T0134500

Lecture Notes in Computer Science 12658

More information about this subseries at http://www.springer.com/series/7412

Alessandro Crimi · Spyridon Bakas (Eds.)

Brainlesion: Glioma, Multiple Sclerosis, Stroke and Traumatic Brain Injuries

6th International Workshop, BrainLes 2020
Held in Conjunction with MICCAI 2020
Lima, Peru, October 4, 2020
Revised Selected Papers, Part I

 Springer

Editors
Alessandro Crimi 🄳
University of Zurich
Zurich, Switzerland

Spyridon Bakas 🄳
University of Pennsylvania
Philadelphia, PA, USA

ISSN 0302-9743   ISSN 1611-3349   (electronic)
Lecture Notes in Computer Science
ISBN 978-3-030-72083-4   ISBN 978-3-030-72084-1   (eBook)
https://doi.org/10.1007/978-3-030-72084-1

LNCS Sublibrary: SL6 – Image Processing, Computer Vision, Pattern Recognition, and Graphics

This Springer imprint is published by the registered company Springer Nature Switzerland AG
The registered company address is: Gewerbestrasse 11, 6330 Cham, Switzerland

in loving memory of Prof. Christian Barillot

Preface

This volume contains articles from the Brain Lesion workshop (BrainLes), as well as (a) the International Brain Tumor Segmentation (BraTS) challenge and (b) the Computational Precision Medicine: Radiology-Pathology Challenge on Brain Tumor Classification (CPM-RadPath). All these events were held in conjunction with the Medical Image Computing and Computer Assisted Intervention (MICCAI) conference on October the 4th 2020 in Lima, Peru, though online due to COVID19 restrictions.

The papers presented describe the research of computational scientists and clinical researchers working on glioma, multiple sclerosis, cerebral stroke, trauma brain injuries, and white matter hyper-intensities of presumed vascular origin. This compilation does not claim to provide a comprehensive understanding from all points of view; however the authors present their latest advances in segmentation, disease prognosis, and other applications to the clinical context.

The volume is divided into four parts: The first part comprises invited papers summarizing the presentations of the keynotes during the full-day BrainLes workshop, the second includes the accepted paper submissions to the BrainLes workshop, the third contains a selection of papers regarding methods presented at the BraTS 2020 challenge, and lastly there is a selection of papers on the methods presented at the CPM-RadPath 2020 challenge.

The content of the first chapter with the three invited papers covers the current state-of-the-art literature on radiomics and radiogenomics of brain tumors, gives a review of the work done so far in detecting and segmenting multiple sclerosis lesions, and last but not least illuminates the clinical gold standard in diagnosis and classification of glioma.

The aim of the second chapter, focusing on the accepted BrainLes workshop submissions, is to provide an overview of new advances of medical image analysis in all of the aforementioned brain pathologies. It brings together researchers from the medical image analysis domain, neurologists, and radiologists working on at least one of these diseases. The aim is to consider neuroimaging biomarkers used for one disease applied to the other diseases. This session did not have a specific dataset to be used.

The third chapter focuses on a selection of papers from the BraTS 2020 challenge participants. BraTS 2020 made publicly available a large (n = 660) manually annotated dataset of baseline pre-operative brain glioma scans from 20 international institutions, in order to gauge the current state of the art in automated brain tumor segmentation using multi-parametric MRI sequences and to compare different methods. To pinpoint and evaluate the clinical relevance of tumor segmentation, BraTS 2020 also included the prediction of patient overall survival, via integrative analyses of radiomic features and machine learning algorithms, and evaluated the algorithmic uncertainty in the predicted segmentations, as noted in: www.med.upenn.edu/cbica/brats2020/.

The fourth chapter contains descriptions of a selection of the leading algorithms participating in the CPM-RadPath 2020 challenge. The "Combined MRI and Pathology

Brain Tumor Classification" challenge used corresponding radiographic and pathologic imaging data towards classifying a cohort of diffuse glioma tumors into three categories. This challenge presented a new paradigm in algorithmic challenges, where data and analytical tasks related to the management of brain tumors were combined to arrive at a more accurate tumor classification. Data from both challenges were obtained from The Cancer Genome Atlas/The Cancer Imaging Archive (TCGA/TCIA) repository, and the Hospital of the University of Pennsylvania.

We heartily hope that this volume will promote further exciting research about brain lesions.

March 2021

Alessandro Crimi
Spyridon Bakas

Organization

Main BrainLes Organizing Committee

Spyridon Bakas   Center for Biomedical Image Computing and
Analytics, University of Pennsylvania, USA

Alessandro Crimi   African Institute for Mathematical Sciences, Ghana

Challenges Organizing Committee

Brain Tumor Segmentation (BraTS) Challenge

Spyridon Bakas   Center for Biomedical Image Computing and
Analytics, University of Pennsylvania, USA

Christos Davatzikos   Center for Biomedical Image Computing and
Analytics, University of Pennsylvania, USA

Keyvan Farahani   Center for Biomedical Informatics and Information
Technology, National Cancer Institute,
National Institutes of Health, USA

Jayashree Kalpathy-Cramer   Massachusetts General Hospital, Harvard University,
USA

Bjoern Menze   Technical University of Munich, Germany

Computational Precision Medicine: Radiology-Pathology (CPM-RadPath) Challenge on Brain Tumor Classification

Spyridon Bakas   Center for Biomedical Image Computing and
Analytics, University of Pennsylvania, USA

Benjamin Aaron Bearce   Massachusetts General Hospital, USA

Keyvan Farahani   Center for Biomedical Informatics and Information
Technology, National Cancer Institute,
National Institutes of Health, USA

John Freymann   Frederick National Lab for Cancer Research, USA

Jayashree Kalpathy-Cramer   Massachusetts General Hospital, Harvard University,
USA

Tahsin Kurc   Stony Brook Cancer Center, USA

MacLean P. Nasrallah   Neuropathology, University of Pennsylvania, USA

Joel Saltz   Stony Brook Cancer Center, USA

Russell Taki Shinohara   Biostatistics, University of Pennsylvania, USA

Eric Stahlberg   Frederick National Lab for Cancer Research, USA

George Zaki   Frederick National Lab for Cancer Research, USA

Program Committee

Meritxell Bach Cuadra	University of Lausanne, Switzerland
Ujjwal Baid	University of Pennsylvania, USA
Jacopo Cavazza	Italian Institute of Technology, Italy
Keyvan Farahani	Center for Biomedical Informatics and Information Technology, National Cancer Institute, National Institutes of Health, USA
Madhura Ingalhalikar	Symbiosis International University, India
Tahsin Kurc	Stony Brook Cancer Center, USA
Jana Lipkova	Harvard University, USA
Sarthak Pati	University of Pennsylvania, USA
Sanjay Saxena	University of Pennsylvania, USA
Anupa Vijayakumari	University of Pennsylvania, USA
Benedikt Wiestler	Technical University of Munich, Germany
George Zaki	Frederick National Lab for Cancer Research, USA

Sponsoring Institution

Center for Biomedical Image Computing and Analytics, University of Pennsylvania, USA

Contents – Part I

Brain Tumor Segmentation

Contents – Part II

**Computational Precision Medicine: Radiology-Pathology
Challenge on Brain Tumor Classification**

Invited Papers

Glioma Diagnosis and Classification: Illuminating the Gold Standard

MacLean P. Nasrallah[(✉)] [iD]

Hospital of the University of Pennsylvania, Philadelphia, PA 19104, USA
maclean.nasrallah@pennmedicine.upenn.edu

Abstract. Accurate glioma classification is essential for optimal patient care, and requires integration of histological and immunohistochemical findings with molecular features, in the context of imaging and demographic information. This paper will introduce classic histologic features of gliomas in contrast to nonneoplastic brain parenchyma, describe the basic clinical algorithm used to classify infiltrating gliomas, and demonstrate how the classification is reflected in the diagnostic reporting structure. Key molecular features include *IDH* mutational status and 1p/19q codeletion. In addition, molecular changes may indicate poor prognosis despite lower grade histology, such as findings consistent with two grade IV infiltrating gliomas: molecular glioblastoma and diffuse midline gliomas. Detailed molecular characterization aids in optimization of treatment with the goal of improved patient outcomes.

Keywords: Glioma · *IDH* mutation · Integrated diagnosis

1 Features of Infiltrating Gliomas

1.1 Introduction

Nonneoplastic brain parenchyma has many constituent cell types and structures, with the cortex composed of both oligodendroglial and astrocytic glial cells, microglial cells, relatively inconspicuous blood vessels, and different types of neurons (Fig. 1a–d). Deep to the cerebral cortex, white matter for the most part lacks neuronal cell bodies and consists of neuronal fibers accompanied by abundant oligodendroglial cells responsible for protecting these processes with sheaths of fatty myelin. The coursing neuronal tracts endow the white matter with a distinctive texture. In contrast, infiltrating gliomas are hypercellular relative to unaffected brain tissue (Fig. 1e–g).

As a result of their infiltrative behavior, these primary brain tumors have no true margins and cannot be entirely surgically resected: tumor cells infiltrate widely throughout the central nervous system. The gradient of dense tumor mass to sparsely infiltrative tumor is captured by radiological imaging. Although they lack contrast enhancement, areas of infiltration are in part detected by abnormal signal on the fluid-attenuated inversion recovery (FLAIR) MRI sequence (Fig. 2a).

Infiltrating gliomas are classified and graded by a combination of histologic and molecular features. Historically, grading has been performed by microscopic assessment

© Springer Nature Switzerland AG 2021
A. Crimi and S. Bakas (Eds.): BrainLes 2020, LNCS 12658, pp. 3–10, 2021.
https://doi.org/10.1007/978-3-030-72084-1_1

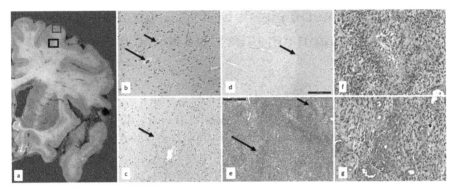

Fig. 1. Two boxes are indicated in a coronal section of normal brain (a). The grey box corresponds to the microscopic image of cortex in (b), with a neuron indicated by the short arrow, and a blood vessel indicated by the longer arrow. A majority of the smaller cells are glial. The red box in (a) corresponds to the area at the border of white matter and cortex and is magnified in (c). The arrow indicates the boundary between cortex on the left, and white matter on the right of the image. The cells in the white matter are predominantly oligodendroglial cells responsible for myelinating the axon tracts. A low power view of normal cortex (left of arrow) and white matter (right of arrow) is shown in (d). In contrast to normal brain parenchyma in (d), a glioblastoma (e) is shown at the same power (scale bars: 500 microns), with pseudopalisading necrosis [short arrow, high power in (f)], and microvascular proliferation [longer arrow, high power in (g)].

according to histologic features. However, ongoing studies demonstrate that molecular features of gliomas are often more important than histologic features for classification, and consequently, for grading a glioma. The Consortium to Inform Molecular and Practical Approaches to CNS Tumor Taxonomy – Not Official WHO (cIMPACT-NOW) works continuously to keep the field of neuropathology up to date with the latest evidence-based science to refine the WHO classifications [1, 2] to ensure that patients receive the best possible care.

1.2 The Importance of *IDH* Mutations in Infiltrating Gliomas

The first question that the pathologist must address when presented with an infiltrating glioma is whether the glioma is characterized by a pathologic *isocitrate dehydrogenase (IDH)* variant (Fig. 3). IDH1 and IDH2 proteins convert isocitrate to 2-ketoglutarate, and in the process produce a metabolic cofactor. When one copy of *IDH* is mutated, mutant IDH protein converts 2-ketoglutarate to the oncometabolite 2-hydroxyglutarate [3]. The cumulative result may be the development of an oligodendroglioma or an astrocytoma.

In addition, astrocytomas often evolve along oncogenic pathways that do not involve changes in one of the *IDH* genes. Although these gliomas develop through separate molecular pathways, *IDH*-mutant and *IDH*-wildtype astrocytomas are indistinguishable by light microscopy. Despite the morphologic similarity of the two tumor types, adult *IDH*-wildtype astrocytomas tend to be significantly more aggressive and portend a much worse prognosis than *IDH*-mutant astrocytomas; therefore, *IDH* mutational status is essential to include in the diagnosis [4].

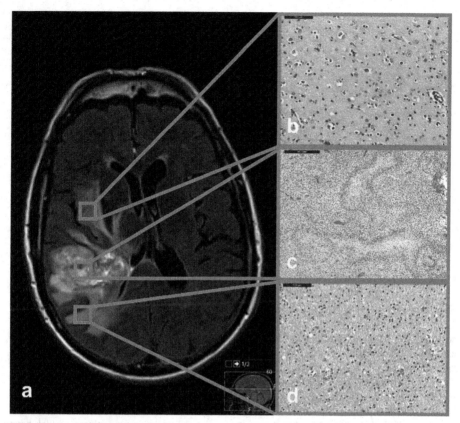

Fig. 2. Axial T2-FLAIR MRI image demonstrates a glioblastoma (a). The mass is hypercellular with pseudopalisading necrosis on microscopic examination (c, scale bar: 500 microns). Although enhancing areas may be circumscribed, the FLAIR sequence highlights the infiltrative nature of the neoplasm, showing signal abnormality in the insula where tumor cells can be seen infiltrating the cortex (b, scale bar: 100 microns). FLAIR abnormality is also prominent in the white matter posterior to the main mass, where edema and infiltrating cells can be seen on histopathology (d, scale bar: 100 microns).

IDH-Mutant Gliomas

IDH-mutant gliomas include two classes of glioma. First, the infiltrating glioma with the best prognosis is the *IDH*-mutant oligodendroglioma, which has a median overall survival on the order of 20 years [5]. An oligodendroglioma is defined molecularly by loss or co-deletion of the chromosomal whole arms 1p and 19q, and often has a mutation in the promoter region of the *telomerase reverse transcriptase* (*TERT*) gene. The latter mutations lead to enhanced transcription factor binding, increased telomerase production and activity, and, consequently, cell immortalization. Morphologically, oligodendroglioma cells have round nuclei and clear perinuclear haloes, often demonstrate a nodular growth pattern, and the tumors are invested by a network of fine vasculature (Fig. 4a).

IDH-mutant astrocytomas are the second class of *IDH*-mutant glioma. *IDH*-mutant astrocytomas have a poorer prognosis than oligodendrogliomas, with median overall

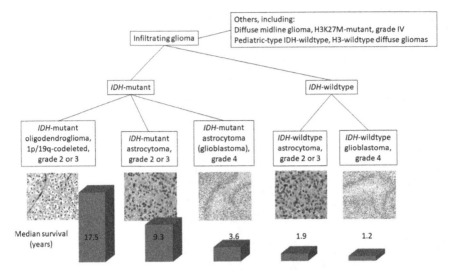

Fig. 3. Classification of infiltrating gliomas. An infiltrating glioma must be tested for an *IDH* variant. Those *IDH*-mutant gliomas that demonstrate 1p/19q codeletion should be classified as oligodendrogliomas. Otherwise, they are classified as *IDH*-mutant astrocytoma. In the absence of an *IDH* mutation, prognosis is worse, as shown by the blue bars, which demonstrate relative median survival time based on survival data from Molinaro *et al.*, 2019 [5]. Given the poor prognosis associated with most lower grade *IDH*-wildtype astrocytomas, a majority of these are considered "molecular glioblastomas" (see text) [6]. Note that the nomenclature is shifting so that grade 4 *IDH*-mutant astrocytomas are not called glioblastomas; this change underlines the distinction of *IDH*-mutant disease from *IDH*-wildtype glioblastomas, despite their identical histology [7]. Finally, other entities must be considered diagnostically in the absence of an *IDH* mutation, such as diffuse midline glioma, which has a poor prognosis, and pediatric-type gliomas, which have relatively good prognoses. (Color figure online)

survival of about 10 years [5]. Molecularly, these gliomas are distinguished from oligodendrogliomas by their lack co-deletion of 1p/19q (Fig. 2). Histologically, *IDH*-mutant astrocytomas have nuclei with irregular borders, and most commonly lack perinuclear haloes and fine vasculature (Fig. 4f). The histologic grade is determined by microscopically assessing mitotic activity, necrosis, and vascular proliferative changes. Increased mitotic activity is required for grade 3 histology, and either necrosis or microvascular proliferation is required for grade 4 histology (demonstrated in Fig. 1f, g).

The basic molecular profiles, including TP53 and ATRX expression, of these two *IDH*-mutant gliomas often clearly differentiate the two tumors at initial pathologic examination, in combination with histological features (Fig. 4). However, codeletion of chromosomal arms 1p and 19q must be demonstrated to render a complete diagnosis of oligodendroglioma. Codeletion may be detected using any of several methods; commonly, two fluorescent *in situ* hybridization (FISH) assays are performed for chromosomes 1 and 19, respectively. Each assay utilizes two probes, one probe specific to the short arm of the chromosome, and a second probe specific to the long arm of the chromosome. In each assay, oligodendroglioma cells will show lack of a probe signal due to the deletion of a chromosomal arm, with the other three probes retained (Fig. 4j).

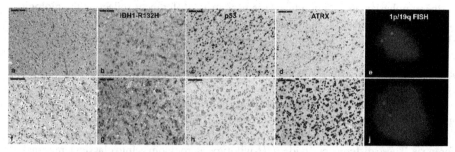

Fig. 4. Histological, immunohistochemical and fluorescent *in situ* (FISH) features of *IDH*-mutant astrocytoma (top row) and oligodendroglioma (bottom row). The infiltrating astrocytoma (a) shows mutant IDH (b), abnormal nuclear p53 protein expression (c), loss of normal ATRX nuclear expression in neoplastic cells (d), and retention of 1p by FISH (e). Retention of 19q would also be seen by FISH. Note that ATRX is retained in nonneoplastic elements, including endothelial cells. The oligodendroglioma demonstrates round nuclei, perinuclear halos, and abundant fine vasculature (f). It shares the IDH mutation (g) with the astrocytoma, but lacks the strong p53 staining (h), and retains normal ATRX (i). However, FISH shows loss of 1p, with only a single green probe present (j), as well as loss of 19q (not shown). Presence of a target on immunohistochemistry is identified by brown staining in the appropriate cellular compartment (scale bars: 100 microns). (Color figure online)

Most infiltrating gliomas may be definitively categorized as either oligodendroglioma or astrocytoma based on histology and molecular features; therefore, the existence of the entity of oligoastrocytoma has been thrown into question. In the 2016 WHO, oligoastrocytoma is listed as a provisional entity [4]. Nevertheless, rare gliomas with dual properties of oligoastrocytoma are reported [8], emphasizing the importance of complete examination and analysis of gliomas in the current age of molecular pathology. However, FISH should not be performed in the absence of an *IDH* mutation, as an oligodendroglioma must have an *IDH* mutation; FISH abnormality in an *IDH*-wildtype astrocytoma is indicative of typical chromosomal abnormalities in these tumors and does not carry the good prognosis associated with an *IDH*-mutant oligodendroglioma [9].

IDH-Wildtype Astrocytomas

In adults, the main and extremely aggressive type of *IDH*-wildtype astrocytoma evolves through gain of a copy of chromosome 7 and loss of chromosome 10, frequently with a mutation in the promoter region of the *TERT* gene, as well as *epidermal growth factor receptor (EGFR)* changes [10, 11]. The *TERT* promoter mutations are the same variants that are seen in oligodendrogliomas. However, patients with *IDH*-wildtype astrocytomas have an overall survival on the order of 1.5 years, dramatically shorter than for *IDH*-mutant tumors [5]. These molecular changes that drive the evolution of *IDH*-wildtype astrocytomas are more indicative of patient outcomes than the histologic grade, which is based on mitotic activity, necrosis and vascular proliferation. Therefore, since 2018, even histologically lower grade *IDH*-wildtype astrocytomas with either *TERT* promoter mutation, *EGFR* gene amplification, or a combination of gain of chromosome 7 and loss of chromosome 10, have been considered grade 4 tumors, essentially equivalent to glioblastoma [6]. These gliomas are referred to as "molecular glioblastoma." To further emphasize and communicate the prognostic differences reflecting the biological differences between *IDH*-mutant and *IDH*-wildtype astrocytomas, the nomenclature has shifted. The term "glioblastoma" is reserved for *IDH*-wildtype grade 4 astrocytomas only. In contrast, diagnoses for *IDH*-mutant astrocytomas are simply qualified by their grade, which is determined by a combination of histologic and additional molecular features (Fig. 3) [12].

IDH-wildtype astrocytomas, including histologic glioblastomas and molecular glioblastomas, are a heterogeneous group [9, 10]. The heterogeneity is seen at multiple levels, with epigenetic, genetic, transcriptomic, and tumor microenvironment differences. These differences divide *IDH*-wildtype astrocytomas into classical, mesenchymal and proneural types [10]. Novel therapies target genetic changes or take advantage of a patient's predisposition to respond to immunotherapies or tumor-treating fields. Although these classifications are not included as part of the WHO clinical diagnosis, complete understanding of the features of glioblastoma allows the design of targeted therapy regimens and the optimal stratification of patients for therapeutic trials.

1.3 Diagnostic Format

As touched on previously, the clinical diagnosis integrates the histology and molecular features in the context of the patient's demographics and imaging features, in a layered format [13]. In a patient report, the top layer will be the final diagnosis, which integrates the subsequent layers of histologic diagnosis, grade and molecular information. For example, an anaplastic oligodendroglioma and an *IDH*-wildtype glioblastoma may show

Table 1. Neuropathology Integrated Diagnosis. Despite identical histologic diagnoses, the two tumors have different integrated diagnoses, based on molecular features.

	Glioma A	Glioma B
Integrated diagnosis	Glioblastoma, *IDH*-wildtype, WHO grade IV	Anaplastic oligodendroglioma, *IDH*-mutant, 1p/19q-codeleted, WHO grade III
Histologic diagnosis	**High-grade glioma with necrosis and microvascular proliferation**	**High-grade glioma with necrosis and microvascular proliferation**
Histologic/molecular grade	Grade 4 (IV)	Grade 3 (III)
Molecular findings	*IDH1/IDH2* wildtype, gain of chromosome 7, loss of chromosome 10, *TERT* promoter mutation	*IDH* mutation, codeletion of 1p/19q, *TERT* promoter mutation, *CIC/FUBP1* mutations

similar histologic findings but are differentiated by their molecular features (Table 1). This format allows complete characterization of each brain tumor, with details that will allow the clinician to customize therapy. The Stupp protocol, involving radiotherapy and temozolomide, remains the standard of therapy [14], but new and diverse options are under investigation and development, with numerous trial opportunities.

1.4 Refining Diagnoses and Improving Patient Care

Just as complete histologic and molecular assessment has revealed the existence of rare dual oligoastrocytomas, careful neuropathologic study continues to identify other gliomas that do not fit neatly into *IDH*-wildtype or *IDH*-mutant categories. Although in the past, a glioma could be diagnosed simply as a glioma, the field has progressed to an understanding of many subtypes of gliomas based on their molecular features and the correlation of these features to outcomes in the context of the patients' demographics. Each tumor must be properly categorized to ensure that the patient receives the appropriate care. For example, an infiltrating astrocytoma in a midline location with a histone mutation in *H3F3A* or *HIST1H3B/C* is defined as a WHO grade 4 diffuse midline glioma regardless of histologic grade [2]. At the other end of the spectrum, in younger patients, an infiltrating glioma is not simply classified as *IDH*-mutant versus *IDH*-wildtype; these patients tend to have alternative genetic findings, including changes in *MYB*, *MYBL1* and *FGFR* genes [12]. Although these tumors are *IDH*-wildtype, they do not possess the poor prognosis seen with adult *IDH*-wildtype astrocytomas, and they are considered low-grade, with precise grade determination requiring additional study and outcomes data. Other gliomas cannot be placed in a known category and may be flagged as "not elsewhere classified" [15]. These are exciting entities that present opportunities for investigation, which will lead to better understanding of the broader field of glioma science, and ultimately improve patient care.

References

1. Louis, D.N., et al.: Announcing cIMPACT-NOW: the consortium to inform molecular and practical approaches to CNS tumor taxonomy. Acta Neuropathol. **133**(1), 1–3 (2016). https://doi.org/10.1007/s00401-016-1646-x
2. Louis, D.N., et al.: WHO Classification of Tumours of the Central Nervous System. International Agency for Research on Cancer (2016)
3. Miller, J.J., Shih, H.A., Andronesi, O.C., Cahill, D.P.: Isocitrate dehydrogenase-mutant glioma: evolving clinical and therapeutic implications. Cancer **123**, 4535–4546 (2017)
4. Louis, D.N., et al.: The 2016 world health organization classification of tumors of the central nervous system: a summary. Acta Neuropathol. **131**(6), 803–820 (2016). https://doi.org/10.1007/s00401-016-1545-1
5. Molinaro, A.M., Taylor, J.W., Wiencke, J.K., Wrensch, M.R.: Genetic and molecular epidemiology of adult diffuse glioma. Nat. Rev. Neurol. **15**, 405–417 (2019)
6. Brat, D.J., et al.: cIMPACT-NOW update 3: recommended diagnostic criteria for "diffuse astrocytic glioma, IDH-wildtype, with molecular features of glioblastoma, WHO grade IV." Acta Neuropathol. **136**(5), 805–810 (2018). https://doi.org/10.1007/s00401-018-1913-0

7. Brat, D.J., et al.: cIMPACT-NOW update 5: recommended grading criteria and terminologies for IDH-mutant astrocytomas. Acta Neuropathol. **139**(3), 603–608 (2020). https://doi.org/10.1007/s00401-020-02127-9

8. Nasrallah, M.L.P., Desai, A., O'Rourke, D.M., Surrey, L.F., Stein, J.M.: A dual-genotype oligoastrocytoma with histologic, molecular, radiological and time-course features. Acta Neuropathol. Commun. **8**, 115 (2020)

9. Brennan, C.W., et al.: The somatic genomic landscape of glioblastoma. Cell **155**, 462–477 (2013)

10. Barthel, F.P., Wesseling, P., Verhaak, R.G.W.: Reconstructing the molecular life history of gliomas. Acta Neuropathol. **135**(5), 649–670 (2018). https://doi.org/10.1007/s00401-018-1842-y

11. Nasrallah, M.P., et al.: Molecular neuropathology in practice: clinical profiling and integrative analysis of molecular alterations in glioblastoma. Acad Pathol. **6**, 2374289519848353 (2019)

12. Louis, D.N., et al.: cIMPACT-NOW update 6: new entity and diagnostic principle recommendations of the cIMPACT-Utrecht meeting on future CNS tumor classification and grading. Brain Pathol. **30**, 844–856 (2020)

13. Hainfellner, J., Louis, D.N., Perry, A., Wesseling, P.: Letter in response to David N. Louis et al, international society of neuropathology-haarlem consensus guidelines for nervous system tumor classification and grading, brain pathology, doi: 10.1111/bpa.12171. Brain Pathol. **24**, 671–672 (2014)

14. Stupp, R., et al.: Radiotherapy plus concomitant and adjuvant temozolomide for glioblastoma. N. Engl. J. Med. **352**, 987–996 (2005)

15. Louis, D.N., et al.: cIMPACT-NOW update 1: not otherwise specified (NOS) and not elsewhere classified (NEC). Acta Neuropathol. **135**(3), 481–484 (2018). https://doi.org/10.1007/s00401-018-1808-0

Multiple Sclerosis Lesion Segmentation - A Survey of Supervised CNN-Based Methods

Huahong Zhang and Ipek Oguz[✉]

Vanderbilt University, Nashville, TN 37235, USA
{huahong.zhang,ipek.oguz}@vanderbilt.edu

Abstract. Lesion segmentation is a core task for quantitative analysis of MRI scans of Multiple Sclerosis patients. The recent success of deep learning techniques in a variety of medical image analysis applications has renewed community interest in this challenging problem and led to a burst of activity for new algorithm development. In this survey, we investigate the supervised CNN-based methods for MS lesion segmentation. We decouple these reviewed works into their algorithmic components and discuss each separately. For methods that provide evaluations on public benchmark datasets, we report comparisons between their results.

Keywords: Multiple Sclerosis · Deep learning · Segmentation · MRI

1 Introduction

Multiple Sclerosis (MS) is a demyelinating disease of the central nervous system. For monitoring the disease course, focal lesion quantification using magnetic resonance imaging (MRI) is commonly used. Recently, deep learning has achieved great success in the computer vision community [44] as well as in medical image analysis tasks, and has been applied to MS lesion segmentation [14,79,91]. Accurate lesion segmentation is valuable for clinical application and subsequent analysis (e.g., [92]).

In this survey, we focus on the methods which are CNN-based, and we review about 100 papers that were published until late October 2020. However, we do not intend to review these papers exhaustively. Instead, we break down the reviewed segmentation pipelines into algorithmic components such as data augmentation and network architecture and compare the representative advances for each component in Sect. 2. This is different from the previous surveys of MS segmentation methods (e.g., [19,41]).

Unsupervised learning methods are not included in this survey since an excellent comprehensive survey of these is already provided by Baur *et al.* [8]. As a general note, supervised methods tend to perform better than unsupervised methods for MS lesion segmentation, given the difficulty of the task [14,17].

To identify the articles to include in this review, we conducted a Google Scholar search using the keywords "Multiple Sclerosis + lesion + segmentation

© Springer Nature Switzerland AG 2021
A. Crimi and S. Bakas (Eds.): BrainLes 2020, LNCS 12658, pp. 11–29, 2021.
https://doi.org/10.1007/978-3-030-72084-1_2

+ neural network". To ensure we include important advances, we also went through the references cited in each reviewed paper. In addition, papers who cite publicly available MS datasets (e.g., [14, 17]) were considered.

2 Review of Methods

2.1 Data Pre-processing

Pre-processing is a common practice in the reviewed papers. It usually includes skull stripping (e.g., BET [73]), bias field correction (e.g., N4ITK [75]), rigid registration and intensity normalization. Rigid registration is used to register between different MRI modalities acquired during a scanning session, between scans of a given patient acquired at different time points, as well as between different subjects of a study or to standard template spaces (e.g., MNI152 [27]).

For intensity normalization, there are two popular approaches: 1) histogram matching (e.g., [10, 78]) and 2) normalizing data to a specific range. For this latter approach, it is common to enforce zero mean and unit variance (whitening-like, e.g., [33, 79, 80]) or fit the intensities into the range [0, 1] (e.g., [25, 31, 76]). Some pipelines (e.g., [10, 78]) also chose to only preserve the values between a range (e.g., 1st and 99th percentile) before normalization to minimize the effect of intensity outliers. Further combinations of these two approaches are also possible. Ravnik et al. [63] argued that in a same-scanner, homogeneous situation, using (whitening-like) intensity normalization, histogram standardization, or both all achieved similar results, and indeed these had no statistically significant improvement over even no normalization at all. But in a multi-scanner, heterogeneous dataset, only using normalization is slightly better than using both, and all performed statistically better than no pre-processing. Other advanced pre-processing techniques (e.g., white stripe [72]) can also be considered.

2.2 Data Representation

For feeding the deep neural network, the input data are usually represented as patches of raw MRI images since whole images are too large. These patches can be 2D, 3D, or any format in between. Choosing the format of patches is an important design decision and will affect the performance of networks. Whether the data is of isotropic resolution needs to be considered for making this decision [36].

With **2D** (slice-based) patches (e.g., [4, 66, 88]), the advantage is that there are far fewer network parameters to train, and therefore, these tend to be not as prone to over-fitting compared to 3D networks. However, contextual information along the third axis is missing. In contrast, **3D** approaches (e.g., [11, 25, 32, 79]) take advantage of the local contextual information, but they do not have any global context. Compared to 2D methods, due to the small size used in 3D patches, the long-distance spatial information cannot be preserved. Further, for processing 3D data, the networks are computationally expensive, need more parameters, and are prone to over-fitting.

Multi-view/2.5D. To obtain a balance between 2D and 3D representations, Birenbaum *et al.* [10] proposed to use multi-view data, in which three orthogonal 2D views passing through the same voxel are input to the network jointly. On the other hand, Aslani *et al.* [4] and Zhang *et al.* [91] used 2.5D by extracting slices from different planes, which are used to train the networks independently from each other. Zhang *et al.* [91] also use "stacked slices" by stacking 2D patches along the third axis, which generates thinner but larger 3D input than normal 3D patches. Using 2.5D and prediction fusion, the networks are able to learn and utilize global information along all three axes.

2.3 Data Preparation

Candidate Extraction. To extract input patches from the raw data, an intuitive way is to move a sliding window voxel-by-voxel throughout the raw MRI volume. However, this approach will generate a lot of similar patches and leads to class imbalance. The strategy used to alleviate these problems is slightly different between fully convolutional network (FCN) and non-FCN methods (as discussed in Sect. 2.4). This distinction is mainly due to the labels of patches for non-FCN methods being single scalars, which are more vulnerable to class imbalance. Furthermore, FCN methods usually solve class imbalance with loss functions.

For **non-FCN** methods, Birenbaum *et al.* [10] proposed to apply a probabilistic WM template and choose the patches whose center is a voxel with high intensity in FLAIR and high probability in WM template. The method does not need lesion maps, so it can be used in both training and test phases to reduce the computation burden. Valverde *et al.* [79,80] randomly under-sample the negative class to make a balanced dataset. Kazancli *et al.* [42] considered strategies of random sampling and sampling around the lesions, with the latter providing better model performance. Ulloa *et al.* [76] augmented data from the lesion class to balance the dataset instead of down-sampling the non-lesion class. Also, they use circular non-uniform sampling, which allows greater contextual information with a radial extension. They further presented a stratified sampling method [77]. Among voxels not labeled as lesions, a portion p of the candidates is extracted from the neighborhood of lesions and the remaining $1 - p$ from the remaining voxels.

For **FCN** methods, Kamnitsas *et al.* [37] extract the patches with a 50% probability of being centered on a foreground or background voxel. Feng *et al.* [25] centered the patches at a lesion voxel with a probability $p = 0.99$. Many methods (e.g., [3,4,66,91]) only extract patches with at least one lesion voxel.

Augmentation. After extracting the patches, data augmentation techniques can be applied. Commonly used augmentations include random flip, random rotation, and random scale. Usually, the rotation is 2D and the angle is $n \times 90°(n \in \mathbb{Z})$. However, a random angle (e.g., [3,10]) or a 3D rotation have also been used. Sometimes random noise and random "bias field" can be added [45]. Additionally, Salem *et al.* [68] suggested synthesizing lesions may be helpful as an augmentation.

Label Processing/Denoising. For training and evaluating models, the label can be a scalar representing whether the central voxel of the input patch is a lesion or not, or it can be the lesion map the same size as the input, depending on the network type (discussed in Sect. 2.4). Either the scalar or the lesion map is extracted from the expert delineations. These unavoidably contain noise, which usually comes from the lesion borders [39]. Also, some datasets (e.g., [14,17]) are delineated by more than one expert, and inter-rater variability needs to be addressed. While a simple majority vote can be used, STAPLE [84] and its variation [2] are very common. On the other hand, a few methods (e.g., [91]) treat delineations from different experts as different samples. In other words, they train networks with the same input patch but different labels.

Even consensus delineations still contain noise that can be further mitigated. Roy *et al.* [66] generated training memberships (labels) by convolving the binary segmentations with a 3×3 Gaussian kernel to get a softer version of boundaries. Kats *et al.* [39] proposed to assign fixed soft labels to the voxels near the lesions by 3D morphological dilation. Even though they define the soft-Dice loss for this purpose, an easier solution may be to generate the soft version of labels at the data preparation stage. Cohen *et al.* [16] further proposed to learn soft labels instead of fixed values, and then apply the soft-STAPLE algorithm [40] to fuse the masks.

Extra Information. Extra information, for example, spatial locations or intermediate results provided by other models, can also be used as input to CNNs. Ghafoorian *et al.* [30] incorporated eight spatial features, with dense features extracted by convolutional layers and fed into fully connected layers. La Rosa *et al.* [46] provided the CNNs with the probability maps generated by a Bayesian partial volume estimation algorithm [24].

2.4   Network Architecture

The network architecture plays a very important role in deep learning. Many works focus on crafting the structure to improve the segmentation performance.

Network Backbones. After Krizhevsky *et al.* [44] won the ImageNet 2012 challenge, Convolutional Neural Networks (**CNN**s) became very popular and have been successfully applied to medical imaging problems [14,17]. For MS lesion segmentation with CNN, the early methods use voxel-wise classification to form the lesion segmentation map [10,42,79]. A typical network of this type consists of a few convolutional layers followed by 2–3 fully connected layers (also called Multilayer Perceptron, MLP) for the final prediction. The input is an image patch, and the networks are trained to classify whether the central voxel of this patch corresponds to a lesion. While these methods have outperformed conventional methods, they have disadvantages: 1) lack of spatial information as only small patches are used; 2) computational inefficiency due to the repetition of similar patches.

Kang *et al.* [38] introduced the fully convolutional neural network (**FCN/FCNN**) for the segmentation task. FCNs do not need to make the lesion prediction by classifying the central voxel of each patch. Instead, they directly generate lesion maps of the same (or similar) size as the input images. However, due to the successive use of convolutional and pooling layers, this approach produces segmentations at a lower resolution. Long et al. [51] preserve the localization information from low-level features and contextual information from high-level features by adding skip connections. Applications of FCNs for MS lesion segmentation include [11,53,66] and some of these methods take advantage of shortcut connections [11,53].

Ronneberger et al. [65] then used a symmetrical u-shape network called **U-Net** to combine features. The network has an encoder-decoder structure and adds shortcut connections between corresponding layers of the two parts. The pooling operations in the encoder are replaced by upsampling operations in the decoder path. Many recent MS segmentation methods [4,25,67] are based on the original U-Net or slight modifications thereof.

Variations of U-Net. For CNNs with voxel-wise prediction or FCN methods that are not U-Net-based, the network structures are quite flexible. As for pipelines that use the U-Net, the most common modification is to introduce some crafted modules (residual block, dense block, attention module, etc.). These new modules can be added to replace the convolutions or between the shortcut connections. Aslani et al. [3,4] presented a U-Net-like structure with convolutions replaced by residual blocks in the encoder path. Hashemi et al. [32] and Zhang et al. [91] adopted the Tiramisu network, which replaces the convolution layers with densely connected blocks (skip connection between layers). Hu et al. [35] presented a context-guided module to expand the perception field and utilize contextual information. Vang et al. [81] augmented the U-Net with the Mask R-CNN framework.

Attention Module. Attention mechanism has been researched in many works for MS lesion segmentation. In general, it can be divided into spatial attention, channel attention, and longitudinal attention. Zhang et al. [89] presented a recurrent slice-wise attention network for 3D CNNs. By performing slice-wise attention from three orientations, they reduced the memory demands for 3D spatial attention. Hu et al. [35] included 3D spatial attention blocks in the decoding stage. Durso-Finley et al. [23] used a saliency-based attention module before the U-Net structure to make the network realize the difference between pre- and post-contrast T1-w images and thus focus on the contrast-enhancing lesions. Hou et al. [34] proposed a cross-attention block that combines spatial attention and channel attention. Zhang et al. [90] extend their folded (slice-wise) attention to the spatial-channel attention module. They use four permutations (corresponding to four dimensions) to build four small sub-affinity matrices to approximate the original affinity matrix. In such a case, the original affinity matrix is regularized as a rank-one matrix and the computational burden is alleviated. Gessert et al. [29] introduced attention-guided interactions to enable effective information exchange between the processing paths of the two time points.

Multi-task Networks. Narayana *et al.* [58] performed segmentation of brain tissue and T2 lesions at the same time. McKinley *et al.* [54] illustrated that the inclusion of additional tissue classes during the segmentation of lesions is helpful for MS segmentation. Duong *et al.* [22] trained the networks with data from many different tasks, making the trained CNN usable for multiple tasks without tuning and thus more applicable in a clinical context.

2.5 Multiple Modalities, Timepoints, Views and Scales

In the context of MS lesion segmentation, the incorporation of multi-modality, multi-timepoints, multi-view and multi-scale data are similar. The data from different sources have to be fused at some point of the pipeline: input, feature map, and/or output. Fusing at the input can be simple concatenation along channels while fusing the output is roughly equivalent to making an ensemble to reach consensus. Fusing the features usually needs parallel paths and interaction between paths, which typically happens in the encoder path in a U-Net-like structure.

Multi-modalities. The commonly used MRI sequences for MS white matter lesion segmentation include T1-weighted (T1-w), T2-weighted (T2w), proton density-weighted (PD-w) and fluid attenuated inversion recovery T2 (FLAIR).

Narayana *et al.* [59] evaluated the performance of U-Net when it is trained with different combinations of modalities on a large multi-center dataset. They concluded that using all the modalities, especially with FLAIR, achieved the best performance. A similar conclusion can be found in [11] and other works that use multiple MRI sequences as input. For fusing the different modalities, Roy *et al.* [66] use parallel pathways for processing different modalities and then concatenate the features along the channels (only once). Aslani *et al.* [4] use parallel encoder paths to process different modalities, and they fuse the different modalities after each convolutional block. Zhang *et al.* [88] also use a similar strategy. Zhang *et al.* [91] fuse the patches from different sequences before feeding into the network.

Multi-modality methods are usually trained on a specific set of modalities and thus require these sequences to be available at the test phase, which can be limiting. To deal with missing modalities, Havaei *et al.* [33] propose to use parallel CNN paths to learn the embeddings of different input sequences into a single latent space for which arithmetic operations are well defined. They randomly drop modalities during training. As such, any subset of available modalities can be used as input at test time. Feng *et al.* [25] also use random "dropout" of modalities but substitute the missing modalities with background values.

Multi-timepoints. Longitudinal studies are common in MS, but the ongoing inflammatory disease activity complicates the analysis of longitudinal MRIs as new lesions can appear and existing lesions can heal between scans. To improve individual **segmentation performance**, Birenbaum *et al.* [10] propose to process the two time-points individually with a Siamese architecture, where the parallel paths share weights, and then concatenate the features for classification.

Denner *et al.* [20] argue that this late-fusion strategy [10] does not properly take advantage of learning from structural changes and they propose two complementary networks for multi-timepoints. The longitudinal network fuses the two time-points early to implicitly use the structural differences. The multi-task network is trained to learn the segmentation with an additional task of deformable registration between the two time-points, which explicitly guides the network to use spatio-temporal information.

To identify **lesion activity**, Placidi *et al.* [62] simply segment the lesions at the two time-points independently and register the previous examination to the current examination to compare the segmentations. However, comparing differences between time-points relies on high similarity between scans and requires highly accurate registration. McKinley *et al.* [55] introduce a segmentation confidence for comparing the lesion segmentation between timepoints. Comparisons are based on the "confident" lesions of each timepoint. Kruger *et al.* [45] fed two timepoints into the same encoder (share weights) and the feature maps are concatenated after each residual block before going to the corresponding decoder block. Salem *et al.* [69] use cascaded networks for detecting new lesions. The first network learns the deformation field between the baseline and follow-up images. The second network takes the two images and the deformation field, and outputs the segmentation. To assist the network in learning to detect new and enlarging T2w lesions, Sepahvand *et al.* [70] illustrate an attention-like mechanism. They multiply the multi-modal MRI at the reference (in contrast to follow-up) with subtraction images, which acts as the attention gate. Then the product is concatenated with the lesion map at the reference to feed the network. Gessert *et al.* [28] propose convolutional gated recurrent units for temporal aggregation. The units are inserted into the bottleneck and skip connections of U-Net. In another work [29], the same team process two timepoints with parallel encoder paths that interact with each other using attention modules. In such a scenario, the attention mechanism functions similarly to masking early time-points.

Multi-views. As discussed in Sect. 2.2, multi-view data can be utilized as data representation between 2D and 3D. To handle multi-view data, Birenbaum *et al.* [10] processed different views by parallel sub-networks and then concatenated the features to feed the fully connected layers. McKinley *et al.* [53] integrate three networks for the three views and the outputs are averaged. Zhang *et al.* [91] use one network for different views, i.e., the network parameters are shared between views. Shachor *et al.* [71] propose a gated model Mixture of Views (MoV) to fuse different views.

Multi-scales. The networks based on FCN and U-Net inherently incorporate multiple scales. However, explicit use of multi-scales may also be useful. For non-FCN networks, Kamnitsas *et al.* [37] propose to use two parallel paths, one for full resolution and another for a lower resolution; these two paths are fused before fully connected layers. As for U-Net-based methods, Wang *et al.* [83] argue that different types of segmentation biases may be generated by networks of different input sizes. To address this issue, they train 3 networks with different input sizes and use another stage for fusing the results. Hou *et al.* [34] feed multi-scale input (original and downsampled patches) to the first three layers and they aggregate the multi-scale outputs from the last three layers to make the final prediction. Hu *et al.* [35] use one input and average the multi-scale outputs.

Others. For training data with annotations from multiple experts, Vaidya *et al.* [78] train two separate networks using the same images but different delineations from two experts. The outputs of the two networks are averaged to get the final prediction. Zhang *et al.* [93] present a segmentation network to estimate the ground truth segmentation and an annotator network to estimate the characteristics of each expert, which can be viewed as a translation of STAPLE to CNN.

No-new-UNet (nnU-Net) [36] is a multi-architecture framework that adaptively chooses an ensemble from 2D, 3D, 3D cascaded networks.

2.6   Loss Functions and Regularization

For MS lesions segmentation, which is usually binary, the most commonly used loss function is the Binary Cross-Entropy (**BCE**, e.g., [3,10]). Other losses such as **L2 loss** (e.g., [11,91]) are also explored. As class imbalance exists within the dataset, the original losses can be weighted based on the probability (prevalence) of each class. The class with the lower probability (i.e., lesion) is compensated with a higher weight. Brosch *et al.* [11] implicitly weight lesion voxels and non-lesion voxels by calculating the **weighted L2 loss** of the lesion voxels and non-lesion voxels. Feng *et al.* [25] used **weighted BCE** with a lesion/non-lesion ratio of 3 to 1. **Focal loss** [50], as the generalization of BCE, was proposed not only to weight the lesion class but also to give more importance to hard examples (e.g., [32,91]).

For FCN-based methods, since the labels/outputs are patches, region-based losses are used to address the intra-patch class imbalance. Milletari *et al.* [56] proposed the **Dice (F_1) loss** to balance between precision and recall equally. **Tversky loss** [32,67] is the generalization of the Dice loss and F_β loss (such that $\beta = 1$ is the Dice loss). The networks trained with higher β have higher recall and lower precision. Based on this property, Ma *et al.* [52] trained individual models with high β to make up an ensemble. Their assumption is that diverse low-precision-high-recall models tend to make different false-positive errors but similar consistent true-positives. Thus the false-positive errors can be canceled out by aggregating the predictions of the ensemble. Further, the **Focal Tversky loss** (e.g., [35]) is the generalization of the Tversky loss. It is similar to Focal loss, which is capable of focusing on mislabeled samples and minority samples by controlling the parameters.

Combining loss functions and introducing domain-specific regularization are helpful in some cases. McKinley *et al.* [53] calculated the 25th percentile of the intensity within the lesion mask and weighted the loss function from these voxels higher than others. Zhang *et al.* [88] proposed to use Generative Adversarial Network (GAN) architecture to provide an additional discriminator-based constraint. They use a combination of BCE loss, Dice loss, L1 loss and GAN-loss.

Additionally, loss functions have been proposed to help the networks with uncertainty analysis [54], domain adaptation [1,3,6] and other goals.

2.7   Implementation

To train the networks, a simple strategy is using fixed epochs. However, early stopping is a common practice to avoid over-fitting. Thus, the training dataset is divided into a fixed training subset and validation subset, or k-fold cross-validation can be utilized [10,32,91]. The "best" models are chosen based on model performance (e.g., Dice score) on the validation set. To provide a fair evaluation, the test set is usually held-out from the training/validation data and different scans of the same patient should not be placed into different datasets.

For optimizing the parameters, stochastic gradient descent (SGD) and its variations are used. To avoid oscillation in local optima, the momentum variable was introduced (e.g., [35].). Further, Nesterov accelerated gradient was proposed to have some prescience about the next update direction (e.g., [33]). To adapt the learning rate to the parameters, Adagrad [21] was proposed. However, Adagrad accumulates the squared gradients in the denominator and leads to monotonically decreasing learning rate. RMSprop and Adadelta [87] are proposed to resolve Adagrad's radically diminishing learning rates. Adaptive Moment Estimation (Adam) [43] is an optimizer with momentum and adaptive learning rates. AMSGrad [64] is a variation of Adam. For MS lesion segmentation, Adadelta (e.g., [3,10,11,53,79,80]) and Adam (e.g., [4,25,42,66]) are most widely used.

2.8   Prediction and Post-processing

Prediction. For non-FCN methods, segmentation is made by classifying all the candidates extracted from the test image voxel-by-voxel. For FCN methods, 2D networks predict the segmentation slices-by-slice, and 3D methods are able to predict the whole image at once. However, Hashemi et al. [32] argue border predictions made by FCN are not as accurate as center voxel predictions, so they propose to predict the results patch-by-patch and then fuse predictions using B-spline weighted soft voting, such that border predictions are given lower weights. For some methods where data from multiple sources (e.g., views) is used or multiple models are trained, a label aggregation step is necessary. Aslani et al. [3,4] and Zhang et al. [91] use majority vote to aggregate labels, but other methods (e.g., STAPLE [84]) can also be considered.

Since the outputs of networks are usually soft predictions (i.e., in the range [0, 1]) indicating the probability of being lesions, the simplest way to get hard predictions is to use a threshold of 0.5 [67]. However, McKinley et al. [55] argue that the scores output by deep networks do not correspond to observed probabilities and are typically overconfident. Brosch et al. [11] and Roy et al. [66] attempt to choose the optimal threshold by maximizing the Dice on the training set.

Post-processing. Many methods attempt to remove false positives from the hard segmentation using post-processing strategies. A common post-processing approach consists of discarding lesions smaller than a volume threshold (e.g., [12]). Vaidya *et al.* [78] use pre-built brain templates to remove predicted lesions outside of white matter, but this strategy is problematic for detecting cortical lesions. Kamnitsas *et al.* [37] use an additional stage of machine learning for post-processing, specifically, a fully-connected Conditional Random Field (CRF), a common strategy. Valverde *et al.* [79, 80] use cascaded networks, in which the first network is trained to recall many possible lesions and the second network refines the output of the first network. Specifically, the training data for the second model is balanced between all the lesion voxels and the random selection of misclassified lesion voxels on the first model. Others [42, 86] use similar strategies.

Nair *et al.* [57] present multiple uncertainty estimates based on Monte Carlo (MC) dropout. Then lesions with high uncertainty can be removed. Their results suggest that uncertainty measures allow choosing superior operating points, compared to only using the network's sigmoid output as a probability.

2.9   Transfer Learning and Domain Adaptation

Transfer learning is an active topic in deep learning, and in the context of MS lesion segmentation, it includes two aspects: 1) using pre-trained models from other domains; 2) applying trained model on different MS datasets (e.g., for clinical use). The latter scenario is also considered as domain adaptation, in which the target task remains the same as the source (i.e., MS lesion segmentation) but the domains (i.e., MRI protocols and therefore image appearance) are different. Multi-task training (Sect. 2.4) is also a form of transfer learning.

Pre-training with Other Domains. The pre-trained blocks (layers) from other domains are typically used to replace the encoder. Brosch *et al.* [11] propose to pre-train the model layer-by-layer with convolutional restricted Boltzmann machines and then apply parameters on both encoder and decoder. Aslani *et al.* [3,4] use the ResNet50 pre-trained on ImageNet as the encoder. Fenneteau *et al.* [26] present a self-supervision method to pre-train the encoder to predict the location of an input patch. However, their results illustrate that pre-training is not helpful. Kruger *et al.* [45] pre-train the encoder path with single time point data and then train the entire network with longitudinal data.

Generalization of Trained Models. For domain adaptation when a few labeled images are available in the target (new) domain, Ghafoorian *et al.* [31] propose to freeze the first few layers of the model trained on the source domain and fine-tune the last few layers. Their results show that using even just 2 images can achieve a good Dice score. Valverde *et al.* [80] propose to freeze all convolutional layers and fine-tune the fully connected layers. A single image for re-training could generate segmentation with human-level performance. Weeda *et al.* [85] further test the one-shot learning proposed by [80] with an independent dataset, and the performance is better than unsupervised methods and is comparable to fully trained supervised methods.

For domain adaptation without any labeled data in the target domain, Baur *et al.* [6] propose to add auxiliary manifold embedding loss for utilizing unlabeled data from target domains. The idea is that latent feature vectors that share the same label (for labeled data) or same noisy prior (unlabeled data) should be similar, and otherwise differ from each other. Baur *et al.* [7] propose to train an auto-encoder for unsupervised anomaly detection in the target domain, and use this unsupervised model to generate artificial labels for jointly training a supervised model with labeled data from the source domain. Ackaouy *et al.* [1] propose a method to perform unsupervised domain adaptation with optimal transport. In the deep learning context, their strategy is implemented as two losses, which ensures the heavily connected source samples and target samples to have similar representations in the latent space and the output, while maintaining good segmentation performance.

Billast *et al.* [9] present domain adaptation with adversarial training. The discriminator is trained to discriminate whether the two input segmentations are from the same scanner, so that the generator learns to map scans from different scanners to the same latent space and thus produce a consistent lesion segmentation. Aslani *et al.* [5] propose a similar idea with a regularization network predicting the feature domain. They use a combination loss of Pearson correlation, randomized cross-entropy and discrete uniform to encourage the latent features to be domain agnostic. Varsavsky *et al.* [82] combine domain adversarial learning and consistency regularization, which enforces invariance to data augmentation.

2.10   Methods for Subtypes of MS Lesions

Most of the methods we have discussed are proposed to segment white matter lesions. Among these, some pipelines focus on detecting contrast-enhancing (CE) lesions since these are indicative of active disease. Gadolinium (Gad) is commonly used in the context of MS. Durso-finley *et al.* [23] and Coronado *et al.* [18] present to detect Gad lesions using pre- and post-contrast T1-weighted images. Brugnara *et al.* [12] propose a network to detect both CE lesions and T2/FLAIR-hyperintense lesions and report the performance separately. On the other hand, radiological monitoring of disease progression also requires detecting new and enlarging T2w lesions, which can be explored by longitudinal approaches (Sect. 2.5, multi-timepoints). La Rosa *et al.* [46] propose to detect the early stage lesions by combining deep neural networks with a shallow model (supervised k-NN with partial volume modeling).

Cortical lesions are also important in MS [13]. La Rosa *et al.* [48] use simplified U-Net to detect both cortical and white matter lesions at 3T MRI. To achieve this, they utilize 3T 3D-FLAIR and magnetization-prepared 2 rapid acquisition with gradient echo (MP2RAGE). They further use 7T MRI (7T MP2RAGE, T2*w echo planar imaging, T2*w gradient recalled echo) for cortical lesion segmentation, which has higher resolution and SNR than 3T [47].

3   Comparison of Experiments and Results

3.1   Datasets

Public Datasets. Currently, the challenge datasets, including the MICCAI 2008[1] [74], ISBI 2015[2] [14], MICCAI (MSSEG) 2016[3] [17] challenges, are widely used. The dataset descriptions can be found on the respective websites. The first two challenges are still (as of November 2020) accepting segmentation submissions on their test dataset, which provides objective comparisons between state-of-the-art MS segmentation methods. Lesjak *et al.* [49] provided a novel public dataset[4], for which three expert raters performed segmentation of WM lesions and reached consensus by several joint sessions. They illustrated that the consensus-based segmentation have better consistency than a single rater's segmentation. It is worth noting that all these public datasets only delineate white matter lesions.

Private Datasets. Using private datasets makes it difficult to compare algorithms but has the advantage of including more subjects than currently available in public datasets. Some proprietary datasets can be of a quite large scale (e.g., 6830 multi-channel MRI [23]). For such large datasets, the "ground truth" labels are usually created by automated or semi-automated algorithms and corrected by experts [23,46].

Narayana *et al.* [60] considered the effect of training data size for training the neural networks. They argue that at least 50 image volumes are necessary to get a meaningful segmentation. But this work does not mention the data augmentations and other advanced techniques for training a network. Based on the results of the ISBI 2015 challenge [14], human-level performance can be achieved by state-of-the-art algorithms with only about 20 images for training.

3.2   Evaluation Metrics

In the task of MS lesion segmentation, the commonly reported metrics include: Dice similarity coefficient (DSC), Jaccard coefficient, absolute volume difference (AVD), average symmetric surface distance (ASSD/SD), true positives rate (TPR, sensitivity, recall), false positives rate (FPR), positive predictive value (PPV, Precision), lesion-wise true positives rate (LTPR) and false positives rate (LFPR).

The above metrics are individually calculated based on each image. Then, the results are aggregated and reported. In addition to reporting mean and standard deviation values, the Wilcoxon signed-rank test is used to statistically test performance differences between methods. Precision-Recall (PR) curve is suitable

---

[1] http://www.ia.unc.edu/MSseg.

[2] https://smart-stats-tools.org/lesion-challenge.

[3] https://portal.fli-iam.irisa.fr/msseg-challenge/overview.

[4] http://lit.fe.uni-lj.si/tools.

for evaluating the performance of the highly unbalanced dataset. Receiver Operating Characteristic (ROC) curve is also used (e.g., [57]). The area under curve (AUC) for these curves is a common aggregation metric. Further, considering the relationship between the model performance and lesion volume, some works divide lesions into groups of different sizes and calculated the metrics (e.g., [83]). Volumes of lesions estimated and the ground truth segmentation can be shown in the correlation (e.g., [4,22,42,66,79]) and Bland-Altman (e.g., [46,54]) plots. A more systematic analysis of algorithm performance can differentiate between correctly detected lesions, nearby lesions merged into one or a single lesion split into many, as well as characterize the performance as a function of lesion size [15,61].

3.3   Results

As previously mentioned, MICCAI 2008 [74] and ISBI 2015 [14] are still accepting submissions and providing the evaluation results on the test dataset, thus serving as objective benchmarks. In this survey, we compare the state-of-the-art methods that have evaluated their performance on these datasets in Table 2 and Table 1.

Table 1. Results on the ISBI 2015 challenge test set. All metrics in percent. DSC: Dice; PPV: Precision; TPR: true positives rate; LTPR: lesion-wise TPR; LFPR: lesion-wise false positives rate; VD: volume difference; SC: total weighted score of other metrics. Code: links to code repositories, if available.

	SC	DSC	PPV	TPR	LFPR	LTPR	VD	Code
Zhang et al. [91]	93.21	64.3	90.8	53.3	12.4	52.0	42.8	Yes
Isensee et al. [36]	92.87	67.9	84.7	60.5	15.9	52.2	36.8	Yes
Hu et al. [35]	92.61	63.4	86.9	52.6	13.4	48.2	39.7	No
Hashemi et al. [32]	92.49	58.4	92.1	45.6	8.7	41.3	49.7	No
Feng et al. [25]	92.41	68.2	78.2	64.5	27.0	60.0	32.6	No
Denner et al. [20]	92.12	64.3	85.9	54.5	19.5	47.1	38.6	Yes
Aslani et al. [4]	92.12	61.1	89.9	49.0	13.9	41.0	45.4	No
Valverde et al. [79]	91.33	63.0	78.7	55.5	15.3	36.7	33.8	Yes
Roy et al. [66]	90.48	52.4	86.6	N/A	11.0	N/A	52.1	Yes
Valverde et al. [80][a]	90.32	57.7	83.1	47.5	18.9	29.7	44.6	Yes
Birenbaum et al. [10]	90.07	62.7	78.9	55.5	49.8	56.8	35.2	No

[a] Trained on other datasets and fine-tune with one sample from this dataset.

From the results, we observe that 3D and 2.5D methods seem to outperform 2D approaches with the development of GPUs. As in Table 1, U-Net-based methods [4,20,32,35,36,91] tend to perform better than non-FCN CNN-based [10,79] and non-U-Net FCN-based [66] methods.

Table 2. Results on the MICCAI 2008 challenge test set. Subscript 1: UNC Rater; 2: CHB Rater. All metrics in percent, except SD in millimeters. VD: volume difference; SD: surface distance; TPR: true positives rate; FPR: false positives rate; SC: total weighted score of other metrics.

	SC	VD_1	SD_1	TPR_1	FPR_1	VD_2	SD_2	TPR_2	FPR_2
Valverde *et al.* [79]	87.1	62.5	5.8	55.5	46.8	40.8	5.2	68.7	46.0
Brosch *et al.* [11]	84.0	63.5	7.4	47.1	52.7	52.0	6.4	56.0	49.8
Havaei *et al.* [33]	83.2	127	7.5	66.1	55.3	68.2	6.6	52.3	61.3

4   Conclusion

In this survey, we explored the advances in different components of supervised CNN MS lesion segmentation methods. Among these, topics including attention mechanism, network designs to combine information from multiple sources, loss functions to handle class imbalance, and domain adaptation are of interest for many researchers.

Acknowledgements. This work was supported, in part, by the NIH grant R01-NS094456 and National Multiple Sclerosis Society award PP-1905-34001.

References

1. Ackaouy, A., et al.: Unsupervised domain adaptation with optimal transport in multi-site segmentation of MS lesions from MRI data. Front. Comput. Neurosci. **14**, 19 (2020)
2. Akhondi-Asl, A., et al.: A log opinion pool based staple algorithm for the fusion of segmentations with associated reliability weights. IEEE Trans. Med. Imaging **33**, 1997–2009 (2014)
3. Aslani, S., Dayan, M., Murino, V., Sona, D.: Deep 2D encoder-decoder convolutional neural network for multiple sclerosis lesion segmentation in brain MRI. In: Crimi, A., Bakas, S., Kuijf, H., Keyvan, F., Reyes, M., van Walsum, T. (eds.) BrainLes 2018. LNCS, vol. 11383, pp. 132–141. Springer, Cham (2019). https://doi.org/10.1007/978-3-030-11723-8_13
4. Aslani, S., et al.: Multi-branch convolutional neural network for multiple sclerosis lesion segmentation. NeuroImage **196**, 1–15 (2019)
5. Aslani, S., et al.: Scanner invariant multiple sclerosis lesion segmentation from MRI. In: ISBI (2020)
6. Baur, C., Albarqouni, S., Navab, N.: Semi-supervised deep learning for FCN. In: Descoteaux, M., Maier-Hein, L., Franz, A., Jannin, P., Collins, D.L., Duchesne, S. (eds.) MICCAI 2017. LNCS, vol. 10435, pp. 311–319. Springer, Cham (2017). https://doi.org/10.1007/978-3-319-66179-7_36
7. Baur, C., et al.: Fusing unsupervised and supervised deep learning for white matter lesion segmentation. In: MIDL (2019)
8. Baur, C., et al.: Autoencoders for unsupervised anomaly segmentation in brain MR images: a comparative study. arXiv:2004.03271 (2020)

9. Billast, M., Meyer, M.I., Sima, D.M., Robben, D.: Improved inter-scanner MS lesion segmentation by adversarial training on longitudinal data. In: Crimi, A., Bakas, S. (eds.) BrainLes 2019. LNCS, vol. 11992, pp. 98–107. Springer, Cham (2020). https://doi.org/10.1007/978-3-030-46640-4_10

10. Birenbaum, A., Greenspan, H.: Longitudinal multiple sclerosis lesion segmentation using multi-view convolutional neural networks. In: Carneiro, G., et al. (eds.) LABELS/DLMIA -2016. LNCS, vol. 10008, pp. 58–67. Springer, Cham (2016). https://doi.org/10.1007/978-3-319-46976-8_7

11. Brosch, T., et al.: Deep 3D conv encoder networks with shortcuts for multiscale feature integration applied to MS lesion segmentation. IEEE Trans. Med. Imaging 35, 1229–1239 (2016)

12. Brugnara, G., et al.: Automated volumetric assessment with ANN might enable a more accurate assessment of disease burden in patients with MS. Eur. Radiol. 30, 2356–2364 (2020)

13. Calabrese, M., et al.: Cortical lesions and atrophy associated with cognitive impairment in relapsing-remitting multiple sclerosis. Arch. Neurol. 66, 1144–1150 (2009)

14. Carass, A., et al.: Longitudinal multiple sclerosis lesion segmentation: resource and challenge. NeuroImage 148, 77–102 (2017)

15. Carass, A., et al.: Evaluating white matter lesion segmentations with refined Sørensen-Dice analysis. Sci. Rep. 10, 1–19 (2020)

16. Cohen, G., et al.: Learning probabilistic fusion of multilabel lesion contours. In: ISBI (2020)

17. Commowick, O., et al.: Objective evaluation of MS lesion segmentation using a data management and processing infrastructure. Sci. Rep. 8, 1–17 (2018)

18. Coronado, I., et al.: Deep learning segmentation of gadolinium-enhancing lesions in multiple sclerosis. Multiple Sclerosis J. (2020)

19. Danelakis, A., et al.: Survey of automated MS lesion segmentation techniques on magnetic resonance imaging. Comput. Med. Imaging Graph. 70, 80–113 (2018)

20. Denner, S., et al.: Spatio-temporal learning from longitudinal data for multiple sclerosis lesion segmentation. arXiv:2004.03675 (2020)

21. Duchi, J., et al.: Adaptive subgradient methods for online learning and stochastic optimization. J. Mach. Learn. Res. (2011)

22. Duong, M.T., et al.: Convolutional neural network for automated flair lesion segmentation on clinical brain MR imaging. AJNR Am. J. Neuroradiol. 40, 1282–1290 (2019)

23. Durso-Finley, J., Arnold, D.L., Arbel, T.: Saliency based deep neural network for automatic detection of gadolinium-enhancing multiple sclerosis lesions in brain MRI. In: Crimi, A., Bakas, S. (eds.) BrainLes 2019. LNCS, vol. 11992, pp. 108–118. Springer, Cham (2020). https://doi.org/10.1007/978-3-030-46640-4_11

24. Fartaria, M.J., Roche, A., Meuli, R., Granziera, C., Kober, T., Bach Cuadra, M.: Segmentation of cortical and subcortical multiple sclerosis lesions based on constrained partial volume modeling. In: Descoteaux, M., Maier-Hein, L., Franz, A., Jannin, P., Collins, D.L., Duchesne, S. (eds.) MICCAI 2017. LNCS, vol. 10435, pp. 142–149. Springer, Cham (2017). https://doi.org/10.1007/978-3-319-66179-7_17

25. Feng, Y., et al.: A self-adaptive network for multiple sclerosis lesion segmentation from multi-contrast MRI with various imaging sequences. In: ISBI (2019)

26. Fenneteau, A., et al.: Learning a CNN on multiple sclerosis lesion segmentation with self-supervision. In: IS&T Electronic Imaging 2020 Symposium (2020)

27. Fonov, V.S., et al.: Unbiased nonlinear average age-appropriate brain templates from birth to adulthood. NeuroImage (2009)

28. Gessert, N., et al.: 4D Deep learning for multiple sclerosis lesion activity segmentation. arXiv:2004.09216 (2020)

29. Gessert, N., et al.: Multiple sclerosis lesion activity segmentation with attention-guided two-path CNNs. Comput. Med. Imaging Graph. **84**, 101772 (2020)

30. Ghafoorian, M., et al.: Location sensitive deep convolutional neural networks for segmentation of white matter hyperintensities. Sci. Rep. **7**, 1–12 (2017)

31. Ghafoorian, M., et al.: Transfer learning for domain adaptation in MRI: application in brain lesion segmentation. In: Descoteaux, M., Maier-Hein, L., Franz, A., Jannin, P., Collins, D.L., Duchesne, S. (eds.) MICCAI 2017. LNCS, vol. 10435, pp. 516–524. Springer, Cham (2017). https://doi.org/10.1007/978-3-319-66179-7_59

32. Hashemi, S.R., et al.: Asymmetric loss functions and deep densely-connected networks for highly-imbalanced medical image segmentation: application to multiple sclerosis lesion detection. IEEE Access **7**, 1721–1735 (2018)

33. Havaei, M., Guizard, N., Chapados, N., Bengio, Y.: HeMIS: hetero-modal image segmentation. In: Ourselin, S., Joskowicz, L., Sabuncu, M.R., Unal, G., Wells, W. (eds.) MICCAI 2016. LNCS, vol. 9901, pp. 469–477. Springer, Cham (2016). https://doi.org/10.1007/978-3-319-46723-8_54

34. Hou, B., et al.: Cross attention densely connected networks for multiple sclerosis lesion segmentation. In: BIBM (2019)

35. Hu, C., et al.: ACU-Net: a 3D attention context u-net for multiple sclerosis lesion segmentation. In: ICASSP (2020)

36. Isensee, F., et al.: NNU-Net: breaking the spell on successful medical image segmentation. arXiv:1904.08128 (2019)

37. Kamnitsas, K., et al.: Efficient multi-scale 3D CNN with fully connected CRF for accurate brain lesion segmentation. Med. Image Anal. **36**, 61–78 (2017)

38. Kang, K., Wang, X.: Fully convolutional neural networks for crowd segmentation. arXiv:1411.4464 (2014)

39. Kats, E., et al.: Soft labeling by distilling anatomical knowledge for improved MS lesion segmentation. In: ISBI (2019)

40. Kats, E., Goldberger, J., Greenspan, H.: A soft STAPLE algorithm combined with anatomical knowledge. In: Shen, D., et al. (eds.) MICCAI 2019. LNCS, vol. 11766, pp. 510–517. Springer, Cham (2019). https://doi.org/10.1007/978-3-030-32248-9_57

41. Kaur, A., et al.: State-of-the-art segmentation techniques and future directions for multiple sclerosis brain lesions. Arch. Comput. Methods Eng. (2020)

42. Kazancli, E., et al.: Multiple sclerosis lesion segmentation using improved convolutional neural networks. In: VISIGRAPP (2018)

43. Kingma, D.P., Ba, J.: Adam: a method for stochastic optimization. arXiv:1412.6980 (2014)

44. Krizhevsky, A., et al.: ImageNet classification with deep CNN. In: NIPS (2012)

45. Krüger, J., et al.: Fully automated longitudinal segmentation of new or enlarged multiple sclerosis lesions using 3D convolutional neural networks. Neuroimage Clin (2020)

46. La Rosa, F., et al.: Shallow vs deep learning architectures for white matter lesion segmentation in the early stages of multiple sclerosis. In: Crimi, A., Bakas, S., Kuijf, H., Keyvan, F., Reyes, M., van Walsum, T. (eds.) BrainLes 2018. LNCS, vol. 11383, pp. 142–151. Springer, Cham (2019). https://doi.org/10.1007/978-3-030-11723-8_14

47. La Rosa, F., et al.: Automated detection of cortical lesions in multiple sclerosis patients with 7T MRI. arXiv:2008.06780 (2020)

48. La Rosa, F., et al.: Multiple sclerosis cortical and WM lesion segmentation at 3T MRI: a deep learning method based on flair and MP2RAGE. Neuroimage Clin. **27**, 102335 (2020)

49. Lesjak, Ž., et al.: A novel public MR image dataset of multiple sclerosis patients with lesion segmentations based on multi-rater consensus. Neuroinformatics **16**, 51–63 (2018)

50. Lin, T.Y., et al.: Focal loss for dense object detection. In: ICCV (2017)

51. Long, J., et al.: Fully convolutional networks for semantic segmentation. In: CVPR (2015)

52. Ma, T., et al.: Ensembling low precision models for binary biomedical image segmentation. arXiv:2010.08648 (2020)

53. McKinley, R., et al.: Nabla-net: a deep dag-like convolutional architecture for biomedical image segmentation. In: Crimi, A., Menze, B., Maier, O., Reyes, M., Winzeck, S., Handels, H. (eds.) BrainLes 2016. LNCS, vol. 10154, pp. 119–128. Springer, Cham (2016). https://doi.org/10.1007/978-3-319-55524-9_12

54. McKinley, R., et al.: Simultaneous lesion and neuroanatomy segmentation in multiple sclerosis using deep neural networks. arXiv:1901.07419 (2019)

55. McKinley, R., et al.: Automatic detection of lesion load change in Multiple Sclerosis using convolutional neural networks with segmentation confidence. Neuroimage Clin. **25**, 102104 (2020)

56. Milletari, F., et al.: V-Net: fully convolutional neural networks for volumetric medical image segmentation. In: 3DV (2016)

57. Nair, T., et al.: Exploring uncertainty measures in deep networks for multiple sclerosis lesion detection and segmentation. Med. Image Anal. **59**, 101557 (2020)

58. Narayana, P.A., et al.: Multimodal MRI segmentation of brain tissue and T2-hyperintense white matter lesions in multiple sclerosis using deep convolutional neural networks and a large multi-center image database. In: CIBEC (2018)

59. Narayana, P.A., et al.: Are multi-contrast magnetic resonance images necessary for segmenting multiple sclerosis brains? A large cohort study based on deep learning. Magn. Reson. Imaging **65**, 8–14 (2020)

60. Narayana, P.A., et al.: Deep-learning-based neural tissue segmentation of MRI in multiple sclerosis: effect of training set size. J. Magn. Reson. Imaging **51**, 1487–1496 (2020)

61. Oguz, I., et al.: Dice overlap measures for objects of unknown number: application to lesion segmentation. In: Crimi, A., Bakas, S., Kuijf, H., Menze, B., Reyes, M. (eds.) BrainLes 2017. LNCS, vol. 10670, pp. 3–14. Springer, Cham (2018). https://doi.org/10.1007/978-3-319-75238-9_1

62. Placidi, G., Cinque, L., Polsinelli, M., Splendiani, A., Tommasino, E.: Automatic framework for multiple sclerosis follow-up by magnetic resonance imaging for reducing contrast agents. In: Ricci, E., Rota Bulò, S., Snoek, C., Lanz, O., Messelodi, S., Sebe, N. (eds.) ICIAP 2019. LNCS, vol. 11752, pp. 367–378. Springer, Cham (2019). https://doi.org/10.1007/978-3-030-30645-8_34

63. Ravnik, D., et al.: Dataset variability leverages white-matter lesion segmentation performance with CNN. In: Medical Imaging 2018: Image Processing (2018)

64. Reddi, S.J., et al.: On the convergence of adam and beyond. arXiv:1904.09237 (2019)

65. Ronneberger, O., Fischer, P., Brox, T.: U-Net: convolutional networks for biomedical image segmentation. In: Navab, N., Hornegger, J., Wells, W.M., Frangi, A.F. (eds.) MICCAI 2015. LNCS, vol. 9351, pp. 234–241. Springer, Cham (2015). https://doi.org/10.1007/978-3-319-24574-4_28

66. Roy, S., et al.: Multiple sclerosis lesion segmentation from brain MRI via fully convolutional neural networks. arXiv:1803.09172 (2018)
67. Salehi, S.S.M., Erdogmus, D., Gholipour, A.: Tversky loss function for image segmentation using 3D fully convolutional deep networks. In: Wang, Q., Shi, Y., Suk, H.-I., Suzuki, K. (eds.) MLMI 2017. LNCS, vol. 10541, pp. 379–387. Springer, Cham (2017). https://doi.org/10.1007/978-3-319-67389-9_44
68. Salem, M., et al.: Multiple sclerosis lesion synthesis in MRI using an encoder-decoder U-Net. IEEE Access **7**, 25171–25184 (2019)
69. Salem, M., et al.: A fully convolutional neural network for new T2-W lesion detection in multiple sclerosis. Neuroimage Clin. **25**, 102149 (2020)
70. Sepahvand, N.M., et al.: CNN detection of new and enlarging multiple sclerosis lesions from longitudinal MRI using subtraction images. In: ISBI (2020)
71. Shachor, Y., et al.: A mixture of views network with applications to multi-view medical imaging. Neurocomputing **374**, 1–9 (2020)
72. Shinohara, R.T., et al.: Statistical normalization techniques for magnetic resonance imaging. NeuroImage Clin. **6**, 9–19 (2014)
73. Smith, S.M.: Fast robust automated brain extraction. Hum. Brain Mapp. **17**, 143–155 (2002)
74. Styner, M., et al.: 3D segmentation in the clinic: a grand challenge II: MS lesion segmentation. Midas J. (2008)
75. Tustison, N.J., et al.: N4ITK: improved N3 bias correction. IEEE Trans. Med. Imaging **29**, 1310–1320 (2010)
76. Ulloa, G., Naranjo, R., Allende-Cid, H., Chabert, S., Allende, H.: Circular non-uniform sampling patch inputs for CNN applied to multiple sclerosis lesion segmentation. In: Vera-Rodriguez, R., Fierrez, J., Morales, A. (eds.) CIARP 2018. LNCS, vol. 11401, pp. 673–680. Springer, Cham (2019). https://doi.org/10.1007/978-3-030-13469-3_78
77. Ulloa, G., Veloz, A., Allende-Cid, H., Allende, H.: Improving multiple sclerosis lesion boundaries segmentation by convolutional neural networks with focal learning. In: Campilho, A., Karray, F., Wang, Z. (eds.) ICIAR 2020. LNCS, vol. 12132, pp. 182–192. Springer, Cham (2020). https://doi.org/10.1007/978-3-030-50516-5_16
78. Vaidya, S., et al.: Longitudinal multiple sclerosis lesion segmentation using 3D convolutional neural networks. In: Proceedings of the 2015 Longitudinal Multiple Sclerosis Lesion Segmentation Challenge (2015)
79. Valverde, S., et al.: Improving automated multiple sclerosis lesion segmentation with a cascaded 3D convolutional neural network approach. NeuroImage **155**, 159–168 (2017)
80. Valverde, S., et al.: One-shot domain adaptation in multiple sclerosis lesion segmentation using convolutional neural networks. Neuroimage Clin. **21**, 101638 (2019)
81. Vang, Y.S., et al.: SynergyNet: a fusion framework for multiple sclerosis brain MRI segmentation with local refinement. In: ISBI (2020)
82. Varsavsky, T., Orbes-Arteaga, M., Sudre, C.H., Graham, M.S., Nachev, P., Cardoso, M.J.: Test-time unsupervised domain adaptation. In: Martel, A.L., et al. (eds.) MICCAI 2020. LNCS, vol. 12261, pp. 428–436. Springer, Cham (2020). https://doi.org/10.1007/978-3-030-59710-8_42
83. Wang, Z., Smith, C.D., Liu, J.: Ensemble of multi-sized FCNs to improve white matter lesion segmentation. In: Shi, Y., Suk, H.-I., Liu, M. (eds.) MLMI 2018. LNCS, vol. 11046, pp. 223–232. Springer, Cham (2018). https://doi.org/10.1007/978-3-030-00919-9_26

84. Warfield, S.K., et al.: Simultaneous truth and performance level estimation (staple): an algorithm for the validation of image segmentation. IEEE Trans. Med. Imaging **23**, 903–921 (2004)

85. Weeda, M., et al.: Comparing lesion segmentation methods in multiple sclerosis. Neuroimage Clin. **24**, 102074 (2019)

86. Xiang, Y., et al.: Segmentation method of multiple sclerosis lesions based on 3D-CNN networks. IET Image Process. **14**, 1806–1812 (2020)

87. Zeiler, M.D.: ADADELTA: an adaptive learning rate method. arXiv:1212.5701 (2012)

88. Zhang, C., et al.: MS-GAN: GAN-based semantic segmentation of multiple sclerosis lesions in brain magnetic resonance imaging. In: DICTA (2018)

89. Zhang, H., et al.: RSANet: recurrent slice-wise attention network for multiple sclerosis lesion segmentation. In: Shen, D., et al. (eds.) MICCAI 2019. LNCS, vol. 11766, pp. 411–419. Springer, Cham (2019). https://doi.org/10.1007/978-3-030-32248-9_46

90. Zhang, H., et al.: Efficient folded attention for 3D medical image reconstruction and segmentation. arXiv:2009.05576 (2020)

91. Zhang, H., et al.: Multiple sclerosis lesion segmentation with tiramisu and 2.5D stacked slices. In: Shen, D., et al. (eds.) MICCAI 2019. LNCS, vol. 11766, pp. 338–346. Springer, Cham (2019). https://doi.org/10.1007/978-3-030-32248-9_38

92. Zhang, H., Bakshi, R., Bagnato, F., Oguz, I.: Robust multiple sclerosis lesion inpainting with edge prior. In: Liu, M., Yan, P., Lian, C., Cao, X. (eds.) MLMI 2020. LNCS, vol. 12436, pp. 120–129. Springer, Cham (2020). https://doi.org/10.1007/978-3-030-59861-7_13

93. Zhang, L., et al.: Learning to segment when experts disagree. In: Martel, A.L., et al. (eds.) MICCAI 2020. LNCS, vol. 12261, pp. 179–190. Springer, Cham (2020). https://doi.org/10.1007/978-3-030-59710-8_18

Computational Diagnostics of GBM Tumors in the Era of Radiomics and Radiogenomics

Anahita Fathi Kazerooni and Christos Davatzikos[✉]

Center for Biomedical Image Computing and Analytics, Department of Radiology,
University of Pennsylvania, Philadelphia 19104, USA
Christos.Davatzikos@pennmedicine.upenn.edu

Abstract. Machine learning (ML) integrated with medical imaging has introduced new perspectives in precision diagnostics of GBM tumors, through radiomics and radiogenomics. This has raised hopes for developing non-invasive and in-vivo biomarkers for prediction of patient survival, tumor recurrence, or molecular characterization, and therefore, encouraging treatments tailored to individualized needs. Characterization of tumor infiltration based on pre-operative multi-parametric magnetic resonance imaging (MP-MRI) scans would help in predicting the loci of future tumor recurrence, and thereby aiding in planning the course of treatment for the patients, such as increasing the resection or escalating the dose of radiation. Specifying molecular properties of GBM tumors and prediction of their changes over time and with treatment would help characterize the molecular heterogeneity of a tumor, and potentially use a respective combination treatment. In this article, we will provide examples of our work on radiomics and radiogenomics, aiming to offer personalized treatments to patients with GBM tumors.

Keywords: Radiomics · Machine learning · Glioblastoma

1 Introduction

Glioblastoma (GBM) is the most common and fatal primary brain tumor. The current standard of care for treatment of patients with GBM tumors involves maximal safe tumor resection followed by radiotherapy and adjuvant temozolomide (TMZ) chemotherapy, and maintenance TMZ therapy for 6–12 months. The standard treatment at the very best provides the patients with a median progression-free survival (PFS) of 6.2–7.5 months and an overall survival (OS) of around 14.6–16.7 months [1, 2]. The patients who tolerate TMZ treatment and do not show tumor progression, can be prescribed to receive tumor-treating fields (TTFields), improving the prognosis only to a median OS of 20.9 months [1, 2]. One of the main reasons for the failure of treatments in GBM patients is marked intra-tumor heterogeneity of GBM tumors, diffuse and immense infiltration of tumor cells in the adjacent brain parenchyma which mainly remain untreated, and resistance of tumor subpopulations to the given therapies.

Medical imaging, and specifically magnetic resonance imaging (MRI), has evolved into an indispensable diagnostic tool in neuro-oncology. It can contribute to personalized

© Springer Nature Switzerland AG 2021
A. Crimi and S. Bakas (Eds.): BrainLes 2020, LNCS 12658, pp. 30–38, 2021.
https://doi.org/10.1007/978-3-030-72084-1_3

patient management by offering patient prognosis, treatment guidance, and monitoring the response of a patient to the therapy, based on specific characteristics of the tumor, manifested with different phenotypes on MRI scans [3]. A mounting body of literature over the past decade has shown that subvisual aspects of GBM tumor heterogeneity can be captured by integrating characteristics that relate to cellular density, neo-angiogenesis, water content, etc. from multi-parametric MRI (MP-MRI) scans [4, 5].

Radiomics is an emerging computational method that combines diverse imaging features through machine learning (ML) modeling into distinctive imaging signatures [6–8]. It can reveal patterns underlying the tumor's progression, response to standard, adjuvant, or novel therapies, and can help to achieve a more personalized medicine for the GBM patients. Imaging phenotypes discovered by radiomics have shown promise in risk stratification, prediction of overall and progression-free survival, disease follow-up (discrimination of true vs pseudo-progression of the disease), characterization of tumor genomics [9], and upfront prediction of the response to treatment.

In this article, we will briefly review the proposed techniques for characterization, prognostication, and treatment planning of the patients with GBM.

2   Patient Prognosis

Upfront prediction of PFS and OS could potentially identify high-risk patients, who are suspected to have a short OS, e.g. of less than 6 months, and enroll them to alternative therapies or palliative care, depending on personal choices. While OS is a primary endpoint for determination of the efficacy of treatment strategies in clinical trials, PFS could serve as a surrogate for OS to overcome several limitations of OS, such as long trial times [10]. Moreover, as patients with GBM have a poor prognosis, with PFS as an early marker of OS, the course of the prescribed treatments for the patients can be modified or changed. In this regard, radiomics has shown potential in stratification of high- and low-risk patients, and prediction of OS or PFS.

In our 2016 study, quantitative imaging features were extracted from pre-operative MP-MRI scans, including pre- and post-contrast T1-weighted images (T1, T1-Gd), T2-weighted (T2), T2 fluid attenuated inversion recovery (FLAIR), dynamic susceptibility contrast-enhanced (DSC)-MRI, and diffusion tensor imaging (DTI). Features of intensity from the MP-MRI scans, volume, location, and growth model parameters were combined by support vector machine (SVM) classification algorithm for stratification of patients with GBM tumors into groups of short, medium, and short survivors. The results suggested that an overall 3-way classification into short/medium/long survivor groups was around 79% in the prospective cohort [11].

As DSC-MRI and DTI sequences are not frequently included with the routine pre-surgical brain tumor protocols at many imaging centers, in a later study, we investigated the accuracy of a predictive model built based on augmented radiomics feature panel (ARFP), including morphology and textural descriptors, extracted only from basic MP-MRI (Bas-mpMRI, comprising of T1, T1-Gd, T2, FLAIR) for stratification of OS risk groups [12]. This predictive model, generated with ARFP from Bas-mpMRI yielded a comparable accuracy to the previous model [11] that was generated with advanced MP-MRI (Adv-mpMRI) scans without using ARFP [12]. It was proposed that Bas-mpMRI and advanced radiomics features can compensate for the lack of Adv-mpMRI [12]. This approach also showed generalizability across different scanners in a multi-center study [13].

In a recent multi-center study, we investigated the role of radiomics in prediction of PFS based on MP-MRI scans, acquired prior to the primary surgery in patients with GBM [14, 15]. A prognostic model was generated with a rich panel of quantitative features of intensity, first-order histogram, texture, morphology, and volume, through an SVM classifier, resulting an AUC = 0.82 for the data from two institutions. This radiomics study was carried out using the publicly available and open-source Cancer Imaging Phenomics Toolkit (CaPTk) software [14].

A few studies have reported the added value of radiomics models to clinical and molecular predictors for risk stratification of patients with GBM tumors. A radiomics study of GBM patients showed a Concordance index (C-index) = 0.696 for prediction of OS by integrating radiomics and clinical variables, i.e. age and Karnofsky performance score (KPS), compared to a C-index = 0.640 with only clinical variables [16]. Similarly, on a larger cohort of patients, it was suggested that a combination of key clinical characteristics, i.e. age, extent of resection, and KPS, with molecular diagnosis, i.e. MGMT methylation status, yielding integrated Brier scores (IBS) of 0.119 and 0.098 for prediction of OS and PFS, respectively, improved to 0.103 and 0.0809 when integrated with radiomics variables. For prediction of PFS, [17]. Another study investigated the improvement of survival prediction using radiomics integrated with clinical and molecular profiles [18]. They found an improvement of risk stratification of GBM patients into low and high survivor groups when radiomics was combined with clinical and molecular variables, denoting an area under the curve (AUC) = 0.78 compared to an AUC = 0.70 for a model based only on clinical and molecular variables [18].

3   Intratumor Heterogeneity and Tumor Recurrence

A hallmark characteristic of GBM is diffuse infiltration into the surrounding brain tissue, extending beyond the hyperintense regions visible on T1-Gd MRI scans into the peritumoral edema, and leading to tumor recurrence. Mapping peritumoral infiltration would augment precision treatment through escalating the radiation therapy dose in densely infiltrated regions, and potentially prolonging survival of the patients [11]. Pattern analysis of MP-MRI scans can reveal infiltration of tumor cells in the peritumoral edema by quantification of spatial heterogeneity in terms of the changes in regional microvasculature, microstructure, and water content [4]. We have developed an imaging signature of tumor infiltration that serves as an early biomarker of the likely location of tumor recurrence [19]. Quantitative features estimated from MP-MRI scans (T1, T1-Gd, T2, FLAIR, DTI, and DSC-MRI) within peritumoral edema were combined using an ML approach to generate predictive maps of tumor recurrence, with an odds ratio of 9.29, AUC of 0.84, sensitivity of 91%, and specificity of 93%. Figure 1 shows examples of infiltration maps generated using pre-operative MP-MRI scans of two patients with GBM tumors and the location of tumor recurrence in the patients.

A common dilemma in evaluating the response of GBM tumors to therapy is differentiation of progressive disease, or true progression (TP), from pseudo-progression (PsP).

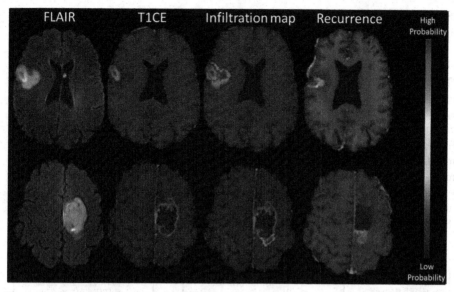

Fig. 1. An illustration of two examples of generating infiltration maps from pre-operative MP-MRI scans. The images on the right side indicate the corresponding recurrence scans of the same patients. As it can be inferred from the images, the predicted infiltration maps are highly predictive of future recurrence.

Radiomics signatures distinguishing between TP and PsP in patients with GBM tumors have been reported including a recent study of our group, on a cohort of GBM patients who underwent second resection due to progressive radiographic changes suspicious for recurrence [20]. A multivariate analysis of deep learning and conventional features from multi-parametric MRI scans was performed, showing an accuracy of 87% for predicting PsP and 84% for predicting TP that compared with similar categories blindly defined by board-certified neuropathologists [20]. In another recent study on a cohort of patients with GBM tumors, a feature learning method based on deep convolutional generative adversarial networks (DCGAN) and AlexNet was implemented to discriminate between PsP and TP, which showed a performance of AUC > 0.90 [21].

4   Radiogenomics

Advances in genomic profiling of tissue specimens in a variety of diseases, especially cancers, has encouraged development of treatments targeted at the genetic makeup of the tumor and paved the way towards personalized treatments and precision medicine [22]. However, several factors, including tumor heterogeneity, sampling error during biopsy, insufficient tissue quality for sequencing, limitations of the sequencing methods, etc., may hinder characterization of tumor genomics. Radiographic imaging phenotypes have shown strong associations with the underlying biology of GBM tumors. Through an ML approach, radiogenomics studies aim to bridge the gap between the two disciplines by generating imaging signatures that represent genetic characteristics or heterogeneity of the tumor. Thereby, radiogenomics signatures can serve as noninvasive biomarkers for tumor genomics or as complementary data for predicting patient prognosis [9].

In our radiogenomics study for generating a signature of EGFRvIII mutation in GBM tumors, we integrated quantitative features derived from MP-MRI through an ML approach that predicted the EGFRvIII mutation status with an accuracy of 87% in a replication cohort [23]. The results suggested that the tumors with EGFRvIII mutations had a propensity to occur in the frontal and temporal regions of the brain, and were associated with higher neovascularization and cell density compared to the wildtype tumors [23]. Figure 2 shows the descriptive characteristics of GBM tumors with EGFRvIII mutation. We further found that heterogeneity in hemodynamic patterns within peritumoral edema, quantified by pattern analysis of perfusion MRI scans, is strongly linked to EGFRvIII mutation status [24]. EGFRvIII mutant tumors displayed a highly infiltrative-migratory phenotype while the wildtype tumors had a confined vascularization within their peritumoral area [24].

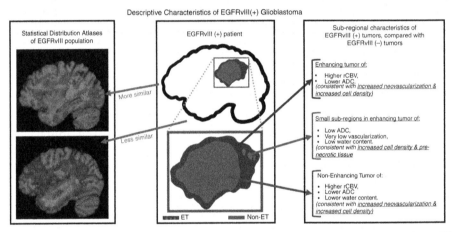

Fig. 2. This image illustrates the characteristics of GBM tumors with mutation in EGFRvIII (EGFRvIII(+)). As indicated, EGFRvIII(+) patients have a fronto-parietal propensity, and show increased neovascularization and cell density [23].

In a study on exploring the synergies between imaging and genomics, we identified three distinct and reproducible imaging subtypes, including rim-enhancing, irregular, and solid, which exhibit clearly different clinical outcome and molecular characteristics, including IDH1, MGMT methylation, EGFRvIII, and transcriptomic molecular subtypes, i.e. classical, mesenchymal, neural, proneural [25]. Our findings signify the importance of precision diagnostics and personalized therapies for patients with GBM tumors.

There are many other notable radiogenomics studies of GBM tumors, which overviewing them is out of scope of this paper. We refer the interested readers to the relevant review papers [9, 22, 26–28].

5 Current Challenges and Future Directions

Despite the promises that radiomics and radiogenomics offer for achieving precision diagnostics in management of GBM tumors, challenges of reproducibility and generalizability of the proposed methods have yet to be tackled for these methods to be translated into clinical applications. First, radiomics is not directly related to biological characteristics, and reproducibility of the features mainly depends on the imaging process, from acquisition, to post-processing and feature extraction. Most radiomics studies are retrospective, therefore, not all aspects of reproducibility such as standardization of image acquisition protocol, intra-patient test-retest repeatability, and across scanner reproducibility can be addressed [29]. This issue complicates generalizability of the radiomics models across different institutions. The clinical trials follow standardized image acquisition guidelines, although the number of data is usually limited and data sharing is restricted due to ownership concerns [9].

Another impedance to reproducible radiomics research is variability in analysis methods, i.e. different results can be achieved for the same data depending on the feature extraction, feature selection and modeling approaches. Lack of clear and comprehensive description of research methodology, including data processing and parametrization of the computational algorithms, and unwillingness or hesitance of the researchers to share their methods with the community mainly due to an understandable reason of intellectual property, further poses a challenge to research reproducibility [9]. In this regard, multiple open-source software toolkits have been developed and made publicly available. To this end, Cancer Imaging Phenomics Toolkit (CaPTk), an imaging analytics suite of open-source software algorithms, has been implemented to facilitate estimation of extensive panels of quantitative imaging phenomics (CIPh) features and integrating them with prognostic models to support a course of tailored treatment strategies for the patients [30]. CaPTk has a three-level functionality design for radiomics analysis. The users can build their image preprocessing pipelines, composed of algorithms for conversion of image format, registration of the scans, segmentation of the desired regions, artifact reduction, and intensity normalization with the tools provided in the first-level functionality. The second level provides general-purpose applications, including extensive set of radiomics features with adjustable parameters for feature extraction, feature selection, and ML model construction. Numerous features representing morphology, volume, intensity, and regional heterogeneity, compliant with the guidelines provided by the Image Biomarker Standardization Initiative (IBSI) [31] to ensure reproducibility and comparability, can be extracted. In the third level, the methods in the first level features have been synthesized into a smaller and meaningful subset of features, combined with second level ML algorithms for specialized applications that aim to support specific clinical applications, including risk stratification of patients, stratification of patients according to their transcriptomic molecular subtypes, predicting genomics of the tumor, etc. [30].

As any other ML approach, radiomics studies require ample and diverse data for learning the underlying patterns of the disease, and to overcome the so-called "curse of dimensionality" problem, i.e. a remarkably larger number of features compared to the number of samples. Clinical data collected at a single institution is usually limited in the number and diversity, thereby, hampering generalization of the ML methods [9]. These challenges motivated the formation of ReSPOND (Radiomics Signatures for PrecisiON Diagnostics) consortium, as an international initiative for machine learning in GBM imaging [32]. ReSPOND is a collaborative effort of over 10 international institutions that aims to gather brain MRI scans from over 3000 de novo GBM patients and develop radiomics biomarkers for personalized prognostication. The main four areas of focus for ReSPOND include prediction of OS and PFS, early prediction of tumor recurrence to help in adopting aggressive treatments of GBMs through an extended resection and escalation of the dose within the peritumoral regions that are suspected of recurrence, differentiation of true tumor progression from pseudo-progression, and prediction of molecular characteristics of the GBM tumors [32, 33].

6 Conclusion

Artificial intelligence, in the forms of radiomics and radiogenomics, has introduced appealing solutions to the current clinical problems for management of GBM tumors and has raised the hopes for accomplishing the purpose of tailoring diagnosis and treatments for the patients at an individual level. Risk stratification of the GBM patients by upfront projection of their OS and PFS, early prediction of tumor recurrence, distinguishing TP from PsP, and prediction of the molecular properties of the tumor and the spatial heterogeneity are among the key applications of radiomics and radiogenomics. Nonetheless, these promising tools face the challenges of reproducibility and generalizability that need to be carefully addressed by the community.

References

1. Stupp, R., Taillibert, S., Kanner, A.A., et al.: Maintenance therapy with tumor-treating fields plus temozolomide vs temozolomide alone for glioblastoma a randomized clinical trial. JAMA – J. Am. Med. Assoc. **314**, 2535–2543 (2015). https://doi.org/10.1001/jama.2015.16669
2. Stupp, R., Taillibert, S., Kanner, A., et al.: Effect of tumor-treating fields plus maintenance temozolomide vs maintenance temozolomide alone on survival in patients with glioblastoma a randomized clinical trial. JAMA – J. Am. Med. Assoc. **318**, 2306–2316 (2017). https://doi.org/10.1001/jama.2017.18718
3. Davatzikos, C., Sotiras, A., Fan, Y., et al.: Precision diagnostics based on machine learning-derived imaging. Magn. Reson. Imaging **64**, 49–61 (2019). https://doi.org/10.1016/j.mri.2019.04.012
4. Kazerooni, A.F., Nabil, M., Zadeh, M.Z., et al.: Characterization of active and infiltrative tumorous subregions from normal tissue in brain gliomas using multiparametric MRI. J. Magn. Reson. Imaging **48**, 938–950 (2018). https://doi.org/10.1002/jmri.25963
5. Fathi Kazerooni, A., Mohseni, M., Rezaei, S., Bakhshandehpour, G., Saligheh Rad, H.: Multi-parametric (ADC/PWI/T2-w) image fusion approach for accurate semi-automatic segmentation of tumorous regions in glioblastoma multiforme. Magn. Reson. Mater. Phys., Biol. Med. **28**(1), 13–22 (2014). https://doi.org/10.1007/s10334-014-0442-7
6. Gillies, R.J., Kinahan, P.E., Hricak, H.: Radiomics: images are more than pictures, they are data. Radiology **278**, 563–577 (2016). https://doi.org/10.1148/radiol.2015151169
7. Kumar, V., Gu, Y., Basu, S., et al.: Radiomics: the process and the challenges. Magn. Reson. Imaging **30**, 1234–1248 (2012). https://doi.org/10.1016/j.mri.2012.06.010

8. Gatenby, R.A., Grove, O., Gillies, R.J.: Quantitative imaging in cancer evolution and ecology. Radiology **269**, 8–15 (2013). https://doi.org/10.1148/radiol.13122697

9. Fathi Kazerooni, A., Bakas, S., Saligheh Rad, H., Davatzikos, C.: Imaging signatures of glioblastoma molecular characteristics: a radiogenomics review. J. Magn. Reson. Imaging **52**, 54–69 (2019). https://doi.org/10.1002/jmri.26907

10. Han, K., Ren, M., Wick, W., et al.: Progression-free survival as a surrogate endpoint for overall survival in glioblastoma: a literature-based meta-analysis from 91 trials. Neuro Oncol. **16**, 696–706 (2014). https://doi.org/10.1093/neuonc/not236

11. Macyszyn, L., Akbari, H., Pisapia, J.M., et al.: Imaging patterns predict patient survival and molecular subtype in glioblastoma via machine learning techniques. Neuro Oncol. **18**, 417–425 (2016). https://doi.org/10.1093/neuonc/nov127

12. Bakas, S., Shukla, G., Akbari, H., Erus, G.: Overall survival prediction in glioblastoma patients using structural magnetic resonance imaging (MRI): advanced radiomic features may compensate for lack of advanced MRI modalities. J. Med. Imaging **7**, 1–18 (2020). https://doi.org/10.1117/1.JMI.7.3.031505

13. Bakas, S., Akbari, H., Shukla, G., et al.: Deriving stable multi-parametric MRI radiomic signatures in the presence of inter-scanner variations: survival prediction of glioblastoma via imaging pattern analysis and machine learning techniques, p. 1057509:8 (2018). https://doi.org/10.1117/12.2293661

14. Fathi Kazerooni, A., Akbari, H., Shukla, G., et al.: Cancer imaging phenomics via CaPTk: multi-institutional prediction of progression-free survival and pattern of recurrence in glioblastoma. JCO Clin. Cancer Inform. 234–244 (2020). https://doi.org/10.1200/cci.19.00121

15. Fathi Kazerooni, A., et al.: NIMG-35. Quantitative estimation of progression-free survival based on radiomics analysis of preoperative multi-parametric MRI in patients with glioblastoma. Neuro Oncol. **21**(Suppl_6), vi168–vi169 (2019). https://doi.org/10.1093/neuonc/noz175.705

16. Kickingereder, P., Burth, S., Wick, A., et al.: Radiomic profiling of glioblastoma: identifying an imaging predictor of patient survival with improved performance over established clinical and radiologic risk models. Radiology **280**, 880–889 (2016). https://doi.org/10.1148/radiol.2016160845

17. Kickingereder, P., Neuberger, U., Bonekamp, D., et al.: Radiomic subtyping improves disease stratification beyond key molecular, clinical, and standard imaging characteristics in patients with glioblastoma. Neuro Oncol **20**, 848–857 (2018). https://doi.org/10.1093/neuonc/nox188

18. Bae, S., Choi, Y.S., Ahn, S.S., et al.: Radiomic MRI phenotyping of glioblastoma: improving survival prediction. Radiology **289**, 797–806 (2018). https://doi.org/10.1148/radiol.2018180200

19. Akbari, H., Macyszyn, L., Da, X., et al.: Imaging surrogates of infiltration obtained via multi-parametric imaging pattern analysis predict subsequent location of recurrence of glioblastoma. Neurosurgery **78**, 572–580 (2016). https://doi.org/10.1227/NEU.0000000000001202

20. Akbari, H., Rathore, S., Bakas, S., et al.: Histopathology-validated machine learning radiographic biomarker for noninvasive discrimination between true progression and pseudoprogression in glioblastoma. Cancer **126**, 2625–2636 (2020). https://doi.org/10.1002/cncr.32790

21. Li, M., Tang, H., Chan, M.D., et al.: DC-AL GAN: pseudoprogression and true tumor progression of glioblastoma multiform image classification based on DCGAN and AlexNet. Med. Phys. **47**, 1139–1150 (2020). https://doi.org/10.1002/mp.14003

22. Pinker, K., Shitano, F., Sala, E., et al.: Background, current role, and potential applications of radiogenomics. J. Magn. Reson. Imaging **47**, 604–620 (2018)

23. Akbari, H., Bakas, S., Pisapia, J.M., et al.: In vivo evaluation of EGFRvIII mutation in primary glioblastoma patients via complex multiparametric MRI signature. Neuro Oncol. **20**, 1068–1079 (2018). https://doi.org/10.1093/neuonc/noy033

24. Bakas, S., Akbari, H., Pisapia, J., et al.: In vivo detection of EGFRvIII in glioblastoma via perfusion magnetic resonance imaging signature consistent with deep peritumoral infiltration: the φ-index. Clin. Cancer Res. **23**, 4724–4734 (2017)

25. Rathore, S., Akbari, H., Rozycki, M., et al.: Radiomic MRI signature reveals three distinct subtypes of glioblastoma with different clinical and molecular characteristics, offering prognostic value beyond IDH1. Sci. Rep. **8**, 1–2 (2018)

26. Zinn, P.O., Mahmood, Z., Elbanan, M.G., Colen, R.R.: Imaging genomics in gliomas. Cancer J. **21**, 225–234 (2015)

27. Smits, M., van den Bent, M.J.: Imaging correlates of adult glioma genotypes. Radiology **284**, 316–331 (2017)

28. Ellingson, B.M.: Radiogenomics and imaging phenotypes in glioblastoma: novel observations and correlation with molecular characteristics. Curr. Neurol. Neurosci. Rep. **15**, 506 (2015)

29. Park, J.E., Kickingereder, P., Kim, H.S.: Radiomics and deep learning from research to clinical workflow: neuro-oncologic imaging. Korean J. Radiol. **21**, 1126–1137 (2020). https://doi.org/10.3348/kjr.2019.0847

30. Davatzikos, C., et al.: Cancer imaging phenomics toolkit: quantitative imaging analytics for precision diagnostics and predictive modeling of clinical outcome. J. Med. Imaging **5**(01), 1 (2018). https://doi.org/10.1117/1.JMI.5.1.011018

31. Zwanenburg, A., Vallières, M., Abdalah, M.A., et al.: The image biomarker standardization initiative: standardized quantitative radiomics for high-throughput image-based phenotyping. Radiology **295**, 328–338 (2020)

32. Davatzikos, C., Barnholtz-Sloan, J.S., Bakas, S., et al.: AI-based prognostic imaging biomarkers for precision neuro-oncology: the ReSPOND consortium. Neuro Oncol. **22**, 886–888 (2020). https://doi.org/10.1093/neuonc/noaa045

33. Davatzikos, C., et al.: NIMG-66. AI-based prognostic imaging biomarkers for precision neurooncology and the respond consortium. Neuro Oncol. **22**(Suppl_2), ii162–ii163 (2020). https://doi.org/10.1093/neuonc/noaa215.679

Brain Lesion Image Analysis

Automatic Segmentation of Non-tumor Tissues in Glioma MR Brain Images Using Deformable Registration with Partial Convolutional Networks

Zhongqiang Liu[1,3], Dongdong Gu[1,2], Yu Zhang[3], Xiaohuan Cao[1], and Zhong Xue[1(✉)]

[1] Shanghai United Imaging Intelligence Co. Ltd., Shanghai, China
gudongdong@hnu.edu.cn, zhong.xue@ieee.org
[2] Hunan University, Hunan, China
[3] Southern Medical University, Guangdong, China

Abstract. In brain tumor diagnosis and surgical planning, segmentation of tumor regions and accurate analysis of surrounding normal tissues are necessary for physicians. Pathological variability often renders difficulty to register a well-labeled normal atlas to such images and to automatic segment/label surrounding normal brain tissues. In this paper, we propose a new registration approach that first segments brain tumor using a U-Net and then simulates missed normal tissues within the tumor region using a partial convolutional network. Then, a standard normal brain atlas image is registered onto such tumor-removed images in order to segment/label the normal brain tissues. In this way, our new approach greatly reduces the effects of pathological variability in deformable registration and segments the normal tissues surrounding brain tumor well. In experiments, we used MICCAI BraTS2018 T1 and FLAIR images to evaluate the proposed algorithm. By comparing direct registration with the proposed algorithm, the results showed that the Dice coefficient for gray matters was significantly improved for surrounding normal brain tissues.

Keywords: Brain tumor · Segmentation · Image recovery · Registration · Partial convolution

1 Introduction

Precise segmentation of tumor regions and quantitative assessment of normal tissues in MR brain images play important roles in diagnosis and surgical planning of glioma and other brain tumors. Since tumors largely influence the brain morphology, it is difficult to accurately segment and label the tissues of the whole brain especially for the normal tissues surrounding the tumor area. Existing brain labeling algorithms include multi-atlas-based [1–3] and deep learning-based [4] methods. However, these algorithms can hardly tackle the tumor images, since most of them are used for normal brains (or brains

---

Z. Liu and D. Gu—These authors contribute equally to this work.

© Springer Nature Switzerland AG 2021
A. Crimi and S. Bakas (Eds.): BrainLes 2020, LNCS 12658, pp. 41–50, 2021.
https://doi.org/10.1007/978-3-030-72084-1_4

without significant morphological variation) and have not considered tumor effects. Therefore, accurate and robust segmentation and labeling of normal tissues in glioma MR brain images is still a challenging task.

Several methods have been proposed to register pathological brain images with a normal brain atlas. In [5], a feature point-based radial mass-effect model is used to first simulate tumor in the atlas and then warp it onto the patient image. Masks of pathologies are utilized in [6] so that the warping of the image regions close to a tumor is only driven by the information of the neighboring normal structures not the tumor region. However, the tumor structures vary considerably large across patients in terms of their location, shape, size, and extension, rendering difficulty in tumor simulation or masking.

In computer vision, it has been shown that recovering the appearance inside a corrupted region is possible by using certain statistics and patterns from the remaining image via deep learning by learning from a large number of samples. With convolutional neural network algorithms such inpainting can amazingly restore the cut-out regions in terms of image appearance, contrast, and even shape [7, 8]. Although only applied to regular corrupted regions such as rectangular bounding boxes, they can be extended to handle the inpainting of any irregular region. Motivated by this technique, it is believed that if the tumor area within a brain MR image is segmented and removed, such inpainting could recover the missing brain tissues, and the registration could be less affected by tumor pathology, thereby giving relatively accurate brain labeling result especially to the surrounding normal tissues of that tumor. Since the location, size, and shape of each brain tumor is different, this kind of method can be adaptive to such variability and greatly reduce the tumor effects in MR brain image registration.

Therefore, in this paper, we propose to use a partial convolution network (PConv-Net) [9–11] to first repair the brain tumor areas from the remaining normal tissues in brain MR images and then register the atlas image onto the subject image. The algorithm first uses a U-Net for tumor segmentation, so that tumors with various sizes and locations can be removed first from the MR images. Then, a PConv-Net is applied to simulate normal tissue images within the cut-out tumor regions. Finally, we perform traditional registration algorithms to warp the atlas onto each recovered subject image to get more reasonable brain labeling result.

Our contributions include: a) a framework for brain tumor image registration; b) a segmentation model to segment brain tumors with various sizes and locations in MR brain images; c) a recovering network based on PConv-Net to recover normal tissue images for brain tumor regions before applying image registration. To our best knowledge, this is the first work for MR tumor brain image registration by applying convolutional neural network inpainting.

To evaluate performance of the proposed registration framework, we use the T1 and FLAIR images from MICCAI BraTS2018 in the experiments [12, 13]. Our method not only adapts to tumor areas with different shapes and sizes but also reduces the computational cost using partial convolutional operation. The registration results from our proposed method can be used to determine the tumor location and provide segmentation of the entire MR brain volume to assist clinical diagnosis and treatment.

Specifically, we used 100 cases for training and 40 cases for testing. The segmentation of BraTS2018 MR FLAIR was used as the ground-truth for segmentation on the corresponding T1 image for training the U-Net tumor segmentation. For training the PConv-Net, the tumor regions from one diseased subjects were extracted and overlaid onto normal subjects to simulate tumor areas. In this way, the ground truth for tumor region recovery using PConv-Net is obtained. After training the PConv-Net, we used 40 cases from our in-house normal brain MR images to test the performance of our PConv network. Finally, the 40 testing images of BraTS2018 were used for tumor segmentation, tumor region recovery, and performing registration to evaluate the performance of the proposed algorithm.

2   Method

The proposed tumor image registration framework is illustrated in Fig. 1. In Fig. 1(a), a U-Net [13] is trained for tumor segmentation. In Fig. 1(b) a PConv-Net is trained to recover normal tissues within a given (tumor) region. Because the ground truth of the tumor region is unknown, to simulate the training data for PConv-Net, we applied artificial "tumor" regions on normal tissues of these images by placing real tumor masks onto normal images after proper registration. After training, we can process a tumor MR brain image by two steps: 1) segment the tumor using the trained U-Net as shown in Fig. 1(a); 2) mask-out the tumor region and recover or simulate normal tissues within the same using PConv-Net as shown in Fig. 1(b). Finally, the SyN registration algorithm [14] was applied to register an normal MR brain atlas to the subject image so that corresponding tissue segmentations and labels can be obtained. Herein, we use MNI152 as the atlas.

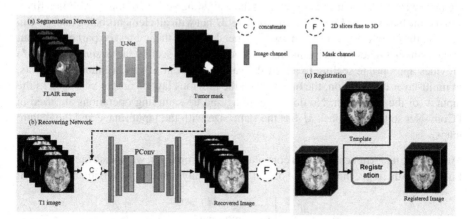

Fig. 1. The framework of the proposed method. (a) Tumor segmentation using U-Net; (b) partial convolutional network (PConv-Net) for tumor-region recovery; (c) image registration using SyN.

2.1   Tumor Segmentation Network

The traditional U-Net is used for tumor segmentation, and herein, we chose to segment the entire tumor/edema region from FLAIR images. The U-Net structure is composed of two parts: encoder and decoder structures. The encoder path contains 3 down-sampling layers and 3 max-pooling layers used for feature extraction, and the decoder path includes 3 up-sampling layers for result recovery. Skip connections are used at each resolution level. In order to minimize tumor boundary intensity propagation effects, we expand the tumor masks using morphological operations before recovery (the element unit size is chosen as 3 mm). The U-Net-based tumor segmentation is implemented in 2D with smoothness constraints across neighboring slices, so the training time and memory are greatly reduced, and we have more training samples. We use the binary cross entropy loss (BCELoss) to measure segmentation errors between the ground truth and the segmentation results of the network.

2.2 Tumor Region Recovery Network

Partial Convolution. To reduce the effect of tumor regions in image registration, a PConv-Net is applied to simulate normal tissues in order to replace the tumor region. For this purpose, we adopt the partial convolution strategy from [15], where convolutional operations are only operated on a certain region of the input image. Mathematically, one PConv layer can be expressed as:

$$
x' = \begin{cases} F^T (X \odot M) \frac{sum(1)}{sum(M)} + b, & \text{if } sum(M) > 0 \\ 0, & \text{otherwise} \end{cases}, \tag{1}
$$

where x' represents the output after convolution of the input image/feature map X. M is a binary mask, which represents the tumor mask in our study. "$\odot$" denotes the element-wise multiplication. F represents the weighting vectors of the convolution filter, and b is the bias vector. 1 has the same shape as M but with all elements being one. As can be seen, the output values of the network depend only on the unmasked part of the input image, which is determined by the shape of the mask. The scaling factor $sum(1)/sum(M)$ provides appropriate scaling to adjust the varying amount of valid (unmasked) inputs. In multi-layer convolution, the output from the previous layer's output x' is used as the input X of the next layer. No down-sampling and up-sampling operations are used in PConv-Net so that the mask M has the same size with the input image and the feature maps.

Mask Update. After each partial convolution operation layer, we update the mask using Eq. (2). For the input mask M, the output m' is actually a threshold version of the input:

$$
m' = \begin{cases} 1, & \text{if } sum(M) > 0 \\ 0, & \text{otherwise} \end{cases}. \tag{2}
$$

This new mask is then fed to the next layer. As the number of network layers increases, the number of pixels in the output mask m' decreases to 0, and the area of the effective area in the output result x' increases. The effect of mask on the overall loss will be smaller.

Loss Functions. The loss functions aim to evaluate how accurate the pixel-wise reconstruction is and how smooth the repaired area transitions into their surrounding tissues.

First, the mean squared error (MSE) loss is used to ensure that the recovered image patch is similar to the ground truth. Given an input image I_{in} masked by the segmented tumor or the initial binary mask M (0 for tumor area and 1 for other regions), the network prediction I_{out} should be close to the ground-truth image I_{gt}, as defined by:

$$L_{masked} = \frac{1}{|I_{gt}|} \left\| (1-M) \odot (I_{out} - I_{gt}) \right\|_1, \qquad (3)$$

$$L_{valid} = \frac{1}{|I_{gt}|} \left\| M \odot (I_{out} - I_{gt}) \right\|_1, \qquad (4)$$

where $|I_{gt}|$ denotes the number of elements in I_{gt}.

The perceptual loss L_{perc} reflects the appearance of recovered areas and measures high-level perceptual and semantic differences between the recovered and the original image regions, which is defined as:

$$L_{perc} = \sum_{p=0}^{P-1} \frac{\left\| \Psi_p^{I_{out}} - \Psi_p^{I_{gt}} \right\|_1}{\left| \Psi_p^{I_{gt}} \right|} + \sum_{p=0}^{P-1} \frac{\left\| \Psi_p^{I_{comp}} - \Psi_p^{I_{gt}} \right\|_1}{\left| \Psi_p^{I_{gt}} \right|}, \qquad (5)$$

where I_{comp} is the raw output image I_{out} but with the non-hole pixels directly set to the ground truth. $\left| \Psi_p^{I_{gt}} \right|$ is the number of elements in $\Psi_p^{I_{gt}}$, and Ψ_p^{I} denotes the activation map of the pth layer of the network. The perceptual loss calculates the L^1 distance between I_{out} and I_{comp} based on autocorrelation (Gram matrix). The style loss terms are introduced in each feature map:

$$L_{style\_out} = \sum_{p=0}^{P-1} \frac{1}{C_p C_p} \left\| K_p \left(\Psi_p^{I_{out}} \right)^T \left(\Psi_p^{I_{out}} \right) - \left(\Psi_p^{I_{gt}} \right)^T \left(\Psi_p^{I_{gt}} \right) \right\|_1, \qquad (6)$$

$$L_{style\_comp} = \sum_{p=0}^{P-1} \frac{1}{C_p C_p} \left\| K_p \left(\Psi_p^{I_{comp}} \right)^T \left(\Psi_p^{I_{comp}} \right) - \left(\Psi_p^{I_{gt}} \right)^T \left(\Psi_p^{I_{gt}} \right) \right\|_1, \qquad (7)$$

where $\Psi(x)_p$ represents the high level features of shape $(H_p W_p) \times C_p$, resulting in a $C_p \times C_p$ Gram matrix, K_p is the normalization factor $K_p = \frac{1}{C_p H_p W_p}$ for the pth selected layer.

The final loss term is the total variation (TV) loss L_{TV}, which is the smoothing penalty on R, and R is the region of 1-pixel dilation of the recovered region:

$$L_{TV} = \sum_{(i,j)\in R,(i,j+1)\in R} \frac{\left\| I_{comp}^{i,j+1} - I_{comp}^{i,j} \right\|_1}{N_{I_{comp}}} + \sum_{(i,j)\in R,(i+1,j)\in R} \frac{\left\| I_{comp}^{i+1,j} - I_{comp}^{i,j} \right\|_1}{N_{I_{comp}}}, \qquad (8)$$

where $N_{I_{comp}}$ is the number of elements in I_{comp}.

The total loss L_{total} is the combination of all the above loss functions [12]:

$$L_{total} = 6L_{masked} + L_{valid} + 0.05L_{perc} + 120(L_{style\_out} + L_{style\_comp}) + 0.1L_{TV}. \qquad (9)$$

The architecture of PConv-Net is similar to that of U-Net. The ReLU layer is used in the encoder phase, and the Leaky ReLU layer is used in the decoder phase. The batch normalization (BN) layer is used in all the layers except the first layer of the encoder and the last layer of the decoder.

3   Results

3.1   Data and Experimental Setting

We used the dataset from MICCAI BraTS2018, including 140 cases to evaluate the performance of the proposed algorithm. The T1-weighted and FLAIR images and the tumor labels were used for tumor segmentation. Since MR scans often display inhomogeneity in intensity distribution caused by the bias fields, we first applied the N4 bias field correction [16] in the preprocessing. Then, to reduce the intensity distribution variability across different subjects, we first applied histogram matching and then normalized each image by subtracting the mean and dividing the standard deviation. The T1 and FLAIR images are already registered, and the tumor masks are combined together to reflect the tumor and edema regions.

To train the segmentation network, we resampled the T1 and FLAIR images and the labels into isotropic volumes with in-plane size 240×240 and 155 slices. We randomly selected 100 images for training, and then used the remaining forty images for testing. The outputs of the network are corresponding tumor masks. For PConv-Net, 100 simulated images (arbitrarily putting simulated tumor masks on normal tissues) were used to train the model for tumor region recovery. Then, we evaluated the performance of PConv-Net using additional 40 T1 images of normal subjects. The original T1 images and simulated tumor masks were used as the inputs, and the tumor-recovered images are obtained. In this way, the ground truth under the simulated tumor masks is known. The PConv-Net was implemented using PyTorch with Adam optimization. The learning rate was initially set to 2e−4 and then fine-tuned using a learning rate of 5e−5. The maximum iterations are 1000 and batch size is 6. The network was trained in one GPU (12G NVIDIA GEFORCE RTX 2080 Ti). Finally, for tumor images, we simulated normal tissues covered by the tumor region and then stacked these repaired 2D slices into 3D images for registration.

3.2   Evaluation of Image Recovery

To evaluate the performance of image recovery, we generated tumor masks on normal images by removing the tissues within the simulated masks and comparing the recovered images with the original images within those masks. Two metrics were used to compare the recovered images with the original images, including the peak signal-to-noise ratio (PSNR) and the structural similarity (SSIM) [17]. PSNR is measured in decibels (dB), and the higher PSNR generally indicates better reconstruction performance. SSIM ranges from 0 to 1, and 1 means perfect recovery.

Figure 2 shows some examples of image recovery results using PConv-Net. The first row shows the original images with generated masks, and the third row displays the recovered images. The original ground-truth images are shown in the second row. We can see from the recovered images that the mask-removed areas were recovered with appearance similar to the real ones. We performed quantitative evaluation by using T1 images with different masks, and obtained PSNR as 42.29 ± 2.39 dB and SSIM as $0.9989 \pm 6.3550\text{E}{-}05$, indicating that our recovered images are very close to the original ones.

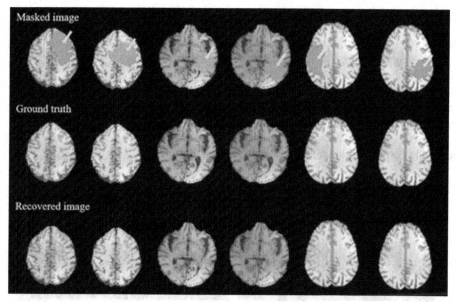

Fig. 2. Examples of recovered images within simulated masks. Top: normal images with simulated masks (yellow arrows); middle: ground truth (original normal images); bottom: recovered images. (Color figure online)

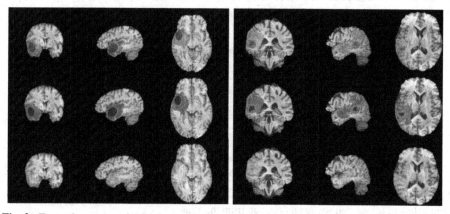

Fig. 3. Example results of PConv-Net for T1 images with glioma. Top: input image; middle: tumor segmented; bottom: tumor-recovered images.

Then, we applied the trained PConv-Net for the real tumor images. Fig. 3 shows two results. The first row is the original T1 images, and the segmented tumors are shown in the second row. By removing all the tumor regions, we can use PConv-Net to recover the images as shown in the third row. Because the tumor boundaries are often not obviously distinctive with the surrounding tissues, we performed a dilation process (using an element unit with size 3mm) on the tumor mask so that the intensities of the tumor areas do not affect the recovery operation. Since the ground truth within the masked area is unknown, only qualitative results can be shown.

3.3 Evaluation of Image Registration

Registration is performed between the T1 template (MNI152) and each tumor-recovered subject image by using SyN. To evaluate the performance of image registration, we used the resultant deformation fields to warp the segmentation of the atlas onto each subject and compared them with the manually corrected segmentation of each subject. Dice indices of grey matter (GM), whiter matter (WM) and cerebrospinal fluid (CSF) were computed for the evaluation. Dice coefficients are defined as $2|A_i \cap B_i|/|A_i| \cup |B_i|$, and i represents the ROI label. A is deformed segmentation from the atlas image, B is the manually corrected segmentation of each subject.

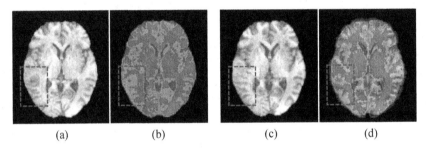

(a)   (b)   (c)   (d)

Fig. 4. A typical result of registration. (a) T1 image with tumor; (b) segmentation result (registering the atlas to the original images); (c) the recovered image using PConv-Net; (d) segmentation result (registering the atlas to the recovered images).

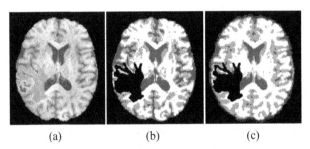

(a)   (b)   (c)

Fig. 5. An example of the segmentation. (a) Original image with tumor; (b) manually corrected segmentation; (c) warped segmentation map from the atlas.

For comparison, we computed the Dice coefficients outside the tumor regions, for evaluating the registration results. The registration is performed with and without tumor region recovery using PConv-Net. Figure 4 shows the comparison results. Figure 4(a) is the input image, and Fig. 4(b) is the segmentation map warped from the atlas after registration. Figure 4(c) is the recovery result of Fig. 4(a) using PConv-Net and Fig. 4(d) is the segmentation map obtained by registering the atlas to the recovered image.

In summary, for all the 40 testing subjects, we extract the tumors and compared Dice between the registered segmentation of the atlas and the manually corrected segmentation. An example of the input and the segmentation images are shown in Fig. 5.

Table 1 shows the quantitative results over 40 testing images. It can be seen that the Dice coefficients were significantly improved for GM ($p < 0.0005$), while there is no significant difference for WM and CSF. Overall the proposed algorithm effectively achieved relatively accurate registration in terms of segmentation Dice.

Table 1. Dice of registration with and without PConv. "*" indicates significant improvement.

	CSF	GM*	WM
Dice within normal region without using PConv	0.577 (±0.022)	0.696 (±0.041)	0.793 (±0.027)
Dice within normal region using PConv	0.569 (±0.015)	0.724 (±0.027)	0.797 (±0.026)

4 Conclusion

A brain MR image registration approach is proposed for registering a normal atlas onto the images with brain tumor. The algorithm is based on recovering missed normal tissues within the tumor regions using a partial convolutional network. Recent learning algorithms suggest that if the tumor area within an MR image can be recovered with normal tissues, the registration could be less affected by pathology, giving relatively accurate registration. Experiments with BraTS2018 tumor images showed the effectiveness of this strategy. We believe that with more accurate segmentation of the normal tissues of tumor image, better assessment or surgical planning can be performed. In the future work, we will further evaluate the performance of labeling the surrounding tissues and study the performance of glioma assessment with clinical datasets.

References

1. Shen, D., Davatzikos, C.: HAMMER: hierarchical attribute matching mechanism for elastic registration. IEEE Trans. Med. Imaging **21**(11), 1421–1439 (2002)
2. Pluim, J.P.W., Maintz, J.B.A., Viergever, M.A.: Mutual-information-based registration of medical images: a survey. IEEE Trans. Med. Imaging **22**(8), 986–1004 (2003)
3. Ou, Y., Akbari, H., Bilello, M., Da, X., Davatzikos, C.: Comparative evaluation of registration algorithms in different brain databases with varying difficulty: results and insights. IEEE Trans. Med. Imaging **33**(10), 2039–2065 (2014)
4. Balakrishnan, G., Zhao, A., Sabuncu, M.R., Dalca, A.V., Guttag, J.: An unsupervised learning model for deformable medical image registration. In: 2018 IEEE/CVF Conference on Computer Vision and Pattern Recognition, Salt Lake City, UT, pp. 9252–9260 (2018)
5. Nowinski, W.L., Belov, D.: Toward atlas-assisted automatic interpretation of MRI morphological brain scans in the presence of tumor. Acad. Radiol. **12**, 1049–1057 (2005)
6. Stefanescu, R., Commowick, O., Malandain, G., Bondiau, P.-Y., Ayache, N., Pennec, X.: Non-rigid atlas to subject registration with pathologies for conformal brain radiotherapy. In: Barillot, C., Haynor, D.R., Hellier, P. (eds.) MICCAI 2004. LNCS, vol. 3216, pp. 704–711. Springer, Heidelberg (2004). https://doi.org/10.1007/978-3-540-30135-6_86

7. Yu, J., Lin, Z., Yang, J.: Generative image inpainting with contextual attention. arXiv preprint arXiv:1801.07892 (2018)

8. Yang, C., Lu, X., Lin, Z., Shechtman, E., Wang, O., Li, H.: High-resolution image inpainting using multi-scale neural patch synthesis. In: 2017 IEEE Conference on Computer Vision and Pattern Recognition (CVPR), Honolulu, HI, pp. 4076–4084 (2017)

9. Menze, B.H., Jakab, A., Bauer, S., et al.: The Multimodal brain tumor image segmentation benchmark (BRATS). IEEE Trans. Med. Imaging **34**(10), 1993–2024 (2015)

10. Bakas, S., Akbari, H., Sotiras, A., et al.: Advancing the cancer genome atlas glioma mri collections with expert segmentation labels and radiomic features. Sci. Data **4**, 170117 (2017). https://doi.org/10.1038/sdata.2017.117

11. Bakas, S., Reyes, M., Jakab, A., et al.: Identifying the best machine learning algorithms for brain tumor segmentation, progression assessment, and overall survival prediction in the BRATS challenge. arXiv:1811.02629 (2018)

12. Liu, G., Reda, F.A., Shih, K.J., Wang, T.-C., Tao, A., Catanzaro, B. : Image inpainting for irregular holes using partial convolutions. arXiv.preprint arXiv:1804.07723 (2018)

13. Ronneberger, O., Fischer, P., Brox, T.: U-net: convolutional networks for biomedical image segmentation. In: Navab, N., Hornegger, J., Wells, W., Frangi, A. (eds.) MICCAI 2004. LNCS, vol. 9351, pp. 234–241. Springer, Cham (2015). https://doi.org/10.1007/978-3-319-24574-4_28

14. Avants, B.B., Epstein, C.L., Grossman, M., Gee, J.C.: Symmetric diffeomorphic image registration with cross correlation: evaluating automated labeling of elderly and neurodegenerative brain. Med. Image Anal. **12**(1), 26–41 (2008)

15. Harley, A.W., Derpanis, K.G., Kokkinos, I.: Segmentation-aware convolutional networks using local attention masks. In: 2017 IEEE International Conference on Computer Vision (ICCV), Venice, pp. 5048–5057 (2017)

16. Tustison, N.J., et al.: N4ITK: improved N3 bias correction. IEEE Trans. Med. Imaging **29**(6), 1310–1320 (2010)

17. Wang, Z., Bovik, A.C., Sheikh, H.R., Simoncelli, E.P.: Image quality assessment: from error visibility to structural similarity. IEEE Trans. Image Process. **13**(4), 600–612 (2004)

Convolutional Neural Network with Asymmetric Encoding and Decoding Structure for Brain Vessel Segmentation on Computed Tomographic Angiography

Guoqing Wu[1], Liqiong Zhang[1], Xi Chen[1], Jixian Lin[2], Yuanyuan Wang[1(✉)], and Jinhua Yu[1(✉)]

[1] Department of Electronic Engineering, Fudan University, Shanghai, China
{yywang,jhyu}@fudan.edu.cn
[2] Minhang Central Hospital, Fudan University, Shanghai, China

Abstract. Segmenting 3D brain vessels on computed tomographic angiography is critical for early diagnosis of stroke. However, traditional filter and optimization-based methods are ineffective in this challenging task due to imaging quality limits and structural complexity. And learning based methods are difficult to be used in this task due to extremely high time consumption in manually labeling and the lack of labelled open datasets. To address this, in this paper, we develop an asymmetric encoding and decoding-based convolutional neural network for accurate vessel segmentation on computed tomographic angiography. In the network, 3D encoding module is designed to comprehensively extract 3D vascular structure information. And three 2D decoding modules are designed to optimally identify vessels on each 2D plane, so that the network can learn more complex vascular structures, and has stronger ability to distinguish vessels from normal regions. What is more, to improve insufficient fine vessel segmentation caused by pixel-wise loss function, we develop a centerline loss to guide learning model to pay equal attention to small vessels and large vessels, so that the segmentation accuracy of small vessels can be improved. Compared to two state-of-the-art approaches, our model achieved superior performance, demonstrating the effectiveness of both whole vessel segmentation and small vessel maintenance.

Keywords: Vessel segmentation · Computed tomographic angiography · Convolutional neural network

1 Introduction

Stroke is the deadliest disease in most developed countries. Rapid and accurate diagnosis is critical to its recovery [1]. Computed tomographic (CT) and computed tomographic angiography (CTA) are the first choices of emergency tools for stroke patients. Segmenting brain vessels quickly and accurately from CTA images and subsequently analyzing blood flow based on them could provide important guidance for the diagnosis and treatment of stroke. However, it is a challenging and time consuming task to precisely segment

© Springer Nature Switzerland AG 2021
A. Crimi and S. Bakas (Eds.): BrainLes 2020, LNCS 12658, pp. 51–59, 2021.
https://doi.org/10.1007/978-3-030-72084-1_5

vessel in CTA due to the complex tree-like structure and the variety of sizes, shapes, and intensities [2]. It will take one radiologist more than 7 h to label one CTA case. In addition, annotations may be error prone [3]. Hence, it comes to no surprise that efficient and effective automatic segmentation techniques are highly desired.

Many recent vessel segmentation methods being proposed are based on supervised learning method. Earlier methods first extracted numerous handcrafted features from local image patches, then built machine learning classification model to divide the patches into vessels and background, and finally aggregated the predictions of local image patches to get global segmentation result [4]. To improve efficiency and effectiveness, an end-to-end fully convolutional network (FCN) method has been widely studied recently. Fu et al. [5] developed FCN model combined with a fully connected Conditional Random Fields (CRFs) to segment vessel. Jin et al. [6] trained a 3D FCN to segment intrathoracic airway on an incompletely labeled dataset, and applied graph-based refinement incorporating fuzzy connectedness segmentation. More recently, a U-type FCN, namely Unet, was proposed to solve the problem of gradient vanishing and fixed receptive fields in segmented networks. Thanks to the benefits from the exquisite skip-connection structure, Unet and its improved versions have achieved encouraging performance in numerous segmentation competitions [7, 8].

Network structure and loss function are the cores of constructing deep convolution network-based vessel segmentation model. In terms of network structure, a desired network for segmenting vessel need to consider the information of coronal plane, axial plane and sagittal plane simultaneously. Currently, there are two commonly used schemes to combine the information of those three planes mentioned above. One is to directly add a dimension on the whole network, that is, to extend 2D image patch convolution to 3D volume convolution. Because this 3D information integration is easy to implement, and 3D convolution-based model has been widely used in tumor and retinal vessel segmentation. The other, also known as 2.5D network, uses three 2D coding networks to encode each plane, and then merges the three decoding results to get the final output. Titinunt et al. [9] proposed a 2.5D network model for hepatic vessel segmentation. Experimental results shown that the performance of this method was improved by 10% compared with 3D network. Jihye et al. [10] proposed an improved 2.5D network model to segment airway. And its results outperformed the results of 15 algorithms of the *2019 airway segmentation challenging*.

In most deep learning-based models, network parameters are trained based on minimizing pixel-wise losses, which measures the pixel level difference between the predicted probability map and the ground truth. Such pixel-wise loss treats every pixel equally. However, due to the large difference in the number of pixels contained in the large and small vessels, pixel-wise loss will guide deep learning models to pay less attention to small vessels than to large vessels. As a result, small vessels are difficult to be effectively maintained in final segmentation results.

In this paper, we propose a convolutional neural network-based approach to accurately segment brain vessels on CTA. Considering the complex tree-like structure, we design a 3D and 2.5D combined network structure, which would be capable of encoding and decoding vessel structure information more comprehensively. In addition, it is actually more difficult to distinguish vessels from three planes independently than from

3D volume. Therefore, we build three decoding networks on the corresponding planes separately, aiming to make the model learn more complex vessel structures and reduce over fitting. In consideration of the correlations among the three planes, we apply 3D convolution to encode the input 3D volume and fuse the three plane decoding results to get the final 3D results. In addition, we develop a centerline loss to improve the ability to maintain small vessels. Compared with pixel-wise loss, such centerline loss would penalize large and small vessels equally. Thus, small vessels would be segmented more effectively. The proposed model was validated on 46 cases, where 28 cases for training, 9 cases for validation and 9 cases for testing. From the view of results, our model achieved the highest Dice coefficient (DSC) and true positive rate (TPR) of 79.73%, 73.96%, respectively, compared with two state-of-the-art methods.

2   Data Acquisition and Preprocessing

The experimental dataset contains 46 CTA volumes provided by the Stroke Center, Anonymous Hospital. All CTA examinations were acquired using ncCT Siemens Sensation 64 with the following scanning parameters: 70 kV, 100 mAs (tube current). The voxel resolution and size of the CTA images are 0.486 * 0.486 * 1 mm^3 and 512 * 512 * 319, respectively. Considering the needs of follow-up stroke diagnosis analysis, we only extract images from 160 to 319 slices which are most likely to have strokes for our experiment. Voxel intensity is clipped by a window [0, 600] (HU) for focusing on vessels. We randomly divide the 46 volumes into 28:9:9 for training, validation and testing, respectively. Each volume was manually labeled by two experienced radiologists on *itk-snap* platform, one of whom was responsible for labeling and the other for correction. As shown in Fig. 1, for more accurate results, the labeling was carried out on the axial plane, the coronal plane and the sagittal plane simultaneously. It took about 7 h to manually label each case.

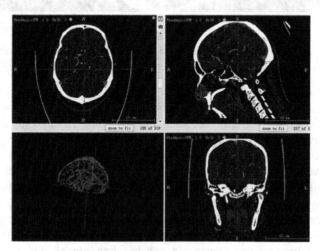

Fig. 1. Vessel labeling on the three planes.

3 Method

3.1 Asymmetric Encoding and Decoding-Based Convolutional Neural Net

Unet models based on 2D, 2.5D and 3D have achieved encouraging performances in tumor, organ and vessel segmentation. Figure 2 shows the brain vessel segmentation results of 3D Unet network, in which Fig. 2(a) is the 3D segmentation result, and Fig. 2(b) and Fig. 2(c) are the results of two different slices on the axial plane. It can be seen that except for the region indicated by the arrow in Fig. 2(a), the other parts have been well divided. However, such over-segmented region can be easily distinguished on the 2D plane shown in Fig. 2(b). Besides, the under-segmented regions that could not be captured in the 3D convolution model can be clearly seen from the 2D plane in Fig. 2(c), because in the process of network training, the loss of 3D Unet guides network to learn the optimal solution in 3D structure. While this process may lose some of optimal solutions in some 2D planes, and thus some features that can be easily distinguished on the 2D plane are difficult to be captured effectively. Comparative experimental results in the literature [9] also validate this interpretation. Therefore, based on the advantages of the 3D Unet network to analyze 3D structures, we develop an asymmetric encoding and decoding-based network framework to enhance network learning ability and segmentation performance.

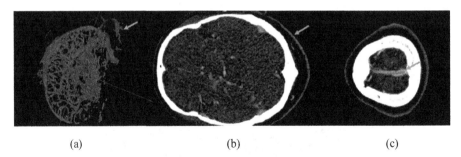

(a)   (b)   (c)

Fig. 2. The segmentation results of 3D Unet model. (a) is the 3D segmentation result. (b) and (c) are the results of two different slices on the axial plane.

As shown in Fig. 3, the asymmetric encoding and decoding-based network consists of three parts: 3D encoding, 2.5D decoding and 3D fusion. 3D encoding part has four convolution layers. Each convolution layer contains two $3 \times 3 \times 3$ convolutions and one rectified linear unit (ReLU) successively. Except for the last convolution layer, each layer is followed by a $2 \times 2 \times 2$ max pooling with strides of two in each dimension. The number of channels of the first convolution layer is set to 20, and the following convolution layer double the number of channels in turn. To satisfy the analysis of 2.5D decoding network, the feature map in bottom layer has done axis exchange of 1 to 2 and 1 to 3, respectively. In the 2.5D decoding network, the three underlying feature maps are decoded on the corresponding 2D plane. In each decoding network, three up-convolution layers with kernel size of 2×2 and stride of 2 are set corresponding to the three max pooling layers. Every up-convolution layer is followed by two 3×3 convolution layer each followed by a ReLU. Each convolution layer halves the number of channels in turn. Shortcut connections from layers of equal resolution in the encoding path provide the essential high-resolution features to the decoding path. Note that to achieve data dimension matching, 3D to 2D and 2D to 3D operations are introduced before and after each 2D convolution layer, respectively. Finally, the 2.5D decoding network gets three 2D plane segmentation results. Through the 2D to 3D operation and the axis transformation operation, the three feature maps of the last layer of the 2.5D decoding networks are spliced in the 3D fusion part. Then, the spliced data are convolved through a $1 \times 1 \times 1$ convolution kernel to get the final 3D segmentation result.

The proposed asymmetric encoding and decoding-based model not only uses the 3D encoding to effectively analyze 3D vascular structure, but also makes the sensitive vascular structure information of each plane be fully utilized in segmentation. In addition, instead of directly stacking and averaging, the 3D fusion part uses an adaptive convolutional network to integrate the decoding information of those three planes. In consequence, our overall design is good for learning complex target structure and helps contribute to discriminate between vessels and background more effectively.

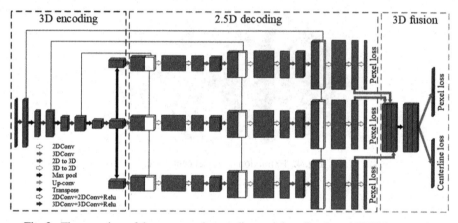

Fig. 3. The overview of the asymmetric encoding and decoding-based CNN framework.

3.2 Centerline Loss Construction

Pixel-wise loss directly compares per-pixel differences between network output and the corresponding ground truth. And each pixel is treated equally under its measure criteria. However, for brain vessel segmentation, due to the highly imbalanced ratio between large and small vessels, small vessels may only get limited attention under the pixel-wise loss function. As a result, many small vessels are missed in segmentation results. To address this, we construct a centerline loss to enforce network model give more penalization on small vessels. As shown in Fig. 3, after fusing decoding information of those three planes, we set up pixel-wise loss and centerline loss respectively to train the network. Because large and small vessels have a close centerline in the same length. Therefore, under the influence of centerline loss, the problem of limited attention to small vessels caused by the imbalance ratio will be effectively solved. In the training phase, the addition of the centerline loss will train the shared learning architecture (the part before fusion network) to learning more features which are favorable for small vessel expression. As a result, more small vessels will be segmented. Hence, the total loss of the proposed model can be written as:

$$loss_{(all)} = \alpha_1 loss_{(axi)} + \alpha_2 loss_{(cor)} + \alpha_3 loss_{(sag)} + \beta loss_{(3D)} + \gamma \, loss_{(3D\_centerline)} \tag{1}$$

where $loss_{(axi)}$, $loss_{(cor)}$, $loss_{(sag)}$, $loss_{(3D)}$ and $loss_{(3D\_centerline)}$ are the losses of axial plane, coronal plane, sagittal plane, 3D fusion loss, and centerline loss, respectively. $\alpha_1, \alpha_2, \alpha_3, \beta$ and γ are the corresponding weights. All the five sub-losses are defined by cross entropy. Figure 4 shows the extracted vessel centerline. In training data construction phase, we apply a skeletonization method [11] to extract centerline. Then we extract centerline patches according to the index of the corresponding image patches.

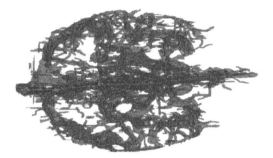

Fig. 4. Extraction of vessel centerline.

3.3 Network Training

We adopted image patch-based training method. 100 3D image patches with size of 128 × 128 × 128 were extracted form per volume. To ensure the positive sample content in the extracted image patch, we adopted a more reasonable patch extraction scheme instead of random extraction. Specifically, we first generated the centerline of vessel, and then extracted 60 patches along the centerline randomly. Next, we extracted the remaining 40 patches from the whole volume randomly. Finally, 2800 patches were extracted from the 28 training cases. In training step, batch size is set to 1. Adam with initial learning rate of 1e−4 is used for model optimize [12]. The weights of loss function, $\alpha_1, \alpha_2, \alpha_3, \beta$ and γ, are set to 0.85, 0.85, 0.85, 0.15 and 0.20, respectively. Training converged after 25–30 epochs. The code of training and testing models were written in Python with Tensorflow, and we ran it on a single NVidia Titan X GPU.

4 Results

In testing phase, the pixel loss branch in 3D fusion part output the final segmentation result. Three evaluation metrics are used [8]: (a) DSC, (b) TPR, (c) Positive predictive value (PPV). Three stated-of-the-art methods are selected for comparison. The first compared method [13] is a graph-cut based method, which has achieved good results in the pulmonary vessels segmentation. The second compared method [6] is a 3D Unet network which has achieved interesting performances in many segmentation competitions. The third compared method [10] is a 2.5D network which has a similar decoding network structure to our method.

Table 1 reports the results of segmentation metrics about the four compared methods, it is observed that our method achieved the highest DSC and TPR of 79.73%, 73.96%, respectively, also with the comparable PPV of 86.71%. Compared with the graph-cut based method [13], the three learning-based methods achieve DSC improvement of at least 4%, which demonstrates the promising performance of learning-based method. The DSC and TPR of our method are 1.55% and 2.9% higher than [6], respectively, which shows that the proposed method not only achieves overall good performance, but also detects more vessel voxels. The segmentation results of method [10] have the smallest standard deviation. With lower DSC, TPR but higher PPV, method [6] segments more conservatively than our method.

Figure 5 shows comparison segmentation results on the three planes. From top to bottom, the three rows show the axial, coronal, and sagittal planes, respectively. From left to right, the five columns show the segmentation of method [13], method [6], method [10], the proposed method and the ground truth. Comparing the regions indicated by the arrow, more small vessels have been correctly segmented by the proposed method than others.

Figure 6 shows 3D visual comparison. The left and right subgraphs show comparison between the proposed method and method [13], method [6] and method [10], respectively. Red regions are detected by both the three compared methods. While the fluorescent green regions are only detected by our method. It is precisely because the combination of multi-dimension network and the centerline loss, our method not only enriches details of segmented large vessels but also segmented small vessels.

Table 1. Comparison of vessel segmentation results (%) on testing dataset.

	Ours			Zhiwei et al. [13]			Jin et al. [6]			Jihye et al. [10]		
	D	T	P	D	T	P	D	T	P	D	T	P
M	**79.7**	**73.9**	86.7	71.0	65.3	78.9	78.1	71.0	**87.4**	75.0	71.6	78.9
S	3.6	5.6	1.7	7.2	10.7	4.1	4.6	7.6	1.9	3.0	3.9	2.5

M and S denote mean and Std., respectively. D, T and P denote DSC, TPR and PPV, respectively.

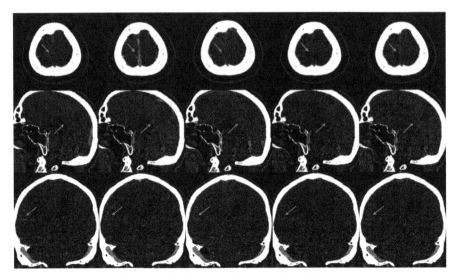

Fig. 5. Comparison results on different planes. The first, second and third rows show the axial, coronal, and sagittal planes, respectively. From left to right, the five columns show the segmentation of method [13], method [6], method [10], the proposed method and the ground truth.

Fig. 6. 3D visual comparison. The left, middle and right subgraphs show the comparison between the proposed method and method [13], method [6] and method [10], respectively.

5   Conclusions

In this paper, we proposed an asymmetric encoding and decoding-based convolutional neural net to segment brain vessels on computed tomographic angiography. Two improvements have been made to improve the overall segmentation results as well as the preservation of small vessels. On the one hand, an asymmetric encoding and decoding-based network framework integrating 3D encoding, 2.5D decoding and 3D fusion is constructed, which could better analyze and mine 2D and 3D sensitive vascular structural features. On the other hand, we proposed a centerline-based loss function, which token into account the adverse effect of the proportion difference of the large and small vessels on small vessels segmentation. Its same penalization for the large and small vessels enforce small vessels to obtain more attention on model training, thus the segmentation effects of small vessels could be improved effectively. Experimental results showed an encouraging performance of the proposed method in terms of visual effects and objective metrics.

Acknowledgement. This study was funded by the National Natural Science Foundation of China (62001119), the Natural Science Foundation and Major Basic Research Program of Shanghai (16JC1420100) and General Program of China Postdoctoral Science Foundation (2019M661363).

References

1. Wu, G., et al.: Early identification of ischemic stroke in noncontrast computed tomography. Biomed. Signal Process. **52**, 41–52 (2019)
2. Qin, Y., et al.: AirwayNet: a voxel-connectivity aware approach for accurate airway segmentation using convolutional neural networks. In: Shen, D., et al. (eds.) MICCAI 2019. LNCS, vol. 11769, pp. 212–220. Springer, Cham (2019). https://doi.org/10.1007/978-3-030-32226-7_24
3. Tschirren, J., Yavarna, T., Reinhardt, J.M.: Airway segmentation framework for clinical environments. In: Brown, M., et al. (eds.) PIA, London, UK, pp. 227–238 (2009)
4. Staal, J., Abràmoff, M.D., Niemeijer, M., Viergever, M.A., van Ginneken, B.: Ridge-based vessel segmentation in color images of the retina. IEEE Trans. Med. Imag **23**(4), 501–509 (2004)
5. Fu, H., Xu, Y., Lin, S., Kee Wong, D.W., Liu, J.: DeepVessel: retinal vessel segmentation via deep learning and conditional random field. In: Ourselin, S., Joskowicz, L., Sabuncu, M.R., Unal, G., Wells, W. (eds.) MICCAI 2016. LNCS, vol. 9901, pp. 132–139. Springer, Cham (2016). https://doi.org/10.1007/978-3-319-46723-8_16
6. Jin, D., Xu, Z., Harrison, A.P., George, K., Mollura, D.J.: 3D convolutional neural networks with graph refinement for airway segmentation using incomplete data labels. In: Wang, Q., Shi, Y., Suk, H.-I., Suzuki, K. (eds.) MLMI 2017. LNCS, vol. 10541, pp. 141–149. Springer, Cham (2017). https://doi.org/10.1007/978-3-319-67389-9_17
7. Ronneberger, O., Fischer, P., Brox, T.: U-net: convolutional networks for biomedical image segmentation. In: Navab, N., Hornegger, J., Wells, W.M., Frangi, A.F. (eds.) MICCAI 2015. LNCS, vol. 9351, pp. 234–241. Springer, Cham (2015). https://doi.org/10.1007/978-3-319-24574-4_28
8. Shen, H., Zhang, J., Zheng, W.: Efficient symmetry-driven fully convolutional network for multimodal brain tumor segmentation. In: Chaker, L., et al. (eds.) ICIP 2017, pp. 3864–3868 (2017)
9. Kitrungrotsakul, T., Han, X.-H., Iwamoto, Y., Foruzan, A.H., Lin, L., Chen, Y.-W.: Robust hepatic vessel segmentation using multi deep convolution network. In: Andrzej, K. (ed.) 2017. Biomedical Applications in Molecular, Structural, and Functional Imaging, SPIE, vol. 10137, pp. 11–16 (2017). https://doi.org/10.1117/12.2253811
10. Yun, J., Park, J., Yu, D., Yi, J., Lee, M.: Improvement of fully automated airway segmentation on volumetric computed tomographic images using a 2.5 dimensional convolutional neural net. MedIA **51**, 13–20 (2019)
11. Lam, L., Lee, S.W., Suen, C.Y.: Thinning methodologies-a comprehensive survey. IEEE TPAMI **14**(9), 869–885 (1992)
12. Nie, D., et al.: Medical image synthesis with deep convolutional adversarial networks. IEEE Trans. Biomed. Eng. **65**, 2720–2730 (2018)
13. Zhai, Z., et al.: Automatic quantitative analysis of pulmonary vascular morphology in CT images. Med. Phys. **46**(9), 3985–3997 (2019)

Volume Preserving Brain Lesion Segmentation

Yanlin Liu[1], Xiangzhu Zeng[2], and Chuyang Ye[1(✉)]

[1] School of Information and Electronics, Beijing Institute of Technology,
Beijing, China
chuyang.ye@bit.edu.cn
[2] Department of Radiology, Peking University Third Hospital, Beijing, China

Abstract. Automatic brain lesion segmentation plays an important role in clinical diagnosis and treatment. *Convolutional neural networks* (CNNs) have become an increasingly popular tool for brain lesion segmentation due to its accuracy and efficiency. CNNs are generally trained with loss functions that measure the segmentation accuracy, such as the cross entropy loss and Dice loss. However, lesion load is a crucial measurement for disease analysis, and these loss functions do not guarantee that the volume of lesions given by CNNs agrees with that of the gold standard. In this work, we seek to address this challenge and propose volume preserving brain lesion segmentation, where a volume constraint is imposed on network outputs during the training process. Specifically, we design a differentiable mapping that approximates the volume of lesions using the segmentation probabilities. This mapping is then integrated into the training loss so that the preservation of brain lesion volume is encouraged. For demonstration, the proposed method was applied to ischemic stroke lesion segmentation, and experimental results show that our method better preserves the volume of brain lesions and improves the segmentation accuracy.

Keywords: Brain lesion segmentation · Convolutional neural network · Volume constraint

1 Introduction

Segmentation of brain lesions can guide clinical diagnosis and treatment strategies. Since manual lesion delineations by experts can be time-consuming and costly, automated brain lesion segmentation methods have been developed. In particular, methods based on *convolutional neural networks* (CNNs) have achieved state-of-the-art segmentation performance for various types of brain lesions, including brain tumors [5,19], ischemic stroke lesions [5], and multiple sclerosis lesions [12]. Typically, CNNs are trained by optimizing loss functions (e.g. the cross entropy loss [15] or Dice loss [11]) that measure the segmentation accuracy. However, these loss functions may not guarantee that the volume of

© Springer Nature Switzerland AG 2021
A. Crimi and S. Bakas (Eds.): BrainLes 2020, LNCS 12658, pp. 60–69, 2021.
https://doi.org/10.1007/978-3-030-72084-1_6

lesions given by CNNs agrees with that of the manual annotation. Since lesion load is often an important measurement for lesion analysis [3,17], such volume difference could introduce undesired bias into the analysis.

To address the issue of volume bias, it is possible to impose a constraint of lesion volume in the training of CNNs. Previous works have explored volume constraints (or area constraints in 2D cases) for different purposes. For example, for situations where pixelwise annotations are not available, Pathak et al. [14] have proposed a constrained CNN to perform weakly supervised semantic segmentation for natural images, where inequality constraints are developed for network training using a rough size estimate of the objects of interest. This method is extended in [7] for weakly supervised 3D medical image segmentation with a more efficient implementation, where the inequality constraints are converted to penalty functions during network training. Similarly in [4], penalty functions for constraints of area are introduced into weakly supervised histopathology image segmentation. The penalty functions for volume constraints have been also applied to semi-supervised and domain adaptation settings [1,6] when rough prior knowledge about segmentation volume is available.

In these existing methods, the segmentation volume in the penalty function is approximated by the sum of segmentation probabilities, so that the loss function is differentiable during training. Although the penalty function for volume constraints is beneficial when the voxelwise annotation is not available and a range of volume is allowed, such volume approximation can be suboptimal in a fully supervised setting, where the precise volume is known for training data. This is because the segmentation probabilities can be smaller than one for lesions and thus underestimate the actual volume, and the underestimation can have a negative effect on network training. Therefore, the incorporation of volume constraints for fully supervised brain lesion segmentation is still an open problem.

In this work, we explore the development of volume constraints for brain lesion segmentation in a fully supervised setting. We propose volume preserving brain lesion segmentation, where we develop a penalty term for volume preservation that can be integrated with typical loss functions when voxelwise annotations are available. Specifically, in the penalty term the segmentation volume is better approximated with a differentiable transformation of segmentation probabilities, and the term penalizes the difference between the volume of annotations and the approximated volume given by the network. For evaluation, the proposed method was applied to the segmentation of ischemic stroke lesions. Experimental results show that the proposed method better preserves the volume of brain lesions and improves the segmentation accuracy.

2   Methods

2.1   Background: Existing Inequality Volume Constraints

The constraint of object size has been incorporated into the training of CNNs to alleviate the lack of pixel/voxel-level annotations [1,7,10,14], e.g., in weakly supervised or semi-supervised settings. In this work, since we consider 3D segmentation of brain lesions, the object size is represented by volume. Suppose we have rough prior knowledge about the volume of the object to segment, where the upper and lower bounds on the volume are a and b, respectively. Then, to introduce the constraint of volume, a constrained optimization problem can be formulated by imposing inequality constraints. For example, in a weakly supervised setting, the minimization problem for each training subject can be expressed as follows [7]

$$\min_{\theta} L(S;\theta) \quad \text{s.t.} \quad a \leq \sum_{i=1}^{n} V_i \leq b, \tag{1}$$

where θ represents network weights, S represents the network output, L is a segmentation loss (e.g., a partial cross-entropy loss), n denotes the number of voxels in the image, and V_i indicates whether the i-th voxel in the network segmentation is foreground (1) or background (0). However, solving Eq. (1) directly is impractical because it is extremely time-consuming to perform Lagrangian-dual optimization for networks with millions of parameters [10,14].

To address the problem, previous works have converted the constraint in Eq. (1) to a penalty term in the loss function, so that the optimization problem can be solved efficiently [1,4,7]. Specifically, the optimization problem becomes

$$\min_{\theta} L(S;\theta) + \lambda F(V;\theta), \tag{2}$$

where F is the penalty term for the volume constraint, V is the volume of the network segmentation ($V = \sum_i V_i$), and λ is a tunable weight. F is positive when the volume of the segmented object is outside $[a, b]$ and zero otherwise, so that volume that agrees with the prior knowledge is encouraged. Note that if V_i is directly used in the loss function in Eq. (2), gradient-based optimization methods cannot be applied for network training. Thus, to ensure that the loss function is differentiable, V_i is approximated by the network output probability S_i at the i-th voxel in existing methods.

2.2   Volume Constraints for Fully Supervised Settings

The inequality volume constraint can be extended for fully supervised settings, where voxel-level annotations are available for each training image. Ideally, the upper and lower volume bounds become the same, which are specified by the voxelwise annotation. In this way, the bias of segmentation volume could be mitigated, and the segmentation accuracy may also be improved. However, due to optimization purposes, the volume $\sum_i V_i$ is approximated by the sum of segmentation probabilities S_i, which can be smaller than one. Thus, an inequality constraint can still be used, which allows the approximated volume to be smaller than the desired volume within a margin. For example, the corresponding penalty term can be designed as follows[1]

---

[1] For brevity, θ is omitted from this point onward.

$$F\left(S\right) = 100 \times \frac{\max\left(0, \sum_{i:S_i>0.5} S_i - \sum_i Y_i\right) + \max\left(0, r\sum_i Y_i - \sum_{i:S_i>0.5} S_i\right)}{\sum_i Y_i}.(3)$$

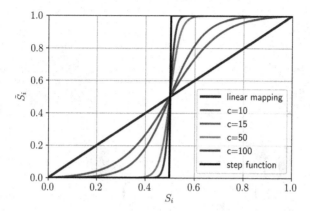

Fig. 1. Curves for different mappings from the segmentation probability S_i to $\tilde{S}_i$ that approximates segmentation volume for a voxel.

Here, $r \in (0,1)$ is a positive constant and Y_i represents the annotation for the i-th voxel. Note that the relative volume difference is considered and a factor of 100 is applied so that this penalty term represents the percentage of relative volume difference. In this term, the approximated segmentation volume that is greater than the annotation volume is penalized, and if the approximated volume is too small, it is also penalized.

This extension for the fully supervised setting is straightforward. However, the volume approximation $\sum_{i:S_i>0.5} S_i$ underestimates the volume, which can still introduce bias despite the use of an inequality constraint. Thus, to address this limitation, we seek to develop an improved penalty term where segmentation volume is better approximated.

Ideally, the relationship between S_i and V_i is a step function, where $V_i = 1$ when $S_i > 0.5$ and $V_i = 0$ when $S_i < 0.5$ (see the red curve in Fig. 1). However, directly using V_i in Eq. (3) is problematic for network training, because the loss is not differentiable with respect to the network weights. The approximation in Eq. (3) can be interpreted as a linear mapping $V_i = S_i$, which underestimates the segmentation volume (see the green curve in Fig. 1). Thus, it is possible to perform a transformation of S_i for better volume approximation, where a compromise is achieved between approximation accuracy and differentiability.

For convenience, we denote the transformed segmentation probability for S_i by $\tilde{S}_i$, which approximates segmentation volume for the i-th voxel. We assume that the transformation of segmentation probabilities should satisfy that: 1) the mapping from S_i to $\tilde{S}_i$ is closer to the step function than the linear mapping is and 2) $\tilde{S}_i = V_i$ when $S_i = 0$ or 1. Thus, the transformation can be designed as an S-shaped mapping:

$$\tilde{S} = \left(\frac{1}{1 + e^{-(S-1/2)c}} - \frac{1}{2} \right) \times \frac{1 + e^{-\frac{1}{2}c}}{1 - e^{-\frac{1}{2}c}} + \frac{1}{2}, \tag{4}$$

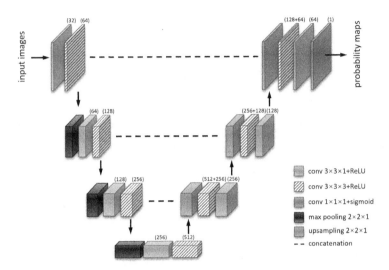

Fig. 2. The 3D U-Net architecture for brain lesion segmentation. Note that the kernel sizes in the slice direction were set to one for some layers, because we used images with thick slices for evaluation.

where c is a tunable parameter that controls the shape of the transformation curve. As shown in Fig. 1, greater c leads to a sharper curve that is closer to the step function. Thus, with a proper selection of c, we can find a better approximation of segmentation volume, where the loss function can still be effectively minimized for network training.

With the transformation of segmentation probabilities, the penalty term in Eq. (3) becomes

$$F(\tilde{S}) = 100 \times \frac{\max\left(0, \sum_{i:\tilde{S}_i>0.5} \tilde{S}_i - \sum_i Y_i\right) + \max\left(0, r\sum_i Y_i - \sum_{i:\tilde{S}_i>0.5} \tilde{S}_i\right)}{\sum_i Y_i}, \tag{5}$$

where $\tilde{S}$ is the map of transformed probabilities. This penalty term can then be integrated with a standard segmentation loss $L(S)$ in a fully supervised setting by minimizing $L(S) + \lambda F(\tilde{S})$.

2.3 Implementation Details

The proposed method is implemented using PyTorch [13]. For demonstration, we selected the 3D U-Net [2] architecture, which has been successfully applied to

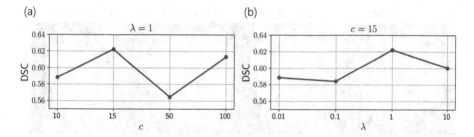

Fig. 3. The impact of hyperparameters: (a) the validation DSCs when $\lambda = 1.0$ and different c was used and (b) the validation DSCs when $c = 15$ and different λ was used.

various tasks of medical image segmentation [9], as our backbone segmentation network.[2] The network structure is shown in Fig. 2. The sum of the Dice loss and weighted cross-entropy loss was used as the segmentation loss. The Adam optimizer [8] was used (the initial learning rate was 0.0001) and the batch size was two. The model was trained for 400 epochs. During the first 200 epochs, the weight for the penalty term was set to zero for warmup. For the lower bound of the volume constraint, we set $r = 0.9$ [6].

For evaluation, we used the *Dice similarity coefficient* (DSC) to measure the segmentation accuracy. In addition, the performance of volume preservation was evaluated by measuring the relative volume error E_{vol}, which is defined as

$$E_{vol} = \frac{\left| \sum_i V_i - \sum_i Y_i \right|}{\sum_i Y_i}. \tag{6}$$

3 Results

3.1 Dataset

For evaluation, the proposed method was applied to ischemic stroke lesion segmentation on *diffusion weighted images* (DWIs). The resolution of the DWIs is $0.96\,\text{mm} \times 0.96\,\text{mm} \times 6.5\,\text{mm}$, and the image dimension is $240 \times 240 \times 21$. We used 150 DWIs annotated by experienced radiologists in the experiment, where 30, 20, and 100 subjects were used for training, validation, and testing, respectively. Skull-stripping was performed using FSL [18], and then the intensities of each input image were normalized to have a zero mean and unit variance [5].

3.2 Evaluation

We first investigated the impact of hyperparameters using the validation set. We set $\lambda = 1.0$ and tested different $c \in \{10, 15, 50, 100\}$. The average DSCs for the validation set are shown in Fig. 3(a), where $c = 15$ corresponds to the best

---

[2] Note that our method can be integrated with arbitrary segmentation networks.

Baseline   Linear   Proposed   Annotation

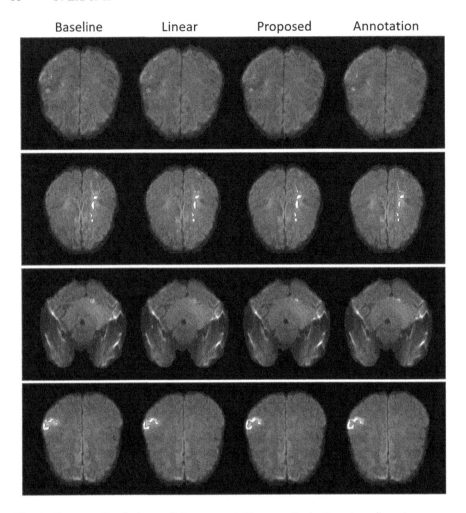

Fig. 4. Cross-sectional views of the segmentation results (red contours) on four representative test subjects. (Color figure online)

validation DSC. Thus, $c = 15$ was selected for the experiment. To show that changing λ does not lead to improved performance, we also fixed $c = 15$ and tested different $\lambda \in \{0.01, 0.1, 1, 10\}$. The corresponding validation DSCs are shown in Fig. 3(b), where we can see that increasing or decreasing λ leads to lower validation DSCs.

The proposed method was then evaluated qualitatively. It was compared with the baseline method, which is the backbone network trained with the sum of the Dice loss and weighted cross-entropy loss and without the volume constraint. In addition, to demonstrate the benefit of the transformation of segmentation probabilities for the volume constraint, the linear volume approximation in Eq. (3) was also considered. Representative cross-sectional views of the segmentation

Table 1. Quantitative segmentation results for the test subjects.

Method	DSC		E_{vol}	
	Mean	Std	Mean	Std
Baseline	0.631	0.251	0.412	0.307
Linear	0.654	0.241	0.384	0.314
Proposed	0.675	0.235	0.378	0.469

results are shown in Fig. 4. In these cases, both linear volume approximation and the proposed method better segmented the small lesions, and the results of the proposed method are more similar to the manual annotations than those of linear approximation are. These results suggest that it is beneficial to introduce the constraint of segmentation volume and that more accurate approximation of segmentation volume is desired.

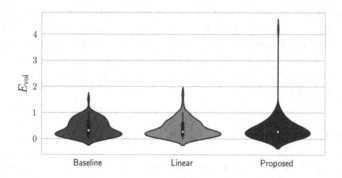

Fig. 5. A violin plot for the distribution of the relative volume error.

The proposed method was next quantitatively evaluated, where the DSC and E_{vol} were computed for the test subjects. The means and *standard deviations* (stds) of the DSC and E_{vol} are shown in Table 1. Compared with the baseline where no volume constraint was imposed during network training, the incorporation of a penalty term for the volume constraint can achieve better performance, as indicated by the higher mean DSC and lower mean E_{vol} of both the linear approximation and the proposed method. Additionally, the proposed method outperforms the linear approximation due to the use of more accurate volume approximation from the segmentation probabilities. We have also plotted the distribution of E_{vol} in Fig. 5. Despite one outlier, there are more small volume errors for the proposed method than for the competing methods. The quantitative results are consistent with the qualitative results and demonstrate the advantage of the proposed method.

4   Discussion

This work has investigated the advantages of volume constraints for brain lesion segmentation and has evaluated the proposed method on ischemic stroke lesions. The segmentation of ischemic stroke lesions is a binary segmentation problem, and it is also possible to extend the constraints to a multi-class segmentation problem, for example, brain tumor segmentation. In this case, the network segments multiple objects, and the volume constraint becomes the sum of the volume constraint of each object.

In addition to the inequality volume constraint, there are other kinds of penalty-based losses. For example, in [16], a generalized loss function based on the Tversky index is proposed, which imposes penalties on false positivity and false negativity. The Tversky loss function alleviates the negative effects caused by data imbalance and achieves better trade-off between precision and recall. It would be interesting to compare the volume constraint loss with the Tversky loss, or combine the two losses in future work.

The proposed method can be evaluated more comprehensively in future work. First, more evaluation metrics, such as the Hausdorff distance, can be included in the evaluation. Second, k-fold cross validation can be considered to show that the proposed method is robust to the selection of training and test data. Third, the proposed method can be evaluated on more diseases, such as brain tumors and multiple sclerosis, to demonstrate that our method is applicable to different types of brain lesions.

5   Conclusion

We have proposed a method for volume preserving brain lesion segmentation. An inequality constraint is developed for the segmentation volume, and the constraint is converted to a penalty term that can be integrated with standard segmentation losses. To reduce the bias of volume underestimation, a better approximation of segmentation volume using transformed segmentation probabilities is proposed and used in the penalty term. Experimental results show that our design of the volume constraint reduces the bias of segmentation volume and improves the segmentation accuracy.

Acknowledgment. This work is supported by the Beijing Natural Science Foundation (L192058 and 7192108) and Beijing Institute of Technology Research Fund Program for Young Scholars.

References

1. Bateson, M., Kervadec, H., Dolz, J., Lombaert, H., Ayed, I.B.: Constrained domain adaptation for segmentation. In: Shen, D., et al. (eds.) MICCAI 2019. LNCS, vol. 11765, pp. 326–334. Springer, Cham (2019). https://doi.org/10.1007/978-3-030-32245-8_37

2. Çiçek, Ö., Abdulkadir, A., Lienkamp, S.S., Brox, T., Ronneberger, O.: 3D U-Net: learning dense volumetric segmentation from sparse annotation. In: Ourselin, S., Joskowicz, L., Sabuncu, M.R., Unal, G., Wells, W. (eds.) MICCAI 2016. LNCS, vol. 9901, pp. 424–432. Springer, Cham (2016). https://doi.org/10.1007/978-3-319-46723-8_49

3. Demeestere, J., et al.: Evaluation of hyperacute infarct volume using ASPECTS and brain CT perfusion core volume. Neurology **88**(24), 2248–2253 (2017)

4. Jia, Z., Huang, X., Chang, I.C., Xu, Y.: Constrained deep weak supervision for histopathology image segmentation. IEEE Trans. Med. Imaging **36**(11), 2376–2388 (2017)

5. Kamnitsas, K., et al.: Efficient multi-scale 3D CNN with fully connected CRF for accurate brain lesion segmentation. Med. Image Anal. **36**, 61–78 (2017)

6. Kervadec, H., Dolz, J., Granger, É., Ben Ayed, I.: Curriculum semi-supervised segmentation. In: Shen, D., et al. (eds.) MICCAI 2019. LNCS, vol. 11765, pp. 568–576. Springer, Cham (2019). https://doi.org/10.1007/978-3-030-32245-8_63

7. Kervadec, H., Dolz, J., Tang, M., Granger, E., Boykov, Y., Ayed, I.B.: Constrained-CNN losses for weakly supervised segmentation. Med. Image Anal. **54**, 88–99 (2019)

8. Kingma, D.P., Ba, J.: Adam: a method for stochastic optimization. arXiv preprint arXiv:1412.6980 (2014)

9. Litjens, G., Kooi, T., Bejnordi, B.E., et al.: A survey on deep learning in medical image analysis. Med. Image Anal. **42**, 60–88 (2017)

10. Márquez-Neila, P., Salzmann, M., Fua, P.: Imposing hard constraints on deep networks: promises and limitations. In: CVPR Workshop on Negative Results (2017)

11. Milletari, F., Navab, N., Ahmadi, S.A.: V-net: fully convolutional neural networks for volumetric medical image segmentation. arXiv preprint arXiv:1606.04797 (2016)

12. Nair, T., Precup, D., Arnold, D.L., Arbel, T.: Exploring uncertainty measures in deep networks for multiple sclerosis lesion detection and segmentation. Med. Image Anal. **59**, 101557 (2020)

13. Paszke, A., Gross, S., Massa, F., et al.: PyTorch: an imperative style, high-performance deep learning library. In: Advances in Neural Information Processing Systems, pp. 8024–8035 (2019)

14. Pathak, D., Krähenbühl, P., Darrell, T.: Constrained convolutional neural networks for weakly supervised segmentation. In: Proceedings of the IEEE International Conference on Computer Vision, pp. 1796–1804 (2015)

15. Ronneberger, O., Fischer, P., Brox, T.: U-Net: convolutional networks for biomedical image segmentation. In: Navab, N., Hornegger, J., Wells, W.M., Frangi, A.F. (eds.) MICCAI 2015. LNCS, vol. 9351, pp. 234–241. Springer, Cham (2015). https://doi.org/10.1007/978-3-319-24574-4_28

16. Salehi, S.S.M., Erdogmus, D., Gholipour, A.: Tversky loss function for image segmentation using 3D fully convolutional deep networks. arXiv preprint arXiv:1706.05721 (2017)

17. Smith, M.C., Byblow, W.D., Barber, P.A., Stinear, C.M.: Proportional recovery from lower limb motor impairment after stroke. Stroke **48**(5), 1400–1403 (2017)

18. Smith, S.M., Jenkinson, M., Woolrich, M.W., et al.: Advances in functional and structural MR image analysis and implementation as FSL. Neuroimage **23**(S1), 208–219 (2004)

19. Zhao, X., Wu, Y., Song, G., Li, Z., Zhang, Y., Fan, Y.: A deep learning model integrating FCNNs and CRFs for brain tumor segmentation. Med. Image Anal. **43**, 98–111 (2018)

Microstructural Modulations in the Hippocampus Allow to Characterizing Relapsing-Remitting Versus Primary Progressive Multiple Sclerosis

Lorenza Brusini[1](✉), Ilaria Boscolo Galazzo[1], Muge Akinci[2],
Federica Cruciani[1], Marco Pitteri[3], Stefano Ziccardi[3], Albulena Bajrami[3],
Marco Castellaro[3], Ahmed M. A. Salih[1], Francesca B. Pizzini[4],
Jorge Jovicich[2], Massimiliano Calabrese[3], and Gloria Menegaz[1]

[1] Department of Computer Science, University of Verona, Verona, Italy
`lorenza.brusini@univr.it`
[2] Center for Mind/Brain Sciences, University of Trento, Trento, Italy
[3] Neurology Unit, Department of Neurosciences, Biomedicine and Movement Sciences, University of Verona, Verona, Italy
[4] Radiology, Department of Diagnostic and Public Health, University of Verona, Verona, Italy

Abstract. Whether gray matter (GM) regions are differentially vulnerable in Relapsing-Remitting and Primary Progressive Multiple Sclerosis (RRMS and PPMS) is still unknown. The objective of this study was to evaluate morphometric and microstructural properties based on structural and diffusion magnetic resonance imaging (dMRI) data in these MS phenotypes, and verify if selective intra-pathological alterations characterise GM structures. Diffusion Tensor Imaging (DTI) and 3D Simple Harmonics Oscillator based Reconstruction and Estimation (3D-SHORE) models were used to fit the dMRI signals, and several features were subsequently extracted from the regional values distributions (e.g., mean, median, skewness). Statistical analyses were conducted to test for group differences and possible correlations with physical disability scores. Results highlighted 3D-SHORE sensitivity to microstructural differences in hippocampus, which was also significantly correlated to physical disability. Conversely, morphometric measurements did not reach any statistical significance. Our study emphasized the potential of dMRI, and in particular the importance of advanced models such as 3D-SHORE with respect to DTI in characterizing the two MS types. In addition, hippocampus has been revealed as particularly relevant in the distinction of RRMS from PPMS and calls for further investigation.

Keywords: Diffusion MRI · SHORE · Multiple sclerosis · Gray matter

---

L. Brusini and I. B. Galazzo—These authors equally contributed as first author to this work.

M. Calabrese and G. Menegaz—These authors equally contributed as last author to this work.

© Springer Nature Switzerland AG 2021
A. Crimi and S. Bakas (Eds.): BrainLes 2020, LNCS 12658, pp. 70–79, 2021.
https://doi.org/10.1007/978-3-030-72084-1_7

1   Introduction

Multiple sclerosis (MS) causes the degeneration of the brain tissue, inducing several debilitating symptoms. Relapsing-Remitting (RRMS) and Primary Progressive Multiple Sclerosis (PPMS) are the two principal MS phenotypes whose clinical manifestations differ in several aspects, but their untimely classification is difficult to establish at disease onset. In particular, while RRMS is characterized by relapses followed by periods of remission, PPMS presents a gradual worsening of symptoms from the outset with a severe progression over time. The clinical and radiological differences between these MS types could imply that distinct mechanisms are at the basis of the onset [1], and that brain regions are differently affected as shown by Huang and colleagues [2]. Recent pathological and MRI studies have highlighted an involvement of gray matter (GM) in the developing of the MS forms [3,4], unlike the tradition that has always described this disease as specific of white matter (WM).

The inherent properties of diffusion MRI (dMRI) make this technique a viable mean to explore the microstructure of the GM tissues and to extract useful information about disease-related modulations. Several approaches can be used to quantify the aforementioned microstructural characteristics [5,6], and are here referred to as compartmental and analytical models. While the latter represents the diffusion signal using objective functions, the former relies on biophysical tissue modelling. Among these models, Neurite Orientation Dispersion and Density Imaging (NODDI) and Composite Hindered and Restricted Model of Diffusion (CHARMED) are the most popular also in clinical settings [7,8]. However, the simplifying model assumptions at the basis of these methods represent their main limitation, which potentially reduces the interpretability and accuracy of the imaging findings [9]. Conversely, analytical models can be ideally applied to any case, including GM. For this reason, we here focus on this second approach relying, in particular, on the characterization of the ensemble average propagator (EAP), which describes the displacement of the water molecules inside the brain in the unit time. Along with the well-known Diffusion Tensor Imaging (DTI) [10], the Simple Harmonic Oscillator-based Reconstruction and Estimation (SHORE) model [11] was adopted. Such a model is able to overcome the DTI limitation implied by the grounding on a Gaussian model to describe the signal in regions with complex topology where this assumption fails (e.g. WM fibers crossings). The indices that can be extracted have been explored in several studies, demonstrating a promising potential for the characterization of brain tissue modulations also in pathological conditions [12–16].

To the best of our knowledge, SHORE indices have never been previously exploited to investigate GM microstructure in MS. Nevertheless, advanced models can offer great advantage over DTI in this context, as shown by several authors. Granberg et al. [17] characterized normal appearing WM and GM using NODDI and DTI, while a detailed comparison across DTI, NODDI, CHARMED and Diffusion Kurtosis Imaging derived metrics was performed in a small cohort of MS patients in [18]. In both studies, advanced measures revealed a greater sensitivity and specificity for uncovering MS neurodegeneration compared to DTI.

Therefore, in this study we investigated whether 3D-SHORE can provide viable indices for differentiating RRMS and PPMS based solely on GM tissue measurements. The challenges of such an intra-pathology study make this exploration particularly relevant, considering also the clinical importance of an early and *ad-personam* treatment that differs in these MS subtypes.

2   Materials and Methods

2.1 Study Participants and MRI Acquisition

Twenty-three RRMS patients (14 females, mean age: 45.8 ± 8.8 years, mean disease duration: 7.6 ± 6.0 years) and twenty-three matched PPMS (14 females, mean age: 45.6 ± 8.6 years, mean disease duration: 8.4 ± 7.3 years) underwent MRI acquisitions on a 3T Philips Achieva scanner. The imaging protocol for the study included 3D T1-weighted Fast Field Echo (T1w), 3D T2-weighted Turbo Spin Echo (T2w), 3D Fluid-Attenuated Inversion Recovery (FLAIR), and two-shells dMRI. The main acquisition parameters for dMRI were: TR/TE = 9300/109 ms, FA = 90°, FOV = 112 × 112 mm^2, 2-mm isotropic resolution, 62 slices, b-values = 700/2000s/mm^2 with 32/64 gradient directions, respectively, and 7 b0 volumes. All patients were recruited in our centre according to their diagnosis based on the McDonald 2010 diagnostic criteria.

The Expanded Disability Status Scale (EDSS) score was measured in all patients. All enrolled subjects gave their written informed consent to participate in the study which was approved by the local Ethical Committee and performed in accordance with the Declaration of Helsinki (2008).

2.2 Signal Modelling and Microstructural Indices

DMRI indices representing microstructural properties of the tissue are derived from the well-known DTI [10,19] and the 3D-SHORE [11,20] models. The EAP can be recovered from the diffusion-weighted signal attenuation $E(q)$ under the narrow pulse assumption [21] via the Fourier relationship:

$$P(\mathbf{r}) = \int_{\mathbf{q} \in R^3} E(\mathbf{q}) e^{i2\pi \mathbf{q}\mathbf{r}} d\mathbf{q} \tag{1}$$

where $P(\mathbf{r})$ is the EAP, indicating the likelihood for a particle to undergo a net displacement $\mathbf{r}$ in the unit time and $\mathbf{q} = q\mathbf{u}$ is the sampling position, with $\mathbf{u}$ being unit vector of the reciprocal space, or $\mathbf{q}$-space.

DTI assumes that the diffusion propagator can be described by a single 3D Gaussian distribution [10,19] from which a 3 × 3 symmetric positive-definite matrix is derived and used to compute the classical tensor-based indices (Fractional Anisotropy [FA] and Mean Diffusivity [MD]). FA provides a measure of the anisotropy of the diffusion process, with higher values reflecting increased directionality of diffusion, independently of the rate of diffusion. MD is calculated as the mean of the eigenvalues and can be considered as a measure of mobility independent from the directionality. Only the shell at b-value $= 700\,\mathrm{s/mm}^2$ was used for the DTI analysis, corresponding to 32 gradient directions.

The more advanced indices explored in this work were calculated by fitting the SHORE model [11,20], that is based on the solutions of the 3D quantum harmonic oscillator in the formulation using the orthonormalized basis:

$$E(\mathbf{q}) = \sum_{l=0,even}^{N_{max}} \sum_{n=l}^{(N_{max}+l)/2} \sum_{m=-l}^{l} c_{nlm}\Phi_{nlm}(\mathbf{q}) \tag{2}$$

In this equation, N_{max} is the maximal order of the functions, $\Phi_{nlm}(\mathbf{q})$ are the functions forming the 3D-SHORE orthonormal basis and are given by:

$$\Phi_{nlm}(\mathbf{q}) = \left[\frac{2(n-l)!}{\zeta^{3/2}\Gamma(n+3/2)}\right]^{1/2} \left(\frac{q^2}{\zeta}\right)^{l/2} \exp\left(\frac{-q^2}{2\zeta}\right) L_{n-l}^{l+1/2}\left(\frac{q^2}{\zeta}\right) Y_l^m(\mathbf{u}) \tag{3}$$

where Γ is the Gamma function and ζ is a scaling parameter determined by the diffusion time and the mean diffusivity [22,23]. For the 3D-SHORE model, the EAP is obtained by plugging Eq. 2 into Eq. 1 [11,22]. Due to the linearity of the Fourier transform, the EAP basis is thus expressed in terms of the same set of coefficients c_{nlm} as the diffusion signal. Six 3D-SHORE-based indices were then derived as in [11]. Return To the Origin Probability (RTOP), Return To the Axis Probability (RTAP) and Return To the Plane Probability (RTPP) can be obtained calculating the integral of the EAP over the whole q-space, along the principal direction of the diffusion process and over the plane defined by the origin and the aforementioned direction, respectively. When the gradient pulses of the sequence are narrow and the diffusion time is long, then these indices are related to the pore geometry where the diffusion process takes place [11,24]. More in detail, RTOP is proportional to the inverse of the volume, whereas RTAP and RTPP are proportional to the inverse of the mean apparent cross-sectional area and length, respectively. Mean Square Displacement (MSD) is considered as another measure of mobility, in fact it is strictly related to MD showing a similar contrast [25]. Finally, Propagator Anisotropy (PA) and Generalized Fractional Anisotropy (GFA) are two measures of anisotropy as FA. The first is calculated as the dissimilarity between the whole propagator and the part of the propagator derived only from isotropic functions of the basis [11]. The second is related to the normalized variance of the Orientation Distribution Function (ODF) that with 3D-SHORE modelling is calculated considering the radial integral of the EAP along a given direction [11,23].

2.3   Image Preprocessing

For dMRI data, image re-sampling, correction for motion, eddy-current and EPI distortions were performed using the Tortoise DIFFPREP pipeline (https://tortoise.nibib.nih.gov/tortoise). Brain extraction and masking were computed using FSL (https://fsl.fmrib.ox.ac.uk/fsl/fslwiki/), along with the rigid-body registration of the T1w image to the mean b0 volume. DTI and 3D-SHORE models were then fitted to preprocessed data (https://dipy.org/).

Concerning structural images, FLAIR images were rigidly registered to the corresponding T1w in each patient by using FSL tools. Lesions maps were automatically derived from the registered FLAIR images using the Lesion Prediction Algorithm (LPA) [26] in the Lesion Segmentation Toolbox (LST) for SPM12 (www.statistical-modelling.de/lst.html). These maps were used to fill the native T1w, resulting in filled images fed into the FreeSurfer software for a complete brain parcellation (http://surfer.nmr.mgh.harvard.edu/). This was projected to the dMRI space by using the previously estimated transformation matrix, and a set of regions of interest (ROIs) known to be highly relevant for the pathology were kept for the subsequent analyses: thalamus (Thal), caudate (Cau), putamen (Put), hippocampus (Hipp), lateral occipital cortex (LOC), lingual gyrus (LgG), pericalcarine cortex (PC), posterior cingulate cortex (PCC), precuneus (Pre), superior-frontal gyrus (SFG), insula (Ins) [3,4,27].

2.4   Features Extraction

Eight microstructural indices were calculated in each subject. The well-known FA and MD [28] were derived from the DTI, while GFA, PA, MSD, RTOP, RTAP and RTPP were estimated from the 3D-SHORE [11,16,24].

The eleven ROIs here considered were used as masks to extract the regional microstructure values for each dMRI index. Starting from these values, for each participant and ROI the following features were calculated: mean, median, mode, skewness, standard deviation and kurtosis. Regarding morphometry, volume (subcortical ROIs) and thickness (cortical ROIs) were derived from Freesurfer. All measures were averaged across the two hemispheres, and volume measures were also normalized by the estimated total intracranial volume (eTIV).

2.5   Statistical Analysis

For each dMRI index, a three-way analysis of variance (ANOVA) was performed to evaluate the differences between all the aforementioned features. Factors were the MS type (STAGE), ROI, distribution feature (FEAT), and their interactions. For each significant interaction, *post-hoc* two sample t-tests were performed and corrected for multiple comparison using the false discovery rate (FDR, $p < 0.05$). For each ROI, differences between morphometric measures were tested through Wilcoxon rank sum tests ($p < 0.05$, FDR-corrected).

Finally, the Spearman correlation coefficient was computed between each MRI-metric and EDSS scores for each ROI separately. FDR was used for correcting for multiple comparisons ($p < 0.05$).

3   Results

ANOVA analyses revealed a significant three-way interaction between STAGE, ROI and FEAT for RTPP ($F_{(50,2904)} = 1.87$, $p = 0.0002$). *Post-hoc* tests highlighted between-group significant differences for RTPP-median and skewness in hippocampus ($p_{FDR} < 0.05$). RTPP values for these two features, expressed as mean ± standard deviation across each group, are reported for all the considered ROIs in Fig. 1. Interestingly, hippocampus also featured the lowest RTPP-median values across all ROIs for both MS types, with a significant between-group difference (PPMS < RRMS, $p_{FDR} < 0.05$). Concerning RTPP-skewness, values were negative for all ROIs except the pericalcarine cortex. In the hippocampus, the RTPP featured higher simmetry for PPMS with respect to RRMS, and this difference was deemed as significant by the ANOVA *post-hoc* analysis (PPMS < RRMS, $p_{FDR} < 0.05$).

Concerning the other dMRI indices, a significant two-way interaction between STAGE and FEAT was found for RTAP ($F_{(5,3024)} = 2.64$, p = 0.0216), and the *post-hoc* tests confirmed a significant between-group difference for RTAP-standard deviation ($p_{FDR} < 0.05$). No significant interactions were found for the remaining indices.

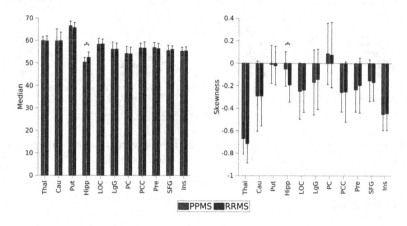

Fig. 1. RTPP features for PPMS and RRMS. For each region, median (left) and skewness (right) values are reported as mean ± standard deviation values across subjects ($^*p_{FDR} < 0.05$).

Regarding morphometric measures, no statistical differences between MS groups could be found in any region as can be observed in Fig. 2.

Figure 3 illustrates the significant correlations between MRI features and EDSS scores. Of note, EDSS resulted in significantly lower EDSS values in PPMS compared to RRMS (two-sample t-test, p < 0.001). RTPP-median and skewness in hippocampus were significantly correlated with EDSS (negatively and positevely, respectively), along with PA-skewness and kurtosis in insula. The highest Spearman correlation coefficient was found for PA-skewness ($r_s = 0.480$, $p_{FDR} = 0.008$).

No significant associations with EDSS could be found for morphometric measures in any region.

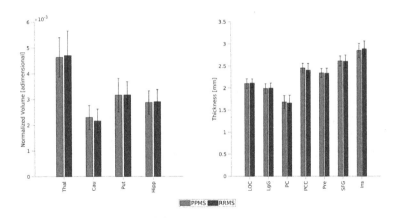

Fig. 2. Morphometric measures for PPMS and RRMS. Volume (left) and thickness (right) are reported as mean ± standard deviation values across subjects.

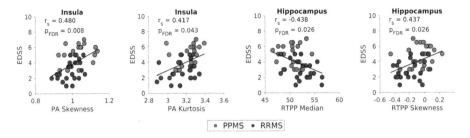

Fig. 3. Significant FDR-corrected correlations. The Spearman correlation coefficient (r_S), and the FDR corrected p-value (p_{FDR}) are also reported.

4   Discussion

In this study, we explored the ability of DTI and 3D-SHORE indices in characterizing GM diffusion patterns in RRMS and PPMS patients, and investigated whether signature of regional microstructural and morphometric differences underpinning the two MS phenotypes could be detected. The differential GM alterations are in agreement with previous studies demonstrating the presence of widespread microstructural modifications encompassing not only WM or MS lesions but also GM [17,18]. In particular, the statistical changes detected by RTPP features demonstrate the advantages provided by advanced models compared to conventional DTI for disentangling complex neurological diseases.

On the basis of our findings, hippocampus seems to play an important role in the microstructure differentiation between pathological types. In particular, the significantly lower RTPP values found in PPMS patients might point to a decreased neuronal density or lower restriction to diffusion [16,22], which has been recently reported as characteristic of MS disease [29]. In addition, a recent study by Koubiyr et al. [30] on subjects with clinically isolated syndrome, usually considered the early clinical sign of MS, demonstrated that hippocampus was the first region to present microstructural alterations and that MD values at baseline enabled to predict its volume loss after 1 year. The significant correlations we found between dMRI features and EDSS further emphasized the relevance of hippocampus in characterizing the MS disease. Rocca et al. [31] deeply discussed the role of this area in brain plasticity and neurogenesis, further supported by pathological studies that showed extensive demyelination, neuronal damage, and synaptic abnormalities in MS patients. Therefore, as also hypothesized by these authors, hippocampus could represent a potential target for the MS early diagnosis and personalized treatment.

Our study also revealed a significant association between physical disability and PA-skewness and kurtosis in insula, which is connected to the hippocampus in its anterior side. This finding is further supported by the role of insula in regulating several physiological functions, such as the viscero-motor and sensitive control. Differently from dMRI features, morphometric measurements did not allow to detect any significant difference or correlation with EDSS scores. Eshaghi et al. [27] recently evaluated the atrophy patterns across the different MS phenotypes in a longitudinal study. They showed that several ROIs, including those here considered, presented similar patterns of volume/thickness loss in RRMS and PPMS. Our results are in line with these findings suggesting that the two considered MS phenotypes might share similar levels of atrophy.

The main limitation of our work is the small number of subjects enrolled in our study. Recruiting new patients could improve the results of statistical tests after FDR correction. Moreover, the intra-pathology nature of our experimental protocol makes more difficult to capture changes differentiating RRMS and PPMS. Involving healthy subjects could highlight which ROIs are more severely compromised from the disease and thus focusing the outcomes. Finally, several other advanced models designed for modeling non-Gaussian signals could be explored in future works. The aim of such an investigation will be to assess the most sensitive derived index for timely capturing modulations defining MS phenotypes.

5   Conclusions

Our cross-sectional study revealed the sensitivity of dMRI features in detecting microstructural differences in PPMS versus RRMS patients, where morphometric measures failed. In particular, 3D-SHORE RTPP in hippocampus was significantly different in the two MS forms and was significantly correlated to physical disability scores. This highlights the importance of using advanced models such as 3D-SHORE and calls for further investigation of the hippocampus, as this ROI seems playing a focal role in the characterization of the MS disease course.

References

1. Lucchinetti, C., Brück, W., Parisi, J., Scheithauer, B., Rodriguez, M., Lassmann, H.: Heterogeneity of multiple sclerosis lesions: implications for the pathogenesis of demyelination. Ann. Neurol. Official J. Am. Neurol. Assoc. Child Neurol. Soc. **47**(6), 707–717 (2000)
2. Huang, W.J., Chen, W.W., Zhang, X.: Multiple sclerosis: pathology, diagnosis and treatments. Exp. Ther. Med. **13**(6), 3163–3166 (2017)
3. Geurts, J.J., Calabrese, M., Fisher, E., Rudick, R.A.: Measurement and clinical effect of grey matter pathology in multiple sclerosis. Lancet Neurol. **11**(12), 1082–1092 (2012)
4. Calabrese, M., et al.: Regional distribution and evolution of gray matter damage in different populations of multiple sclerosis patients. PLoS ONE **10**(8), e0135428 (2015)
5. Alexander, D.C., Dyrby, T.B., Nilsson, M., Zhang, H.: Imaging brain microstructure with diffusion MRI: practicality and applications. NMR Biomed. **32**(4), e3841 (2019)
6. Novikov, D.S., Fieremans, E., Jespersen, S.N., Kiselev, V.G.: Quantifying brain microstructure with diffusion MRI: theory and parameter estimation. NMR Biomed. **32**(4), e3998 (2019)
7. Assaf, Y., Basser, P.J.: Composite hindered and restricted model of diffusion (charmed) MR imaging of the human brain. Neuroimage **27**(1), 48–58 (2005)
8. Zhang, H., Schneider, T., Wheeler-Kingshott, C.A., Alexander, D.C.: NODDI: practical in vivo neurite orientation dispersion and density imaging of the human brain. Neuroimage **61**(4), 1000–1016 (2012)
9. Lampinen, B., Szczepankiewicz, F., Mårtensson, J., van Westen, D., Sundgren, P.C., Nilsson, M.: Neurite density imaging versus imaging of microscopic anisotropy in diffusion MRI: a model comparison using spherical tensor encoding. Neuroimage **147**, 517–531 (2017)
10. Basser, P.J., Mattiello, J., LeBihan, D.: Estimation of the effective self-diffusion tensor from the NMR spin echo. J. Magn. Reson. Ser. B **103**(3), 247–254 (1994)
11. Özarslan, E., et al.: Mean apparent propagator (map) MRI: a novel diffusion imaging method for mapping tissue microstructure. Neuroimage **78**, 16–32 (2013)
12. Avram, A.V., et al.: Clinical feasibility of using mean apparent propagator (map) MRI to characterize brain tissue microstructure. Neuroimage **127**, 422–434 (2016)
13. Brusini, L., et al.: Assessment of mean apparent propagator-based indices as biomarkers of axonal remodeling after stroke. In: Navab, N., Hornegger, J., Wells, W.M., Frangi, A.F. (eds.) MICCAI 2015. LNCS, vol. 9349, pp. 199–206. Springer, Cham (2015). https://doi.org/10.1007/978-3-319-24553-9_25
14. Brusini, L., et al.: Ensemble average propagator-based detection of microstructural alterations after stroke. Int. J. Comput. Assist. Radiol. Surg. **11**(9), 1585–1597 (2016)
15. Ma, K., et al.: Mean apparent propagator-MRI: a new diffusion model which improves temporal lobe epilepsy lateralization. Eur. J. Radiol. 108914 (2020)
16. Boscolo Galazzo, I., Brusini, L., Obertino, S., Zucchelli, M., Granziera, C., Menegaz, G.: On the viability of diffusion MRI-based microstructural biomarkers in ischemic stroke. Front. Neurosci. **12**, 92 (2018)
17. Granberg, T., et al.: In vivo characterization of cortical and white matter neuroaxonal pathology in early multiple sclerosis. Brain **140**(11), 2912–2926 (2017)

18. De Santis, S., et al.: Characterizing microstructural tissue properties in multiple sclerosis with diffusion MRI at 7 T and 3 T: the impact of the experimental design. Neuroscience **403**, 17–26 (2019)
19. Basser, P.J., Mattiello, J., LeBihan, D.: MR diffusion tensor spectroscopy and imaging. Biophys. J. **66**(1), 259–267 (1994)
20. Özarslan, E., Koay, C., Shepherd, T., Blackb, S., Basser, P.: Simple harmonic oscillator based reconstruction and estimation for three-dimensional q-space MRI (2009)
21. Stejskal, E.O., Tanner, J.E.: Spin diffusion measurements: spin echoes in the presence of a time-dependent field gradient. J. Chem. Phys. **42**(1), 288–292 (1965)
22. Zucchelli, M., Brusini, L., Méndez, C.A., Daducci, A., Granziera, C., Menegaz, G.: What lies beneath? Diffusion EAP-based study of brain tissue microstructure. Med. Image Anal. **32**, 145–156 (2016)
23. Merlet, S.L., Deriche, R.: Continuous diffusion signal, EAP and ODF estimation via compressive sensing in diffusion MRI. Med. Image Anal. **17**(5), 556–572 (2013)
24. Zucchelli, M., Fick, R.H.J., Deriche, R., Menegaz, G.: Ensemble average propagator estimation of axon diameter in diffusion MRI: implications and limitations. In: 2016 IEEE 13th International Symposium on Biomedical Imaging (ISBI), pp. 465–468 (2016)
25. Wu, Y.C., Alexander, A.L.: Hybrid diffusion imaging. Neuroimage **36**(3), 617–629 (2007)
26. Schmidt, P., et al.: An automated tool for detection of flair-hyperintense white-matter lesions in multiple sclerosis. Neuroimage **59**(4), 3774–3783 (2012)
27. Eshaghi, A., et al.: Progression of regional grey matter atrophy in multiple sclerosis. Brain **141**(6), 1665–1677 (2018)
28. Pierpaoli, C., Basser, P.J.: Toward a quantitative assessment of diffusion anisotropy. Magn. Reson. Med. **36**(6), 893–906 (1996)
29. Carassiti, D., Altmann, D., Petrova, N., Pakkenberg, B., Scaravilli, F., Schmierer, K.: Neuronal loss, demyelination and volume change in the multiple sclerosis neocortex. Neuropathol. Appl. Neurobiol. **44**(4), 377–390 (2018)
30. Koubiyr, I., et al.: Differential gray matter vulnerability in the 1 year following a clinically isolated syndrome. Front. Neurol. **9**, 824 (2018)
31. Rocca, M.A., et al.: The hippocampus in multiple sclerosis. Lancet Neurol. **17**(10), 918–926 (2018)

Symmetric-Constrained Irregular Structure Inpainting for Brain MRI Registration with Tumor Pathology

Xiaofeng Liu[1]([✉]), Fangxu Xing[1], Chao Yang[2], C.-C. Jay Kuo[3], Georges El Fakhri[1], and Jonghye Woo[1]

[1] Gordon Center for Medical Imaging, Department of Radiology, Massachusetts General Hospital and Harvard Medical School, Boston, MA 02114, USA
`xliu61@mgh.harvard.edu`
[2] Facebook Artificial Intelligence, Boston, MA 02142, USA
[3] Ming Hsieh Department of Electrical and Computer Engineering, University of Southern California, Los Angeles, CA 90007, USA

Abstract. Deformable registration of magnetic resonance images between patients with brain tumors and healthy subjects has been an important tool to specify tumor geometry through location alignment and facilitate pathological analysis. Since tumor region does not match with any ordinary brain tissue, it has been difficult to deformably register a patient's brain to a normal one. Many patient images are associated with irregularly distributed lesions, resulting in further distortion of normal tissue structures and complicating registration's similarity measure. In this work, we follow a multi-step context-aware image inpainting framework to generate synthetic tissue intensities in the tumor region. The coarse image-to-image translation is applied to make a rough inference of the missing parts. Then, a feature-level patch-match refinement module is applied to refine the details by modeling the semantic relevance between patch-wise features. A symmetry constraint reflecting a large degree of anatomical symmetry in the brain is further proposed to achieve better structure understanding. Deformable registration is applied between inpainted patient images and normal brains, and the resulting deformation field is eventually used to deform original patient data for the final alignment. The method was applied to the Multimodal Brain Tumor Segmentation (BraTS) 2018 challenge database and compared against three existing inpainting methods. The proposed method yielded results with increased peak signal-to-noise ratio, structural similarity index, inception score, and reduced L1 error, leading to successful patient-to-normal brain image registration.

Keywords: Brain tumor · Registration · Image inpainting · Irregular structure · Symmetry · Contextual learning · Deep learning

---

X. Liu and F. Xing—Contribute Equally.

© Springer Nature Switzerland AG 2021
A. Crimi and S. Bakas (Eds.): BrainLes 2020, LNCS 12658, pp. 80–91, 2021.
https://doi.org/10.1007/978-3-030-72084-1_8

1   Introduction

In brain imaging studies, magnetic resonance imaging (MRI) as a noninvasive tool is widely used to provide information on the brain's clinical structure, tissue anatomy, and functional behaviors [4,28]. When multiple datasets from a population of interest are involved, to establish a comparable framework in which similarity and variability in the tissue structure can be evaluated, deformable image registration between subjects are often used to achieve inter-subject alignment [37]. Brain tumor is a common type of disorder diagnosed using medical imaging [35]. However, tumors in MRI tend to cause difficulties with deformable registration: 1) Tumor regions have no matching structure in a normal brain, nullifying the basic mathematical assumptions made for regular image registration methods and subsiding their performance; 2) Expansion of tumor regions often alters its peripheral structure, causing the whole image to become asymmetric with distorted hemispheres or ventricles; and 3) The locations of tumors are sometimes irregularly scattered around the whole brain, causing inconsistencies when matching multiple tumor spots [10].

There has been a great deal of work that tackles patient-to-normal tissue registration in a traditional way [19,38]. Especially, for small tumor cases, Dawant et al. [9] introduced a tumor seed and Cuadra et al. [8] extended it with a tumor growth model to drive the registration process. For larger tumors, Mohamed et al. [27] used a biomechanical model of tumor-induced deformation to generate a similar tumor image from the normal image. Since then many methods have been focusing on tumor growth simulations to facilitate symmetry computation [14,40]. More traditional methods are summarized in [37]. In this work, we propose a new image inpainting method—i.e., a restorative method that treats tumor as defective holes in an ideal image and reconstructs them with synthetic normal tissue. The synthesized brain can be processed with regular deformable registration and the tumor region will eventually be re-applied after being mapped to the new space.

Traditional inpainting methods are either diffusion-based or patch-based with low-level features [2,3,5,7,12,32]. These prior approaches usually perform poorly in generating semantically meaningful contents and filling in large missing regions [21]. Recently developed learning-based inpainting methods usually use generative adversarial networks (GANs) to learn image semantics and infer contents in the missing region [15,31,36,39]. In the brain tumor application, difficulties 2) and 3) need to be addressed specifically. Starting from the initial context encoder deep learning method [31], Liu et al. [20] updated the mask and convolution weights in each layer to handle irregular holes. However, it is challenging for these 1-step inpainting solutions to address the large holes with complicated texture. Song et al. [36] proposed a multi-step framework to refine the results with patch-swap, but its coarse inpainting module does not fit for multiple irregular holes. Moreover, the above methods are designed for general image cases and do not involve priors such as brain anatomy and physiology.

In this work, we propose a novel multi-step inpainting method capable of making fine-grained prediction within irregular holes with feature patch-wise

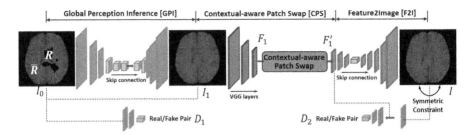

Fig. 1. Overview of the proposed network architecture. GPI is used for coarse inference and VGG is used for extracting the feature map. The patch-swap layer propagates high frequency information from the boundary to the hole. F2I translates to a complete, high-resolution image further constrained with symmetric loss.

conditional refinement. It also incorporates a symmetry constraint to explicitly exploit the quasi-symmetry property of the human brain for better structure understanding. Deformable registration is applied between inpainted patient images and normal controls whose deformation field is then used to deform original patient data into the target space, achieving patient-to-normal registration.

2   Methods

Given a brain MRI slice I_0 with tumor, the goal is to replace the pathological regions with normal brain appearances. The incomplete input I_0 is composed of R and $\overline{R}$, representing the removed pathological region (the hole) and the remaining normal region (boundary or context), respectively. Mathematically, the task is to generate a new, complete image I with plausible contents in $\overline{R}$.

Following the basic idea of contextual-based image inpainting [36], our framework consists of three sequential modules: global perception inference (GPI), context-aware patch swapping (CPS), and feature-to-image translator (F2I). The intuition behind the multi-step operation is that direct learning of the distribution of high dimensional image data is challenging. Thus using a coarse generation followed by a refinement scheme can increase the inpainting performance [36]. Our network architecture is shown in Fig. 1.

2.1   Global Perception Inference

The input to the GPI network I_0 is a $1 \times 240 \times 240$ image with irregular holes. Its output is a coarse prediction I_1. Considering the potential irregular distribution of tumor locations, the rectangular hole generation module used in [36] is not applicable. Therefore, we first adopt the GPI network structure from the image-to-image translation network proposed in [17], which consists of 4×4 convolutions with skip connections in order to concatenate different features from each encoder layer and the corresponding decoder layer. We slightly modify the size of each layer since only single channel T1-weighted MRI is used in this task.

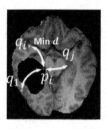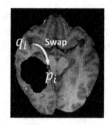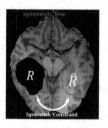

Fig. 2. Illustration of the patch-swap operation (left) and symmetry constraint (right). Patch-swap is implemented in the FCN-based VGG's feature space to search for the most similar boundary 1×1 feature patch with minimum $d(p, q)$.

The GPI module is explicitly trained using the L_1 reconstruction loss, which is important for stabilizing the adversarial training [23]. It can be formulated as

$$\mathcal{L}_1(I_1, I_{gt}) = \| I_1 - I_{gt} \|_1, \tag{1}$$

where I_1 and I_{gt} are the rough inpainting result of GPI and the ground truth, respectively.

The second objective is the adversarial loss based on GANs [24], which can be defined as:

$$\mathcal{L}_{adv} = \max_{D_1} \mathbb{E}[\log(D_1(I_0, I_{gt})) + \log(1 - D_1(I_0, I_1))]. \tag{2}$$

Here, a pair of images are input to the discriminator D_1 as is the setting of adversarial training. The incomplete image I_0 and the original image I_{gt} are the real pair, and the incomplete image I_0 and the prediction I_1 are the fake pair.

During training, the overall loss function is given by $\mathcal{L}_{GPI} = \lambda_1 \mathcal{L}_1 + \lambda_2 \mathcal{L}_{adv}$, where λ_1 and λ_1 are the balancing hyperparameters for the two losses.

2.2 Context-Aware Patch Swapping

We use I_1 as input to the CPS network which is implemented in two phases. First, I_1 is encoded as F_1 by a fully convolutional network (FCN) using the pre-trained VGG network as in [36]. Then the patch-swap operation is applied to propagate the texture from $\bar{R}$ to R while maintaining the high frequency information in R [22].

r and $\bar{r}$ denote the regions in F_1 corresponding to R and $\bar{R}$ in I_1, respectively. For each 1×1 neural patch[1] p_i of F_1 overlapping with r, the closest-matching neural patch in $\bar{r}$, indexed by q_i, is found using the following cross-correlation metric

$$d(p, q) = \frac{< p, q >}{\| p \| \cdot \| q \|}, \tag{3}$$

---

[1] In the inpainting community, the 1×1 patch (in a feature map) is a widely used concept. The output of F1 $\in \mathbb{R}^{256 \times 60 \times 60}$, while the original image is $240 \times 240 \times 1$; therefore a 1×1 area in a feature map is not considered as a pixel.

where p_i is replaced by q_i. We first swap each patch in r with its most similar patch in $\bar{r}$, followed by averaging overlapping patches. The output is then a new feature map F_1'. This process is illustrated in Fig. 2 left.

2.3   Feature-to-Image Translator

Next, we use the F2I network to learn the mapping from the swapped feature map to a complete and vivid image, which has a U-Net style generator. The input to the U-Net is a feature map extracted by the FCN-based VGG network. The generator consists of seven convolution layers and eight deconvolution layers, where the first six corresponding deconvolutional and convolutional layers are connected through skip connections. The output is a complete $1 \times 240 \times 240$ image. In addition, the F2I network comprises a patch-GAN based discriminator D_2 for adversarial training. However, the input to D_2 is a pair of an image and its feature map in contrast to the GPI network.

In practice, we follow [36] that uses the ground truth as training input. Specifically, the feature map $F_{gt} = \text{vgg}(I_{gt})$ is the input to the patch-swap layer followed by using the swapped feature $F_{gt}' = \text{patch\_swap}(F_{gt})$ to train the F2I model. $F_1' = \text{patch\_swap}(F_1)$ is still used as input for inference, since I_{gt} is not accessible at test time. Of note, using different types of input for both training and testing is not a common practice in training a machine learning model. However, its effectiveness in inpainting has been demonstrated in [36]. Similar to [42], the robustness can be further improved by sampling from both the ground truth and the GPI prediction.

The first objective is the perceptual loss defined on the entire image between the final output I and the ground truth I_{gt}:

$$\mathcal{L}_{perceptual}(I, I_{gt}) = \| vgg(I) - vgg(I_{gt}) \|_2 . \tag{4}$$

This perceptual loss has been widely used in many tasks [6,11,13,18] as it corresponds better with human perception of similarity [41].

The adversarial loss is defined by the discriminator D_2, which can be expressed as:

$$\mathcal{L}_{adv} = \max_{D_2} \mathbb{E}[\log(D_2(F_{gt}', I_{gt})) + \log(1 - D_2(F_{gt}', I))], \tag{5}$$

where the real and fake pairs for adversarial training are (F_{gt}', I_{gt}) and (F_{gt}', I), respectively.

2.4   Quasi-Symmetry Constraint

While the brain is not exactly symmetrical w.r.t. the mid-sagittal plane, there is a large degree of symmetry between left and right hemispheres in the brain which we call the "quasi-symmetry property" [29,33]. As such, using this anatomical symmetry constraint on the generated images can mitigate the ill-posed inpainting task and further improve performance especially for large hole cases. The symmetry loss is given by

$$\mathcal{L}_{sym}(I) = \mathbb{E} \parallel I_R - I_{\hat{R}} \parallel_2, \tag{6}$$

where R and $\hat{R}$ are the hole and its mirrored region as shown in Fig. 2 right.

Therefore, we can easily transfer the appearance of the normal brain tissue to the corresponding tumor part by teaching the network to recover the lost information from the mirrored side. Note that the brains used in our experiments are coarsely aligned on their mid-sagittal planes. More importantly, our technique is robust against any potential misalignments, since the resolution of the feature space is 60×60, while the input is 240×240, with the down-sampling of the max-pooling operation, the deep neural networks, in general, are robust against small rotation [25]. Besides, the deep neural network can tackle this simple rotation. With the symmetry constraint, the overall loss for the F2I translation network is defined as:

$$\mathcal{L}_{F2I} = \lambda_3 \mathcal{L}_{perceptual} + \lambda_4 \mathcal{L}_{adv} + \lambda_5 \mathcal{L}_{sym}, \tag{7}$$

where λ_3, λ_4, and λ_5 are the balancing hyperparameters for different losses. Considering the brain is not strictly symmetrical w.r.t. the mid-sagittal plane, we usually choose a relatively small weight λ_5 for $\mathcal{L}_{sym}$.

3   Experiments and Results

The proposed method was validated both qualitatively and quantitatively on the T1 modality of Brain Tumor Segmentation (BraTS) 2018 database[2]. From a total of 210 patients each with ˜150 slices, we randomly selected 16 patients for testing and the remaining subjects were used for training in a subject independent manner. Training was performed on four NVIDIA TITAN Xp GPUs with the PyTorch deep learning toolbox [30], which took about 5 h.

The normal slices without tumors in the training set were selected to train our network. Since tumors in BraTS data can occur in different spatial locations, our network is capable of familiarizing with the normal appearance in different slices. We randomly chose the irregular tumor segmentation labels in our training set as training masks.

---

[2] https://www.med.upenn.edu/sbia/brats2018/data.html.

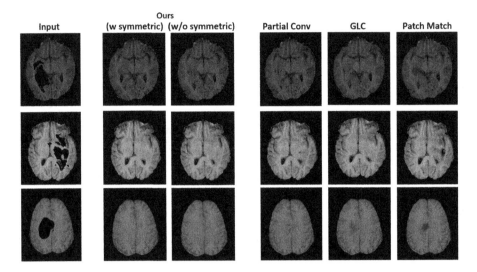

Fig. 3. An ablation study of our symmetry constraint and the comparison with the other inpainting methods.

The process of computing cross-correlation for all the neural patch pairs between the hole and the remaining region (e.g., boundary) is computationally prohibitive. To alleviate this, the strategy in [6,36] was used to speed up computation via paralleled convolution. In practice, processing one feature map only took about 0.1 s.

In order to match the absolute value of each loss, we set different weights for each part. For the training of GPI, we set weight $\lambda_1 = 10$ and $\lambda_2 = 1$. Adam optimizer was used for training. The learning rate was set at $lr_{GPI} = 1e{-}3$ and $lr_{D_1} = 1e{-}4$ and the momentum was set at 0.5. When training the F2I network, we set $\lambda_3 = 10$, $\lambda_4 = 3$ and $\lambda_5 = 1$. For the learning rate, we set $lr_{F2I} = 2e{-}4$ and $lr_{D_2} = 2e{-}4$. Same as the GPI module, the momentum was set as 0.5.

The inpainting results of various cases are shown in Figs. 3, 4, and 5. The proposed network can deal with incomplete data from different unseen patients, different slice positions, and arbitrary shape and number of holes.

Comparisons with the other inpainting methods are shown in Fig. 3. Our proposed method using context-aware inpainting [36] shows superior performance over the other methods as visually assessed. In addition, an ablation study to evaluate the contribution of the symmetry constraint is illustrated in Fig. 3. Of note, the inpainting quality was further improved using the symmetry constraint plus marginal training cost without additional testing cost. This is partly attributed to the use of the context and quasi-symmetry property of the brain.

For quantitative evaluation, we manually generated holes with random size and positions on normal slices of the testing subjects. Therefore, the ground truth is known. The inpainted images were expected to have sharp and realistic looking textures, be coherent with $\bar{R}$, and look similar to its corresponding

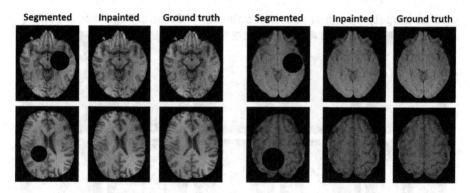

Fig. 4. Inpainting results comparing to the ground truth.

Table 1. Numerical comparison of four methods using BraTS 2018 testing set. Note that smaller mean L1 error and larger SSIM mean error indicate higher similarity.

Methods	Mean L1 error ↓	SSIM ↑	PSNR ↑	Inception Score ↑
Patch-match [3]	445.8	0.9460	29.55	9.13
GLC [16]	432.6	0.9506	30.34	9.68
Partial Conv [20]	373.2	0.9512	33.57	9.77
Proposed	292.5	0.9667	34.26	10.26
Proposed+symmetry	**254.8**	**0.9682**	**34.52**	**10.58**

ground truth. Our results are illustrated in Fig. 4. The proposed method generated visually satisfying results. Table 1 lists numerical comparisons between the proposed approach, Patch-match [3], GLC [16], and Partial Conv [20]. We note that the compared inpainting baselines [16,20] are based on the 1-step framework. We used four quality measurements to assess the performance: mean L1 error, structural similarity index (SSIM), peak signal-to-noise ratio (PSNR), and inception score [34]. We directly computed the mean L1 error and SSIM over the holes, while the incepetion score is measured on the completed I.

Finally, Fig. 5 and Table 2 show the results of deformable registration using the ANTs SyN method [1] with normalized cross-correlation as a similarity metric. As for the target atlas, we used a T1-weighted brain atlas constructed using healthy subjects from the OASIS database [26]. The result was evaluated using mutual information (MI) computed only in normal tissues to achieve a fair comparison (tumor masks were used to exclude the tumor region). Direct patient-to-normal registration was affected by the existence of tumor, thus reducing the MI score even in normal tissues. This was corrected by using the inpainted volume as registration input, yielding improved or equal MI scores on every subject tested. The mean of MI was improved from 0.3097 to 0.3129.

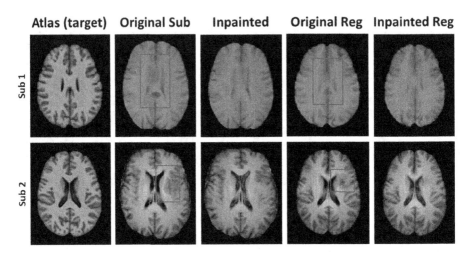

Fig. 5. Deformable registration of two brain tumor subjects to a brain atlas: direct registration vs. inpainted registration. Tumors are marked in red boxes. (Color figure online)

Table 2. Mutual information between registered brain volumes and the brain atlas on ten test subjects using direct patient registration and inpainted volume registration.

Methods	Sub1	Sub2	Sub3	Sub4	Sub5	Sub6	Sub7	Sub8	Sub9	Sub10
Direct registration	0.303	0.311	0.308	0.324	0.315	0.299	0.309	0.303	0.317	0.308
Inpainted registration	0.309	0.311	0.309	0.324	0.316	0.304	0.312	0.313	0.320	0.312

4   Conclusion

This paper presented an inpainting network that replaces the pathological tumor regions with normal brain appearances, targeting patient-to-normal deformable registration. The challenges lie in irregular brain tumor distribution. The two-stage inpainting scheme utilized both the complete and segmented samples, producing the refined results based on pixel-wise semantic relevance. Our experimental results demonstrate that the proposed method surpassed the comparison methods, which can be used for the registration between healthy subjects and tumor patients.

Acknowledgements. This work was supported by NIH R01DE027989, R01DC018511, R01AG061445, and P41EB022544.

References

1. Avants, B.B., Tustison, N.J., Song, G., Cook, P.A., Klein, A., Gee, J.C.: A reproducible evaluation of ants similarity metric performance in brain image registration. Neuroimage **54**(3), 2033–2044 (2011)

2. Ballester, C., Bertalmio, M., Caselles, V., Sapiro, G., Verdera, J.: Filling-in by joint interpolation of vector fields and gray levels. IEEE Trans. Image Process. **10**(8), 1200–1211 (2001)
3. Barnes, C., Shechtman, E., Finkelstein, A., Goldman, D.B.: Patchmatch: a randomized correspondence algorithm for structural image editing. ACM Trans. Graph. **28**(3), 24-1 (2009)
4. Bauer, S., Wiest, R., Nolte, L.P., Reyes, M.: A survey of MRI-based medical image analysis for brain tumor studies. Phys. Med. Biol. **58**(13), R97 (2013)
5. Bertalmio, M., Sapiro, G., Caselles, V., Ballester, C.: Image inpainting. In: Proceedings of the 27th Annual Conference on Computer Graphics and Interactive Techniques, pp. 417–424 (2000)
6. Chen, T.Q., Schmidt, M.: Fast patch-based style transfer of arbitrary style. arXiv preprint arXiv:1612.04337 (2016)
7. Criminisi, A., Pérez, P., Toyama, K.: Region filling and object removal by exemplar-based image inpainting. IEEE Trans. Image Process. **13**(9), 1200–1212 (2004)
8. Cuadra, M.B., et al.: Atlas-based segmentation of pathological brains using a model of tumor growth. In: Dohi, T., Kikinis, R. (eds.) MICCAI 2002. LNCS, vol. 2488, pp. 380–387. Springer, Heidelberg (2002). https://doi.org/10.1007/3-540-45786-0_47
9. Dawant, B., Hartmann, S., Pan, S., Gadamsetty, S.: Brain atlas deformation in the presence of small and large space-occupying tumors. Comput. Aided Surg. **7**(1), 1–10 (2002)
10. DeAngelis, L.M.: Brain tumors. N. Engl. J. Med. **344**(2), 114–123 (2001)
11. Dosovitskiy, A., Brox, T.: Generating images with perceptual similarity metrics based on deep networks. In: Advances in Neural Information Processing Systems, pp. 658–666 (2016)
12. Efros, A.A., Freeman, W.T.: Image quilting for texture synthesis and transfer. In: Proceedings of the 28th Annual Conference on Computer Graphics And Interactive Techniques, pp. 341–346 (2001)
13. Gatys, L.A., Ecker, A.S., Bethge, M.: Image style transfer using convolutional neural networks. In: 2016 IEEE Conference on Computer Vision and Pattern Recognition (CVPR), pp. 2414–2423. IEEE (2016)
14. Gooya, A., Biros, G., Davatzikos, C.: Deformable registration of glioma images using EM algorithm and diffusion reaction modeling. IEEE Trans. Med. Imaging **30**(2), 375–390 (2010)
15. Iizuka, S., Simo-Serra, E., Ishikawa, H.: Globally and locally consistent image completion. ACM Trans. Graph. (TOG) **36**(4), 107 (2017)
16. Iizuka, S., Simo-Serra, E., Ishikawa, H.: Globally and locally consistent image completion. ACM Trans. Graph. (ToG) **36**(4), 1–14 (2017)
17. Isola, P., Zhu, J.Y., Zhou, T., Efros, A.A.: Image-to-image translation with conditional adversarial networks. In: Proceedings of the IEEE Conference On Computer Vision and Pattern Recognition, pp. 1125–1134 (2017)
18. Johnson, J., Alahi, A., Fei-Fei, L.: Perceptual losses for real-time style transfer and super-resolution. In: Leibe, B., Matas, J., Sebe, N., Welling, M. (eds.) ECCV 2016. LNCS, vol. 9906, pp. 694–711. Springer, Cham (2016). https://doi.org/10.1007/978-3-319-46475-6_43
19. Lamecker, H., Pennec, X.: Atlas to image-with-tumor registration based on demons and deformation inpainting (2010)
20. Liu, G., Reda, F.A., Shih, K.J., Wang, T.C., Tao, A., Catanzaro, B.: Image inpainting for irregular holes using partial convolutions. In: Proceedings of the European Conference on Computer Vision (ECCV), pp. 85–100 (2018)

21. Liu, H., Jiang, B., Xiao, Y., Yang, C.: Coherent semantic attention for image inpainting. In: Proceedings of the IEEE International Conference on Computer Vision, pp. 4170–4179 (2019)

22. Liu, X., et al.: Permutation-invariant feature restructuring for correlation-aware image set-based recognition. In: Proceedings of the IEEE International Conference on Computer Vision, pp. 4986–4996 (2019)

23. Liu, X., Kumar, B.V., Ge, Y., Yang, C., You, J., Jia, P.: Normalized face image generation with perceptron generative adversarial networks. In: 2018 IEEE 4th International Conference on Identity, Security, and Behavior Analysis (ISBA), pp. 1–8. IEEE (2018)

24. Liu, X., et al.: Feature-level Frankenstein: eliminating variations for discriminative recognition. In: Proceedings of the IEEE Conference on Computer Vision and Pattern Recognition, pp. 637–646 (2019)

25. Marcos, D., Volpi, M., Tuia, D.: Learning rotation invariant convolutional filters for texture classification. In: ICPR (2016)

26. Marcus, D.S., Wang, T.H., Parker, J., Csernansky, J.G., Morris, J.C., Buckner, R.L.: Open access series of imaging studies (OASIS): cross-sectional MRI data in young, middle aged, nondemented, and demented older adults. J. Cogn. Neurosci. **19**(9), 1498–1507 (2007)

27. Mohamed, A., Zacharaki, E.I., Shen, D., Davatzikos, C.: Deformable registration of brain tumor images via a statistical model of tumor-induced deformation. Med. Image Anal. **10**(5), 752–763 (2006)

28. Oishi, K., Faria, A.V., Van Zijl, P.C., Mori, S.: MRI Atlas of Human White Matter. Academic Press (2010)

29. Oostenveld, R., Stegeman, D.F., Praamstra, P., van Oosterom, A.: Brain symmetry and topographic analysis of lateralized event-related potentials. Clin. Neurophysiol. **114**(7), 1194–1202 (2003)

30. Paszke, A., et al.: Automatic differentiation in pytorch (2017)

31. Pathak, D., Krahenbuhl, P., Donahue, J., Darrell, T., Efros, A.A.: Context encoders: feature learning by inpainting. In: Proceedings of the IEEE Conference on Computer Vision and Pattern Recognition, pp. 2536–2544 (2016)

32. Prados, F., et al.: Fully automated patch-based image restoration: application to pathology inpainting. In: Crimi, A., Menze, B., Maier, O., Reyes, M., Winzeck, S., Handels, H. (eds.) BrainLes 2016. LNCS, vol. 10154, pp. 3–15. Springer, Cham (2016). https://doi.org/10.1007/978-3-319-55524-9_1

33. Raina, K., Yahorau, U., Schmah, T.: Exploiting bilateral symmetry in brain lesion segmentation. arXiv preprint arXiv:1907.08196 (2019)

34. Salimans, T., Goodfellow, I., Zaremba, W., Cheung, V., Radford, A., Chen, X.: Improved techniques for training GANs. In: Advances in Neural Information Processing Systems, pp. 2234–2242 (2016)

35. Sartor, K.: MR imaging of the brain: tumors. Eur. Radiol. **9**(6), 1047–1054 (1999)

36. Song, Y., et al.: Contextual-based image inpainting: infer, match, and translate. In: Proceedings of the European Conference on Computer Vision (ECCV), pp. 3–19 (2018)

37. Sotiras, A., Davatzikos, C., Paragios, N.: Deformable medical image registration: a survey. IEEE Trans. Med. Imaging **32**(7), 1153–1190 (2013)

38. Tang, Z., Wu, Y., Fan, Y.: Groupwise registration of MR brain images with tumors. Phys. Med. Biol. **62**(17), 6853 (2017)

39. Yang, C., Song, Y., Liu, X., Tang, Q., Kuo, C.C.J.: Image inpainting using block-wise procedural training with annealed adversarial counterpart. arXiv preprint arXiv:1803.08943 (2018)

40. Zacharaki, E.I., Shen, D., Lee, S.K., Davatzikos, C.: Orbit: a multiresolution framework for deformable registration of brain tumor images. IEEE Trans. Med. Imaging **27**(8), 1003–1017 (2008)

41. Zhang, R., Isola, P., Efros, A.A., Shechtman, E., Wang, O.: The unreasonable effectiveness of deep features as a perceptual metric. arXiv preprint arXiv:1801.03924 (2018)

42. Zheng, S., Song, Y., Leung, T., Goodfellow, I.: Improving the robustness of deep neural networks via stability training. In: Proceedings of the IEEE Conference on Computer Vision and Pattern Recognition, pp. 4480–4488 (2016)

Multivariate Analysis is Sufficient for Lesion-Behaviour Mapping

Lucas Martin[1], Julie Josse[2], and Bertrand Thirion[1(✉)]

[1] Inria, CEA, Université Paris Saclay, Palaiseau, France
bertrand.thirion@inria.fr
[2] Inria, Sophia-Antipolis, Paris, France

Abstract. Lesion-behaviour mapping aims at predicting individual behavioural deficits, given a certain pattern of brain lesions. It also brings fundamental insights on brain organization, as lesions can be understood as interventions on normal brain function. We focus here on the case of stroke. The most standard approach to lesion-behaviour mapping is mass-univariate analysis, but it is inaccurate due to correlations between the different brain regions induced by vascularisation. Recently, it has been claimed that multivariate methods are also subject to lesion-anatomical bias, and that a move towards a causal approach is necessary to eliminate that bias. In this paper, we reframe the lesion-behaviour brain mapping problem using classical causal inference tools. We show that, in the absence of additional clinical data and if only one region has an effect on the behavioural scores, suitable multivariate methods are sufficient to address lesion-anatomical bias. This is a commonly encountered situation when working with public datasets, which very often lack general health data. We support our claim with a set of simulated experiments using a publicly available lesion imaging dataset, on which we show that adequate multivariate models provide state-of-the art results.

Keywords: Lesion-behaviour mapping · Multivariate methods · Causal inference

1 Introduction

Lesion-behaviour mapping aims at predicting individual behavioural impacts of brain lesions, such as those induced by stroke. Based on large-scale datasets including brain images and corresponding deficits, this mapping can be used to assess the critical impact of brain territories on behaviour. Yet, this remains a complex endeavour [14]. Traditionally, univariate methods, such as voxel-based lesion-symptom mapping, have been used for this purpose [2]. However, such methods are subject to the topographical bias induced by brain vascularization,

---

Electronic supplementary material The online version of this chapter (https://doi.org/10.1007/978-3-030-72084-1_9) contains supplementary material, which is available to authorized users.

© Springer Nature Switzerland AG 2021
A. Crimi and S. Bakas (Eds.): BrainLes 2020, LNCS 12658, pp. 92–100, 2021.
https://doi.org/10.1007/978-3-030-72084-1_9

i.e. brain regions that are irrigated by the same artery often die together in the case of stroke. In turn, this induces a correlation between the lesion status of different brain regions, which can lead to spurious effects being detected [12].

Later on, multivariate methods that incorporate the lesion status of every brain region in a single model have been introduced [16]. While these methods were thought to be able to overcome topographical bias, this notion has recently been challenged [17], based on numerical experiments involving support vector regression. It has also been argued that only causal inference methods would be able to overcome topographical bias.

In this work, we tackle the question of which multivariate methods are suited for this type of inference. First, we notice that inference based on multivariate models is hard due to the high dimensionality of the problem, and the covariance structure between different brain regions [14]. Specific methods, such as desparsified Lasso [19], are required for accurate inference. Second, we point out the possible inadequacy of linear models, given that deficits can result from a complex combination of lesions [7].

We then discuss in detail the expected benefits of causal inference tools: those are limited unless additional clinical data are available together with the lesion data. Numerical experiments are performed by simulating behavioral deficits based on a publicly available lesion database [15]. These support our claims, and also replicate the results reported in [17].

2   Multivariate Methods Considered

Let us consider the case where the only data available to us are segmented lesion maps, typically reduced to lesion occurrence in a predefined set of regions of interest (ROIs) of a given brain atlas, together with behavioural scores highlighting some deficits, but no other clinical data. This is a commonly encountered situation when working with public datasets.

The outcome is a given behavioural score $\mathbf{y}$, observed in n subjects, and the potential causes are the lesion status of the different brain regions $(\mathbf{X}_j)_{j=1..p}$. Multivariate statistical inference consists in finding which variables are predictive of the deficit, given the status of the other regions.

If we assume a linear model, the reference method is the desparsified LASSO [19], namely a de-biased version of the lasso, in which the model weights $\beta_{DLASSO}(\lambda)$ are computed from the regular Lasso weights $\beta_{LASSO}(\lambda)$ as follows:

$$\beta_{DLASSO}(\lambda) = \beta_{LASSO}(\lambda) + \frac{1}{n}\widehat{\Sigma^{-1}}\mathbf{X}^T(\mathbf{y} - \mathbf{X}\beta_{LASSO}(\lambda))$$

where λ is the regularization parameter of the LASSO, and $\widehat{\Sigma^{-1}}$ is an estimate of the inverse covariance matrix of the model. Explicit formulae for the covariance matrix of $\beta_{DLASSO}(\lambda)$, allowing the computation of reliable confidence intervals.

The desparsified LASSO is well-suited to the problem of lesion-behaviour mapping, because it gives reliable confidence intervals on model weights. Moreover, it takes into account the correlations between brain regions, while allowing inference in the case where $p \geq n$.

We also consider non-linear multivariate models, such as random forests with permutation feature importance [3], and random forests with approximated Shapley values [11]. Support vector regression is investigated as well, in order to relate our findings to those of [17], and as a baseline commonly used in the literature [20].

3   Causal Analysis of the Problem

Causal analysis imposes to first split the problem into several cases, depending on whether only a single, or multiple ROIs affect the behavioural scores.

Confounding Bias and Backdoor Paths. Confounding variables are variables that have a causal effect on potential causes and outcome. Those can spuriously enhance the performance of predictive machine learning algorithms, but hamper their usefulness when trying to infer brain-behaviour (or in our case, lesion-behaviour) relationships [4]. The type of bias induced by these confounding variables is called *confounding bias*.

Confounding bias occurs whenever there exists an unblocked *backdoor path* between the cause and the outcome. A backdoor path is a path in the undirected causal graph that does not include the direct edge from the cause to the outcome. See the graph in Fig. 1(a) for a token example. Backdoor paths are blocked by colliders, that are vertices with two incoming edges: $\rightarrow Z \leftarrow$, or by conditioning on vertices that are not colliders.

Classical causal inference methods aim at eliminating confounding bias by conditioning on an appropriate set of observed variables. For a more in-depth presentation of this topic, the reader can consult Sect. 3.3 of [13]. In particular, these methods make the assumption that no unobserved confounding variable exists. Therefore, in the absence of observed confounders, classical causal inference methods should not have any advantage over appropriate multivariate methods.

Single ROI Case. In Fig. 1(b), the middle causal graph links the observed or known variables in a simplified situation with only two brain regions denoted A and B. As brain vascularization induces correlations between regions, we choose to represent it in the causal graph. As can be seen from the middle graph, there are no observed confounding variables in this case, because there are no backdoor paths between the behavioural scores and region A. Causal inference methods should not yield improved performance over proper multivariate inference methods.

Multiple ROI Case. In the case where multiple brain regions linked by vasculature have a causal effect on behavioural scores, a backdoor path opens up as shown on the right causal graph in Fig. 1(c), hence causal inference methods may be necessary. However, the magnitude of the confounding bias induced by this backdoor path is unknown, and may be small. In turn, this may confer limited performance improvements with respect to causal inference methods.

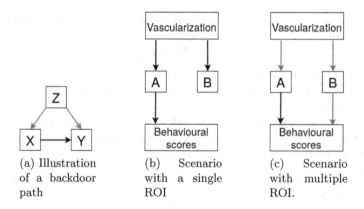

(a) Illustration of a backdoor path

(b) Scenario with a single ROI

(c) Scenario with multiple ROI.

Fig. 1. The sample causal graph on the left illustrates the concept of a backdoor path. X is the cause, Y the outcome, and Z the confounding variable. The middle causal graph represents the known variables in the single ROI case. The right causal graph represents the multiple ROI case. A and B are two different brain regions irrigated by the same artery. Backdoor paths are highlighted in red. (Color figure online)

4   Experiments

It is important to notice that there is no ground truth available for such problems, hence we have to rely on simulations to control the behavior of inference tools.

Dataset Used. Similar to the experiments done in [17], we used lesion maps from the publicly available LESYMAP dataset (https://github.com/dorianps/ LESYMAP) [15], which includes left hemisphere stroke lesions from 131 patients. We partition the brain volumes into 403 regions using the parcellation atlas provided with the LESYMAP dataset.

Generative Model of Simulated Outcomes. We simulate behavioural scores using the status of several regions pairs of the provided atlas. We considered regions to be lesioned if more than 60 % of their voxels were lesioned. The figures in this paper showcase the results obtained using regions 100 and 101. The same experiments were performed on several other region pairs ({100, 101}, {108, 114}, {109, 114}, {79, 108}, {80, 108}), and similar results were obtained. These are available in the supplementary materials. We partition the brain volumes into 403 regions using the parcellation atlas provided with the LESYMAP dataset. Because the LESYMAP dataset only contains left hemisphere stroke lesions, of the 403 previously mentioned regions, only 179 present a lesion in at least one patient. We keep these regions and discard the rest. The region pairs with the highest number of subjects showing a lesion in both regions were picked. The rationale behind this is to pick the regions most susceptible to topographical bias. Indeed, in the 131 subjects, 49 had a lesion in region 100, 45 a lesion in region 101, and 41 a lesion in both. Figure 2 shows the location of these regions in the brain.

We then simulate behavioural scores using a simple linear model with additive Gaussian noise.

$$\mathbf{y}^i = \phi(\mathbf{X}^i) + \varepsilon^i$$

where $\mathbf{y}^i$ is the behavioural score for subject i, $\mathbf{X}^i$ represents the lesion status across regions for subject i, ϕ is a function mapping this lesion status to deficits, and $\varepsilon^i \sim \mathcal{N}(0, \sigma)$.

In all our experiments, $\sigma = 1$. All random variables ε^i are i.i.d. We use four scenarii of simulation for behavioural scores, that are based upon real lesion-behaviour interactions documented in the literature [7]

1. Single ROI scenario: $\phi(\mathbf{X}) = \mathbf{X}_j$, the jth brain region lesion status (e.g., $j = 101$), i.e. $\phi(\mathbf{X}) = \mathbf{X}_j = 1$ if region j is lesioned, 0 otherwise.
2. OR scenario: $\phi(\mathbf{X}) = \text{OR}(\mathbf{X}_j, \mathbf{X}_k)$, with e.g. j = 100, k = 101.
3. AND scenario: $\phi(\mathbf{X}) = \text{AND}(\mathbf{X}_j, \mathbf{X}_k)$, with e.g. j = 100, k = 101.
4. Sum scenario: $\phi(\mathbf{X}) = \mathbf{X}_j + \mathbf{X}_k$, with e.g. j = 100, k = 101.

Models. We compare two causal models to various multivariate models. For the first causal model, we use Bayesian Additive Regression Trees (abbreviated BART) [8], which is a state-of-the-art model for the estimation of average treatment effects [6]. We fit one BART model per atlas region, using the lesion status of that region as the treatment variable, and the lesion status of other regions as potential confounding covariates. We then take the estimated average treatment effect output by each BART model to be the effect on the behavioural scores of lesion presence in each region. We use the `bartCause` R package (https://rdrr.io/github/vdorie/bartCause/) for our experiments.

The second causal model is a doubly robust AIPW model (abbreviated DR) [9], where the two response surfaces were modeled by random forests. Following the same procedure as BART, we fit one model per region, and get the average treatment effect as output from the DR model.

The multivariate models we use are:

1. Support vector regression (SVR), using the `scikit-learn` package
2. Desparsified LASSO [19] (DLASSO), using a custom `Python` implementation
3. Random forests with permutation feature importance (RF), using the `scikit-learn` package

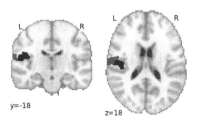

Fig. 2. Location in the brain of the two ROIs used in our simulations. Region 101 is colored red and region 100 is colored blue. (Color figure online)

4. Random forests with Shapley additive explanations [11] (abbreviated RF+SHAP), using the SHAP package (https://github.com/slundberg/shap)

Assessing Model Performance. Each model gives a score per ROI corresponding to its effect on the behavioural scores. The scores are the model weights for SVR and DLASSO, average treatment effects for BART and DR, feature importance for RF, and approximated Shapley values for RF+SHAP. For the models that give standard deviation estimates on their score (BART and DLASSO), we compute Z-scores using these values, and calculate precision-recall curves from the Z-scores.

For the other models, we robustly fit a Gaussian distribution to their scores (denoted $(\mathbf{w}_j, j = 1 \ldots p)$), by taking $m = \mathrm{median}(\mathbf{w}_j)$ as a mean parameter, and $\sigma^2 \propto \mathrm{mad}(\mathbf{w}_j)^2)$ (where mad() stands for mean absolute deviation) as variance parameter. We then compute the following statistic $Z_j = \frac{\mathbf{w}_j - m}{\sigma}$, which we call pseudo-Z-scores, and calculate precision-recall curves from the pseudo Z-scores.

Finally, we take the area under each precision-recall curve (AUC) as a final measure of model performance. The hyperparameters of each model are optimized using grid search and cross-validation, and the hyperparameters which yield the best predictive performance are picked.

This procedure is repeated over 50 bootstrap runs to obtain Fig. 3. We also provide results averaged over all considered region pairs in Table 1.

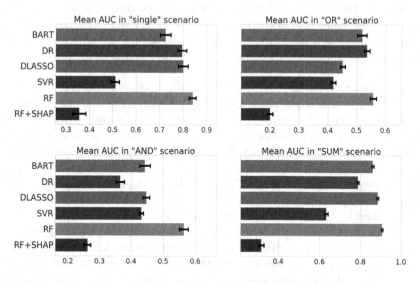

Fig. 3. Area under the precision-recall curve for our 6 models under the four simulation scenarii. Signal to noise ratio is equal to 1. Results are averaged over 50 bootstrap runs.

Table 1. Mean AUC for each model under each scenario, averaged across all region pairs. The winning model in each scenario is highlighted in bold.

	BART	DR	DLASSO	SVR	RF	RF+SHAP
Single ROI	0.69	0.75	0.81	0.46	**0.81**	0.28
OR	0.50	**0.53**	0.50	0.34	0.53	0.17
AND	0.42	0.37	0.34	0.36	**0.45**	0.24
SUM	0.83	0.80	**0.86**	0.60	0.85	0.26

Results. The RF model performs very well across all scenarii, being either the best or very close to the best. In each scenario, we see that SVR performs worse than the other models. This is consistent with the findings of [17], and suggests that support vector regression is not a good model for inference. We also see that RF+SHAP perform poorly across all scenarii. The SHAP approximation method makes the hypothesis that model features are independent, and it is well known that it does not perform adequately when this hypothesis is violated [1].

In our case, the model features are the brain regions, which are heavily correlated because of topographical bias. Therefore the poor performance of RF+SHAP is unsurprising. It is worth noting that other approximations for Shapley values exist, but are prohibitively computationally expensive when considering the size of our problem.

While the DLASSO model performs very well in the single ROI and SUM scenarii, it performed significantly worse in the AND and OR scenarii. This is because DLASSO is a linear model, and fails to accurately represent the non-linear interactions between lesions and behavioural scores in the AND and OR scenarii.

For the region pair $\{100, 101\}$ displayed in Fig. 3, causal models underperform in the AND scenario when compared to RF. However, this is not necessarily the case across all region pairs, as shown in the supplementary materials (see region pair $\{108, 114\}$). Both causal models (DR and BART) posit an underlying additive model, where $\mathbf{y} = f(\mathbf{X}) + \tau \mathbf{W} + \varepsilon$ where τ is the causal effect and $\mathbf{W}$ the treatment variable (here, the lesion status of the region investigated for causal effect). However, in the AND and OR scenario, the underlying causal model is effectively $y = f(\mathbf{X}) + \tau \phi(\mathbf{W}, \mathbf{W}') + \varepsilon$, where $(\mathbf{W}, \mathbf{W}')$ represents treatments on two regions and ϕ is not additive, which hampers causal models.

Additionally, the strength of the confounding introduced by the presence of multiple ROIs may vary between region pairs and scenarii, and may compensate more or less well for this impairment. Overall, we notice that causal models were not always the best-performing ones. These numerical experiments suggest that multivariate models are still sufficient for lesion-behaviour mapping in the proposed framework.

5 Discussion

5.1 Outlook

Through a simple causal analysis of the lesion-behaviour mapping problem, we show that in the case where a single region affects behavioural scores, and no other clinical data is observed, there are no observed confounding variables. Therefore, traditional causal inference methods that assume no unobserved confounders should not perform better than multivariate methods that have good inference capabilities (random forests, desparsified LASSO). We illustrate this through our experiments based on documented lesion-behaviour interactions. although confounding variables exist in the case where multiple regions affect behavioural scores, we also show empirically that appropriate multivariate methods still perform adequately. As the absence of clinical data is almost always the case when working with public brain lesion datasets, appropriate multivariate methods are good enough in the cases created through standard simulations. The generalization to more complex scenarii, where non-linear combinations of several regions would cause the deficits, is an important future direction.

5.2 Future Work

Causal Inference with Unobserved Confounders. Among recent developments in the causal inference literature, methods that deal with unobserved confounders and do away with the assumption of strong ignorability have been proposed. Examples include [10] and [18]. Although some of these methods are still the subject of debate [5], we believe that they could be an interesting basis for a causal approach to lesion-behaviour mapping when no other clinical data are observed, as is the case in publicly available datasets.

Causal Inference with Additional Clinical Data. In the case where additional clinical data are available, additional confounding variables such as age might be observed. In that case, we conjecture that traditional causal inference methods that make the assumption of strong ignorability would perform better than multivariate methods, as they could effectively eliminate confounding bias.

Acknowledgements. This project has received funding from the European Unions Horizon 2020 research and innovation programme under grant agreement No 826421 (TVB-Cloud).

References

1. Aas, K., Jullum, M., Løland, A.: Explaining individual predictions when features are dependent: more accurate approximations to shapley values (2019)
2. Bates, E., et al.: Voxel-based lesion-symptom mapping. Nat. Neurosci. **6**, 448–50 (2003). https://doi.org/10.1038/nn1050
3. Breiman, L.: Random forests. Mach. Learn. **45**(1), 5–32 (2001)

4. Chyzhyk, D., Varoquaux, G., Thirion, B., Milham, M.: Controlling a confound in predictive models with a test set minimizing its effect, pp. 1–4 (2018). https://doi.org/10.1109/PRNI.2018.8423961
5. D'Amour, A.: Comment: reflections on the deconfounder (2019). https://arxiv.org/abs/1910.08042
6. Dorie, V., Hill, J., Shalit, U., Scott, M., Cervone, D.: Automated versus do-it-yourself methods for causal inference: lessons learned from a data analysis competition. Stat. Sci. **34**(1), 43–68 (2019). https://doi.org/10.1214/18-STS667
7. Godefroy, O., Duhamel, A., Leclerc, X., Saint Michel, T., Hénon, H., Leys, D.: Brain-behaviour relationships. Some models and related statistical procedures for the study of brain-damaged patients. Brain **121**(Pt 8), 1545–1556 (1998)
8. Hill, J.: Bayesian nonparametric modeling for causal inference. J. Comput. Graph. Stat. **20**, 217–240 (2011). https://doi.org/10.1198/jcgs.2010.08162
9. Kang, J.D.Y., Schafer, J.L.: Demystifying double robustness: a comparison of alternative strategies for estimating a population mean from incomplete data. Stat. Sci. **22**(4), 523–539 (2007)
10. Louizos, C., Shalit, U., Mooij, J.M., Sontag, D., Zemel, R., Welling, M.: Causal effect inference with deep latent-variable models. In: Guyon, I., et al. (eds.) Advances in Neural Information Processing Systems 30, pp. 6446–6456. Curran Associates, Inc. (2017). http://papers.nips.cc/paper/7223-causal-effect-inference-with-deep-latent-variable-models.pdf
11. Lundberg, S.M., Lee, S.I.: A unified approach to interpreting model predictions. In: Guyon, I., et al. (eds.) Advances in Neural Information Processing Systems 30, pp. 4765–4774. Curran Associates, Inc. (2017). http://papers.nips.cc/paper/7062-a-unified-approach-to-interpreting-model-predictions.pdf
12. Mah, Y.H., Husain, M., Rees, G., Nachev, P.: Human brain lesion-deficit inference remapped. Brain **137**(Pt 9), 2522–2531 (2014)
13. Pearl, J.: Causality: Models, Reasoning and Inference, 2nd edn. Cambridge University Press, Cambridge (2009)
14. Price, C.J., Hope, T.M., Seghier, M.L.: Ten problems and solutions when predicting individual outcome from lesion site after stroke. Neuroimage **145**(Pt B), 200–208 (2017)
15. Pustina, D., Avants, B., Faseyitan, O.K., Medaglia, J.D., Coslett, H.B.: Improved accuracy of lesion to symptom mapping with multivariate sparse canonical correlations. Neuropsychologia **115**, 154–166 (2018)
16. Smith, D.V., Clithero, J.A., Rorden, C., Karnath, H.O.: Decoding the anatomical network of spatial attention. Proc. Natl. Acad. Sci. **110**(4), 1518–1523 (2013)
17. Sperber, C.: Rethinking causality and data complexity in brain lesion-behaviour inference and its implications for lesion-behaviour modelling. Cortex **126**, 49–62 (2020)
18. Wang, Y., Blei, D.M.: The blessings of multiple causes. J. Am. Stat. Assoc. **114**(528), 1574–1596 (2019)
19. Zhang, C.H., Zhang, S.: Confidence intervals for low-dimensional parameters in high-dimensional linear models. J. R. Stat. Soc.: Ser. B (Stat. Methodol.) **76** (2011). https://doi.org/10.1111/rssb.12026
20. Zhang, Y., Kimberg, D., Coslett, H., Schwartz, M., Wang, Z.: Multivariate lesion-symptom mapping using support vector regression. Hum. Brain Mapp. **35** (2014). https://doi.org/10.1002/hbm.22590

Label-Efficient Multi-task Segmentation Using Contrastive Learning

Junichiro Iwasawa[1(✉)], Yuichiro Hirano[2], and Yohei Sugawara[2]

[1] The University of Tokyo, Tokyo, Japan
jiwasawa@ubi.s.u-tokyo.ac.jp
[2] Preferred Networks, Tokyo, Japan
{hirano,suga}@preferred.jp

Abstract. Obtaining annotations for 3D medical images is expensive and time-consuming, despite its importance for automating segmentation tasks. Although multi-task learning is considered an effective method for training segmentation models using small amounts of annotated data, a systematic understanding of various subtasks is still lacking. In this study, we propose a multi-task segmentation model with a contrastive learning based subtask and compare its performance with other multi-task models, varying the number of labeled data for training. We further extend our model so that it can utilize unlabeled data through the regularization branch in a semi-supervised manner. We experimentally show that our proposed method outperforms other multi-task methods including the state-of-the-art fully supervised model when the amount of annotated data is limited.

Keywords: Multi-task learning · Brain tumor segmentation · Semi-supervised learning

1 Introduction

For precision medicine, it is imperative that interpretation and classification of medical images is done quickly and efficiently; however, it is becoming a major hurdle due to the shortage of clinical specialists who can provide informed clinical diagnoses. Automated segmentation can not only save physicians' time but can also provide accurate and reproducible results for medical analysis. Recent advances in convolutional neural networks (CNN) have yielded state-of-the-art segmentation results for both 2D and 3D medical images [11,12,15], a significant step toward fully automated segmentation. However, this level of performance is only possible when sufficient amount of labeled data is available.

J. Iwasawa—This work was done when J.I. worked at Preferred Networks as an intern and part-time researcher.

Electronic supplementary material The online version of this chapter (https://doi.org/10.1007/978-3-030-72084-1_10) contains supplementary material, which is available to authorized users.

A. Crimi and S. Bakas (Eds.): BrainLes 2020, LNCS 12658, pp. 101–110, 2021.
https://doi.org/10.1007/978-3-030-72084-1_10

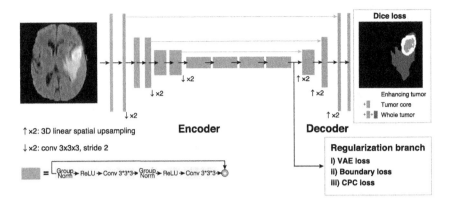

Fig. 1. Schematic image of the encoder-decoder based network combined with a regularization branch. The green block represents a ResBlock. (Color figure online)

Furthermore, obtaining annotations from medical experts is both expensive and time-consuming, despite its importance for training CNNs. Thus, methods that utilize small labeled datasets have been explored extensively. Specifically, multi-task learning has been considered as an efficient method for small data, since parameter sharing for both the main segmentation task and regularization subtask could reduce the risk of overfitting [12,16]. Although subtasks in multi-task learning have been extensively investigated, we still lack a systematic understanding of the impact of subtasks on the main segmentation model, especially in the low labeled data regime.

Contrastive learning based approaches have recently shown state-of-the-art performance in image classification with small amount of labels [3,5,7,19]. In this work, we integrated a contrastive learning based approach as a subtask of a multi-task segmentation model. However, its applicability in segmentation tasks, especially medical image segmentation is yet to be explored [5]. Here, we systematically assess the performance of a brain tumor segmentation problem for three different multi-task models including our proposed method. The main contributions can be summarized as follows[1]:

- We propose a novel method for tumor segmentation by utilizing contrastive learning as a subtask for the main segmentation model.
- We experimentally show that our proposed method, combined with a semi-supervised approach which utilizes unlabeled data, could enhance segmentation performance when the amount of labeled data is small.

---

[1] Our implementation is available at github.com/pfnet-research/label-efficient-brain-tumor-segmentation.

2   Methods

2.1   Encoder-Decoder Network with Regularization Branches

Figure 1 is the schematic image of our model. The model constitutes an encoder-decoder architecture with a regularization branch starting from the encoder output. The encoder and decoder are composed of ResNet [6]-like blocks (ResBlock) with skip-connections between the encoder and decoder [11,12,15]. We adopted the encoder-decoder architecture from Myronenko (2018) [12], and each Res-Block consists of two $3 \times 3 \times 3$ convolution layers with group normalization [21].

The loss function of the model consists of two terms: $\mathcal{L}_{\text{total}} = \mathcal{L}_{\text{Dice}} + \mathcal{L}_{\text{branch}}$, where $\mathcal{L}_{\text{Dice}}$ is a softmax Dice loss applied to the decoder output p_{pred}, compelling it to match the ground truth label p_{true}. The Dice loss [11] is given by,

$$\mathcal{L}_{\text{Dice}}\left(p_{\text{true}}, p_{\text{pred}}\right) = 1 - \frac{2 \times \sum_i p_{\text{true},i} p_{\text{pred},i}}{\sum_i \left(p_{\text{true},i}^2 + p_{\text{pred},i}^2\right) + \epsilon}, \tag{1}$$

where $\epsilon = 10^{-7}$ is a constant preventing zero division. $\mathcal{L}_{\text{branch}}$ is the loss applied to the output of the regularization branch and is dependent on the architecture of the branch.

To explore the regularization effects of the multi-task learning, we compared the performance of different subtasks for the regularization branch on the encoder endpoint. Based on the type of information it uses to calculate the loss function, subtasks for multi-task learning can be categorized into the following classes:

1. Subtasks that use the input X itself, without using any additional information, such as a decoder-like branch attempting to reconstruct the original input X [12].
2. Subtasks that attempt to predict a transformed feature of the label y, such as boundary-aware networks that predict the boundary of the given label [4,13].
3. Subtasks that compel the encoder to obtain certain representations of the input by predicting low or high-level features from the input X, such as tasks predicting the angles of rotated images [14]. Our proposed method using contrastive predictive coding (CPC) [7,19] would also be classified in this class.

To investigate a wide spectrum of subtasks, we implemented three different types of regularization branches that use either a variational autoencoder (VAE) loss, a boundary loss, or a CPC loss. The description of each branch and loss is provided subsequently (the architectures are shown in supplementary materials).

Variational Autoencoder Branch. The purpose of the VAE branch is to guide the encoder to extract an efficient low-dimensional representation for the input [12]. The encoder output is first fed to a linear layer to reduce its dimension to 256 (128 to represent the mean μ, and 128 to represent the standard deviation (SD) σ). Accordingly, a sample is drawn from a 128-dimension Gaussian distribution with the given mean and SD. This sample is processed by several upsizing layers, where the number of features is reduced by a factor of two using $1 \times 1 \times 1$ convolution, and the spatial dimension is doubled using 3D bilinear sampling, so that the final output size matches the input size. The output is fed to the VAE loss ($\mathcal{L}_{\mathrm{VAE}}$) which is given by $\mathcal{L}_{\mathrm{branch}} = \mathcal{L}_{\mathrm{VAE}} = 0.1 \times (\mathcal{L}_{\mathrm{rec}} + \mathcal{L}_{\mathrm{KL}})$, where $\mathcal{L}_{\mathrm{rec}}$ is the L2 loss between the output of the VAE branch and the input image, and $\mathcal{L}_{\mathrm{KL}}$ is the Kullback-Leibler divergence between the two normal distributions, $\mathcal{N}(\mu, \sigma^2)$ and $\mathcal{N}(0, 1)$.

Boundary Attention Branch. The boundary attention branch aims to regularize the encoder by extracting information for predicting the boundaries of the given labels [4, 13, 20]. We prepared the boundary labels by applying a 3D Laplacian kernel to the ground truth binary labels. The attention layer first upsizes the output of the encoder, and concatenates it with the feature map from the encoder with the same spatial dimensions. Accordingly, a $1 \times 1 \times 1$ convolution is applied to the concatenated sample, followed by a sigmoid function that yields an attention map. The output of each attention layer is given by an element-wise multiplication of the attention map and the input to the layer. This operation is repeated until the spatial dimension matches that of the model input.

The loss for the boundary attention branch is given by $\mathcal{L}_{\mathrm{branch}} = \mathcal{L}_{\mathrm{boundary}} = \mathcal{L}_{\mathrm{Dice}}(b_{\mathrm{pred}}, b_{\mathrm{true}}) + \mathcal{L}_{\mathrm{edge}}$, where $\mathcal{L}_{\mathrm{Dice}}(b_{\mathrm{pred}}, b_{\mathrm{true}})$ is the Dice loss between the branch output (b_{pred}) and boundaries of the ground truth label (b_{true}). $\mathcal{L}_{\mathrm{edge}}$ is given by a weighted binary cross entropy loss:

$$\mathcal{L}_{\mathrm{edge}} = -\beta \sum_{j \in y_+} \log P\left(b_{\mathrm{pred},j} = 1 | X, \theta\right) - (1 - \beta) \sum_{j \in y_-} \log P\left(b_{\mathrm{pred},j} = 0 | X, \theta\right), \tag{2}$$

where X, θ, y_+ and y_- denote the input, model parameters, and the boundary and non-boundary voxels, respectively. β, the ratio of the non-boundary voxels to the total number of voxels is introduced to handle the imbalance of the boundary and non-boundary voxels.

Contrastive Predictive Coding Branch. Self-supervised learning, where a representation is learned from unlabeled data by predicting missing input data based on other parts of the input, has become a promising method for learning representations; it is useful for downstream tasks such as classification [7, 9] and segmentation [22]. Recently, CPC has been proposed as a self-supervised method that can outperform the fully supervised methods in ImageNet classification tasks in the small labeled data regime [7]. Despite its performance in classification tasks, the effectiveness of CPC in segmentation tasks is yet to be explored. Here, we incorporated the CPC architecture into the encoder-decoder structure as a regularization branch and investigated its performance with regard to medical image segmentation.

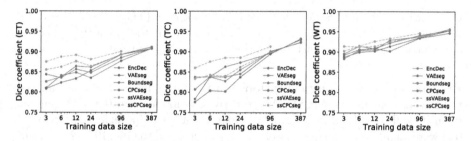

Fig. 2. Comparison of the Dice scores of the test predictions by the proposed models with different training data sizes. For the ssVAEseg and ssCPCseg, the training data size refers to the number of labeled images.

First, we divided the input image into $32 \times 32 \times 32$ overlapping patches with a 16-voxel overlap (resulting in an $8 \times 8 \times 7$ grid for the brain tumor dataset). Each divided patch is individually encoded to a latent representation using the encoder in the encoder-decoder architecture and spatially mean-pooled into a single feature vector $z_{i,j,k}$. Here, it should be noted that the visual field of the encoder when using the CPC branch is $32 \times 32 \times 32$ pixels, which is smaller than that with the other regularization branches due to the initial image division. Accordingly, eight layered ResBlocks f_{res8} and a linear layer W were applied to the upper half of the feature vectors, $\hat{z}_{i,j,k_{low}} = W\left(f_{res8}\left(z_{i,j,k_{up}}\right)\right)$, to predict the lower half of the feature vectors $z_{i,j,k_{low}}$. The predictions were evaluated based on the CPC loss (i.e. the InfoNCE [19]),

$$\mathcal{L}_{\text{branch}} = \mathcal{L}_{\text{CPC}} = -\sum_{i,j,k_{low}} \log \frac{\exp\left(\hat{z}_{i,j,k_{low}}^T z_{i,j,k_{low}}\right)}{\exp\left(\hat{z}_{i,j,k_{low}}^T z_{i,j,k_{low}}\right) + \sum_l \exp\left(\hat{z}_{i,j,k_{low}}^T z_l\right)}, \quad (3)$$

where the negative samples $\{z_l\}$ are randomly taken from the other feature vectors which are encoded from different patches of the same image.

3   Experiments and Results

Brain Tumor Dataset. We used the brain tumor dataset provided by the Medical Segmentation Decathlon [17], which is a subset of the data used in the 2016 and 2017 Brain Tumor Image Segmentation (BraTS) challenges [1,2,10]. This dataset includes multimodal 3D magnetic resonance imaging (MRI) scans from patients diagnosed with either glioblastoma or lower-grade glioma with an image size of $240 \times 240 \times 155$. For the experiments, we randomly split the 484 labeled data into the training (387), validation (48), and test set (49).

The dataset has three different labels corresponding to different tumor sub-regions (i.e., necrotic and non-enhancing parts of the tumor (NCR & NET), the peritumoral edema (ED), and the enhancing tumor (ET)). Following the BraTS challenge, we evaluated the model's prediction accuracy using the three nested structures of these sub-regions: ET, tumor core (TC: ET+NCR+NET), and the whole tumor (WT: TC+ED). The output of the network was set to three channels to predict each of the three nested structures above.

Preprocessing and Augmentation. All the input images were normalized to have zero mean and unit SD. Accordingly, a random scale (0.9, 1.1), random intensity shift ($-0.1, 0.1$ of SD), and random axis mirror flip (all axes, probability 0.5) were applied. All the inputs were randomly cropped to $(160, 192, 128)$, except when utilizing the CPC branch ($(144, 144, 128)$ in this case), due to the graphics processing unit (GPU) memory limitation. Note that the random cropping was performed without respect to the location of the brain region.

Comparison of Different Regularization Branches. We implemented our network in Chainer [18] and trained it on eight Tesla V100 32GB GPUs. We used a batch size of eight, and the Adam optimizer [8] with an initial learning rate of $\alpha_0 = 10^{-4}$ that was further decreased according to $\alpha = \alpha_0(1 - n/N)^{0.9}$, where n, N denotes the current epoch and the total number of epochs, respectively.

To compare the different regularization effects, we measured the segmentation performances of four models: the encoder-decoder alone (EncDec), EncDec with a VAE branch (VAEseg), EncDec with a boundary attention branch (Boundseg), and EncDec with a CPC branch (CPCseg). The performance was evaluated using the mean Dice and 95th-percentile Hausdorff distance of the ET, TC, and WT. To evaluate the regularization effect with varied amounts of labeled data, we also evaluated each model's performance with the training data size reduced to 3–96. The results are given in Fig. 2 and Table 1, S4.

It can first be observed that no regularization branch consistently outperforms the others. Furthermore, it can be observed that in some cases the EncDec that has no regularization branch had the highest mean Dice score among the

Table 1. Performance of EncDec, VAEseg, Boundseg, CPCseg (trained with six labeled data), and ssVAEseg, ssCPCseg (trained with six labeled and 381 unlabeled data). Evaluation was done using the mean Dice and 95th percentile Hausdorff distance. The performance of VAEseg trained with 387 labels is shown for comparison.

Test data	Dice			Hausdorff distance (mm)		
	ET	TC	WT	ET	TC	WT
VAEseg (387 labels)	0.9077	0.9323	0.9536	3.6034	10.2344	8.3895
EncDec (6 labels)	0.8412	0.8383	0.9144	11.4697	20.12	24.3726
VAEseg (6 labels)	0.8234	0.8036	0.8998	14.3467	22.4926	17.9775
Boundseg (6 labels)	0.8356	0.8378	0.9041	17.0323	27.2128	25.8112
CPCseg (6 labels)	0.8374	0.8386	0.9057	10.2839	14.9661	15.0633
ssVAEseg (6 labels)	0.8626	0.8425	0.9131	9.1966	**12.5302**	14.8056
ssCPCseg (6 labels)	**0.8873**	**0.8761**	**0.9151**	**8.7092**	16.0947	**12.3962**

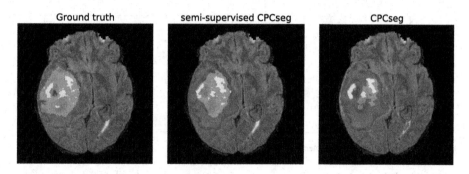

Fig. 3. Prediction results of the semi-supervised CPCseg (24 labeled + 363 unlabeled) and the CPCseg (24 labeled). The union of purple, orange, and yellow corresponds to the WT, orange plus yellow corresponds to the TC, and yellow corresponds to the ET. (Color figure online)

fully supervised models (Table 1). However, it should also be noted that multi-task models tended to outperform the EncDec when using all 387 labeled data (Table S4). These results imply that the regularization branches using labeled data have a limited effect on the segmentation performance when the amount of labeled data is small. Interestingly, the VAEseg, the state-of-the-art model in the BraTS 2018 challenge [12], was not necessarily the best model for various training data sizes. This was surprising, although the dataset we used in this study slightly differs from that used in the BraTS 2018 challenge. Our results suggest that the VAE subtask would not always be the optimal approach to brain tumor segmentation tasks.

Semi-supervised Multi-task Learning. Typically, unlabeled data is readily accessible in large quantities, compared to annotated data. This naturally leads to the question: is it possible to utilize unlabeled data to guide the segmentation model at small data regimes? To answer this question, we focused on the VAEseg and CPCseg, because they do not require labels to optimize the regularization branch. For training, we used all the 387 unlabeled data from the training set, and varied the number of labels used. To utilize the unlabeled data, we devised a semi-supervised update method wherein the model could be updated using only $\mathcal{L}_{\text{branch}}$ when the image had no label, and by $\mathcal{L}_{\text{Dice}} + \mathcal{L}_{\text{branch}}$, otherwise. This update method lets the encoder (and regularization branch) learn representations from both labeled and unlabeled data. The segmentation results for the semi-supervised VAEseg (ssVAEseg) and semi-supervised CPCseg (ssCPCseg) are shown in Fig. 2, 3 and Table 1. It can be observed that the semi-supervised methods outperform their fully supervised counterparts. In addition, the ssCPCseg outperformed all the other regularization methods, including the fully supervised state-of-the-art model, VAEseg, in the small labeled data regime. This tendency was most apparent for the ET and TC. For example, the ssCPCseg using six labels exhibited a 6% decrease in the Dice score for TC compared to

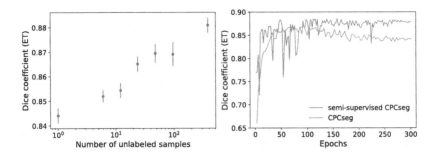

Fig. 4. The left panel shows the dependency of the number of unlabeled data on the segmentation accuracy for the ssCPCseg when using six labeled data. The right panel show learning curves for the test data of the brain tumor dataset. The learning curves for the CPCseg (24 labeled) and ssCPCseg (24 labeled + 363 unlabeled) are shown.

the VAEseg using 387 labels, while other methods using six labels exhibited a 10–14% decrease (Table 1). We speculate that this is because the areas of the ET and TC are smaller, compared to that of the WT, and thus, provide less supervision signals per sample to the model. Our results imply that difficult and non-trivial subtasks such as CPC, as well as unlabeled data, can be exploited to achieve state-of-the-art performance when the amount of annotated data is limited.

To explore the importance of the number of unlabeled data to the segmentation performance, we compared the ET Dice score of the ssCPCseg trained using six labeled and different amounts of unlabeled data (Fig. 4). It can be observed that the Dice score increases monotonically with the number of unlabeled data. However, it should be noted that the Dice score seems to increase linearly with the log of unlabeled data, indicating that the number of data required to improve accuracy would increase exponentially. We also investigated the effect of unlabeled data on the learning behavior of the model. As can be seen in Fig. 4, the semi-supervised CPCseg not only produced more accurate results but was also more robust to overfitting. This might be because the encoder needs to map good representations for both the labeled and unlabeled data to optimize the CPC objective. Overall, our results imply that utilizing unlabeled data could both enhance and stabilize the model's performance.

4   Discussion and Conclusion

In this work, we systematically investigated the effectiveness of different regularization subtasks for medical image segmentation. Our experiments on the brain tumor dataset showed that utilizing unlabeled data through the regularization branch improved and stabilized the performance of segmentation models when the number of labeled data was small. Especially, our proposed ssCPCseg outperformed other methods including the state-of-the-art fully supervised model in the small labeled data regime. In previous works, CPC has been used for self-supervised pre-training for image classification tasks with unlabeled images [7,19]. Our work is the first to show the effectiveness of CPC as a regularization subtask for image segmentation by utilizing both unlabeled and labeled images, providing a novel direction for label efficient segmentation. It should also be noted that ssCPCseg achieved particularly higher Dice scores than the other methods for tumor sub-regions with small area size (i.e. ET and TC) whose structure varies across the data and provides limited supervision signals to the model. Although our results are based on a single dataset, we believe that our method could be applicable to various targets in the field since target labels for medical images are often relatively small, and have widely varied structures.

It is generally expensive to obtain annotations for 3D medical images. On the other hand, large number of unlabeled images are often available. Thus, our semi-supervised method should have wide applicability to medical image segmentation tasks. However, it should be noted that all the MRI scans in the brain tumor dataset are normalized using a reference atlas; furthermore, they have the same size and voxel resolution. Therefore, we have not been able to evaluate the segmentation performance when the quality of the unlabeled data varies. An important future work is to evaluate the model's segmentation performance when it is fed unlabeled images from different modalities and domains. We believe that our systematic study provides important designing principles for segmentation models, leading to more cost-efficient medical image segmentation.

Acknowledgments. J.I. was supported by the Grant-in-Aid for JSPS Fellows JP18J21942.

References

1. Bakas, S., et al.: Advancing the cancer genome Atlas glioma MRI collections with expert segmentation labels and radiomic features. Sci. Data **4**, 170117 (2017)
2. Bakas, S., et al.: Identifying the best machine learning algorithms for brain tumor segmentation, progression assessment, and overall survival prediction in the BRATS challenge. arXiv e-prints arXiv:1811.02629 (2018)
3. Chen, T., Kornblith, S., Norouzi, M., Hinton, G.: A simple framework for contrastive learning of visual representations. arXiv e-prints arXiv:2002.05709 (2020)
4. Hatamizadeh, A., Terzopoulos, D., Myronenko, A.: End-to-end boundary aware networks for medical image segmentation. In: Suk, H.-I., Liu, M., Yan, P., Lian, C. (eds.) MLMI 2019. LNCS, vol. 11861, pp. 187–194. Springer, Cham (2019). https://doi.org/10.1007/978-3-030-32692-0_22
5. He, K., Fan, H., Wu, Y., Xie, S., Girshick, R.: Momentum contrast for unsupervised visual representation learning. In: The IEEE/CVF Conference on Computer Vision and Pattern Recognition (CVPR) (2020)
6. He, K., Zhang, X., Ren, S., Sun, J.: Identity mappings in deep residual networks. In: Leibe, B., Matas, J., Sebe, N., Welling, M. (eds.) ECCV 2016. LNCS, vol. 9908, pp. 630–645. Springer, Cham (2016). https://doi.org/10.1007/978-3-319-46493-0_38
7. Hénaff, O.J., et al.: Data-efficient image recognition with contrastive predictive coding. arXiv e-prints arXiv:1905.09272 (2019)

8. Kingma, D.P., Ba, J.: Adam: a method for stochastic optimization. In: 3rd International Conference on Learning Representations, ICLR 2015, San Diego, CA, USA, Conference Track Proceedings (2015)

9. Kolesnikov, A., Zhai, X., Beyer, L.: Revisiting self-supervised visual representation learning. In: IEEE Conference on Computer Vision and Pattern Recognition, CVPR 2019, Long Beach, CA, USA, pp. 1920–1929 (2019)

10. Menze, B.H., et al.: The multimodal brain tumor image segmentation benchmark (BRATS). IEEE Trans. Med. Imaging **34**(10), 1993–2024 (2015)

11. Milletari, F., Navab, N., Ahmadi, S.: V-net: fully convolutional neural networks for volumetric medical image segmentation. In: Fourth International Conference on 3D Vision, 3DV 2016, Stanford, CA, USA, pp. 565–571 (2016)

12. Myronenko, A.: 3D MRI brain tumor segmentation using autoencoder regularization. In: Crimi, A., Bakas, S., Kuijf, H., Keyvan, F., Reyes, M., van Walsum, T. (eds.) BrainLes 2018. LNCS, vol. 11384, pp. 311–320. Springer, Cham (2019). https://doi.org/10.1007/978-3-030-11726-9_28

13. Myronenko, A., Hatamizadeh, A.: 3D kidneys and kidney tumor semantic segmentation using boundary-aware networks. arXiv e-prints arXiv:1909.06684 (2019)

14. Noroozi, M., Favaro, P.: Unsupervised learning of visual representations by solving Jigsaw puzzles. In: Leibe, B., Matas, J., Sebe, N., Welling, M. (eds.) ECCV 2016. LNCS, vol. 9910, pp. 69–84. Springer, Cham (2016). https://doi.org/10.1007/978-3-319-46466-4_5

15. Ronneberger, O., Fischer, P., Brox, T.: U-net: convolutional networks for biomedical image segmentation. In: Navab, N., Hornegger, J., Wells, W.M., Frangi, A.F. (eds.) MICCAI 2015. LNCS, vol. 9351, pp. 234–241. Springer, Cham (2015). https://doi.org/10.1007/978-3-319-24574-4_28

16. Ruder, S.: An overview of multi-task learning in deep neural networks. arXiv e-prints 1706.05098 (2017)

17. Simpson, A.L., et al.: A large annotated medical image dataset for the development and evaluation of segmentation algorithms. arXiv e-prints arXiv:1902.09063 (2019)

18. Tokui, S., et al.: Chainer: a deep learning framework for accelerating the research cycle. In: Proceedings of the 25th ACM SIGKDD International Conference on Knowledge Discovery & Data Mining, KDD 2019, Anchorage, AK, USA, pp. 2002–2011 (2019)

19. van den Oord, A., Li, Y., Vinyals, O.: Representation learning with contrastive predictive coding. arXiv e-prints arXiv:1807.03748 (2018)

20. Vaswani, A., et al.: Attention is all you need. In: Advances in Neural Information Processing Systems 30: Annual Conference on Neural Information Processing Systems 2017, Long Beach, CA, USA, pp. 5998–6008 (2017)

21. Wu, Y., He, K.: Group normalization. In: Ferrari, V., Hebert, M., Sminchisescu, C., Weiss, Y. (eds.) ECCV 2018. LNCS, vol. 11217, pp. 3–19. Springer, Cham (2018). https://doi.org/10.1007/978-3-030-01261-8_1

22. Zhou, Z., et al.: Models genesis: generic autodidactic models for 3D medical image analysis. In: Shen, D., et al. (eds.) MICCAI 2019. LNCS, vol. 11767, pp. 384–393. Springer, Cham (2019). https://doi.org/10.1007/978-3-030-32251-9_42

Spatio-Temporal Learning from Longitudinal Data for Multiple Sclerosis Lesion Segmentation

Stefan Denner[1], Ashkan Khakzar[1], Moiz Sajid[1], Mahdi Saleh[1], Ziga Spiclin[3], Seong Tae Kim[1(✉)], and Nassir Navab[1,2]

[1] Computer Aided Medical Procedures, Technical University of Munich, Munich, Germany
`seongtae.kim@tum.de`
[2] Computer Aided Medical Procedures, Johns Hopkins University, Baltimore, USA
[3] Faculty of Electrical Engineering, University of Ljubljana, Ljubljana, Slovenia

Abstract. Segmentation of Multiple Sclerosis (MS) lesions in longitudinal brain MR scans is performed for monitoring the progression of MS lesions. We hypothesize that the spatio-temporal cues in longitudinal data can aid the segmentation algorithm. Therefore, we propose a multi-task learning approach by defining an auxiliary self-supervised task of deformable registration between two time-points to guide the neural network toward learning from spatio-temporal changes. We show the efficacy of our method on a clinical dataset comprised of 70 patients with one follow-up study for each patient. Our results show that spatio-temporal information in longitudinal data is a beneficial cue for improving segmentation. We improve the result of current state-of-the-art by 2.6% in terms of overall score ($p < 0.05$). Code is publicly available (https://github.com/StefanDenn3r/Spatio-temporal-MS-Lesion-Segmentation).

Keywords: Longitudinal analysis · MS lesion segmentation

1 Introduction

Multiple Sclerosis (MS) is a neurological disease characterized by damage to myelinated nerve sheaths (demyelination) and is a potentially disabling disease of the central nervous system. The affected regions appear as focal lesions in the white matter [22] and Magnetic Resonance Imaging (MRI) is used to visualize and detect the lesions [9]. MS is a chronic disease, therefore longitudinal MRI patient studies are conducted to monitor the progression of the disease. Accurate lesion segmentation in the MRI scans is important to quantitatively assess response to treatment [21] and future disease-related disability progression [25]. However, manual segmentation of MS lesions in MRI volumes is time-consuming, prone to errors and intra/inter-observer variability [6].

S. Denner and A. Khakzar—First two authors contributed equally to this work.
S. T. Kim and N. Navab—Share senior authorship.

© Springer Nature Switzerland AG 2021
A. Crimi and S. Bakas (Eds.): BrainLes 2020, LNCS 12658, pp. 111–121, 2021.
https://doi.org/10.1007/978-3-030-72084-1_11

Several studies have proposed automatic methods for MS lesion segmentation in MRI scans [1,2,11,12,26,29]. Valverde et al. [26] proposed a cascade of two 3D patch-wise convolutional networks, where the first network provides candidate voxels to the second network for final lesion prediction. Hashemi et al. [12] introduced an asymmetric loss, similar to the Tversky index, which is supposed to tackle the problem of high class imbalance in MS lesion segmentation by achieving a better trade-off between precision and recall. Whereas the previous two approaches worked with 3D input, Aslani et al. [2] proposed a 2.5D slice-based multimodality approach, where they use a single branch for each modality. They trained their network with slices from all plane orientations (axial, coronal, sagittal). During inference, they merge those 2D binary predictions to a single lesion segmentation volume by applying a majority vote. Zhang et al. [29] also proposed a 2.5D slice-based approach, but they concatenated all modalities instead of processing them in multiple branches. In contrast, they utilize a separate model for each plane orientation.

However, none of these works use the data from multiple time-points. The work of Birenbaum et al. [5] is the only method that processes longitudinal data. Birenbaum et al. [5] proposed a siamese architecture, where input patches from two time-points are given to separate encoders that share weights and subsequently the encoders' outputs are concatenated and fed into subsequent CNN to predict the class of pixel of interest. Birenbaum et al. [5] set the direction for using longitudinal data and opens up a line of opportunities for future work. However, their work does not extensively investigate the potential of using information from longitudinal data. Specifically, their proposed late-fusion of features does not properly take advantage of learning from structural changes.

In this paper, we initially propose an improved baseline methodology over that of Birenbaum et al. [5] by employing an early fusion of multimodal longitudinal data, which allows for proper capturing of the differences between inputs from different time points, as opposed to the late fusion of data proposed in [5]. Our main contribution is *proposing a multitask learning framework by adding an auxiliary deformable registration task to the segmentation model.* The two tasks share the same encoder, but two separate decoders are assigned for each task, hence the learned features for predicting the deformation are shared with the segmentation task. The notion of joint registration and segmentation itself is previously proposed in [28] (Deep Atlas), however, the methodology uses two different segmentation and registration networks where there is no feature sharing in between the models, and the output of segmentation model for two different and single-time point scans are corrected by the deformation and vice versa. On the contrary, our proposed approach aims to use the learned features for the registration task *explicitly* in the segmentation task, and to our knowledge this is the first work which applies such a model to longitudinal data. We hypothesize that structural changes of lesions through time are valuable cues for the model to detect these lesions, and we evaluate our approaches on a clinical dataset including 70 patients with one follow-up study for each patient. We compare our methods to the state-of-the-art works on MS lesion segmentation [12,29], and the previous longitudinal approach [5]. Moreover, we adapt the joint registration and segmentation methodology of [28] to longitudinal data and the problem of MS lesion segmentation, and compare it to our own methodology to investigate how explicitly incorporating spatio-temporal features can improve the segmentation.

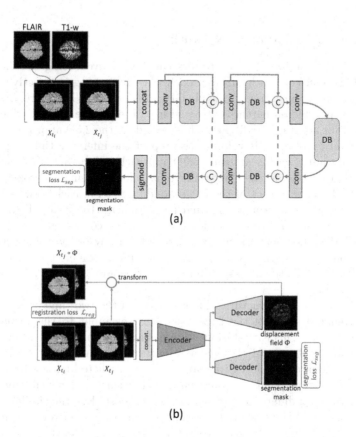

Fig. 1. Our proposed methods: (a) Baseline Longitudinal Network: longitudinal scans are concatenated and given to the segmentation model to implicitly use the structural differences (b) Multitask Longitudinal Network: The network is trained with an auxiliary task of deformable registration between two longitudinal scans, to explicitly guide the network toward using spatio-temporal changes.

2   Methodology

This section describes our approaches for incorporating spatio-temporal features into the learning pipeline of a neural network. We hypothesize that structural changes of lesions between the longitudinal scans are valuable cues for detecting these lesions. Note that the aim is not to model *how* the lesions deform or change, but to find *what* has changed and to use that information to improve segmentation. To this aim, we propose a baseline neural network that improves on the methodology of [5] by proposing early-fusion of input data and subsequently introduce our multitask learning approach with deformable registration approach.

2.1 Baseline Longitudinal Network

It is shown that state-of-the-art results can be achieved using the 2.5D approach [29]. For the case of the 3D approach, it is challenging to directly process a full 3D volume by the current available GPU memory [20]. Therefore, 3D models are usually operated on patches extracted from the volume [12,27], which limits the context for accurate prediction. Thus, we adopt the 2.5D approach [2,19,29] for segmentation of 3D MR volumes (we report the inference time required for the segmentation of each scan using 2.5D approach in Sect. 3.2). For each voxel, segmentation is done on the three orthogonal slices crossing the voxel of interest. The probability output of the corresponding pixel in each view is averaged and thresholded to determine the final prediction for the voxel. To segment a given slice, we use a fully convolutional and densely connected neural network (Tiramisu) [13]. The network receives a slice from any of the three orthogonal views and outputs a segmentation mask. To account for different modalities (T1-w, FLAIR), we stack the corresponding slices from all modalities and feed them to the network.

In order to use the structural changes between the two time-points, we give the concatenated scans of the two-time points as input to the segmentation network (Fig. 1a). This early-fusion of inputs allows the network filters to capture the minute structural changes at all layers leading to the bottleneck, as opposed to the late fusion of Birenbaum et al. [5], where high-level representations from each time point are concatenated. The early fusion's effectiveness for learning structural differences can be further supported by the similar architectural approaches in the design of deformable registration networks [3,4].

2.2 Multitask Learning with Deformable Registration

In this section we describe our approach involving the augmentation of the segmentation task with an auxiliary deformable registration task. We aim to explicitly use the structural change information between the two longitudinal scans. In longitudinal scans, only specific structures such as MS lesions change substantially. Deformable registration is defined to learn a deformation field between two instances. We therefore propose augmenting our baseline longitudinal segmentation model with a deformable registration loss. We hypothesize that this would further guide the network towards using structural differences between the inputs of two different time points. Note that the longitudinal scans are already rigidly registered in the pre-processing step, therefore the deformation field only reflects the structural differences of lesions.

The resulting network (Fig. 1b) consists of a shared encoder followed by two decoders used to generate the specific outputs for the two tasks. One head of the network is associated with generating the segmentation mask and the other one with deformation field map. The encoder-decoder architecture here is that of Tiramisu [13], and two decoders for registration and segmentation are architecturally equivalent. The deformable registration task is trained without supervision or ground truth registration data. This is rather trained self-supervised and by reconstructing one scan from the other which helps adding additional generic information to the network. The multi task loss is defined as:

$$\mathcal{L} = \mathcal{L}_{seg} + \mathcal{L}_{reg} \tag{1}$$

A common pitfall in multi task learning is the imbalance of different tasks which leads to under-performance of multitask learning compared to single tasks. To solve this one needs to normalize loss functions or gradients flow [8]. Here we use the same type of loss function for both tasks. Specifically we use MSE loss, which is used for both registration and segmentation problems. We use a CNN based deformable registration methodology similar to VoxelMorph [4], but adapted to 2D inputs and using Tiramisu architecture (Fig. 1b). The registration loss ($\mathcal{L}_{reg}$) is defined as:

$$\mathcal{L}_{reg} = \mathcal{L}_{sim}(X_{t_i}, X_{t_j} \circ \Phi) + \lambda \mathcal{L}_{smooth}(\Phi) \tag{2}$$

where Φ is the deformation field between inputs X_{t_i} and X_{t_j} (t_i and t_j denote the time-points). $X_{t_j} \circ \Phi$ is the warping of X_{t_j} by Φ, and $\mathcal{L}_{sim}$ is the loss imposing the similarity between X_{t_i} and warped version of X_{t_j}. $\mathcal{L}_{smooth}$ is regularization term to encourage Φ to be smooth. We use MSE loss for $\mathcal{L}_{sim}$, and for the smoothness term $\mathcal{L}_{smooth}$, similar to [4] we use a diffusion regularizer on the spatial gradients of the displacement field.

3   Experiment Setup

3.1   Datasets and Preprocessing

The clinical dataset [10, 15] consists of 1.5T and 3T MR images. Follow-up images of 70 MS patients were acquired. Images are 3D T1-w MRI scans and 3D FLAIR scans with approximately 1 mm isotropic resolution. The MR scans were preprocessed initially by applying non-local mean-based image denoising [16] and the N4 bias correction [24]. From the preprocessed T1-weighted (T1) image the brain mask was extracted by a multi-atlas label fusion segmentation method [7], which employed 50 manually segmented T1 MR brain images of age-matched healthy subjects. The atlases were aligned to the brain masked and preprocessed T1 by a nonlinear B-spline registration method [14]. Using the same registration method, the corresponding preprocessed FLAIR images were aligned to the preprocessed T1. MS lesions were manually annotated by an expert rater. The data of 40 patients were used as a training set (30 patients for training the model and 10 patients for validation). The remaining data of 30 patients were used as an independent test set.

3.2   Implementation Details

The encoders and decoders of our architectures are based on FC-DenseNet57 [13]. We used Adam optimizer with AMSGrad [18] and a learning rate of 1e–4. We use a single model with shared weights for all plane orientations. Since our approaches are 2.5D, we average and threshold the probability output predictions of all plane orientations. The inference time for a whole 3D volume with a resolution of $224 \times 224 \times 224$ is 8.29 s on a 8 GB GPU. Our Multitask Longitudinal Network has about 2 million parameters. PyTorch 1.4 [17] is used for neural network implementation.

3.3   Evaluation Metrics

For evaluating our methods, we use Dice Similarity Coefficient (DSC), Positive Predictive Value (PPV), Lesion-wise True Positive Rate (LTPR), Lesion-wise False Positive Rate (LFPR) and Volume Difference (VD). To consider the overall effect of different metrics, we adopt an *Overall Score* in a similar way to MS lesion segmentation challenges [6,23] as follows:

$$
\begin{aligned}
OverallScore = 0.125 DSC + 0.125 PPV + 0.25 \left(1 - VD\right) \\
+ 0.25 LTPR + 0.25 \left(1 - LFPR\right).
\end{aligned}
\tag{3}
$$

3.4   Method Comparisons

In this section we elaborate upon the methods used for comparison and the naming scheme used throughout the paper. Henceforth we call the single time point segmentation models as *static* models. All models are based on the FC-DenseNet-57 architecture to provide comparability.

Static Network (Zhang et al. [29]): The approach from Zhang et al. [29] feeds three consecutive slices of one 3D scan to the network. It uses one model for each plane orientation, i.e. weights are not shared along the different plane orientations.

Static Network (Asymmetric Dice Loss [12]): Hashemi et al. [12] proposed an asymmetric dice loss which is applied on the FC-DenseNet-57. $\beta = 1.5$.

Deep Atlas [28]: Two separate FC-DenseNet-57 which are jointly trained. One model is trained on the deformation task, the other on is trained on segmentation.

Longitudinal Siamese Network [5]: We implement the longitudinal siamese model [5] with a FC-DenseNet-57 [13]. As in [5], we use the late fusion which combines the features of different time point scans in the bottleneck layer.

Baseline Static Network (Ours): Similar to Zhang et al. [29], our model is based on FC-DenseNet-57 and uses only a *single* time-point. Our model is trained with all three plane orientations (axial, coronal, and sagittal).

Baseline Longitudinal Network (Ours): Our proposed longitudinal model (Sect. 2.1).

Multitask Longitudinal Network (Ours): Our multitask network (Sect. 2.2).

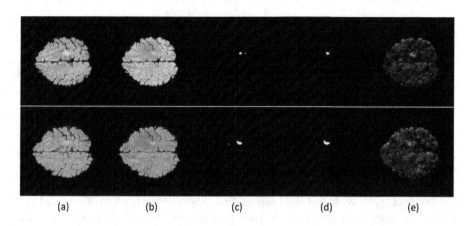

(a)	(b)	(c)	(d)	(e)

Fig. 2. Visualisation of MS lesion's structural change in two longitudinal MR FLAIR scans. Each row presents data from one patient. (a) is the scan from the first time-point, (b) is the scan from the follow up study, (c) visualizes ground truth of MS lesions on the follow up image, (d) shows the predicted segmentation mask of Multitask Longitudinal Network on the follow up image, (e) represents the predicted displacement field between the two scans using the registration module of our multitask method.

Table 1. Comparison of different methodologies. Our methods are shown in bold letters. For LFPR and VD, lower is better.

Method	DSC ↑	PPV ↑	LTPR ↑	LFPR ↓	VD ↓	Overall score ↑
Multitask longitudinal network	**0.695**	0.771	0.680	0.212	0.221	**0.745**
Baseline longitudinal network	0.694	0.752	0.654	0.227	0.227	0.731
Baseline static network	0.684	0.762	0.647	0.250	0.247	0.718
Longitudinal Siamese Network [5]	0.684	**0.777**	0.614	**0.194**	0.245	0.726
Static network (Zhang et al. [29])	0.684	0.761	0.604	0.223	0.263	0.710
Static network (asymmetric dice loss [12])	0.690	0.648	**0.752**	0.346	0.336	0.685
Deep atlas [28]	0.656	0.701	0.652	0.260	**0.180**	0.723

4   Results and Discussion

We first compare our Baseline Static Network (Sect. 3.4) to the state-of-the-art static segmentation approaches [12,29]. Table 1 shows that our Baseline Static Network, which is inspired by the method of [29] achieves an overall score of 0.718 and performs similar to our implementation of [29] which achieves 0.710. The difference between the methods is that we use one single network for all orthogonal views, and the results show that using a single network and thus, fewer parameters achieves similar but slightly better results. We also report the

Table 2. Statistical significance analysis of performance improvements of our Multitask Longitudinal Network over other methods in terms of the overall score by paired t-test.

Comparison with	Mean difference ± standard error	95% CI	p-value
Baseline longitudinal network	0.0141 ± 0.0047	[0.0047, 0.0236]	0.0040
Baseline static network	0.0267 ± 0.0053	[0.0161, 0.0372]	<0.0001
Longitudinal siamese network [5]	0.0187 ± 0.0048	[0.0091, 0.0282]	0.0002
Static network (Zhang et al. [29])	0.0389 ± 0.0055	[0.0280, 0.0499]	<0.0001
Static network (Asymmetric Dice Loss [12])	0.0603 ± 0.0144	[0.0315, 0.0891]	0.0001
Deep atlas [28]	0.0223 ± 0.0094	[0.0034, 0.0412]	0.0218

results of asymmetric Dice loss of [12] on our static model which improves the DSC, however hurts the Overall Score.

We proceed by evaluating our Baseline Longitudinal Network against our Baseline Static Network and the Longitudinal Siamese Network of [5]. In Table 1 we observe that both longitudinal models improve the overall score. We also observe that our Baseline Longitudinal Network which uses early fusion of input data improves the score further. In Sect. 2.1, we stated that the early-fusion of inputs allows the network to capture the structural differences between the inputs better than late-fusion. The comparison between our Longitudinal Network and Longitudinal Siamese Network [5] which only differ in how they fuse the inputs, serves as an ablation experiment for this claim.

To illustrate the behavior of our Multitask Longitudinal Network, we visualize the segmentation mask and the displacement field in Fig. 2. The displacement field shows *what* has changed. In Fig. 2, the colors in the displacement encode the direction of the field at any point, and the brightness signifies the magnitude of displacement. As can be seen, the areas corresponding to MS lesions have high brightness indicating that the deformable registration model has captured the change of MS lesions. As shown in Table 1, by exploiting the spatio-temporal information from deformable registration, the segmentation performances were improved compared to the Baseline Longitudinal Network. It is also observed that the Deep Atlas [28] methodology while improving over the static approaches, achieves inferior results compared to all longitudinal approaches.

To verify that the performance improvements of our Multitask Longitudinal Network are statistically significant, we conducted further analysis of the models with paired t-test on the overall score. The paired t-test provides a statistical evaluation of the performance differences between models. Table 2 shows the results of the statistical significance analysis for differences of overall score. The multitask learning framework significantly improves the overall score compared with longitudinal network ($p = 0.004$). The improvements from our Multitask Longitudinal Network compared with previous methods [5,12,28,29] are also statistically significant $p < 0.05$.

5   Conclusion

In this work, we investigated the utilization of spatio-temporal information in longitudinal brain MR data to improve the segmentation of MS lesions. We proposed a novel multitask formulation where an auxiliary unsupervised deformable registration task is adopted. We evaluated our approaches on a clinical dataset comprising of 70 patients with one follow-up scan for each patient. Our evaluations against state-of-the-art MS lesion segmentation works confirm that incorporating spatio-temporal information into segmentation models improves the segmentation performance. Furthermore, we showed transferring previous work on joint registration and segmentation to longitudinal data achieves inferior results compared to our methodology as we explicitly incorporate spatio-temporal features into our model. In future work, our proposed methodology can be extended to other longitudinal medical studies to improve segmentation.

Acknowledgements. The authors acknowledge the financial support for this work by Siemens Healthineers and Munich Center for Machine Learning (MCML). Ziga Spiclin was supported by the Slovenian Research Agency (research core funding No. P2-0232, and research grant No. J2-2500).

References

1. Andermatt, S., Pezold, S., Cattin, P.C.: Automated segmentation of multiple sclerosis lesions using multi-dimensional gated recurrent units. In: Crimi, A., Bakas, S., Kuijf, H., Menze, B., Reyes, M. (eds.) BrainLes 2017. LNCS, vol. 10670, pp. 31–42. Springer, Cham (2018). https://doi.org/10.1007/978-3-319-75238-9_3
2. Aslani, S., Dayan, M., Storelli, L., Filippi, M., Murino, V., Rocca, M.A., Sona, D.: Multi-branch convolutional neural network for multiple sclerosis lesion segmentation. NeuroImage **196**, 1–15 (2019)
3. Balakrishnan, G., Zhao, A., Sabuncu, M.R., Dalca, A.V., Guttag, J.: An unsupervised learning model for deformable medical image registration. In: Proceedings of the IEEE Computer Society Conference on Computer Vision and Pattern Recognition (2018). https://doi.org/10.1109/CVPR.2018.00964
4. Balakrishnan, G., Zhao, A., Sabuncu, M.R., Guttag, J., Dalca, A.V.: VoxelMorph: a learning framework for deformable medical image registration. IEEE Trans. Med. Imaging (2019). https://doi.org/10.1109/TMI.2019.2897538
5. Birenbaum, A., Greenspan, H.: Longitudinal multiple sclerosis lesion segmentation using multi-view convolutional neural networks. In: Carneiro, G., et al. (eds.) LABELS/DLMIA -2016. LNCS, vol. 10008, pp. 58–67. Springer, Cham (2016). https://doi.org/10.1007/978-3-319-46976-8_7
6. Carass, A., et al.: Longitudinal multiple sclerosis lesion segmentation: resource and challenge. NeuroImage **148**, 77–102 (2017)
7. Cardoso, M.J., et al.: Geodesic information flows: spatially-variant graphs and their application to segmentation and fusion. IEEE Trans. Med. Imaging **34**(9), 1976–1988 (2015)
8. Chen, Z., Badrinarayanan, V., Lee, C.Y., Rabinovich, A.: GradNorm: gradient normalization for adaptive loss balancing in deep multitask networks. In: 35th International Conference on Machine Learning, ICML 2018 (2018)

9. Compston, A., Coles, A.: Multiple sclerosis (2008). https://doi.org/10.1016/S0140-6736(08)61620-7
10. Galimzianova, A., Pernuš, F., Likar, B., Špiclin, Ž: Stratified mixture modeling for segmentation of white-matter lesions in brain MR images. NeuroImage **124**, 1031–1043 (2016)
11. Ghafoorian, M., Platel, B.: Convolutional neural networks for MS lesion segmentation, method description of diag team. In: Proceedings of the 2015 Longitudinal Multiple Sclerosis Lesion Segmentation Challenge, pp. 1–2 (2015)
12. Hashemi, S.R., Salehi, S.S.M., Erdogmus, D., Prabhu, S.P., Warfield, S.K., Gholipour, A.: Asymmetric loss functions and deep densely-connected networks for highly-imbalanced medical image segmentation: application to multiple sclerosis lesion detection. IEEE Access **7**, 1721–1735 (2018)
13. Jégou, S., Drozdzal, M., Vazquez, D., Romero, A., Bengio, Y.: The one hundred layers tiramisu: fully convolutional densenets for semantic segmentation. In: Proceedings of the IEEE Conference on Computer Vision and Pattern Recognition Workshops, pp. 11–19 (2017)
14. Klein, S., Staring, M., Murphy, K., Viergever, M.A., Pluim, J.P.W.: elastix: a toolbox for intensity-based medical image registration. IEEE Trans. Med. Imaging **29**(1), 196–205 (2010)
15. Lesjak, Ž, et al.: A novel public MR image dataset of multiple sclerosis patients with lesion segmentations based on multi-rater consensus. Neuroinformatics **16**(1), 51–63 (2018)
16. Manjón, J.V., Coupé, P., Buades, A., Louis Collins, D., Robles, M.: New methods for MRI denoising based on sparseness and self-similarity. Med. Image Anal. **16**(1), 18–27 (2012). https://doi.org/10.1016/j.media.2011.04.003. http://www.sciencedirect.com/science/article/pii/S1361841511000491
17. Paszke, A., et al.: Pytorch: An imperative style, high-performance deep learning library. In: Advances in Neural Information Processing Systems, pp. 8024–8035 (2019)
18. Reddi, S.J., Kale, S., Kumar, S.: On the convergence of Adam and beyond (2018)
19. Roth, H.R., et al.: A new 2.5D representation for lymph node detection using random sets of deep convolutional neural network observations. In: Golland, P., Hata, N., Barillot, C., Hornegger, J., Howe, R. (eds.) MICCAI 2014. LNCS, vol. 8673, pp. 520–527. Springer, Cham (2014). https://doi.org/10.1007/978-3-319-10404-1_65
20. Roy, A.G., Conjeti, S., Navab, N., Wachinger, C., Initiative, A.D.N., et al.: Quick-NAT: a fully convolutional network for quick and accurate segmentation of neuroanatomy. NeuroImage **186**, 713–727 (2019)
21. Stangel, M., Penner, I.K., Kallmann, B.A., Lukas, C., Kieseier, B.C.: Towards the implementation of 'no evidence of disease activity' in multiple sclerosis treatment: the multiple sclerosis decision model. Therap. Adv. Neurol. Disord. **8**(1), 3–13 (2015)
22. Steinman, L.: Multiple sclerosis: A coordinated immunological attack against myelin in the central nervous system (1996). https://doi.org/10.1016/S0092-8674(00)81107-1
23. Styner, M., et al.: 3D segmentation in the clinic: a grand challenge II: MS lesion segmentation. Midas J. **2008**, 1–6 (2008)
24. Tustison, N.J., et al.: N4ITK: improved N3 bias correction. IEEE Trans. Med. Imaging **29**(6), 1310–1320 (2010)
25. Uher, T., et al.: Combining clinical and magnetic resonance imaging markers enhances prediction of 12-year disability in multiple sclerosis. Multiple Sclerosis **23**(1), 51–61 (2017)

26. Valverde, S., et al.: Improving automated multiple sclerosis lesion segmentation with a cascaded 3D convolutional neural network approach. NeuroImage **155**, 159–168 (2017)

27. Wachinger, C., Reuter, M., Klein, T.: DeepNAT: deep convolutional neural network for segmenting neuroanatomy. NeuroImage **170**, 434–445 (2018)

28. Xu, Z., Niethammer, M.: DeepAtlas: joint semi-supervised learning of image registration and segmentation. In: Shen, D., et al. (eds.) MICCAI 2019. LNCS, vol. 11765, pp. 420–429. Springer, Cham (2019). https://doi.org/10.1007/978-3-030-32245-8_47

29. Zhang, H., et al.: Multiple sclerosis lesion segmentation with tiramisu and 2.5D stacked slices. In: Shen, D., et al. (eds.) MICCAI 2019. LNCS, vol. 11766, pp. 338–346. Springer, Cham (2019). https://doi.org/10.1007/978-3-030-32248-9_38

MMSSD: Multi-scale and Multi-level Single Shot Detector for Brain Metastases Detection

Hui Yu[1], Wenjun Xia[1], Yan Liu[2(✉)], Xuejun Gu[3], Jiliu Zhou[1], and Yi Zhang[1]

[1] College of Computer Science, Sichuan University, Chengdu 610065, China
[2] College of Electrical Engineering, Sichuan University, Chengdu 610065, China
liuyan77@scu.edu.cn
[3] Department of Radiation Oncology, University of Texas Southwestern
Medical Center, Dallas, TX, USA

Abstract. Stereotactic radio surgery (SRS) is the preferred treatment for brain metastases (BM), in which the delineation of metastatic lesions is one of the critical steps. Taking into consideration that the BM always have clear boundary with surrounding tissues but very small volume, the difficulty of delineation is object detection instead of segmentation. In this paper, we presented a novel lesion detection framework, called Multi-scale and Multi-level Single Shot Detector (MMSSD), to detect the BM target accurately and effectively. In MMSSD, we took advantage of multi-scale feature maps, while paid more attention on the shallow layers for small objects. Specifically, first we only preserved the applicable large-and-middle-scale features in SSD, then generated new feature representations by multi-level feature fusion module, and finally made predictions on those feature maps. The proposed MMSSD framework was evaluated on the clinical dataset, and the experiment results demonstrated that our method outperformed existing popular detectors for BM detection.

Keywords: Brain metastases detection · Small objects · Single shot detector

1 Introduction

Brain metastases (BM) are malignant tumors that metastasize into the cranial cavity from other areas of the body [1], like lung, breast, alimentary canal, etc. In recent clinical practice, the stereotactic radio surgery (SRS) has become the preferred treatment for BM, which delivers radiation to target tumor in fewer high-dose treatments to avoid side-effects on surrounding normal tissues, unlike fractional treatment of traditional radiation therapy [2]. Due to that, the precise delineation of metastatic lesions plays an extremely vital role to ensure the curative effect of SRS treatment. Besides, as a one-day procedure, SRS treatment planning should be finished in a short time. Currently, the treatment planning is mainly designed on the contrasted enhanced T1-weighted (T1c) magnetic resonance imaging (MRI), with manual delineation from well-trained radiation oncologists. Same as other manual jobs, it faces the subjective inconsistency as well as the labor-and-time-consuming problem. Therefore, there is an urgent need for clinical practice to find the BM target accurately and fast.

© Springer Nature Switzerland AG 2021
A. Crimi and S. Bakas (Eds.): BrainLes 2020, LNCS 12658, pp. 122–132, 2021.
https://doi.org/10.1007/978-3-030-72084-1_12

Generally, automatic segmentation methods are employed to find BM targets since they can localize the lesion and extract the boundary simultaneously. Several popular segmentation approaches [3–7], including level-set [3, 4], fuzzy clustering [5, 6] and machine learning [7] based ones, are easy to suffer from parameter tuning, complex calculation and manual design of features problems when applied on BM segmentation, since most BM have small sizes and various locations [1]. Currently, deep learning (DL) is superior to most traditional segmentation methods on BM segmentation [8–10], especially convolutional neural networks (ConvNets [11]) based methods due to their hierarchical feature learning capability. They treat segmentation as a pixelwise classification problem ignoring sematic understanding, so that generate large numbers of false-positives [9]. In fact, BM treated by SRS always have clear boundaries and small sizes so it becomes an easy problem to delineate the contour once the target is found. Thereby the challenge in finding BM is more a detection mission instead of segmentation [9]. To address that, we adopt a detection framework in this paper and treat the 3D space as a composition of 2D orthogonal (i.e., axial, sagittal and coronal) planes.

Single shot detector (SSD) approach is developed for object detection [12, 13]. Different from two-stage detectors [14, 15], SSD places anchors (i.e., prior boxes) densely and directly classifies as well as regresses each anchor. It adopts multi-scale feature maps and multi-scale-and-aspect-ratio priors, which are the most important improvements for performance. The pyramidal features within ConvNets help SSD to detect multi-scale objects. The large-scale features are used for relatively small objects, while small-scale features mainly for large objects. Inspired by SSD architecture, we try to develop a modified SSD network for BM detection task. The newly designed model will pay more attention on the shallower layer features to achieve better performance on the small BM targets.

In this paper, we present a novel brain metastases detection framework, named Multi-scale and Multi-level Single Shot Detector (MMSSD). First, we preserve the large-and-middle-scale features used in SSD. Second, we generate fusion features by multi-level feature fusion modules. Finally, we make predictions on those multi-scale feature representations for BM detection. Our proposed approach was validated on the clinical dataset and outperformed popular object detectors. We also conducted ablation experiments to demonstrate the effectiveness of each component in our framework.

2   Methodology

As we analyzed before, the large-scale features are more suitable for object localization and small-scale features mainly contribute to sematic classification. The essential of BM detection is localization while it is relatively insensitive for classification in the case of only having one class label of tumor, so we develop SSD with more focusing on the shallower layers. Figure 1 overviews the proposed pipeline of MMSSD.

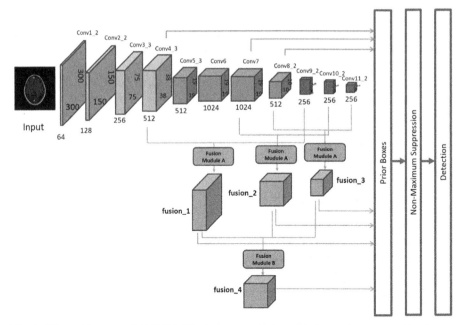

Fig. 1. The architecture of MMSSD. We take an input in the axial view as the example. Predictions are made on the parts of original features in SSD (light blue arrows) and new fusion features (light purple arrows) simultaneously. (Color figure online)

2.1 Multi-scale Feature Maps for BM Detection

As shown in Fig. 1, SSD takes VGG16 [12] as the backbone network instead of the deeper ConvNet (e.g. ResNet [16] or DenseNet [17]). Deeper ConvNets are not designed for small object localization and inference speed promotion [13]. It makes predictions on six multi-scale feature maps (Conv4_3, Conv7, Conv8_2, Conv9_2, Conv10_2 and Conv11_2). However, large-and-middle scale features (Conv4_3, Conv7, and Conv8_2), instead of small-scale ones (Conv9_2-Conv11_2) with large receptive fields, are mainly responsible for small object detection [13]. Therefore, in our task we keep them for direct prediction. To further improve the performance of SSD, we design multi-level feature fusion modules, which makes full use of fine-grained information in shallower layers and avoids losing semantic information from deep layers. In addition, we adopt symmetrical connections in feature fusion for sharing the feature structure.

Particularly, in the first-level feature fusion, we up-sample Conv9_2, Conv10_2 and Conv11_2 and merge them with Conv4_3, Conv7 and Conv8_2 to generate new fusion features fusion_1, fusion_2 and fusion_3 respectively. They replace the original Conv9_2, Conv10_2 and Conv11_2 of SSD. It is necessary to mine hidden feature information, thus we design the second-level feature fusion to continue generating more descriptive feature representation fusion_4 based on the fusion_1, fusion_2 and fusion_3. The more details will be described in the following section.

In brief, we totally have 7 feature representations for prediction at different scales, including 4 fusion features (fusion_1, fusion_2, fusion_3 and fusion_4) and 3 original

SSD feature maps (Conv4_3, Conv7 and Conv8_2). Those features also form the feature pyramids in both lateral and longitudinal directions.

2.2 Multi-level Feature Fusion Modules

In order to take full advantage of the feature pyramid in SSD, we develop multi-level feature fusion strategy to fuse the information provided in top and bottom-and-middle parts of the pyramid, as illustrated in Fig. 1. The main idea is to utilize the semantic understanding in the deep layers while compensate the spatial knowledge with the shallower layers.

For Fusion Module A, it achieves the first-level feature fusion with symmetrical connections (i.e., left and right). Note that the feature maps should have the same size and channel when we perform element-wise operation among them. Fusion Module A is formulated as:

$$F_A = Conv_2\big(Conv_1(f_{low}) + DeConv(f_{high})\big) \tag{1}$$

Where f_{low} and f_{high} represents large-and-middle-scale features and small-scale features; $Conv_1$ and $Conv_2$ are two different convolution layers, involving ReLU and L2 normalization operation, respectively; $DeConv$ indicates a set of deconvolution operations.

We take the generation of fusion_1 as an example. In order to fuse Conv4_3 and Conv9_2, we up-sample the spatial resolution of Conv9_2 through three deconvolution operations and produce an output map having the same size and channel with Conv4_3 of $512 \times 38 \times 38$. Figure 2 describes the detailed process of fusing Conv4_3 and Conv9_2. The symmetric connections enable fusion_2 and fusion_3 to follow the identical fusion principle.

For Fusion Module B, it completes the second-level feature fusion with symmetrical connections (i.e., left, middle and right) to fully mine hidden feature information. Fusion Module B is formulated as:

$$F_B = Conv_2\big(Max(f_{f1}) + Conv_1(f_{f2}) + DeConv(f_{f3})\big) \tag{2}$$

Where f_{f1}, f_{f2} and f_{f3} represents fusion features, i.e., fusion_1, fusion_2 and fusion_3, respectively; $Conv_1$ and $Conv_2$ are two different convolution layers, involving ReLU and L2 normalization operation as Eq. (1); Max is the max-pooling operation; $DeConv$ indicates a set of deconvolution operations.

Based on the result of Fusion Module A, fusion_1 and fusion_3 need to down-sample and up-sample separately to obtain output having the same size and channel with the fusion_2. Then merge them together through element-wise operation to produce a whole fusion feature (fusion_4). Figure 3 showcases the details of Fusion Module B.

3 Experiments and Results

3.1 Dataset and Data Processing

We evaluated our framework on a subset of clinical dataset proposed in [8] and 95 patients were randomly selected. All the patients were undertaken SRS treatment and

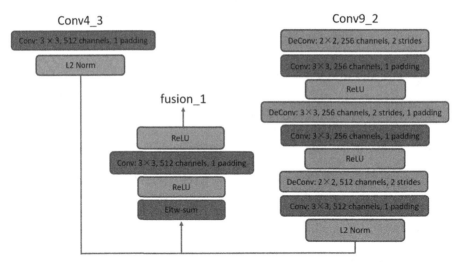

Fig. 2. The details of generating fusion_1. We use different colors to emphasize different operations. (Color figure online)

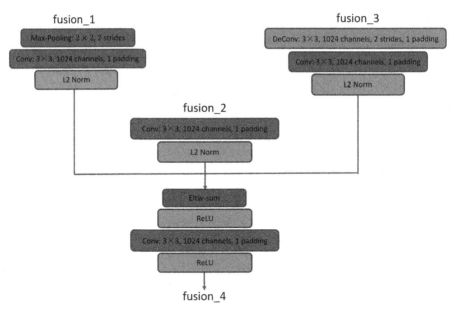

Fig. 3. The details of Fusion Module B. We use different colors to emphasize different operations. (Color figure online)

the T1c sequences were manually delineated by the physicians following the standard clinical protocols. The clinical dataset has the following characteristics: 1) the BM are always solid lesions with diameters less than 1cm, and have ball-like shape as well as clear boundary with surrounding tissues; 2) the BM are mainly caused by hematogenous dissemination so that easily occur in multiple sites simultaneously. Figure 4 describes the slice examples of the clinical dataset in three orthogonal planes.

In our work, we trained an individual MMSSD for each of the three axes independently. Therefore, we divided the 3D MRI volumes of all patients into the 2D slices in each orthogonal view and obtained corresponding 2933, 2985 and 2977 slices having the presence of BM. Totally, there were 8799, 8955 and 8931 lesions in the axial, sagittal and coronal plane, respectively. We allocated the training and testing datasets with ratio of 7:3. In addition, we adopted the VOC (i.e., XML files) data format for BM detection. The XML files mainly contained the dataset paths, bounding boxes and corresponding object class labels. The raw clinical dataset only had segmentation masks, so for our detection task we generated the ground truths of bounding boxes by parsing segmentation labels. We implemented our framework with the PyTorch library under the environment of Spyder.

3.2 Training

We trained the whole framework directly without the warm-up phase of supervised learning in a minibatch of size 8. The SGD optimizer was utilized for optimizing the detection network. We set the initial learning rate of SGD as 3e-4 and divided it by 0.2 every 100 epochs for a total of 200 epochs. Other parameters for MMSSD were shown as follows.

Prior Boxes. One of the most important changes in SSD is to set the different scales and aspect ratios of prior boxes for objects with various sizes and shapes. Same as SSD, the vital step in MMSSD is to densely generate multiple prior boxes from each feature map. The scale s_k indicates the ratio of the box area to the feature map area, and aspect ratio $a_r \in \left\{1, 2, 3, \frac{1}{2}, \frac{1}{3}\right\}$ is used to constrain the shape of the box. In addition, the scale s_k reveals a linear decreasing trend that with the decreasing of feature map size, the corresponding s_k linearly increases, which is formulated as:

$$s_k = s_{min} + \frac{s_{max} - s_{min}}{m - 1}k - 1, k \in [1, m] \tag{3}$$

Where $s_{min} = 0.2$ and $s_{max} = 0.9$ follow [12]; m is the index number of feature maps used in MMSSD.

The height and width of the prior box on the kth feature map could be determined through $w_k^{\alpha} = s_k \sqrt{\alpha_r}$ and $h_k^{\alpha} = s_k / \sqrt{\alpha_r}$. For fusion_1, fusion_2, fusion_3 and fusion_4, their corresponding scales and aspect ratios were consistent with Conv4_3, Conv7, Conv8_2 and Conv7 in SSD, respectively. In each feature map cell, the number of prior boxes also matches with the aspect ratio. For example, if a cell needs to generate six prior boxes, the corresponding aspect ratio should be set to $\left[1, 2, 3, \frac{1}{2}, \frac{1}{3}\right]$. In this work, each feature cell of Conv4_3, Conv7 and fusion_1 predicted four prior boxes, and the others had six prior boxes. Therefore, we generated 18528 prior boxes for BM detection totally.

Loss Function. In general object detectors, the training objective is the weighted sum between localization loss and confidence loss, which is formulated as:

$$L(x, c, l, g) = L_{conf}(x, c) + \alpha L_{loc}(x, l, g) \tag{4}$$

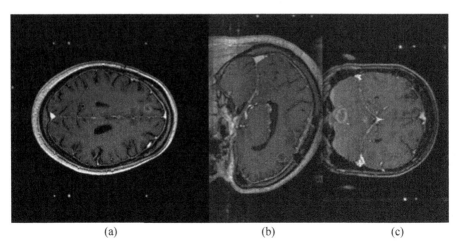

(a)   (b)   (c)

Fig. 4. The slice examples of the clinical dataset in three orthogonal views. (a) shows a slice in the axial view, (b) shows a slice in the sagittal view, and (c) shows a slice in the coronal view.

Where c, l and g represents the confidence coefficient, predicted box and ground truth box, respectively; $L_{conf}(\cdot)$ is the cross-entropy loss widely used in binary-classification; $L_{loc}(\cdot)$ is the smooth L1 loss commonly in regression; α is the weighted factor.

3.3 Experiment Results

We compared our framework (MMSSD) with the other well-performed object detectors including Faster R-CNN [14], YOLO-3 [18], CenterNet [19] and the baseline (SSD) [12]. We used the average precision (AP) metric to measure the performance of those methods on the clinical dataset. The detection results in terms of AP metric and training time are presented in Table 1. In natural object detection, Faster-RCNN usually has good performance at the cost of time, but failed in the BM detection. The reason may be that sizes of BM are much smaller than objects in natural images. YOLO is one-stage detector as well as SSD as and CenterNet, so they ran far faster than Faster-RCNN. we adopted an advanced YOLO-3 that adjusted the network structure and also utilized multi-scale features for object detection, and its performance was still not as good as SSD. CenterNet is the anchor-free detector based on key-point estimation and has achieved state-of-the-art detection performance for natural images, while it only rivaled SSD in this task. This also implied that it was effective to choose SSD model as our baseline. Finally, our method delivered the best results of 20.2%, 14.6% and 18.5% AP in the axial, sagittal and coronal view separately, with relatively slower inference speed than SSD because of more prior boxes. Figure 5 shows some visual results of MMSSD.

We conducted a set of ablation experiments to evaluate the effectiveness of the large-and-middle scale features and multi-level feature fusion module: (i) baseline (SSD), (ii) SSD without deep layers (Conv9_2, Conv10_2 and Conv11_2), (iii) SSD without shallower layers (Conv4_3, Conv7 and Conv8_2), (iv) our proposed method (MMSSD) without Fusion Module B (fusion_4) and (v) MMSSD. Table 2 establishes the results

Table 1. Comparison results of different methods.

Method	Axial		Sagittal		Coronal	
	AP(%)	Time(h)	AP(%)	Time(h)	AP(%)	Time(h)
Faster-RCNN	<10	>100	<10	>100	<10	>100
YOLO-3	<10	<6	<10	<6	< 10	<6
CenterNet	14.5	>12	13.5	>12	12.3	>12
SSD	14.6	<5	11.2	<5	15.2	<5
MMSSD	**20.2**	<6	**14.6**	<6	**18.5**	<6

in terms of *AP* metric. The results of (ii) and (iii) compared with (i) proved that large-and-middle features from shallower layers did contribute more to small object detection. There was little influence on the performance while we directly removed the predictions from deep layers. However, the performance underwent sharp dropping with only deep layers, especially in the axial plane. Comparing (i) with (iv), (iv) with (v), we also observed further improvement by 3.5% vs. 2.1%, 1.6% vs.1.8% and 0.9% vs. 2.4% in three axes. This demonstrated that Fusion Module A and Fusion Module B both served to improve BM detection.

Table 2. Comparison results of the ablation experiment.

Condition	Axial	Sagittal	Coronal
SSD	14.6%	11.2%	15.2%
SSD without deep layers	13.3%	10.6%	14.5%
SSD without shallower layers	<10%	<10%	<10%
MMSSD without Fusion Module B	18.1%	12.8%	16.1%
MMSSD	**20.2%**	**14.6%**	**18.5%**

3.4 Discussion

The challenge in finding BM is more a detection issue instead of segmentation since BM treated by SRS always have clear boundaries and small sizes. Therefore, in this paper we took advantage of the detection framework to find the BM. In fact, with the development of object detection, some successful ConvNets-based methods in the field of natural images also have demonstrated impressive results on medical imaging analysis [7, 15, 20]. Inspired by SSD, we proposed an advanced framework of MMSSD, which enjoyed two advantages. It can be seen from Table 1 that MMSSD delivered better performance in terms of *AP* metric with high inference speed. Our approach utilized shallower layer features for BM localization, that is to say, we retained the large-and-middle-scale features in the original SSD and generated new fusion features combing

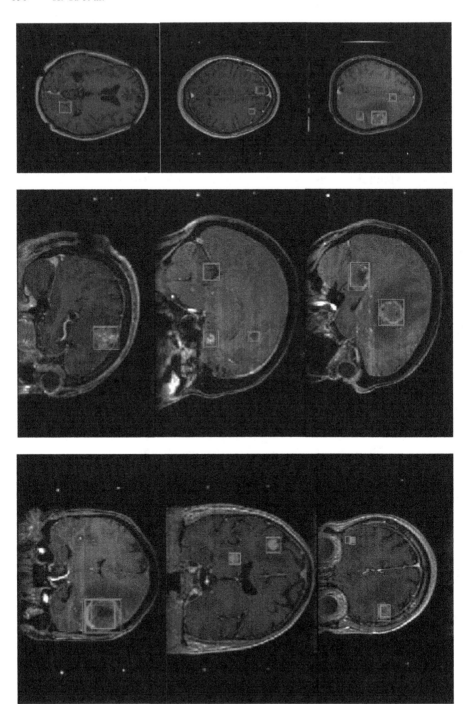

Fig. 5. Some visual results. The red boxes indicate the ground truths and the green ones indicate the predictions. The first row shows the results in the axial view, the second row shows the results in the sagittal view, and the last row shows the results in the coronal view. (Color figure online)

fine-grained details with rich sematic information to make predictions. The results in Table 2 also evaluated the effectiveness of the shallower layer features and multi-level feature fusion modules for small object detection.

4 Conclusion

We present a Multi-scale and Multi-level Single Shot Detector for BM detection. To address small target issue, we pay more attention on large-and-middle feature maps that are better for localization. The fusion features combing rich sematic information with fine-grained details are more descriptive. Experiments conducted on the clinical dataset demonstrate the effectiveness of our method. The shallow layers only start from Conv4_3 to extract features for detection, other larger-scale features, such as Conv3_3 and Conv2_2, should be explored. Therefore, how to design prior boxes to balance inference speed and performance will be our future work. In addition, seen from the visual results, there are still many false-positives, which will need to be settled urgently in the future work as well.

Acknowledgments. Publications of this article were sponsored by the National Science Foundation of China under Grant 61902264 and by the Key Research and Development projects in Sichuan Province under Grant 2019YFS0125.

References

1. Patchell, R.A.: The management of brain metastases. Cancer Treat. Rev. **29**(6), 533–540 (2003)
2. Landoni, V., Pinzi, V., Gomellini, S., et al.: 1539 poster tumor control probability in stereotactic radio-surgery for brain metastases. Radiother. Oncol. **99**(1), 572 (2011)
3. James, A.S.: Level Set Methods and Fast Marching Methods. Cambridge University Press, Cambridge (1999)
4. Li, C., Xu, C., Gui, C., et al.: Level set evolution without re-initialization: a new variational formulation. In: Proceedings of 2005 IEEE Computer Society Conference on Computer Vision and Pattern Recognition, pp. 430–436. CVPR, Washington (2005)
5. Arakeri, M.P., Ram Mohana Reddy, G.: Efficient fuzzy clustering based approach to brain tumor segmentation on MR images. In: Das, V.V., Thankachan, N. (eds.) CIIT 2011. CCIS, vol. 250, pp. 790–795. Springer, Heidelberg (2011). https://doi.org/10.1007/978-3-642-25734-6_141
6. Rastgarpour, M., Shanbehzadeh, J.: A new kernel-based fuzzy level set method for automated segmentation of medical images in the presence of intensity in homogeneity. Comput. Math. Methods Med. **1**, 231–239 (2014)
7. Kamnitsas, K., Ledig, C., Newcombe, V.F.J., et al.: Efficient multi-scale 3D CNN with fully connected CRF for accurate brain lesion segmentation. Med. Image Anal. **36**, 61–78 (2017)
8. Liu, Y., Stojadinovic, S., Hrycushko, B., et al.: A deep convolutional neural network-based automatic delineation strategy for multiple brain metastases stereotactic radiosurgery. PLoS ONE **12**(10), 1–7 (2017)
9. Charron, O., Lallement, A., Jarnet, D., et al.: Automatic detection and segmentation of brain metastases on multimodal MR images with a deep convolutional neural network. Comput. Biol. Med. **95**, 43–54 (2018)

10. Grøvik, E., Yi, D., Iv, M., et al.: Deep learning enables automatic detection and segmentation of brain metastases on multi-sequence MRI. J. Magn. Reson. Imaging **3**, 1–9 (2019)

11. Lecun, Y., Boser, B., Denker, J.S., et al.: Back-propagation applied to handwritten zip code recognition. Neural Comput. **1**(4), 541–551 (2014)

12. Liu, W., Anguelov, D., Erhan, D., Szegedy, C., Reed, S., Fu, C.-Y., Berg, A.C.: SSD: single shot multibox detector. In: Leibe, B., Matas, J., Sebe, N., Welling, M. (eds.) ECCV 2016. LNCS, vol. 9905, pp. 21–37. Springer, Cham (2016). https://doi.org/10.1007/978-3-319-464 48-0_2

13. Cui, L., Ma, R., Lv, P., et al.: MDSSD: multi-scale deconvolutional single shot detector for small objects. arXiv preprint arXiv: 1805.07009 (2018)

14. Ren, S., He, K., Girshick, R., et al.: Faster R-CNN: towards real-time object detection with region proposal networks. IEEE Trans. Pattern Anal. Mach. Intell. **39**(6), 1137–1149 (2017)

15. Xu, X., Zhou, F., Liu, B., et al.: Efficient multiple organ localization in CT image using 3D region proposal network. IEEE Trans. Med. Imaging **38**(8), 1885–1898 (2019)

16. He, K., Zhang, X., Ren, S., et al.: Deep residual learning for image recognition. In: Proceedings of 2016 IEEE Computer Society Conference on Computer Vision and Pattern Recognition, pp. 770–778. CVPR, Washington (2016)

17. Huang, G., Liu, Z., Laurens, V.D.M., et al.: Densely connected convolutional networks. In: Proceedings of 2016 IEEE Computer Society Conference on Computer Vision and Pattern Recognition, pp. 1–9. CVPR, Washington (2016)

18. Redmon, J., Farhadi, A.: YOLOv3: an incremental improvement. arXiv preprint arXiv: 1804.02767 (2018)

19. Duan, K., Bai, S., Xie, L., et al.: CenterNet: keypoint triplets for object detection. In: Proceedings of 2019 IEEE Computer Society Conference on Computer Vision and Pat-tern Recognition, p. 10. CVPR, Long Beach (2019)

20. De, V.B., Wolterink, J., De Jong, P., et al.: ConvNet-based localization of anatomical structures in 3D medical image. IEEE Trans. Med. Imaging **36**(7), 1470–1481 (2017)

Unsupervised 3D Brain Anomaly Detection

Jaime Simarro Viana[1,2(✉)], Ezequiel de la Rosa[1,3], Thijs Vande Vyvere[1,4], David Robben[1,5,6], Diana M. Sima[1], and CENTER-TBI Participants and Investigators

[1] icometrix, Research and Development, Leuven, Belgium
jaime.simarro@icometrix.com
[2] Erasmus Joint Master in Medical Imaging and Applications, University of Girona, Girona, Spain
[3] Department of Computer Science, Technical University of Munich, Munich, Germany
[4] Department of Radiology, Neuroradiology Division, Antwerp University Hospital and University of Antwerp, Antwerp, Belgium
[5] Medical Image Computing (MIC), ESAT-PSI, Department of Electrical Engineering, KU Leuven, Leuven, Belgium
[6] Medical Imaging Research Center (MIRC), KU Leuven, Leuven, Belgium

Abstract. Anomaly detection (AD) is the identification of data samples that do not fit a learned data distribution. As such, AD systems can help physicians to determine the presence, severity, and extension of a pathology. Deep generative models, such as Generative Adversarial Networks (GANs), can be exploited to capture anatomical variability. Consequently, any outlier (i.e., sample falling outside of the learned distribution) can be detected as an abnormality in an unsupervised fashion. By using this method, we can not only detect expected or known lesions, but we can even unveil previously unrecognized biomarkers. To the best of our knowledge, this study exemplifies the first AD approach that can efficiently handle volumetric data and detect 3D brain anomalies in one single model. Our proposal is a volumetric and high-detail extension of the 2D f-AnoGAN model obtained by combining a state-of-the-art 3D GAN with refinement training steps. In experiments using non-contrast computed tomography images from traumatic brain injury (TBI) patients, the model detects and localizes TBI abnormalities with an area under the ROC curve of ∼75%. Moreover, we test the potential of the method for detecting other anomalies such as low quality images, pre-processing inaccuracies, artifacts, and even the presence of post-operative signs (such as a craniectomy or a brain shunt). The method has potential

CENTER-TBI participants and investigators are listed at the end of the supplementary material.

Electronic supplementary material The online version of this chapter (https://doi.org/10.1007/978-3-030-72084-1_13) contains supplementary material, which is available to authorized users.

A. Crimi and S. Bakas (Eds.): BrainLes 2020, LNCS 12658, pp. 133–142, 2021.
https://doi.org/10.1007/978-3-030-72084-1_13

for rapidly labeling abnormalities in massive imaging datasets, as well as identifying new biomarkers.

Keywords: Unsupervised learning · Anomaly detection · Deep generative networks · 3D GAN · Biomarker discovery

1   Introduction

Supervised deep learning techniques have shown outstanding performance in a wide diversity of medical imaging tasks, and can even outperform radiologists in areas such as lung cancer detection [2] or breast tumor identification [24]. However, these techniques require large annotated databases, which are expensive and time-consuming to obtain [23]. Furthermore, manual annotations often are disease-specific and do not always cover the wide range of abnormalities that can be present in a scan [1,22]. In contrast, *unsupervised* learning models are capable of discovering patterns from label-free databases. A current challenge in this field is unsupervised *anomaly detection (AD)*. AD is the task of identifying test data that does not fit the data distribution seen during training [21]. In clinical practice, AD represents a crucial step. Physicians learn the normal anatomical variability and they recognize anomalies by implicitly comparing to normal cases or healthy surrounding areas. As such, many AD models identify abnormalities in an unconstrained fashion by mimicking this human behavior.

State of the Art. Deep generative models, such as Variational Auto-encoders (VAEs) [13] and Generative Adversarial Networks (GANs) [8], are able to generate synthetic images that capture the variability of the training images. Thus, if a deep generative model is trained over lesion-free data, anomalies could be discovered by detecting samples that do not fit this lesion-free variability. For AD in retina images, Schlegl et al. [21] suggest that a GAN trained on healthy images should not be able to reconstruct abnormalities. In that work, a slow iterative optimization algorithm is used to find the GAN's latent space projection of a given image. To make this mapping technique faster, Schlegl et al. [20] propose f-AnoGAN, which replaces the iterative algorithm with an encoder network. In brain imaging, most recent AD work has focused on 2D axial images. Baur et al. [5] use a combination of a spatial VAE and an adversarial network for delineating multiple sclerosis lesions in MR images. You et al. [25] detect brain tumors using a Gaussian Mixture VAE with restoration of the latent space, while Pawlowski et al. [18] use Bayesian Auto-encoders to detect traumatic brain injury lesions. In a very recent comparative study on brain AD, the performance of f-AnoGAN is remarkable in diverse datasets [4]. All of these 2D-based approaches have several drawbacks: i) they do not consider volumetric information and, consequently, they do not effectively handle the complex brain anatomy; ii) they have to consider the whole brain image since there is no prior information of the anomaly localization; iii) they require multiple models for evaluating an entire scan.

Contributions of This Work. We propose, to the best of our knowledge, the first 3D brain anomaly detector. This model effectively handles complex brain structures and provides reliable 3D reconstructions based on brain anatomy. The present work is inspired by the 2D f-AnoGAN architecture. However, the proposed methodology differs in several aspects from f-AnoGAN: i) the network learns from *volumetric information* creating 3D image reconstructions; ii) the architecture is enhanced by using a modified version of a state-of-the-art 3D GAN; iii) a new training step is proposed to deal with the lack of details in reconstruction images. We show the AD capability of our proposed method in two independent traumatic brain injury (TBI) datasets. Besides, we evaluate its potential for AD in postsurgical cases and poor quality scans.

2   Methods

Database. For devising and validating the approach, we use non-contrast computed tomography (NCCT) data of traumatic brain injury patients. TBI includes a vast spectrum of pathoanatomical anomalies that may affect any brain region. Two independent datasets are used for our experiments:

* **CENTER-TBI.** The collaborative European NeuroTrauma Effectiveness Research in Traumatic Brain Injury (CENTER-TBI) project include a database collection of NCCT images [16]. The study protocol was approved by the national and local ethics committees for each participating center. Informed consent, including use of data for other research purposes, was obtained in each subject according to local regulations. Patient data was de-identified and coded by means of a Global Unique Patient Identifier. In this multi-center, multi-scanner, longitudinal study, all the NCCT images of TBI patients were visually reviewed and the abnormal findings were reported in a structured way by an expert panel. We retrieve a selection of images from a centralized imaging repository that stores the data collected and sent by the different sites. This dataset includes brain images without NCCT abnormal findings by expert review ($n = 637$ total scans) and manually annotated TBI scans ($n = 102$) with abnormal NCCT findings.
* **PhysioNet.** The model is also tested on the publicly available database, online at the PhysioNet repository[1] [7,10,11]. This dataset includes 37 subjects without NCCT abnormal findings and 33 TBI patients.

The training of the model is performed over $\sim$80% ($n = 532$) of the CENTER-TBI data without abnormal NCCT findings. As test sets we use CENTER-TBI (remaining 20%, $n = 105$ and all TBI cases with abnormal findings) and PhysioNet (entire database).

Preprocessing. All scans undergo the following preprocessing steps:

1. NCCT images are registered to the MNI space with an affine transformation.
2. An automatic quality control process is performed using the FDA approved **icobrain** TBI software [12]. Highly corrupted images are automatically discarded.

---

[1] https://physionet.org/content/ct-ich/1.3.1/.

3. Using the same software, a skull-stripping operation is performed.
4. Black boundaries caused by the application of the brain mask are removed.
5. After a Gaussian smoothing, images are resized to $64 \times 64 \times 64$ using linear interpolation.
6. A soft tissue windowing $[-20, 100 \text{ HU}]$ is performed, similarly as in [17].
7. The images are globally min-max normalized between -1 and 1.

Training Strategy. As proposed in f-AnoGAN, the model framework is composed of a GAN and an encoder network. These networks are trained in a multi-step training strategy where brain images without NCCT abnormal findings are used. Then, these trained models are able to detect anomalies using an anomaly score. The training strategy is divided into the following training steps:

1. GAN Training. The GAN (Fig. 1) training is based on a competitive game between two networks: the *generator* network (G) and the *discriminator* network (D). During training, G maximizes the probability of D making a mistake, while D maximizes the probability of correctly predicting the real and generated samples. Equation 1 shows the objective function for parametrizing the model.

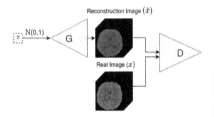

$$\min_{G} \max_{D} \mathbb{E}_{x \sim \mathbb{P}_r}[\log D(x)] + \mathbb{E}_{\tilde{x} \sim \mathbb{P}_g}[\log(1 - D(\tilde{x}))] \quad (1)$$

Equation 1: Original GAN loss function. Where $\mathbb{P}_r$ is the real data distribution and $\mathbb{P}_g$ is the model distribution defined by $\tilde{x} = G(z)$. The input z of G is sampled from a Gaussian distribution, $N(\mu = 0, \sigma^2 = 1)$.

Fig. 1. 3D GAN networks architecture.

The main challenge in 3D generation is the mode collapse problem [15]. Thus, we use Wasserstein-1 distance (also called Earth-Mover) [3] in our GAN loss. Moreover, the gradient penalty [9] is also included in order to increase training stability. The resulting discriminator and generator loss functions (respectively L_D and L_G) are hence as follows:

$$L_D = \mathbb{E}_{\tilde{x} \sim \mathbb{P}_g}[D(\tilde{x})] - \mathbb{E}_{x \sim \mathbb{P}_r}[D(x)] + \lambda \mathbb{E}_{\hat{x} \sim \mathbb{P}_{\hat{x}}}[(\|\nabla_{\hat{x}} D(\hat{x})\|_2 - 1)^2] \quad (2)$$
$$L_G = -\mathbb{E}_{\tilde{x} \sim \mathbb{P}_g}[D(\tilde{x})] \quad (3)$$

where $\mathbb{P}_{\hat{x}}$ is sampled uniformly along straight lines between a pair of points sampled from $\mathbb{P}_r$ and $\mathbb{P}_g$ and λ is a weighting parameter.

2. Encoder Training. Once the adversarial training is completed, G knows how to map from the latent space (z) to an image ($\tilde{x}$), $G(z) = z \rightarrow \tilde{x}$. However, the representation of a given image in the latent space is unknown. The *encoder* network (E) makes this mapping, $E(x) = x \to z$. As shown in Fig. 2, the weights of G and D remain frozen while only the weights of E are optimized. Results show that the E network exploits proper latent space representations and, therefore, G outcomes good reconstructions without requiring a forced constraint over z. The E network is optimized by minimizing L_E, a weighted sum of: *image space loss (L_{img}) and discriminator feature space loss (L_{feat})* (see Eq. 4–6). The use of this feature space is suggested by [21] and is inspired by the feature matching technique [19].

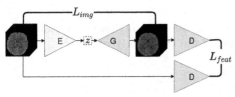

Fig. 2. Image space loss and discriminator feature space loss. Networks in blue do not change their weights during encoder training phase.

$$L_{img} = \frac{1}{n} \|x - \tilde{x}\|^2 \qquad (4)$$

$$L_{feat} = \frac{1}{m} \|f(x) - f(\tilde{x})\|^2 \qquad (5)$$

$$L_E = L_{img} + \kappa \cdot L_{feat} \qquad (6)$$

Equation (4–6): n is the number of voxels in the image, m is the dimensionality of the discriminator feature space, $f(x)$ the activation on the intermediate layer of D and κ is a weighting parameter.

3. Techniques for Improving The Performance. After preliminary experiments, a lack of details in the reconstructed images is observed. Therefore, we propose a new learning step that provides explicit learning feedback of the vast information that a 3D image contains. This extra training step provides a fine-tuning of the networks weights rather than a full model training from scratch. Empirical results show that optimizing the weights of E and G while minimizing $L_{Encoder}$ is the most convenient training strategy (see supplementary material Table S2).

Anomaly Score. The anomaly score quantifies the deviation of test images and corresponding reconstruction [20]. Note that the reconstructions are generated by considering only the distribution of the data used for training, i.e., NCCT images without any radiological findings. Therefore, this anomaly score can be interpreted as a distance metric between the input image and the learned anatomical variability. As presented in f-AnoGAN, the anomaly score for a given image is obtained using the function shown in Eq. 6. Thresholding the anomaly score provides a global classification of an image as abnormal or not. AD performance can thus be evaluated using a ROC analysis of the anomaly score.

Model Architecture. A state-of-the-art brain 3D GAN architecture [15] is used as foundation for the AD model. We add a hyperbolic tangent activation in the last encoder. We also increase the original latent space dimension to 2500. Architecture details are shown in Fig. 3.

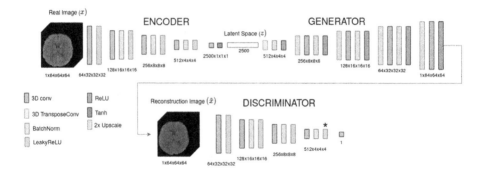

Fig. 3. Architecture details. Dimensions of each feature maps are shown for each block. The asterisk (*) denotes the intermediate layer of the D used in $f(x)$.

3   Results and Discussion

Comparison with 2D f-AnoGAN. Axial slices within a middle brain range (ensuring similar anatomical structures) are randomly selected to train a 2D model. In order to have a fair comparison in terms of AD of TBI abnormalities, images with abnormalities that are not located in the selected slice are discarded. In our experiments, the 3D model outperforms the 2D one by 4% in the area under the ROC curve.

AD Performance. In Fig. 4 we show a comparison of ROC curves for the different TBI datasets. The AD performance reaches $\sim 75\%$ area under the ROC in both databases. At the Youden index of the ROC curve over the combined datasets, the model has a 70.75% of accuracy, 54.07% of recall, and 86.66% of specificity. Another operating point could be chosen, depending on the clinically desired balance between sensitivity and specificity.

AD Performance by Lesion Type. Subjects from both datasets with at least one of the following hematomas are used to evaluate the performance of the model: *epidural*, *subdural*, and *intraparenchymal*. If the model detects an anomalous case having one of these lesions, this is counted as a detection, no matter if other lesions are also present. The model performance is similar across lesion types, performing slightly better for subdural hematomas (see Fig. 5).

Qualitative Results and Anomaly Localization. The method is able to localize anomalies through a voxel-wise subtraction of the original image x and its reconstruction $\tilde{x}$. Figure 6 exemplifies the most common cases in anomaly localization. *i) Case without abnormal findings:* In the second row of the figure, we can appreciate reliable reconstructions of the input images. No relevant region can be considered as a lesion in the voxel-wise error image (third row). *ii) Undetected TBI lesions:* Tiny lesions can be missed inside the anatomical variability, so the reconstruction image matches with the original one, making the TBI lesion hardly detectable. *iii) Detected TBI lesions:* If lesions fall outside the learned

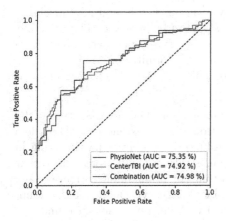
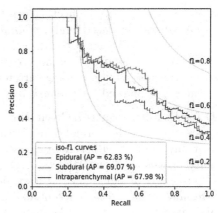

Fig. 4. Comparison of ROC curves for the different datasets. AUC: Area under the ROC curve.

Fig. 5. Comparison of Precision Recall curve for each TBI lesion. AP: Average precision.

distribution, the model will not be able to reconstruct this region and it will select the closest representation that has been learned, which could be thought of as a *healthy* brain representation. Hence, lesions are well localized (see blue arrows). Refer to supplementary material Fig. S1-S3 to visualize examples in different anatomical planes.

AD for Biomarker Discovery. The proposed unsupervised learning model is capable of detecting unknown/unlabeled abnormalities. Figure 6-c shows an epidural hematoma that has been labeled. It can be noticed that the lesion

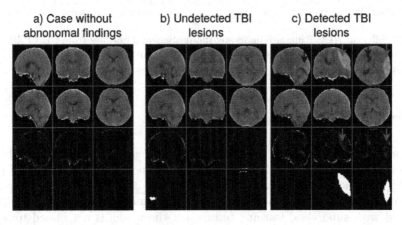

Fig. 6. Anomalous region localization using voxel-level error. First row: original image after preprocessing (x). Second row: reconstruction image ($\tilde{x}$). Third row: voxel-level error image. Fourth row: ground truth lesion segmentation. Arrows indicate the anomalies detected by the model; the blue ones show a labeled anomaly in the database while orange ones show an unlabeled anomaly. (Color figure online)

introduces a *mass effect* affecting nearby structures: the lateral ventricles are compressed and displaced, and a *midline shift* can be observed. The mass effect is not labeled in the dataset and, hence, no supervised approach would detect it. However, the proposed method overcomes this limitation, highlighting and locating this anomaly (see orange arrows). This property of detecting unlabeled anomalies can be used for *biomarker discovery*.

AD for Quality Control. Given that an anomaly is defined as any type of data unrepresented by the normal data distribution, we can extend our AD model to detect any kind of *outlier* sample. We evaluate its potential for detecting low quality images (such as artifacts, wrong registrations, and wrong orientations) and post-surgical signs (such as a craniectomy or brain shunt). Figure 7 shows the results of this proof of concept application. Images with anomalies have much higher anomaly scores than the distribution without any radiological findings.

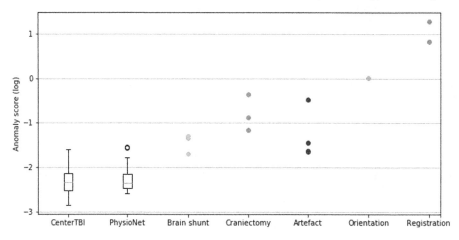

Fig. 7. Different NCCT images: images without abnormal findings from CENTER-TBI and PhysioNet, followed by various post-surgical pathologies and low quality images. Logarithm of the anomaly scores are used for better visualization.

4   Conclusion

The present model is, to our knowledge, the first feasible attempt for 3D brain AD screening in a real-world scenario. We overcome several limitations of previous approaches: i) our model handles volumetric information; ii) it is capable of detecting a wide variety of abnormalities in an unconstrained manner; iii) in contrast with supervised learning techniques, the model is not biased towards expert annotations; iv) the model offers good generalization capabilities, providing database-invariant anomaly scores; v) the voxel-wise error image localizes abnormalities, increasing the model interpretability.

In our experiments, the GPU memory limited the input data resolution. As future perspectives, we consider working with higher resolution images

($>64 \times 64 \times 64$), which would help to detect small anomalies and, hence, improve the model performance. This improvement will reduce the difference between unsupervised TBI detection performance and supervised learning models such as [6,14]. Besides, the voxel-error image could be extended to perform 3D anomaly segmentation (i.e., generating anomaly masks). Also, we want to extend our model to work with MR images, which is much more challenging than working with NCCT. In MR, the trained GAN should capture both the anatomical brain variability and the intrinsic MR scans variability. In addition, the model could be improved by taking into consideration demographic variables such as age.

Acknowledgments. JSV received an Erasmus+ scholarship from the Erasmus Mundus Joint Master Degree in Medical Imaging and Applications (MAIA), a programme funded by the Erasmus+ programme of the European Union (EU grant 20152491). This project received funding from the European Union's Horizon 2020 research and innovation program under the Marie Sklodowska-Curie grant agreement TRABIT No 765148. DR is supported by an innovation mandate of Flanders Innovation & Entrepreneurship (VLAIO). Data used in preparation of this manuscript were obtained in the context of CENTER-TBI, a large collaborative project with the support of the European Union 7th Framework program (EC grant 602150). Additional funding was obtained from the Hannelore Kohl Stiftung (Germany), from OneMind (USA) and from Integra LifeSciences Corporation (USA).

We thank Charlotte Timmermans and Nathan Vanalken for performing the manual TBI segmentations.

References

1. Alaverdyan, Z., Jung, J., Bouet, R., Lartizien, C.: Regularized siamese neural network for unsupervised outlier detection on brain multiparametric magnetic resonance imaging: application to epilepsy lesion screening. Med. Image Anal. **60**, 101618 (2020)
2. Ardila, D., et al.: End-to-end lung cancer screening with three-dimensional deep learning on low-dose chest computed tomography. Nat. Med. **25**(6), 954–961 (2019)
3. Arjovsky, M., Chintala, S., Bottou, L.: Wasserstein generative adversarial networks. In: Proceedings of the 34th International Conference on Machine Learning, vol. 70, pp. 214–223. PMLR (2017)
4. Baur, C., Denner, S., Wiestler, B., Albarqouni, S., Navab, N.: Autoencoders for unsupervised anomaly segmentation in brain MR images: a comparative study. arXiv preprint arXiv:2004.03271 (2020)
5. Baur, C., Wiestler, B., Albarqouni, S., Navab, N.: Deep autoencoding models for unsupervised anomaly segmentation in brain MR images. In: Crimi, A., Bakas, S., Kuijf, H., Keyvan, F., Reyes, M., van Walsum, T. (eds.) BrainLes 2018. LNCS, vol. 11383, pp. 161–169. Springer, Cham (2019). https://doi.org/10.1007/978-3-030-11723-8_16
6. Chilamkurthy, S., et al.: Deep learning algorithms for detection of critical findings in head CT scans: a retrospective study. Lancet **392**(10162), 2388–2396 (2018)
7. Goldberger, A.L., et al.: PhysioBank, PhysioToolkit, and PhysioNet: components of a new research resource for complex physiologic signals. Circulation **101**(23), e215–e220 (2000)

8. Goodfellow, I., et al.: Generative adversarial nets. In: Advances in Neural Information Processing Systems, pp. 2672–2680 (2014)
9. Gulrajani, I., Ahmed, F., Arjovsky, M., Dumoulin, V., Courville, A.C.: Improved training of Wasserstein GANs. In: Advances in Neural Information Processing Systems, pp. 5767–5777 (2017)
10. Hssayeni, M.D.: Computed tomography images for intracranial hemorrhage detection and segmentation (version 1.3.1). PhysioNet (2020). https://doi.org/10.13026/4nae-zg36
11. Hssayeni, M.D., Croock, M.S., Salman, A.D., Al-khafaji, H.F., Yahya, Z.A., Ghoraani, B.: Intracranial hemorrhage segmentation using a deep convolutional model. Data **5**(1), 14 (2020)
12. Jain, S., et al.: Automatic quantification of Computed Tomography features in acute traumatic brain injury. J. Neurotrauma **36**(11), 1794–1803 (2019)
13. Kingma, D.P., Welling, M.: Auto-encoding variational bayes. arXiv preprint arXiv:1312.6114 (2013)
14. Kuo, W., Häne, C., Mukherjee, P., Malik, J., Yuh, E.L.: Expert-level detection of acute intracranial hemorrhage on head computed tomography using deep learning. Proc. Natl. Acad. Sci. **116**(45), 22737–22745 (2019)
15. Kwon, G., Han, C., Kim, D.: Generation of 3D brain MRI using auto-encoding generative adversarial networks. In: Shen, D., et al. (eds.) MICCAI 2019. LNCS, vol. 11766, pp. 118–126. Springer, Cham (2019). https://doi.org/10.1007/978-3-030-32248-9_14
16. Maas, A.I., et al.: Collaborative european neurotrauma effectiveness research in traumatic brain injury (CENTER-TBI) a prospective longitudinal observational study. Neurosurgery **76**(1), 67–80 (2015)
17. Monteiro, M., et al.: TBI lesion segmentation in head CT: impact of preprocessing and data augmentation. In: Crimi, A., Bakas, S. (eds.) BrainLes 2019. LNCS, vol. 11992, pp. 13–22. Springer, Cham (2020). https://doi.org/10.1007/978-3-030-46640-4_2
18. Pawlowski, N., et al.: Unsupervised lesion detection in brain CT using Bayesian convolutional autoencoders. In: Medical Imaging with Deep Learning (2018)
19. Salimans, T., Goodfellow, I., Zaremba, W., Cheung, V., Radford, A., Chen, X.: Improved techniques for training GANs. In: Advances in Neural Information Processing Systems, pp. 2234–2242 (2016)
20. Schlegl, T., Seeböck, P., Waldstein, S.M., Langs, G., Schmidt-Erfurth, U.: f-AnoGAN: fast unsupervised anomaly detection with generative adversarial networks. Med. Image Anal. **54**, 30–44 (2019)
21. Schlegl, T., Seeböck, P., Waldstein, S.M., Schmidt-Erfurth, U., Langs, G.: Unsupervised anomaly detection with generative adversarial networks to guide marker discovery. In: Niethammer, M. (ed.) IPMI 2017. LNCS, vol. 10265, pp. 146–157. Springer, Cham (2017). https://doi.org/10.1007/978-3-319-59050-9_12
22. Seeböck, P., et al.: Exploiting epistemic uncertainty of anatomy segmentation for anomaly detection in retinal OCT. IEEE Trans. Med. Imaging **39**(1), 87–98 (2019)
23. Seeböck, P., et al.: Unsupervised identification of disease marker candidates in retinal OCT imaging data. IEEE Trans. Med. Imaging **38**(4), 1037–1047 (2018)
24. Stower, H.: AI for breast-cancer screening. Nat. Med. **26**(2), 163 (2020)
25. You, S., Tezcan, K.C., Chen, X., Konukoglu, E.: Unsupervised lesion detection via image restoration with a normative prior. In: Medical Imaging with Deep Learning, vol. 102, pp. 540–556. PMLR (2019)

Assessing Lesion Segmentation Bias of Neural Networks on Motion Corrupted Brain MRI

Tejas Sudharshan Mathai[1]([✉]), Yi Wang[1,2], and Nathan Cross[2]

[1] Philips Healthcare, Bothell, WA 98201, USA
[2] Department of Radiology, University of Washington, Seattle, WA 98195, USA

Abstract. Patient motion during the magnetic resonance imaging (MRI) acquisition process results in motion artifacts, which limits the ability of radiologists to provide a quantitative assessment of a condition visualized. Often times, radiologists either "see through" the artifacts with reduced diagnostic confidence, or the MR scans are rejected and patients are asked to be recalled and re-scanned. Presently, there are many published approaches that focus on MRI artifact detection and correction. However, the key question of the bias exhibited by these algorithms on motion corrupted MRI images is still unanswered. In this paper, we seek to quantify the bias in terms of the impact that different levels of motion artifacts have on the performance of neural networks engaged in a lesion segmentation task. Additionally, we explore the effect of a different learning strategy, curriculum learning, on the segmentation performance. Our results suggest that a network trained using curriculum learning is effective at compensating for different levels of motion artifacts, and improved the segmentation performance by ~9%–15% ($p < 0.05$) when compared against a conventional shuffled learning strategy on the same motion data. Within each motion category, it either improved or maintained the dice score. To the best of our knowledge, we are the first to quantitatively assess the segmentation bias on various levels of motion artifacts present in a brain MRI image.

Keywords: MRI · Motion · Segmentation · Deep learning · Bias

1 Introduction

Gliomas are a family of neoplasms of the brain that includes the devastating and most common tumor of the brain, glioblastoma [1–4]. The most aggressive types of gliomas associated with low patient survival rates are the high grade gliomas (HGG), such as glioblastoma, which are extremely infiltrative, spreading extensively through surrounding tissue. Surgery usually focuses on areas of highest grade and debulking of tumor to improve short term symptoms and quality of life

---

T. S. Mathai and Y. Wang—Equal contribution.

A. Crimi and S. Bakas (Eds.): BrainLes 2020, LNCS 12658, pp. 143–156, 2021.
https://doi.org/10.1007/978-3-030-72084-1_14

[1,4]. Magnetic Resonance Imaging (MRI) is the standard approach to diagnose and follow gliomas [2] particularly over the course of their management. Unfortunately, imaging of gliomas, and HGG in particular, is limited by their varied and similar appearance to other neoplasms, irregular shapes, and heterogeneous histologic grade throughout their volume making it difficult to characterize and quantify [4–10]. Different contrasts, such as T1, T1 with contrast (T1ce), T2, and Fluid Attenuation Inversion Recovery (FLAIR), are usually acquired to visualize the tissue properties and the extent of the tumor [4]. Furthermore, varying MRI scanners and exam protocols are used at different institutions to acquire these scans [4,6]. Given the morbidity surrounding HGGs, there is increasing interest in quantitatively describing these lesions by identifying and segmenting the tumor and its sub-components in MRI images.

Further confounding analysis is patient motion, the predominant source of image degradation in clinical practice affecting 10%–42% of brain exams [12, 17]. Degradation of the MR images can potentially be identified at the time of exam, but this would require almost 20% of all MR exams to be repeated [12]. In some cases, the entire MR exam can be rejected and the patient will need to be recalled and re-scanned, resulting in tremendous financial and time costs to the healthcare provider [12]. Moreover, radiologists usually attempt to "see-through" [17] motion artifacts to diagnose the underlying condition, but if the quality of the MR image is severely corrupted, the accurate localization of the tumor boundary is hindered. Many techniques have been proposed in a plethora of prior work to segment brain MR images [4], but convolutional neural network approaches are the state of the art in this domain [4–11]. Similarly, many schemes have also been proposed for prospective or retrospective [13–16] motion correction. Deep learning-based approaches for motion correction [17–21] have also shown promise in improving the image quality. Some approaches have enabled the segmentation performance of a network to improve post-correction [21]. We focus on the deep learning approaches in this paper.

The Multimodal Brain Tumor Segmentation Challenge [4] has been predominantly used by prior deep learning algorithms, and it provides a dataset of multi-sequence, pre- and post-operative, HGG and low grade gliomas (LGG), neuroradiologist-segmented brain MRIs. These prior lesion segmentation methods utilize all available MR sequences (e.g. T1, T2, T1ce, FLAIR), but in clinical practice, radiologists focus on the slices through the mass of a few sequences (e.g. T1ce or FLAIR) for sizing of the tumor and refer to other sequences for confirmation. Prior research has shown that the addition of a small amount of Gaussian noise to the training data degraded the segmentation performance of the top performing models from the BraTS challenge by 10%–15% [10]. Attempts to make robust motion insensitive models have augmented the training set with images that incorporate simulated motion [7,21]. But, they do not quantify the algorithmic performance across different severities of motion corruption: minimal, mild, moderate, severe. Thus, the impact and segmentation performance relative to the spectrum of motion artifacts is not characterized.

In this paper, we assess the bias caused by a gamut of simulated motion on the segmentation performance (in terms of dice score) of neural networks using the BraTS 2019 dataset. We use five previously published neural networks - UNet with Attention [26], UNet 2.5D [21], LSTM UNet [27], LSTM ResNet34 [28], and LSTM CorNet [29] and quantify their segmentation bias. We focus on the task of HGG segmentation in only T1ce sequences affected by different levels of motion corruption. We postulate that a curriculum learning [7,23] strategy, wherein an easier task (minimal) is first learned followed by progressively harder tasks (mild, moderate, severe), helps these networks converge faster and improves their segmentation performance. The intent is to evaluate and quantify segmentation performance of these algorithms relative to the severity of motion artifact, and show that an alternative learning strategy improves performance.

Contribution. 1) For a lesion segmentation task, we characterize the bias caused by different levels of motion artifacts on the segmentation performance of neural networks. 2) We propose a curriculum learning strategy as a way to overcome this bias and improve the segmentation performance.

2   Methods

Data. In this study, the 2019 Brain Tumor Segmentation challenge [4] dataset was used for experimentation. The dataset is split into training, validation, and testing sets. The training set contains 259 HGG and 76 LGG subjects whose corresponding ground truth labels were provided. The ground truth masks include four labels: 1 - necrotic (NCR) and non-enhancing (NET) tumor, 2 - edema (ED), 4 - enhancing tumor (ET), and 0 - normal tissue and background. Each case contained four MRI sequences: T1-weighted, T1-weighted contrast-enhanced (T1ce), T2-weighted and FLAIR. The cases were pre-processed before segmentation, including co-registration to an anatomical template, interpolation to an isotropic resolution ($1 \, \text{mm}^3$) and skull stripping. The resulting sequences had dimensions of $240 \times 240 \times 155$ pixels. In this study, we focus on the tumor core (labels 1 and 4) of the T1ce sequence, which is crucial to clinical tumor size estimation.

Motion Simulation. The BraTS dataset is not representative of a real clinical scenario where data is often limited, noisy, and labels are sparse or incorrect [10]. To mimic a real clinical setting, we utilized only the 259 HGG T1ce sequences from subjects in the training set, and divided them into four motion categories: minimal (64), mild (64), moderate (64), and severe (67). In each motion category, we further subdivided the data into train/validation/testing splits. For the first three motion categories, each split contained 38/6/20 cases, while the split for the severe motion category contained 40/6/21 cases. To study the bias introduced by motion, artifacts were simulated on T1ce sequences assuming a rigid body translation and rotation model as shown in Fig. 2. The severity of the artifact depended heavily on the timing and amplitude of the movement during MRI signal sampling. Motion was artificially introduced in the frequency domain (k-space): 1) Minimal (no changes to 64 cases), 2) Mild (± 2 px translation/$\pm 1°$ rotation applied on 64 cases), 3) Moderate (± 3 px translation/$\pm 2°$ rotation applied on 64 cases), 4) Severe (± 4 px translation/$\pm 3°$ rotation applied on 67 cases). The motion corrupted k-space data can be considered a linear combination of two versions of k-space: (1) k-space before motion and (2) k-space after simulated motion event (rotation, translation). When the combined k-space data was converted back as images, varying degrees of simulated motion artifact were evident. To more closely approximate the visual appearance of motion artifact seen in the clinical setting, a simulated skull was added back to the skull-stripped data which produced the typical ringing, blurring and ghosting appearances.

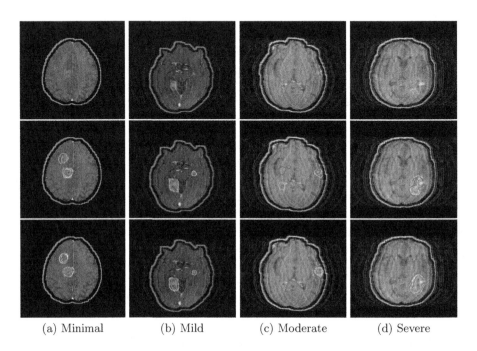

(a) Minimal   (b) Mild   (c) Moderate   (d) Severe

Fig. 1. Qualitative results of the five different networks on images from test datasets in the (a) minimal, (b) mild, (c) moderate and (d) severe motion categories respectively. The second and third rows show the segmentation results of the networks trained with the shuffled learning and curriculum learning strategies respectively. Magenta: ground truth label, Yellow: UNet Attention, Green: UNet 2.5D, Cyan: LSTM UNet, Orange: LSTM ResNet, Brown: LSTM CorNet. (Color figure online)

Baseline Comparisons. Currently, radiologists center their tumor sizing efforts over a few slices in particular sequences (e.g. T1ce or FLAIR) and refer to the other sequences for confirmation. As prior work has been focused on volumetric brain lesion segmentation, we paid attention to other popular models that utilized either 1- or 3- slice(s) for segmenting the lesion. We extend the prior work by testing the performance of five previously proposed neural networks: 1) UNet with attention [26], 2) UNet 2.5D [21], 3) LSTM UNet [27], 4) LSTM ResNet34

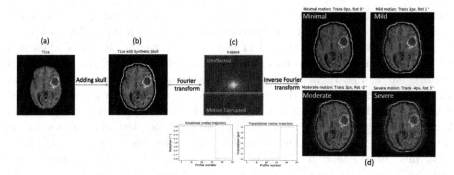

Fig. 2. Simulation of rigid body translation and rotation on T1ce images are shown. (a) Original T1ce, (b) T1ce with synthetic skull, (c) Combined k-space signal of unaffected k-space (blue) and motion corrupted k-space (yellow). Example motion trajectories are plotted as a function of k-space profile number. (d) T1ce with four levels of simulated motion artifacts: minimal, mild, moderate and severe. The motion simulation was applied only to the T1ce images, and the underlying ground truth labels were left unchanged. (Color figure online)

[28], and 5) LSTM CorNet [29]. These network architectures follow the basic encoder-decoder design [25]. UNet with attention [26] used attention gates to identify salient image regions and suppress noisy responses. The LSTM-based networks [27–29] incorporated a convLSTM decoder structure and other modifications, such as residual [30] and dilated convolutions [31] etc., to exploit the extracted contextual features. The UNet 2.5D [21] used three T1ce slices as the network input, while the other networks used a single T1ce slice.

Learning Strategy. Prior work on MR image segmentation has utilized a shuffled learning strategy during training that relied on randomly presenting an individual case to the network from the training set [4–6,9,10]. Within the context of motion corruption, this strategy does not account for the varying degrees of motion severity. Instead of a randomly shuffled dataset order, a curriculum-based strategy [23,32] was used in this work to present the network with easier datasets (e.g. minimal) first, and gradually feeding it with data of increasing motion severity (mild, moderate, severe). Only the order of datasets presented to the network was changed, and the time complexity of learning remained the same as with shuffled learning.

3   Experiments

Design. We quantitatively evaluated the lesion segmentation performance of each network using the BraTS 2019 data. The first experiment computed the upper limit of performance by using the networks to segment sequences without artifacts or simulated skull (blue bar in Fig. 3). Second, the networks were trained and tested on data where the skull was simulated (gold bar in Fig. 3). Third, the networks that were trained on artifact-free data containing the simulated skull were tested on artifact simulated data with the skull (green bar in Fig. 3). Fourth, the networks were trained on data containing the simulated skull and motion artifacts, and tested on data containing the skull and motion artifacts (orange bar in Fig. 3). In the aforementioned experiments, a shuffled learning strategy was used, i.e., all the training data fed to the network were shuffled with no regard to the motion severity. In the last experiment, a curriculum learning strategy was introduced during training time; datasets were fed to the networks during training in the order of increasing motion severity i.e., minimal, mild, moderate, severe. The training data contained both the simulated skull and motion artifacts, and the networks were tested on data containing the skull and motion artifacts (pink bar in Fig. 3).

Pre-processing. Each image in the training sequence was normalized [6,7,10] by subtracting its mean value and dividing by its standard deviation of the intensities within the brain region. After normalization, the images were padded to 256×256 pixels for training the network.

Data Augmentation. In addition to the motion simulation, heavy data augmentation was conducted in the form of flipping (horizontal and vertical), gamma adjustment, Gaussian noise addition, Gaussian blurring, Median blurring, Bilateral blurring, cropping, and affine transformations.

Training. A grid search across the parameter space was conducted for each network on a $4\times$ sub-sampled dataset to identify the best hyper-parameters given the input data. We trained each network for 30 epochs with the training being terminated early if the validation loss did not improve for 7 epochs. The networks were trained with the dice loss [28] as it has been experimentally proven to be less sensitive to the class imbalance problem. In most cases, grid search yielded chosen parameters that were the same as those proposed in the original papers with the exception of the batch size, which varied from 4–16 T1ce images per batch. The reader is referred to the original papers for full implementation details. The ADAM optimizer [33] was used for optimizing the loss function, and the final pixel level probabilities were classified using the softmax function. All experiments were run on a workstation running Ubuntu 16.04LTS, a NVIDIA Titan Xp GPU, and the average inference time on a test slice was 44.7 ms for UNet Attention, 49.1 ms for UNet 2.5D, 58.8 ms for LSTM UNet, 86.3 ms for LSTM ResNet34, and 96.2 ms for LSTM CorNet.

Post-processing. In contrast to the prior approaches, no post-processing was done on the predicted segmentation masks. The predicted labels were directly compared against the ground truth masks, and the dice score was computed.

4 Results and Discussion

Results Across All Motion Categories. As seen in Fig. 1, qualitative results are presented for all the models on test dataset images organized by motion category. Shuffled learning tends to under-segment the tumor (false negatives) and in some cases generates false positives, contributing to a lower dice score. The spread of the dice scores across the test datasets for the five networks are presented in Fig. 3, and the associated mean and standard deviations of the dice scores are listed in Table 1. From the experimental designs in Sect. 3, these networks have been trained with different learning strategies and data categories. Table 2 lists the p-values from statistical analysis using ANOVA that was conducted on dice scores between different learning strategies and motion categories for each network.

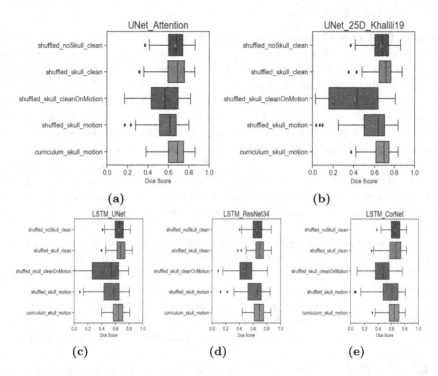

Fig. 3. Each plot displays the spread of the dice scores obtained using different learning strategies and data collections across all the testing set for the (a) UNet with attention model, (b) UNet 2.5D model, (c) LSTM UNet model, (d) LSTM ResNet34 model, (e) LSTM CorNet model. (Color figure online)

From Fig. 3, adding the simulated skull to the data did not change the segmentation performance when compared against data without the skull across the neural networks (blue vs gold bars, $p > 0.05$). However, the performance degraded significantly when a network trained with the shuffled learning strategy and clean data was tested on motion data (gold vs green bars, $p \ll 0.05$); e.g., the dice scores of the UNet Attention network dropped by >12% when the network trained on clean data was tested on motion corrupted data. This observation holds true across the four other networks, as shown in Table 1 and Fig. 3. Also consistent with prior work [21], a 23% mean improvement in performance across all networks was seen with incorporating motion data during training. Figure 3 also demonstrates that curriculum learning compensates for the motion corruption. From Table 1, the results indicate that the curriculum learning strategy provided a 9%–15% improvement when compared against the shuffled learning strategy on the same motion data (orange vs pink bars, $p < 0.05$). Across all five neural networks, the resulting segmentations from curriculum learning were comparable to the corresponding cases where the networks were trained on clean data (gold vs pink bars, $p > 0.05$) as seen from the dice scores in Table 1 and p-values in Table 2 (last row).

Results for Each Motion Category. Figure 4 shows the performance of networks trained with different learning strategies on the individual motion categories. For each network, the dice scores of networks trained with a shuffled learning strategy on clean data are also shown for reference. Table 3 shows the statistical significance of the two learning strategies on each motion category; following a test of normality, either a paired t-test or the Wilcoxon signed rank test was used. Figure 5 depicts the training loss and the validation loss as a function of epochs for the UNet Attention network: (a) shuffled learning strategy on clean data, (b) shuffled learning strategy on motion data, and (c) curriculum learning strategy on motion data. In contrast to the shuffled learning, curriculum learning exhibits a relatively stable convergence rate with a lower loss. Other observations that were made for each motion category include:

- **UNet Attention:** From Fig. 4(a) and Table 3, curriculum learning improved the segmentation performance across motion categories (p < 0.05).
- **UNet 2.5D:** From Fig. 4(b) and Table 3, curriculum learning improved the dice score significantly for the minimal, moderate and severe category (p < 0.05). No significant difference was found between the two learning strategies for the mild motion category (p = 0.257). Curriculum learning either improved or maintained the segmentation performance by yielding higher or comparable dice scores.
- **LSTM UNet:** As seen in Table 3, curriculum learning improved the segmentation performance across all motion categories, except for mild motion. However, it resulted in lower standard deviations in the mild category in contrast to shuffled learning as shown in Fig. 4(c).
- **LSTM ResNet34:** As seen in Fig. 4(d) and Table 3, significantly higher dice scores were obtained with curriculum learning for minimal and severe motion categories (p < 0.05), but there is no statistically significant difference with the other motion groups. Curriculum learning did provide comparable dice scores in the mild motion category, and a higher mean dice score and a lower standard deviation in the moderate category.
- **LSTM CorNet:** Curriculum learning improved the segmentation performance across all motion categories as shown in Fig. 4(e) and Table 3.

Table 1. Comparison of the dice score for each neural network trained with different learning strategies and data collections.

	UNet attention	UNet 2.5D	LSTM UNet	LSTM ResNet34	LSTM CorNet
Shuffled noSkull Clean	0.67 ± 0.1	0.67 ± 0.1	0.65 ± 0.09	0.66 ± 0.1	0.65 ± 0.09
Shuffled Skull Clean	0.66 ± 0.12	0.7 ± 0.1	0.67 ± 0.09	0.69 ± 0.09	0.64 ± 0.12
Shuffled Skull CleanOnMotion	0.55 ± 0.16	0.4 ± 0.24	0.46 ± 0.22	0.48 ± 0.17	0.46 ± 0.14
Shuffled Skull Motion	0.58 ± 0.13	0.58 ± 0.17	0.54 ± 0.16	0.62 ± 0.15	0.56 ± 0.17
Curriculum Skull Motion	0.67 ± 0.1	0.67 ± 0.1	0.64 ± 0.1	0.68 ± 0.09	0.63 ± 0.11

Table 2. Statistical significance of networks trained with different learning strategies across all motion categories at a 5% significance level. s = shuffled, c = curriculum, S = Skull, C = Clean, M = Motion, n = no

			UNet attention	UNet 2.5D	LSTM UNet	LSTM ResNet34	LSTM CorNet
s_{nSC}	vs.	s_{SC}	0.665	0.068	0.211	0.058	0.724
s_{SC}	vs.	s_{SM}	0.0001	0.0001	0.0007	0.001	0.001
s_{SM}	vs.	c_{SM}	0.0001	0.0001	0.0001	0.001	0.004
s_{SC}	vs.	c_{SM}	0.525	0.125	0.059	0.688	0.508

Table 3. Statistical comparison of the learning strategies (shuffled vs. curriculum) used for training different networks on each motion category with $\alpha = 0.05$.

	UNet attention	UNet 2.5D	LSTM UNet	LSTM ResNet34	LSTM CorNet
Minimal	0.0005	4.91e−5	0.0002	71e−5	0.0029
Mild	8.81e−6	0.257	0.232	0.473	0.0012
Moderate	8.41e−8	0.007	0.003	0.056	0.0045
Severe	6.81e−5	0.012	6.81e−5	0.038	0.0003

Discussion. From Fig. 3 and Tables 1 and 2, when contrasted against curriculum learning, shuffled learning on motion corrupted data resulted in significantly inferior segmentation performance. Potential reasons include an unbalanced set of data for training (64 cases without added motion; 195 cases with added motion), and weight oscillation during training. From Fig. 5, curriculum learning overcame the weight oscillations [23] with the data being fed to a network in the order of increasing motion corruption. Overall, the curriculum learning strategy helped to improve the segmentation dice score for all the neural networks that were tested. An in-depth analysis suggested that curriculum learning either improved or maintained the segmentation dice score across the different motion categories for all the networks. While the least improvements in dice scores were seen with the mild motion category, improvements were visualized in the other (minimal, moderate and severe) motion categories across all the networks tested. Our results suggest that the curriculum learning is most beneficial to the UNet with Attention model.

In general, there is a tendency of curriculum learning to overestimate the volume of the mass as illustrated in Fig. 1. While a T1ce image was corrupted by simulating motion using rotations, translations, blurring, and ringing artifacts,

Fig. 4. Dice scores for the (a) UNet with attention, (b) UNet 2.5D, (c) LSTM UNet, (d) LSTM ResNet34, (e) LSTM CorNet across the various motion categories. The first two error bars show the shuffled learning strategy trained with/without motion corrupted data respectively, and the last error bar displays the curriculum learning strategy trained with motion corrupted data. Curriculum learning either enhanced or maintained the segmentation performance across the different motion categories.

the underlying ground truth label was not altered in the same way; due to MR physics, applying the same motion simulation to the ground truth label would not ascertain a 1-1 correspondence with the image. Thus, there was a slight mismatch between the motion corrupted image and its corresponding ground truth label. As a network is fed these images during training and minimizes the dice loss function, it has to consider the label mismatch. Therefore, it accounted for the motion and over-segmented the lesion by generating false positives, such that the predicted labels are slightly larger than the underlying ground truth. This caused a drop in the dice scores. Moreover, the dice scores from our study are not as high as those from the BraTS challenge studies, possibly due to training being done

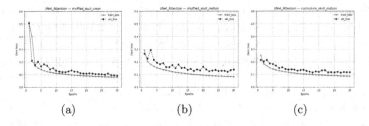

(a)   (b)   (c)

Fig. 5. Comparison of the training loss (red) and validation loss (blue) for the UNet architecture with attention mechanism with (a) shuffled learning strategy on clean data, (b) shuffled learning strategy on motion data, and (c) curriculum learning strategy on motion data. (Color figure online)

on a single image contrast with only 1/3 slice(s). Furthermore, no additional post-processing methods were performed to rectify the predicted segmentation, which has been shown to significantly enhance the dice scores [4–10]. We would like to point out that the main purpose of this paper was not to achieve the highest segmentation performance. Instead, these results help to quantify the effects of motion corruption, which is commonly encountered in clinical imaging, on the segmentation performance of a variety of common deep learning models.

Our observation suggests that for a network trained on motion corrupted data for the task of lesion segmentation and not motion correction, it might be necessary to shrink the prediction while interpreting clinical cases. Potentially, conventional image analysis techniques could be used to erode the labels further down to the lesion margin. While human vision is different and may be tempered by a variety of other information than just the pixels on the screen, our work provides some insight into the effect of motion artifacts on the radiologist interpretations of lesion size. 3D volumetric techniques have been shown to more accurately assess changes in glioblastoma [34], but it is rarely performed due to the cumbersome segmentation process and limited tools. Furthermore, as radiologists size lesions often over the course of a busy clinical day, there is a need for accurate 2D/2.5D automated segmentation techniques. If automated techniques are to be applied to generate volumetric measurements of lesions, our results show that models will need adequate experience with motion corrupted data. Since motion is an ubiquitous issue in clinical imaging, it will also behoove those analyzing model performance to ensure that results are robust to motion artifacts, or at least identify it as a factor and fail gracefully.

5   Conclusion and Future Work

Volumetric assessment of neoplasms are more accurate and clinically relevant. However, in a busy clinical practice, this is rarely performed due to limited tools and the tedious mass segmentation process. Automated tools can fill this need, but they need to be robust to the commonplace artifacts degrading clinical imaging, such as the ubiquitous motion artifact. In this paper, we analyzed the performance of neural networks tasked with segmenting lesions in motion corrupted brain MRI images, and quantified the bias exhibited by them on individual motion categories. We also explored a different learning strategy in curriculum learning, and demonstrated that it is effective in compensating for the effects of simulated motion. It either improved or maintained the segmentation performance across all five networks and four motion severity categories. To the best of our knowledge, this is the first quantitative assessment of segmentation bias from motion artifacts on brain MRI tumor measurement. In the future, we plan to correct for the motion and segment the MR image in a multi-task fashion, and explore curriculum learning further with bootstrapping.

Acknowledgements. We would like to thank Karsten Sommer, Axel Saalbach, Ekta Walia, Chris Martel, Prashanth Pai, and Shawn Stapleton for their comments during discussions related to this work.

References

1. Marsh, J., et al.: Current status of immunotherapy and gene therapy for high-grade gliomas. J Clin. Oncol. **20**(1), 43–48 (2013)
2. Wen, P., et al.: Updated response assessment criteria for high-grade gliomas: response assessment in neuro-oncology working group. J Clin. Oncol. **28**(11), 1963–1972 (2010)
3. Mazzara, G., et al.: Brain tumor target volume determination for radiation treatment planning through automated MRI segmentation. Int. J Rad. Oncol. Biol. Phys. **59**(1), 300–312 (2004)
4. Bakas, S., et al.: Identifying the best machine learning algorithms for brain tumor segmentation, progression assessment, and overall survival prediction in the BRATS challenge. arXiv (2018)
5. Myronenko, A.: 3D MRI brain tumor segmentation using autoencoder regularization. In: Crimi, A., Bakas, S., Kuijf, H., Keyvan, F., Reyes, M., van Walsum, T. (eds.) BrainLes 2018. LNCS, vol. 11384, pp. 311–320. Springer, Cham (2019). https://doi.org/10.1007/978-3-030-11726-9_28
6. Isensee, F., Kickingereder, P., Wick, W., Bendszus, M., Maier-Hein, K.H.: No new-net. In: Crimi, A., Bakas, S., Kuijf, H., Keyvan, F., Reyes, M., van Walsum, T. (eds.) BrainLes 2018. LNCS, vol. 11384, pp. 234–244. Springer, Cham (2019). https://doi.org/10.1007/978-3-030-11726-9_21
7. Zhou, C., Chen, S., Ding, C., Tao, D.: Learning contextual and attentive information for brain tumor segmentation. In: Crimi, A., Bakas, S., Kuijf, H., Keyvan, F., Reyes, M., van Walsum, T. (eds.) BrainLes 2018. LNCS, vol. 11384, pp. 497–507. Springer, Cham (2019). https://doi.org/10.1007/978-3-030-11726-9_44
8. Zhou, C., Ding, C., Lu, Z., Wang, X., Tao, D.: One-pass multi-task convolutional neural networks for efficient brain tumor segmentation. In: Frangi, A.F., Schnabel, J.A., Davatzikos, C., Alberola-López, C., Fichtinger, G. (eds.) MICCAI 2018. LNCS, vol. 11072, pp. 637–645. Springer, Cham (2018). https://doi.org/10.1007/978-3-030-00931-1_73
9. McKinley, R., Meier, R., Wiest, R.: Ensembles of densely-connected CNNs with label-uncertainty for brain tumor segmentation. In: Crimi, A., Bakas, S., Kuijf, H., Keyvan, F., Reyes, M., van Walsum, T. (eds.) BrainLes 2018. LNCS, vol. 11384, pp. 456–465. Springer, Cham (2019). https://doi.org/10.1007/978-3-030-11726-9_40

10. Muller, S., et al.: Robustness of brain tumor segmentation. arXiv (2019)
11. Hu, X., et al.: Brain SegNet: 3D local refinement network for brain lesion segmentation. BMC Med. Imaging **20**, 17 (2020)
12. Andre, J., et al.: Toward quantifying the prevalence, severity, and cost associated with patient motion during clinical MR examinations. J. Am. Coll. Radiol. **12**, 689–695 (2015)
13. Ooi, M., et al.: Prospective real-time correction for arbitrary head motion using active markers. Mag. Res. Med. **62**(4), 943–954 (2009)
14. Pipe, J., et al.: Motion correction with PROPELLER MRI: application to head motion and free-breathing cardiac imaging. Mag. Res. Med. **42**, 963–969 (1999)
15. Kober, T., et al.: Head motion detection using FID navigators. Mag. Res. Med. **66**(1), 135–143 (2011)
16. Godenschweger, F., et al.: Motion correction in MRI of the brain. Phys. Med. Biol. **61**(5), R32 (2018)
17. Sommer, K., et al.: Correction of motion artifacts using a multiscale fully convolutional neural network. AJNR **41**, 416–423 (2020)
18. Sommer, K., et al.: Correction of motion artifacts using a multi-resolution fully convolutional neural network. In: ISMRM (2018)
19. Duffy, B., et al.: Retrospective correction of motion artifact affected structural MRI images using deep learning of simulated motion. In: MIDL (2018)
20. Pawar, K., et al.: Motion correction in MRI using deep convolutional neural network. In: ISMRM (2018)
21. Khalili, N., Turk, E., Zreik, M., Viergever, M.A., Benders, M.J.N.L., Išgum, I.: Generative adversarial network for segmentation of motion affected neonatal brain MRI. In: Shen, D., et al. (eds.) MICCAI 2019. LNCS, vol. 11766, pp. 320–328. Springer, Cham (2019). https://doi.org/10.1007/978-3-030-32248-9_36
22. Shaw, R., et al.: A k-space model of movement artefacts: application to segmentation augmentation and artefact removal. IEEE Trans. Med. Imaging **39**, 2881–2892 (2020)
23. Bengio, Y., et al.: Curriculum learning. In: ICML, pp. 41–48 (2009)
24. Chen, C., et al.: Realistic adversarial data augmentation for MR image segmentation. arXiv (2020)
25. Ronneberger, O., Fischer, P., Brox, T.: U-net: convolutional networks for biomedical image segmentation. In: Navab, N., Hornegger, J., Wells, W.M., Frangi, A.F. (eds.) MICCAI 2015. LNCS, vol. 9351, pp. 234–241. Springer, Cham (2015). https://doi.org/10.1007/978-3-319-24574-4_28
26. Oktay, O., et al.: Attention U-net: learning where to look for the pancreas. In: MIDL (2018)
27. Arbelle, S., et al.: Microscopy cell segmentation via convolutional LSTM networks. In: IEEE ISBI, pp. 1008–1012 (2019)
28. Milletari, F., Rieke, N., Baust, M., Esposito, M., Navab, N.: CFCM: segmentation via coarse to fine context memory. In: Frangi, A.F., Schnabel, J.A., Davatzikos, C., Alberola-López, C., Fichtinger, G. (eds.) MICCAI 2018. LNCS, vol. 11073, pp. 667–674. Springer, Cham (2018). https://doi.org/10.1007/978-3-030-00937-3_76
29. Mathai, T.S., Gorantla, V., Galeotti, J.: Segmentation of vessels in ultra high frequency ultrasound sequences using contextual memory. In: Shen, D., et al. (eds.) MICCAI 2019. LNCS, vol. 11765, pp. 173–181. Springer, Cham (2019). https://doi.org/10.1007/978-3-030-32245-8_20
30. He, K., et al.: Deep residual learning for image recognition. In: IEEE CVPR, pp. 770–778 (2016)

31. Koltun, V., et al.: Multi-scale context aggregation by dilated convolutions. In: ICLR (2016)
32. Oksuz, I., et al.: Automatic CNN-based detection of cardiac MR motion artefacts using k-space data augmentation and curriculum learning. Med. Image Anal. **55**, 136–147 (2019)
33. Kingma, D., et al.: Adam: a method for stochastic optimization. In: ICLR (2015)
34. Dempsey, M.: Measurement of tumor "size" in recurrent malignant glioma: 1D, 2D, or 3D? AJNR **26**(4), 770–776 (2005)

Estimating Glioblastoma Biophysical Growth Parameters Using Deep Learning Regression

Sarthak Pati[1,2,6], Vaibhav Sharma[3], Heena Aslam[3], Siddhesh P. Thakur[1,2,6], Hamed Akbari[1,2], Andreas Mang[4], Shashank Subramanian[5], George Biros[5], Christos Davatzikos[1,2], and Spyridon Bakas[1,2,6(✉)]

[1] Center for Biomedical Image Computing and Analytics, University of Pennsylvania, Philadelphia, PA, USA
sbakas@upenn.edu
[2] Department of Radiology, Perelman School of Medicine, University of Pennsylvania, Philadelphia, PA, USA
[3] Department of Electrical and Electronics Engineering, Aligarh Muslim University, Aligarh, Uttar Pradesh, India
[4] Department of Mathematics, University of Houston, Houston, TX, USA
[5] Oden Institute of Computational Engineering and Sciences, The University of Texas at Austin, Austin, TX, USA
[6] Department of Pathology and Laboratory Medicine, Perelman School of Medicine, University of Pennsylvania, Philadelphia, PA, USA

Abstract. Glioblastoma (*GBM*) is arguably the most aggressive, infiltrative, and heterogeneous type of adult brain tumor. Biophysical modeling of GBM growth has contributed to more informed clinical decision-making. However, deploying a biophysical model to a clinical environment is challenging since underlying computations are quite expensive and can take several hours using existing technologies. Here we present a scheme to accelerate the computation. In particular, we present a deep learning (*DL*)-based logistic regression model to estimate the GBM's biophysical growth in seconds. This growth is defined by three tumor-specific parameters: 1) a diffusion coefficient in white matter (*Dw*), which prescribes the rate of infiltration of tumor cells in white matter, 2) a mass-effect parameter (*Mp*), which defines the average tumor expansion, and 3) the estimated time (*T*) in number of days that the tumor has been growing. Preoperative structural multi-parametric MRI (*mpMRI*) scans from $n = 135$ subjects of the TCGA-GBM imaging collection are used to quantitatively evaluate our approach. We consider the mpMRI intensities within the region defined by the abnormal FLAIR signal envelope for training one DL model for each of the tumor-specific growth parameters. We train and validate the DL-based predictions against parameters derived from biophysical inversion models. The average Pearson correlation coefficients between our DL-based estimations and the biophysical parameters are 0.85 for Dw, 0.90 for Mp, and 0.94 for T, respectively.

C. Davatzikos and S. Bakas—Equally contributing senior author.

© Springer Nature Switzerland AG 2021
A. Crimi and S. Bakas (Eds.): BrainLes 2020, LNCS 12658, pp. 157–167, 2021.
https://doi.org/10.1007/978-3-030-72084-1_15

This study unlocks the power of tumor-specific parameters from biophysical tumor growth estimation. It paves the way towards their clinical translation and opens the door for leveraging advanced radiomic descriptors in future studies by means of a significantly faster parameter reconstruction compared to biophysical growth modeling approaches.

Keywords: Deep learning · Regression · Glioblastoma · Brain tumor · Biophysical growth model

1   Introduction

Glioblastoma (***GBM***), being the most common, aggressive, and infiltrative adult brain tumor, has unfavorable prognosis [1,2]. Recent advances in clinical GBM care has not made a huge difference to patient prospects [3]. The highly infiltrative nature of GBM render its recurrence essentially a guaranteed process [4,5] and its management varies drastically on a case-by-case basis. Integration of computational imaging, biophysical modeling, and machine learning has the potential to provide valuable tools to aid clinical diagnosis and decision making in a consistent and reproducible way [6], and by that possibly dramatically improve treatment planning for patients diagnosed with GBM [7–17].

In particular, biophysical modelling has the potential to become an indispensable tool to provide additional insights into GBM progression and development, and hence improve clinical decision making resulting in further improving quality of life for patients [6]. For example, in Fig. 1, the growth characteristics of the tumor would help stratifying the tumor into "proliferating" (tumor predominantly growing by hyperplasia and hypertrophy that push surrounding normal tissues), "infiltrating" (tumor predominantly growing by invasion and replacing surrounding tissues without mass effects) or "necrotic" (tumor is predominantly stagnant). Previous works [6,18–28] have used rigorous mathematical modelling to automatically estimate parameters that define the physical characteristics of tumor progression. In the present work, we consider the following main parameters:

1. *Diffusion coefficient in white matter (Dw)*. This parameter, also termed as *diffusivity*, controls the rate at which tumor cells infiltrate white matter [29].
2. *Mass-effect parameter (Mp)*. This parameter captures the mechanical deformation of the brain parenchyma caused as a function of the tumor growth [30], and defines the average tumor expansion. Clinically, this parameter quantifies how much the surrounding tissues deform because of the force exerted by the tumor's expansion.
3. *Estimated time (T)*. This parameter defines the number of days the tumor has been expanding [31].

In this study, we propose a deep learning (***'DL'***) based regression model (see Sect. 2.3) to glean multi-parametric insight into the data by *training* on the parameter values extracted using Boosted GLioma Image SegmenTation

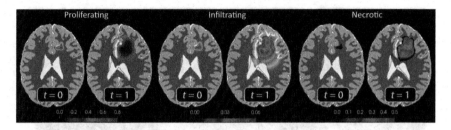

Fig. 1. Tumor proliferation across two time-points. These figures showcase the different types of clinical diagnoses that arise from the considered tumor growth parameters.

and Registration (GLISTRboost[1]) , which can then be used to approximate the aforementioned parameters during the *inference* phase. Our results (Sect. 3) show that this approach compares favorably to the current state-of-the-art, and is significantly more efficient during inference. This reduction in execution time is a critical for deploying our methodology in a clinical setting.

2 Methods

2.1 Data

The Cancer Imaging Archive (TCIA) [32] has released The Cancer Genome Atlas Glioblastoma Multiforme (TCGA-GBM) collection [33], which contains clinically-acquired multi-parametric MRI (mpMRI) scans for all patients of the collection. TCIA Analysis Results repository also offers a curated version of this collection, extended by corresponding segmentation labels and imaging features for all ($n = 135$) the pre-operative structural scans of the TCGA-GBM collection [34]. These mpMRI explicitly refer to native T1-weighted (T1) and post-contrast T1-weighted (T1Gd) scans, T2-weighted (T2), and T2 Fluid-Attenuated Inversion Recovery (FLAIR) scans. Furthermore, the included imaging features, also contain estimates for the parameters of interest, namely, Dw, Mp, and T. We use these parameter estimates along with the image intensities present in the abnormal FLAIR signal envelope to train the proposed DL model.

2.2 Pre-processing

To guarantee the homogeneity of the dataset, we applied the same pre-processing pipeline across all mpMRI scans. All the raw DICOM scans obtained from TCIA are converted to the NIfTI [35] file format. Subsequently, we followed the protocol for pre-processing as defined in the International Brain Tumor Segmentation (BraTS) challenge [36–38]. Specifically, each patient's T1Gd scan was rigidly registered to a common anatomical atlas of $240 \times 240 \times 155$ image size, and resampled to an isotropic resolution of $1\,\mathrm{mm}^3$ [39]. The remaining scans of each patient

---

[1] https://www.med.upenn.edu/cbica/sbia/glistrboost.html.

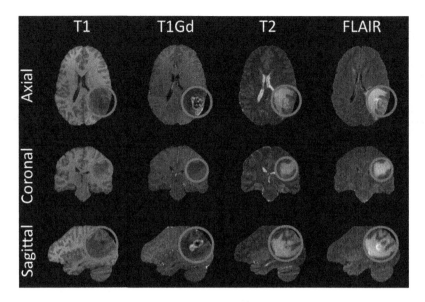

Fig. 2. Images of pipeline described in Sect. 2.2. The region of pathology used to train the DL model is highlighted using a red circle. (Color figure online)

(T1, T2, FLAIR) were subsequently rigidly co-registered to the same patient's resampled T1Gd scan. All the registrations were done using Greedy ([2]) [40] - a CPU-based C++ implementation of the greedy diffeomorphic registration algorithm and is integrated into the ITK-SNAP ([3]) segmentation software [41], as well as the Cancer Imaging Phenomics Toolkit (CaPTk ([4])) [42–44]. Following registration to a common anatomical atlas, and resampling to an isotropic resolution of 1mm^3, we perform instance-level normalization, where the intensity of each modality of each individual subject is normalized to zero mean and unit variance. This process is also known as Z-scoring, after which we performed skull stripping using BrainMaGe ([5]) [45] (see Fig. 2). We use the z-score normalized images during the downstream analysis.

All intensities of the whole tumor, defined by the whole abnormal FLAIR signal envelope, for each of the mpMRI modalities are concatenated in a 1D vector. This results in a stacked 2D matrix, where the number of rows define the intensities picked from all voxel included in the whole tumor region, and the number of columns equates to the number of the input mpMRI modalities, i.e., 4. This 2D matrix forms the data of each patient used in our proposed approach. This process is done over the entire dataset and, with the exception of 20% of

---

[2] https://github.com/pyushkevich/greedy.

[3] http://www.itksnap.org.

[4] https://www.cbica.upenn.edu/captk.

[5] https://github.com/CBICA/BrainMaGe/.

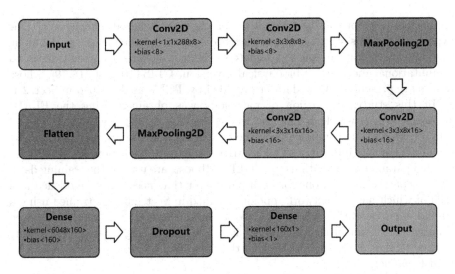

Fig. 3. The base model used to train for all 3 parameters.

the data held out for final performance evaluation, forms the complete *training input data.*

2.3   Network Topology

Once the *training input data* is constructed, we train one 2D DL model for each of the three parameters that are to be predicted. Each DL model consisted of two sets of convolution layers, each set followed by a single max-pooling layer. After feature reduction using the convolution/max-pooling layers, a fully connected flatten layer is used to flatten the output from the last max pooling layer. The flatten layer is followed by a fully connected dense layer having 160 nodes (which ensures a fixed number of inputs going into the final dense layer, regardless of the number of intensities) and followed by a 20% dropout (see Fig. 3). The model is trained using an adaptive gradient algorithm with an initial learning rate of 1E–2, and the model minimization metric used is normalized root mean squared error (NRMS), which is defined as follows:

$$NRMS = \sqrt{\frac{1}{N}\sum_{i=1}^{N}(x_i - y_i)}, \tag{1}$$

where x_i, y_i denotes the original and predicted observations indexed with i, respectively, and N represents the total number of samples.

2.4   Experimental Design

The current state-of-the-art approaches for personalized brain tumor modeling and parameter estimation consider optimization formulations with multiple, tightly coupled partial differential equations as constraints [6]. These types of approaches pose formidable computational and mathematical challenges [6]. Our hypothesis for this study is that a regression-based machine learning technique would be able to capture these parameters using the outputs of a widely-accepted computational method for biophysical inversion, GLISTRboost [18,19], whose results are presented in the dataset provided by TCIA, as described in Sect. 2.1.

In this study, the region of pathology was obtained using the FLAIR abnormal signal, which defines the whole tumor, i.e., the peritumoral edematous/invaded tissue combined with the enhancing and the necrotic parts of the tumor. The image intensities of all mpMRI scans from this region, along with the original parameter estimations from GLISTRboost, are used as the training data. Three separate models, one for each parameter that needs to be modelled and each of which has the network topology defined in Sect. 2.3, was trained using a 10-fold cross validation scheme [46]. We utilized a nested cross-validation based training i) in favor of reproducibility, ii) while attempting to avoid over-fitting to the training data, and iii) to tune the network hyper-parameters in a more robust manner.

2.5   Evaluation Metric

Following the literature on similar predictive modelling and classification tasks [47,48], we have used the Pearson's correlation coefficient (r_p) [49] to evaluate the efficacy of the network. These measures are defined as follows:

$$r_p = \frac{\sum_{i=1}^{N}(x_i - \bar{x})(y_i - \bar{y})}{\sqrt{\sum_{i=1}^{N}(x_i - \bar{x})^2}\sqrt{\sum_{i=1}^{N}(y_i - \bar{y})^2}}, \tag{2}$$

where x_i, y_i denotes the original and predicted observations indexed with i, respectively, N represents the total number of observations, and $\bar{x}, \bar{y}$ are the sample means for the original and predicted values, respectively.

3   Results

Following the experimental design described in Sect. 2.4, we used the training data of the fold with the best accuracy (defined as the lowest NRMS loss) to validate against a hold-out set of $n = 20$ subjects. Notably this hold out set was excluded from the cross-validated training phase. We estimated the Pearson's correlation coefficient between the original and predicted values. The experimental results for all the 10 folds (see Fig. 4) show that the DL models were able to approximate the parameters reasonably well.

From Fig. 4, it is evident that the model's performance is best for Mp, with the best correlation score of $r_p = 0.99$. The median correlation scores for Mp and Dw are $r_p = 0.89$, and $r_p = 0.889$, respectively. In terms of stability, the model that predicts T performs best, with a median correlation score of $r_p = 0.92$. This highlights the need for more rigorous validation with increased amount of diverse datasets and possibly using more sophisticated modelling techniques, such as the one described in [50].

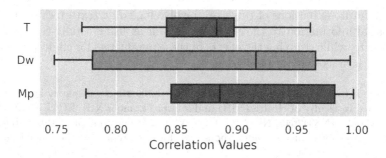

Fig. 4. Pearson's correlation coefficient for original and predicted values for the growth parameters under consideration across 10 training folds.

4 Discussion

This study compares the outputs from an existing mathematically rigorous formulation of a biophysical modelling technique [19] provided by TCIA [32–34] with that of the proposed DL model, trained solely on image intensities and those pre-existing estimates of tumor growth parameters. Using a 2D convolution based architecture for the DL network, combined with a normalized mean squared error loss (see Sect. 2.3 for details), and intensity values from the pathological region of the dataset, we were able to get favorable results for data not seen by the model during training, as show in Sect. 3.

The most notable difference was that for the inference phase for predicting the growth parameters, where the DL-based method took about 15–30 s on a GPU (NVIDIA Titan Xp), and 2–5 min when executed on a CPU architecture (Intel i7-8700K). This is in contrast to the mathematical formulation (GLISTRboost [18]), which would take about 5 h on the same CPU. This significant reduction in run-time renders such methods for parameter estimation potentially translatable to clinical applications.

The results presented in this explorative study show that the proposed framework yields a good accompanying tool along with sophisticated mathematical modelling techniques, where the latter is used to generate the training data. We note that our framework is generic; introducing more complicated (realistic) mathematical models of tumor progression alongside with efficient numerical methods for their solution into the training phase forms the basis of our current research. Additionally, we note that this method is limited in capability and confined by the input training data, and more work is needed to ensure further analysis of the generalizability of the proposed method when compared with biophysical tumor growth modelling techniques.

We envision an integrated computational framework that augments clinical imaging data in a consistent and reproducible way by estimating tumor growth model parameters for individual patients, without the burden of extreme computational footprint, with the potential to be clinically translated and hence aid clinical decision making, towards ultimately improving clinical outcome. Towards this end, future extensions of this work involve including texture characteristics and spatial information to the network, as well as using a fully-connected 3D architecture, while studying the effects of these parameters for clinically-relevant outcomes such as survival prediction.

Acknowledgments. Research reported in this publication was partly supported by the National Institutes of Health (NIH) under award number NIH/NCI:U01CA242871, NIH/NINDS:R01NS042645, NIH/NCI:U24CA189523. The content of this publication is solely the responsibility of the authors and does not represent the official views of the NIH.

References

1. Ostrom, Q.T., Rubin, J.B., Lathia, J.D., Berens, M.E., Barnholtz-Sloan, J.S.: Females have the survival advantage in glioblastoma. Neuro Oncol. **20**(4), 576 (2018)
2. Herrlinger, U., et al.: Lomustine-temozolomide combination therapy versus standard temozolomide therapy in patients with newly diagnosed glioblastoma with methylated MGMT promoter (CeTeG/NOA-09): a randomised, open-label, phase 3 trial. The lancet **393**(10172), 678–688 (2019)
3. Hou, L.C., Veeravagu, A., Hsu, A.R., Victor, C.: Recurrent glioblastoma multiforme: a review of natural history and management options. Neurosurg. Focus **20**(4), E3 (2006)
4. Akbari, H., et al.: Imaging surrogates of infiltration obtained via multiparametric imaging pattern analysis predict subsequent location of recurrence of glioblastoma. Neurosurgery **78**(4), 572–580 (2016)
5. Fathi Kazerooni, A., et al.: Cancer imaging phenomics via CaPTk: multi-institutional prediction of progression-free survival and pattern of recurrence in glioblastoma. JCO Clin. Cancer Inf. **4**, 234–244 (2020)
6. Mang, A., Bakas, S., Subramanian, S., Davatzikos, C., Biros, G.: Integrated biophysical modeling and image analysis: application to neuro-oncology. Ann. Rev. Biomed. Eng. **22**, 309–341 (2020)
7. Gutman, D.A., et al.: MR imaging predictors of molecular profile and survival: multi-institutional study of the TCGA glioblastoma data set. Radiology **267**(2), 560–569 (2013)
8. Gevaert, O., et al.: Glioblastoma multiforme: exploratory radiogenomic analysis by using quantitative image features. Radiology **273**(1), 168–174 (2014)
9. Jain, R., et al.: Outcome prediction in patients with glioblastoma by using imaging, clinical, and genomic biomarkers: focus on the nonenhancing component of the tumor. Radiology **272**(2), 484–493 (2014)
10. Aerts, H.J.: The potential of radiomic-based phenotyping in precision medicine: a review. JAMA Oncol. **2**(12), 1636–1642 (2016)
11. Bilello, M., et al.: Population-based MRI atlases of spatial distribution are specific to patient and tumor characteristics in glioblastoma. NeuroImage: Clinical **12**, 34–40 (2016)
12. McNitt-Gray, M., et al.: Standardization in quantitative imaging: a multicenter comparison of radiomic features from different software packages on digital reference objects and patient data sets. Tomography **6**(2), 118 (2020)

13. Bakas, S., et al.: Overall survival prediction in glioblastoma patients using structural magnetic resonance imaging (MRI): advanced radiomic features may compensate for lack of advanced MRI modalities. J. Med. Imaging **7**(3), 031505 (2020)
14. Zwanenburg, A., et al.: The image biomarker standardization initiative: standardized quantitative radiomics for high-throughput image-based phenotyping. Radiology **295**(2), 328–338 (2020)
15. Bakas, S., et al.: In vivo detection of EGFRvlll in glioblastoma via perfusion magnetic resonance imaging signature consistent with deep peritumoral infiltration: the φ-index. Clin. Cancer Res. **23**(16), 4724–4734 (2017)
16. Binder, Z.A., et al.: Epidermal growth factor receptor extracellular domain mutations in glioblastoma present opportunities for clinical imaging and therapeutic development. Cancer Cell **34**(1), 163–177 (2018)
17. Akbari, H., et al.: In vivo evaluation of EGFRvlll mutation in primary glioblastoma patients via complex multiparametric MRI signature. Neuro Oncol. **20**(8), 1068–1079 (2018)
18. Bakas, S., et al.: GLISTRboost: combining multimodal MRI segmentation, registration, and biophysical tumor growth modeling with gradient boosting machines for glioma segmentation. In: Crimi, A., Menze, B., Maier, O., Reyes, M., Handels, H. (eds.) BrainLes 2015. LNCS, vol. 9556, pp. 144–155. Springer, Cham (2016). https://doi.org/10.1007/978-3-319-30858-6_13
19. Hogea, C., Davatzikos, C., Biros, G.: An image-driven parameter estimation problem for a reaction-diffusion glioma growth model with mass effects. J. Math. Biol. **56**(6), 793–825 (2008)
20. Ostrom, Q.T., et al.: Cbtrus statistical report: primary brain and central nervous system tumors diagnosed in the united states in 2008–2012. Neuro-oncology **17**(suppl_4), iv1–iv62 (2015)
21. Konukoglu, E., et al.: Image guided personalization of reaction-diffusion type tumor growth models using modified anisotropic eikonal equations. IEEE Trans. Med. Imaging **29**(1), 77–95 (2009)
22. Gooya, A., Biros, G., Davatzikos, C.: Deformable registration of glioma images using EM algorithm and diffusion reaction modeling. IEEE Trans. Med. Imaging **30**(2), 375–390 (2010)
23. Scheufele, K., Mang, A., Gholami, A., Davatzikos, C., Biros, G., Mehl, M.: Coupling brain-tumor biophysical models and diffeomorphic image registration. Comput. Methods Appl. Mech. Eng. **347**, 533–567 (2019)
24. Menze, B.H., et al.: A generative approach for image-based modeling of tumor growth. In: Székely, G., Hahn, H.K. (eds.) IPMI 2011. LNCS, vol. 6801, pp. 735–747. Springer, Heidelberg (2011). https://doi.org/10.1007/978-3-642-22092-0_60
25. Wang, C.H., et al.: Prognostic significance of growth kinetics in newly diagnosed glioblastomas revealed by combining serial imaging with a novel biomathematical model. Can. Res. **69**(23), 9133–9140 (2009)
26. Geremia, E., Menze, B.H., Prastawa, M., Weber, M.-A., Criminisi, A., Ayache, N.: Brain tumor cell density estimation from multi-modal MR images based on a synthetic tumor growth model. In: Menze, B.H., Langs, G., Lu, L., Montillo, A., Tu, Z., Criminisi, A. (eds.) MCV 2012. LNCS, vol. 7766, pp. 273–282. Springer, Heidelberg (2013). https://doi.org/10.1007/978-3-642-36620-8_27
27. Jackson, P.R., Juliano, J., Hawkins-Daarud, A., Rockne, R.C., Swanson, K.R.: Patient-specific mathematical neuro-oncology: using a simple proliferation and invasion tumor model to inform clinical practice. Bull. Math. Biol. **77**(5), 846–856 (2015)

28. Ivkovic, S., et al.: Direct inhibition of myosin ii effectively blocks glioma invasion in the presence of multiple motogens. Mol. Biol. Cell **23**(4), 533–542 (2012)

29. Wong, K.C., Summers, R.M., Kebebew, E., Yao, J.: Tumor growth prediction with reaction-diffusion and hyperelastic biomechanical model by physiological data fusion. Med. Image Anal. **25**(1), 72–85 (2015)

30. Clatz, O., et al.: Realistic simulation of the 3-D growth of brain tumors in MR images coupling diffusion with biomechanical deformation. IEEE Trans. Med. Imaging **24**(10), 1334–1346 (2005)

31. Rahman, M.M., Feng, Y., Yankeelov, T.E., Oden, J.T.: A fully coupled space-time multiscale modeling framework for predicting tumor growth. Comput. Methods Appl. Mech. Eng. **320**, 261–286 (2017)

32. Clark, K., et al.: The cancer imaging archive (TCIA): maintaining and operating a public information repository. J. Digit. Imaging **26**(6), 1045–1057 (2013)

33. Scarpace, L., et al.: Radiology data from the cancer genome atlas glioblastoma multiforme [TCGA-GBM] collection. Cancer Imaging Arch. **11**(4), 1 (2016)

34. Bakas, S., et al.: Segmentation labels and radiomic features for the pre-operative scans of the TCGA-GBM collection. The cancer imaging archive. Nat. Sci. Data **4**, 170117 (2017)

35. Cox, R., et al.: A (sort of) new image data format standard: Nifti-1: we 150. Neuroimage **22** (2004). https://nifti.nimh.nih.gov/nifti-1/documentation/hbm_nifti_2004.pdf

36. Menze, B.H., et al.: The multimodal brain tumor image segmentation benchmark (BRATS). IEEE Trans. Med. Imaging **34**(10), 1993–2024 (2014)

37. Bakas, S., et al.: Advancing the cancer genome atlas glioma MRI collections with expert segmentation labels and radiomic features. Scientific data **4**, 170117 (2017)

38. Bakas, S., et al.: Identifying the best machine learning algorithms for brain tumor segmentation, progression assessment, and overall survival prediction in the brats challenge. arXiv preprint arXiv:1811.02629 (2018)

39. Rohlfing, T., Zahr, N.M., Sullivan, E.V., Pfefferbaum, A.: The SRI24 multichannel atlas of normal adult human brain structure. Hum. Brain Mapp. **31**(5), 798–819 (2010)

40. Yushkevich, P.A., Pluta, J., Wang, H., Wisse, L.E., Das, S., Wolk, D.: Fast automatic segmentation of hippocampal subfields and medial temporal lobe subregions in 3 Tesla and 7 Tesla T2-weighted MRI. Alzheimer's Dementia **7**(12), P126–P127 (2016)

41. Yushkevich, P.A., et al.: User-guided 3D active contour segmentation of anatomical structures: significantly improved efficiency and reliability. Neuroimage **31**(3), 1116–1128 (2006)

42. Davatzikos, C., et al.: Cancer imaging phenomics toolkit: quantitative imaging analytics for precision diagnostics and predictive modeling of clinical outcome. J. Med. Imaging **5**(1), 011018 (2018)

43. Rathore, S., et al.: Brain cancer imaging phenomics toolkit (brain-CaPTk): an interactive platform for quantitative analysis of glioblastoma. In: Crimi, A., Bakas, S., Kuijf, H., Menze, B., Reyes, M. (eds.) BrainLes 2017. LNCS, vol. 10670, pp. 133–145. Springer, Cham (2018). https://doi.org/10.1007/978-3-319-75238-9_12

44. Pati, S., et al.: The cancer imaging phenomics toolkit (CaPTk): technical overview. In: Crimi, A., Bakas, S. (eds.) BrainLes 2019. LNCS, vol. 11993, pp. 380–394. Springer, Cham (2020). https://doi.org/10.1007/978-3-030-46643-5_38

45. Thakur, S., et al.: Brain extraction on MRI scans in presence of diffuse glioma: multi-institutional performance evaluation of deep learning methods and robust modality-agnostic training. NeuroImage **220**, 117081 (2020)

46. Allen, D.M.: The relationship between variable selection and data agumentation and a method for prediction. Technometrics **16**(1), 125–127 (1974)
47. Kather, J.N., et al.: Deep learning can predict microsatellite instability directly from histology in gastrointestinal cancer. Nat. Med. **25**(7), 1054–1056 (2019)
48. Kuo, C.-C., et al.: Automation of the kidney function prediction and classification through ultrasound-based kidney imaging using deep learning. NPJ Digit. Med. **2**(1), 1–9 (2019)
49. Student: Probable error of a correlation coefficient. Biometrika, pp. 302–310 (1908)
50. Mang, A., et al.: SIBIA-GLS: scalable biophysics-based image analysis for glioma segmentation. In: The Multimodal Brain Tumor Image Segmentation Benchmark (BRATS), MICCAI (2017)

Bayesian Skip Net: Building on Prior Information for the Prediction and Segmentation of Stroke Lesions

Julian Klug[1,2]([⊠]) [ID], Guillaume Leclerc[3], Elisabeth Dirren[1],
Maria Giulia Preti[2][ID], Dimitri Van De Ville[2][ID], and Emmanuel Carrera[1][ID]

[1] Stroke Research Group, Department of Clinical Neurosciences,
University Hospital and Faculty of Medicine, Geneva, Switzerland
`julian.klug@etu.unige.ch`
[2] Medical Image Processing Laboratory, Institute of Bioengineering,
Ecole Polytechnique Fédérale de Lausanne (EPFL), Lausanne, Switzerland
[3] Massachusetts Institute of Technology, Cambridge, USA

Abstract. Perfusion CT is widely used in acute ischemic stroke to determine eligibility for acute treatment, by defining an ischemic core and penumbra. In this work, we propose a novel way of building on prior information for the automatic prediction and segmentation of stroke lesions. To this end, we reformulate the task to identify differences from a prior segmentation by extending a three-dimensional Attention Gated Unet with a skip connection allowing only an unchanged prior to bypass most of the network. We show that this technique improves results obtained by a baseline Attention Gated Unet on both the Geneva Stroke Dataset and the ISLES 2018 dataset.

Keywords: Prior · Stroke · Convolutional neural network · Medical image segmentation

1 Introduction

Ischemic stroke is a leading cause of mortality and disability worldwide [5]. In the last decade, advances in stroke imaging have enabled a more targeted approach to treatment and reperfusion. Perfusion CT (pCT) is widely used in acute ischemic stroke to determine eligibility for treatment by providing perfusion parameter maps informing about voxelwise cerebral blood flow, cerebral blood volume, transit time and time to maximum of the residue function (CBF, CBV, MTT and Tmax respectively) [8]. Necrotic tissue defining the ischemic core, as well as hypoperfused but still salvageable tissue, the ischemic penumbra, are commonly delineated by threshold-based tools [6]. These techniques have shown to be of great benefit to select patients for treatment in several clinical trials [2,21]. However, accurate segmentation of the ischemic core on pCT, as well as prediction of the final infarct is challenging, and diffusion weighed magnetic resonance imaging (MRI) remains the gold standard.

© Springer Nature Switzerland AG 2021
A. Crimi and S. Bakas (Eds.): BrainLes 2020, LNCS 12658, pp. 168–180, 2021.
https://doi.org/10.1007/978-3-030-72084-1_16

With the advent of deep convolutional neural networks (CNNs), significant improvements in stroke prediction and segmentation have been made [18]. Current methods take as input acute imaging sequences and learn a tissue outcome prediction function from labelled data obtained from follow-up imaging. Many currently presented models rely on a Unet backbone [9,10,15,18,28], allowing for multi-scale localisation and contextualisation through an encoder-decoder structure with skip-connections. In the setting of the ISLES 2018 challenge, the best performing submissions achieved a Dice score ranging from 0.48 to 0.51 for the segmentation of the ischemic core on perfusion CT [1,12,17]. Current methods for the prediction of the final lesion often use a combination of perfusion MRI and diffusion weighed imaging leading to a Dice score of up to 0.53 [20,30]. Predicting the final lesion from perfusion CT alone has been attempted through the use of baseline perfusion images and infarct growth estimations [16,26] with Dice scores of up to 0.48.

Incorporating prior knowledge has proven useful in many medical image segmentation tasks [22]. Indeed, the inclusion of a prior can simplify a model's task by reducing the amount of information to be learnt. Atlas models, distance boundaries, shape and topology specifications as well as edge polarity have been successfully used as regularisation terms in region growing segmentation methods. Recent work has used prior constraints in the form of adjacency, boundary or learned anatomical conditions [23].

In acute ischemic stroke, tissue of the penumbra is progressively recruited into the core, contributing to the growth of the lesion. Previous models have successfully used prior manual ischemic core and penumbra segmentations to obtain a representation of ischemic stroke growth directions [16]. Using standardized thresholds for the automated segmentation of the ischemic core could potentially leverage the strengths of a largely clinically validated model. Integrating this information as a prior can be used as a starting point to either refine the segmentation of the ischemic core or to predict the final lesion. In this work we aim to efficiently integrate prior information obtained by a standard threshold segmentation of the ischemic core directly into the commonly used Attention Gated 3D Unet [23,27,30].

2   Materials and Methods

2.1   Data

Geneva Stroke Dataset. This dataset comprises acute pCT images and final lesion labels of 144 patients who have benefitted from treatment by thrombectomy and/or thrombolysis as described in prior work [13]. For every subject a full-volume CBF, CBV, MTT, Tmax and non-contrast CT (NCCT) image is available. Manually annotated labels for the final infarct have been obtained from follow-up MRI in the sub-acute phase by expert neurologists. Briefly, a model has to learn the prediction of the final infarct after treatment.

ISLES 2018 Dataset. The Ischemic Stroke Lesion Segmentation 2018 dataset contains pCT images and acute lesion labels of patients before undergoing treatment [12,17]. Every subject has the main slices containing the acute lesion of CBF, CBV, MTT, Tmax and NCCT sequences. The gold-standard ischemic cores were defined on the subsequent MRI, performed in the acute setting and manually annotated. We used the publicly available training subset comprising 94 data points. In this setting, a model has to learn the segmentation of stroke lesions before treatment.

2.2   Data Pre-processing

We normalize by subtracting the mean and dividing by the standard deviation all input sequences, which are subsequently scaled to a 0–1 range. To ensure divisibility by 16 necessary for our Unet architecture we pad by 0 along every dimension. All data used for training is augmented ad hoc by applying random flip, random elastic transform, random shift, random scale and Gaussian noise [25]. Every individual transformation has a 50% probability of being applied on each batch.

Ischemic core is commonly defined as relative CBF < 0.3 (rCBF). In our work we have chosen a slightly more inclusive threshold at 0.38 for greater sensitivity. rCBF is defined as relative (by ratio) to the mean CBF of the contralateral hemisphere after smoothing with a 3D Gaussian kernel of 2 voxels width [7]. Cerebral spinal fluid (CSF) is segmented on NCCT images by applying a 5th percentile threshold on voxels bounded between 0 and 100 Hounsfield units (HU). A segmentation of the skull is obtained with the bet2 algorithm of the FMRIB Software Library (FSLv.5) [11,19] on NCCT images with the same bounds. Finally, major blood vessels are segmented by applying a 99th percentile threshold on CBF images. The resulting CSF, skull and vessel segmentations were then extended slightly by binary dilation with a spheroid structuring element of width 2 voxels. They were then removed from the initial ischemic core segmentations. A similar pipeline is implemented by the commercially available RAPID software package (RAPID, Ischemaview Stanford University, Stanford, USA) and widely used in clinical practice [29]. The final ischemic core segmentation is defined as prior segmentation (Fig. 1).

2.3   Network Architecture

Attention Gated Unet. A 3D Unet with attention gated skip connections is used as the baseline model. This encoder-decoder architecture makes use of contextual information by down-sampling input volumes. Spatial details are then recovered by up-sampling. The information bottleneck is classically overcome by skip connections spanning between down-sampling and up-sampling layers of corresponding scale to provide more contextual information to the model [31]. Attention gates applied to the skip connections extract features from coarse scale to highlight salient regions on a finer scale [24]. Every convolutional module is composed of two layers with a 3D-convolution with a $3 \times 3 \times 3$ kernel followed by

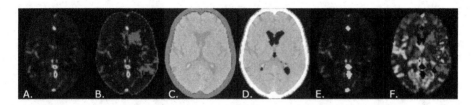

Fig. 1. From left to right: axial view of CBF (A.), thresholded relative CBF at 0.38 overlaid in magenta on CBF (B.), skull segmentation (C.) and ventricle segmentation (D.) overlaid in blue on NCCT, vascular noise filter in orange (E.) and the final ischemic core segmentation in magenta (F.) overlaid on CBF of a representative subject of the Geneva Stroke Dataset. The final threshold-based ischemic core segmentation is subsequently used as prior. (Color figure online)

batch-normalisation and a Rectified linear unit (ReLU) as activation function. Four convolutional modules are used respectively during down- and up-sampling with an additional central module. Deep supervision is used to enhance the transition between feature space and semantically relevant segmentation [14].

Bayesian Skip for Attention Gated Unet. The above described Unet creates large receptive fields by successive down-sampling of input information to model relationships at a gradually coarser scale. It remains however challenging to reduce false-positive rates for small and patchy segmentations of varying shape. To address this issue, we propose to allow selected input channels defined as prior to bypass the main part of the model, before being reintegrated at the final layer. The network thus effectively learns to model the difference relative to the prior given all inputs. To this end we add an additional skip connection spanning the network from its input to its final layer, termed bayesian skip. There, it is integrated by Method A) summation with the output or Method B) by convolution with the output. In Method B) a $1 \times 1 \times 1$ convolution is used to reduce computational overhead. The network can thus be described as the function $U(x)$ with $u(x)$ defined as the skipped part of the network and p the prior. Note that at any given step, $p \in x$ as the prior is not removed from the input. Thus, for the baseline model $U(x) = u(x)$. The Methods A) and B) can be respectively described by Eqs. 1 and 2.

$$U(x)_A = u(x) + p \tag{1}$$

$$U(x)_B = u(x) * p \tag{2}$$

Subsequently, $u(x)$ learns to fit the difference between $U(x)$ and p in Method A) and the deconvolution in Method B) (Fig. 2).

2.4 Experimental Setup

We rearrange our data in a three-way split, with 70% used for training, 15% for validation and 15% for testing. Volumes of size $96 \times 96 \times 96$ and $256 \times 256 \times 22$

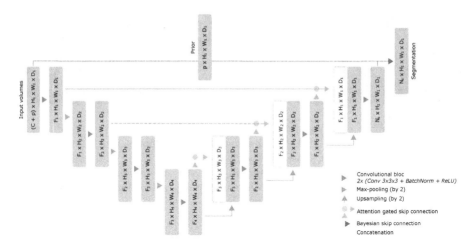

Fig. 2. A bloc diagram of the proposed Bayesian Skip Net, building on the previously designed Attention Gated Unet. The bayesian skip connection enables the prior to bypass the network unchanged. The prior is then reintegrated with the output of the final layer via Method A) summation or Method B) convolution.

are used respectively for the Geneva Stroke Dataset and the ISLES 2018 Dataset. We use a batch-size of 2 and 4 respectively for the datasets, dictated by computational limitations. Stochastic gradient descent is used with an initial learning rate of 0.0001 and Nesterov momentum. Weight decay is used as regularisation. The learning rate is reduced by a factor of 2 every 200 epochs. We optimize the commonly employed smooth Sorensen-Dice loss defined as follows:

$$L_{Dice} = 1 - \frac{2|X \cap Y| + \epsilon}{|X| + |Y| + \epsilon} \tag{3}$$

where a smoothing factor ϵ of 0.01 is used. Early stopping on the validation loss is used with a patience of 20 after 200 epochs. The best model is selected based on best Dice score on the validation set and is subsequently evaluated on the independent test set. No hyperparameter optimisation is attempted as absolute results are not the objective of this work. A baseline Attention Gated Unet, as well as two Attention Gated Unets with bayesian skip integrated through Method A) or B) respectively, have been trained and evaluated using CBF, CBV, MTT, Tmax and NCCT sequences, as well as the prior segmentation as input channels of the Geneva Stroke Dataset. We then validated the experiments for the baseline network and Method B) on the ISLES 2018 Dataset. All computations were done on a single Nvidia Tesla P100 GPU (Nvidia Corporation, Santa Clara, California, USA). Our Pytorch implementation for the proposed architecture is publicly available *here*.

Impact of Prior Quality. To evaluate the impact of the quality of the prior segmentation on the performance of the Bayesian Skip Net, we computed the correlation between the Dice score obtained by the prior alone and the score obtained by the model making use of the prior using Pearson's correlation coefficient.

We further evaluated the performance of the proposed model given a degraded prior. To this aim we added random binary noise where $p(1) = 0.002$ on the prior segmentation used for training and testing. On average, this results in a noisy prior in which 25% of the total volume corresponds to randomly segmented voxels. For both datasets, a Bayesian Skip Net was trained and tested using the resulting noisy segmentation as prior.

3   Results

Experimental results on validation and test splits for the Geneva Stroke Dataset and ISLES 2018 Dataset are reported respectively in Tables 1 and 2. Overall, the Bayesian Skip Net with Method B) performs better than the baseline Attention Gated Unet on both datasets. It consistently achieves better Dice and precision scores on the Geneva Stroke Dataset as well as on the test split of the ISLES 2018 Datatset. On the ISLES 2018 validation split a performance similar to baseline is achieved. The baseline model mostly remains superior in terms of recall. All evaluated methods outperform the prior segmentation. Furthermore, the proposed method greatly speeds up the training process, as the Bayesian Skip Net with Method B) achieves convergence after about 300 epochs, compared to a baseline convergence of 450 epochs when evaluated on the ISLES 2018 dataset as shown in Fig. 3. The proposed model with Method A) does not yield any increase in performance and was therefore not further evaluated on the second dataset.

Correlation between the quality of the prior used and Dice scores obtained by the proposed model with Method B) on the test splits of both datasets are shown in Fig. 4. For both the Geneva Stroke Dataset (Pearson's $R = 0.82$, $p < 10^{-5}$) and the ISLES 2018 Dataset (Pearson's $R = 0.76$, $p < 10^{-2}$), Dice scores obtained by the prior and by the Bayesian Skip Net correlate strongly.

Evaluation of the Bayesian Skip Net with Method B) given a prior with and without added noise on the test splits for the Geneva Stroke Dataset and ISLES 2018 Dataset are reported in Table 3. On both tasks, adding noise to the prior slightly reduces model performance. When comparing with the baseline Unet, the proposed model achieves comparable or better results even when using a noisy prior.

Example predictions are shown in Fig. 5. A case-by-case visual analysis of the predicted lesions of all evaluated methods revealed that the proposed model consistently produced more precise segmentations and was less likely to produce false positive results than the baseline method. A selective analysis of cases where both models failed to segment the lesion revealed small infarcts along the midline, in the brainstem or the cerebellum. The prior segmentation did not include any of these lesions.

Table 1. Experimental results on test and validation splits for the Geneva Stroke Dataset. The results are reported in terms of mean and standard deviation for Dice score, precision and recall. The proposed Unet with bayesian skip with Method A) and B) is benchmarked against the baseline Attention Gated Unet. The prior segmentation's performance is reported as reference. Best model results in bold.

Data Split	Method	Dice	Precision	Recall
Validation	Prior	0.125 ± 0.135	0.149 ± 0.128	0.171 ± 0.194
	Unet	0.270 ± 0.215	0.265 ± 0.299	**0.404** ± 0.336
	Unet+Method A	0.246 ± 0.206	0.221 ± 0.247	0.300 ± 0.299
	Unet+Method B	**0.292** ± 0.211	**0.348** ± 0.333	0.294 ± 0.257
Test	Prior	0.099 ± 0.110	0.109 ± 0.116	0.119 ± 0.163
	Unet	0.192 ± 0.156	0.189 ± 0.235	**0.271** ± 0.284
	Unet+Method A	0.181 ± 0.154	0.132 ± 0.187	0.278 ± 0.324
	Unet+Method B	**0.212** ± 0.136	**0.289** ± 0.333	0.188 ± 0.219

Table 2. Experimental results on test and validation splits for the ISLES 2018 Dataset compared with the top three submissions in the ISLES 2018 challenge. The results are reported in terms of mean and standard deviation for Dice score, precision and recall. The proposed Unet with bayesian skip with Method B) is benchmarked against the baseline Attention Gated Unet. The prior segmentation's performance, as well as the top three submissions of the ISLES 2018 challenge are reported as reference. Best model results in bold.

Data Split	Method	Dice	Precision	Recall
Validation	Prior	0.189 ± 0.174	0.214 ± 0.181	0.236 ± 0.230
	Unet	**0.417** ± 0.072	**0.433** ± 0.304	**0.419** ± 0.264
	Unet+Method B	0.415 ± 0.073	0.431 ± 0.301	0.411 ± 0.264
Test	Prior	0.296 ± 0.256	0.251 ± 0.222	0.374 ± 0.307
	Unet	0.524 ± 0.182	0.532 ± 0.354	0.560 ± 0.312
	Unet+Method B	**0.552** ± 0.195	**0.561**± 0.238	**0.573** ± 0.292
ISLES 2018 Challenge	Song et al. [28]	0.51 ± 0.31	0.55 ± 0.36	0.55 ± 0.34
	Liu et al. [15]	0.49 ± 0.31	0.56 ± 0.37	0.53 ± 0.33
	Chen et al. [9]	0.48 ± 0.31	0.59 ± 0.38	0.46 ± 0.33

Table 3. Impact of noise added to the prior on the performance of the Bayesian Skip Net with Method B). Experimental results on the test splits of the Geneva Stroke Dataset and the ISLES 2018 Dataset with and without noise added to the prior are reported in terms of mean and standard deviation for Dice score, precision and recall.

Data set	Prior quality	Dice	Precision	Recall
Geneva Stroke Dataset	Standard	0.212 ± 0.136	0.289 ± 0.333	0.188 ± 0.219
	Noisy	0.191 ± 0.165	0.204 ± 0.277	0.208 ± 0.244
ISLES 2018 Dataset	Standard	0.552 ± 0.195	0.561 ± 0.238	0.573 ± 0.292
	Noisy	0.538 ± 0.205	0.565 ± 0.282	0.565 ± 0.271

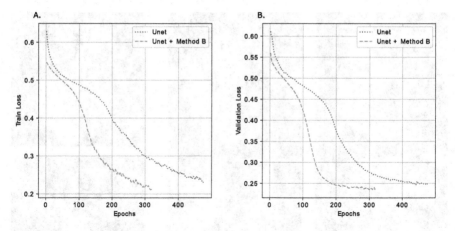

Fig. 3. Training (A.) and validation (B.) loss during training of the baseline Attention Gated Unet (blue) and the proposed Bayesian Skip Net with Method B) (orange) on the ISLES 2018 Dataset. Efficient integration of the prior ensures faster training. Both models tend to overfit on validation and test data. This is common in deep learning and generally does not prevent generalisation on test data [4]. (Color figure online)

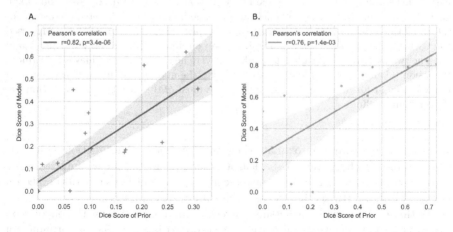

Fig. 4. Correlation of Dice scores obtained by the prior segmentation and the Bayesian Skip Net with Method B) on the test splits of the Geneva Stroke Dataset (A.) and the ISLES 2018 Dataset (B.) For both datasets, the performance of the model strongly correlates with the quality of the prior.

4   Discussion and Conclusion

In this paper, we present a novel bayesian skip connection as a way to take into account a prior in medical image segmentation. By reintegrating prior knowledge at the end of the proposed model, we explicitly let the rest of the network approximate the divergence from this prior. Having the same information at their disposal, both the baseline model and the model with the bayesian skip connec-

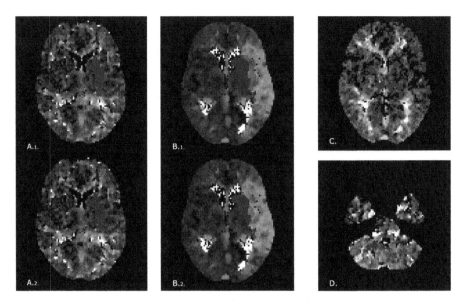

Fig. 5. Lesion labels, in blue, and model predictions, in red, projected on Tmax axial views for four subjects (A-D). The left and middle images show the predicted segmentation by the baseline Attention Gated Unet (upper subimage, 1.) and by the proposed Unet with bayesian skip with Method B) (lower subimage, 2.) for two patients (A and B). The two images on the right show subjects were neither method achieves to segment the lesion (C and D). Small lesions located along the midline and in the posterior fossa remain difficult for both models and account for most of the variability in performance. (Color figure online)

tion should be able to approximate the desired segmentation function. However, the amount of information to be learnt is smaller in the latter, improving its learning process. This results in faster convergence and better performance on both the Geneva Stroke Dataset and the ISLES 2018 dataset.

Within our experimental setting, bayesian skip connections with convolution (Method B) but not with summation (Method A) improve the performance of an Attention Gated Unet when a prior segmentation is given as input for a task of ischemic lesion prediction and segmentation. The prior used in our setting is associated with its own uncertainty. This can be modelled by the convolution but not by a summation explaining the superiority of Method B) relative to Method A).

The overall gain in performance is mainly driven by increased precision as the proposed model focuses on regions already segmented in the prior, thus reducing its false-positive rate. However, this comes at the cost of a lower recall value. The results reported here further show that the proposed method greatly speeds up model convergence which is crucial for rapid model iteration and reduces computational costs.

The Bayesian Skip Net compares favourably with the best performing models of the ISLES 2018 challenge. Although the analysis is somewhat limited as the test set used for the challenge is not publicly available and does not correspond exactly to the same subjects used for the test split in our experiments, this comparison on a same dataset suggests that our proposed model achieves state-of-the art performance for the segmentation of the ischemic core.

The comparison of results for the prediction of the final lesion from perfusion CT is more complicated as datasets referenced across the literature are heterogenous and not open-source. Unsurprisingly, the Bayesian Skip Net outperforms earlier general linear models with receptive fields on the Geneva Stroke Dataset [13]. Robben et al. use the raw perfusion CT signal as input to a CNN and report a Dice score of 0.48 on the dataset of the MRCLEAN study compromising globally bigger lesions (mean lesion volume [interquartile range]: 78 cm^3 [21-121] vs. 29.4 cm^3 [2.5-37.6]) [26]. This highlights the importance of lesion volume as a component of model performance and could suggest a potential loss of information due to the conversion from raw perfusion CT images to perfusion maps (Tmax, MTT, CBF, CBV). In a small sample of the TRAVESTROKE dataset, Lucas et al. obtain a Dice score of 0.46 by modulating a deformation model obtained from manual segmentations of the ischemic penumbra and core [16]. Although manual segmentations are not sustainable in clinical practice, this underscores the importance of the quality of the prior used and suggests that adding a threshold-based segmentation of the ischemic penumbra to the prior might yield further gains in performance.

We show that our model is relatively resistant to added noise as it maintains similar or greater performance than the baseline Unet when presented with a degraded prior. Nonetheless, the gain in performance of our model strongly correlates with the quality of the prior given as input. For the prediction of the final lesion, it also depends on the overlap between ischemic core and final lesion that strongly depends on clinical intervention. In this work, we choose to use our own open-source implementation of the ischemic core segmentation to ensure reproducibility. However, this could be improved by using commercially available software [3,29]. Moreover, for the prediction of the final lesion, a model segmenting the ischemic core such as the one proposed for the ISLES 2018 dataset, could be used to provide a prior of greater quality.

All models evaluated showed great inter-case variability and weak performance on small lesions located along the midline, as well as on lesions in the posterior fossa which remain difficult to detect. The prior segmentation failed to cover the same lesions and its integration could therefore not improve final performance. Moreover, results varied substantially between the two datasets, with greater scores achieved on the ISLES 2018 dataset. This can be explained by a more complicated task in the Geneva Stroke Dataset which consists of predicting a segmentation of the final infarct obtained several days after treatment. Greater inter-subject variability in this dataset mainly stems from a greater range of treatment administered to the patients, as well as highly divergent radiological outcomes.

In a clinical context, medical models rarely stand in isolation but are integrated with a prior representation of the problem. The segmentation and prediction of infarct evolution in acute ischemic stroke are challenging tasks for humans and machines alike. Threshold-based ischemic core segmentation has been largely validated and is commonly used as a starting point in this setting. Our proposed model can effectively leverage this prior segmentation to enhance the prediction of radiological outcome.

References

1. ISLES: Ischemic Stroke Lesion Segmentation Challenge (2018). http://www.isles-challenge.org/
2. Albers, G.W., et al.: Thrombectomy for Stroke at 6 to 16 Hours with Selection by Perfusion Imaging. New England J. Med. **378**(8), 708–718 (2018). https://doi.org/10.1056/NEJMoa1713973. Massachusetts Medical Society, publisher eprint
3. Shalini, A., et al.: Cerebral blood flow predicts the infarct core. Stroke **50**(10), 2783–2789 (2019). https://doi.org/10.1161/STROKEAHA.119.026640. https://www.ahajournals.org/doi/10.1161/STROKEAHA.119.026640. Publisher: American Heart Association
4. Belkin, M., Hsu, D., Ma, S., Mandal, S.: Reconciling modern machine learning practice and the bias-variance trade-off. arXiv:1812.11118 [cs, stat] (2019)
5. Benjamin Emelia, J., et al.: Heart disease and stroke statistics–2019 update: a report from the American Heart Association. Circulation, **139**(10), e56–e528 (2019). https://doi.org/10.1161/CIR.0000000000000659. https://www.ahajournals.org/doi/10.1161/CIR.0000000000000659. Publisher: American Heart Association
6. Campbell, B.C.V., Khatri, P.: Stroke. Lancet **396**(10244), 129–142 (2020). https://doi.org/10.1016/S0140-6736(20)31179-X. https://www.thelancet.com/journals/lancet/article/PIIS0140-6736(20)31179-X/abstract. Publisher: Elsevier
7. Campbell, B.C.V., et al.: Cerebral blood flow is the optimal CT perfusion parameter for assessing infarct core. Stroke, **42**(12), 3435–3440 (2011). https://doi.org/10.1161/STROKEAHA.111.618355. https://www.ahajournals.org/doi/10.1161/strokeaha.111.618355. Publisher: American Heart Association
8. Carrera, E., Wintermark, M.: Imaging-based selection of patients for acute stroke treatment: is it ready for prime time? Neurology **88**(24), 2242–2243 (2017). https://doi.org/10.1212/WNL.0000000000004051
9. Chen, Y., Li, Y., Zheng, Y.: Ensembles of modalities fused model for ischemic stroke lesion segmentation, p. 1
10. Clèrigues, A., Valverde, S., Bernal, J., Freixenet, J., Oliver, A., Lladó, X.: Acute ischemic stroke lesion core segmentation in CT perfusion images using fully convolutional neural networks. Comput. Biol. Med. **115**, 103487 (2019). https://doi.org/10.1016/j.compbiomed.2019.103487. http://www.sciencedirect.com/science/article/pii/S0010482519303555
11. Jenkinson, M., Pechaud, M., Smith, S.: BET2 - MR-Based Estimation of Brain, Skull and Scalp Surfaces, p. 1
12. Kistler, M., Bonaretti, S., Pfahrer, M., Niklaus, R., Büchler, P.: The virtual skeleton database: an open access repository for biomedical research and collaboration. J. Med. Internet Res. **15**(11) (2013). https://doi.org/10.2196/jmir.2930. https://www.ncbi.nlm.nih.gov/pmc/articles/PMC3841349/

13. Klug, J., et al.: Integrating regional perfusion CT information to improve prediction of infarction after stroke. J. Cerebr. Blood Flow Metab. (2020). https://doi.org/10.1177/0271678X20924549. https://journals.sagepub.com/doi/10.1177/0271678X20924549. Publisher: SAGE PublicationsSage UK: London, England

14. Lee, C.Y., Xie, S., Gallagher, P., Zhang, Z., Tu, Z.: Deeply-supervised nets. arXiv:1409.5185 [cs, stat] (2014)

15. Liu, P.: Stroke lesion segmentation with 2D novel CNN pipeline and novel loss function. In: Crimi, A., Bakas, S., Kuijf, H., Keyvan, F., Reyes, M., van Walsum, T. (eds.) BrainLes 2018. LNCS, vol. 11383, pp. 253–262. Springer, Cham (2019). https://doi.org/10.1007/978-3-030-11723-8_25

16. Lucas, C., Aulmann, L., Kemmling, A., Madany Mamlouk, A., Heinrich, M.: Estimation of the principal ischaemic stroke growth directions for predicting tissue outcomes, pp. 69–79 (2020). https://doi.org/10.1007/978-3-030-46640-4_7

17. Maier, O., et al.: ISLES 2015 - A public evaluation benchmark for ischemic stroke lesion segmentation from multispectral MRI. Med. Image Anal. **35**, 250–269 (2017). https://doi.org/10.1016/j.media.2016.07.009

18. Kim, M., Patrick, T., Greg, Z.: Artificial intelligence applications in stroke. Stroke **51**(8), 2573–2579 (2020). https://doi.org/10.1161/STROKEAHA.119.027479. https://www.ahajournals.org/doi/10.1161/STROKEAHA.119.027479. Publisher: American Heart Association

19. Muschelli, J., Ullman, N.L., Mould, W.A., Vespa, P., Hanley, D.F., Crainiceanu, C.M.: Validated automatic brain extraction of head CT images. Neuroimage **114**, 379–385 (2015). https://doi.org/10.1016/j.neuroimage.2015.03.074

20. Nielsen, A., Hansen, M.B., Tietze, A., Mouridsen, K.: Prediction of tissue outcome and assessment of treatment effect in acute ischemic stroke using deep learning. Stroke **49**(6), 1394–1401 (2018). https://doi.org/10.1161/STROKEAHA.117.019740. https://www.ahajournals.org/doi/full/10.1161/strokeaha.117.019740. Publisher: American Heart Association

21. Nogueira, R.G., et al.: DAWN trial investigators: thrombectomy 6 to 24 hours after stroke with a mismatch between deficit and infarct. N. Engl. J. Med. **378**(1), 11–21 (2018). https://doi.org/10.1056/NEJMoa1706442

22. Nosrati, M.S., Hamarneh, G.: Incorporating prior knowledge in medical image segmentation: a survey. arXiv:1607.01092 [cs] (2016)

23. Oktay, O., et al.: Anatomically constrained neural networks (ACNNs): application to cardiac image enhancement and segmentation. IEEE Trans. Med. Imaging **37**(2), 384–395 (2018). https://doi.org/10.1109/TMI.2017.2743464. Conference Name: IEEE Transactions on Medical Imaging

24. Oktay, O., et al.: Attention U-Net: learning where to look for the pancreas, p. 10

25. Péez-García, F., Sparks, R., Ourselin, S.: TorchIO: a Python library for efficient loading, preprocessing, augmentation and patch-based sampling of medical images in deep learning. arXiv:2003.04696 [cs, eess, stat] (2020)

26. Robben, D., et al.: Prediction of final infarct volume from native CT perfusion and treatment parameters using deep learning. Med. Image Anal. **59**, 101589 (2020). https://doi.org/10.1016/j.media.2019.101589. http://www.sciencedirect.com/science/article/pii/S136184151930129X

27. Schlemper, J., et al.: Attention gated networks: learning to leverage salient regions in medical images. Med. Image Anal. **53**, 197–207 (2019). https://doi.org/10.1016/j.media.2019.01.012. http://www.sciencedirect.com/science/article/pii/S1361841518306133

28. Song, T., Huang, N.: Integrated extractor, generator and segmentor for ischemic stroke lesion segmentation. In: Crimi, A., Bakas, S., Kuijf, H., Keyvan, F., Reyes, M., van Walsum, T. (eds.) BrainLes 2018. LNCS, vol. 11383, pp. 310–318. Springer, Cham (2019). https://doi.org/10.1007/978-3-030-11723-8_31

29. Straka, M., Albers, G.W., Bammer, R.: Real-time diffusion-perfusion mismatch analysis in acute stroke. J. Magn. Reson. Imaging: JMRI, **32**(5), 1024–1037 (2010). https://doi.org/10.1002/jmri.22338. https://www.ncbi.nlm.nih.gov/pmc/articles/PMC2975404/

30. Yu, Y., et al.: Use of deep learning to predict final ischemic stroke lesions from initial magnetic resonance imaging. JAMA Netw. Open **3**(3), e200772–e200772 (2020). https://doi.org/10.1001/jamanetworkopen.2020.0772. https://jamanetwork.com/journals/jamanetworkopen/fullarticle/2762679.   Publisher: American Medical Association

31. Çiçek, Ö., Abdulkadir, A., Lienkamp, S.S., Brox, T., Ronneberger, O.: 3D U-Net: learning dense volumetric segmentation from sparse annotation. In: Ourselin, S., Joskowicz, L., Sabuncu, M.R., Unal, G., Wells, W. (eds.) MICCAI 2016. LNCS, vol. 9901, pp. 424–432. Springer, Cham (2016). https://doi.org/10.1007/978-3-319-46723-8_49

Brain Tumor Segmentation

Brain Tumor Segmentation Using Dual-Path Attention U-Net in 3D MRI Images

Wen Jun, Xu Haoxiang[(⊠)], and Zhang Wang

School of Information and Software Engineering, University of Electronic
Science and Technology of China, Chengdu, China
201822090510@str.uestc.edu.cn

Abstract. Semantic segmentation plays an essential role in brain tumor diagnosis and treatment planning. Yet, manual segmentation is a time-consuming task. That fact leads to hire the Deep Neural Networks to segment brain tumor. In this work, we proposed a variety of 3D U-Net, which can achieve comparable segmentation accuracy with less graphic memory cost. To be more specific, our model employs a modified attention block to refine the feature map representation along the skip-connection bridge, which consists of parallelly connected spatial and channel attention blocks. Dice coefficients for enhancing tumor, whole tumor, and tumor core reached 0.752, 0.879 and 0.779 respectively on the BRATS- 2020 valid dataset.

Keywords: Brain tumor segmentation · U-Net · 3D convolution

1 Introduction

Comprising about 30% of all intracranial tumors, gliomas are one of the most common type of intracranial tumor with a highly variable clinical prognosis, and only one-fifth of gliomas are benign. Gliomas could lead to various symptoms, such as headaches, vomiting, seizures, and cranial nerve disorders. According to WHO's classification, the gliomas have four grades. Grade I and II are the Low-Grade Gliomas, namely LGG, which bring lower threatens to patients. Likewise, Grade III and IV are the High-Grade Gliomas, namely HGG, which bring higher threat. Gliomas could be divided into several components: the enhancing tumor (ET), tumor core (TC), and the whole tumor (WT). The TC subregion describes the bulk of the tumor and is usually removed. The TC subregion entails the ET, along with the fluid-filled and the solid parts of the tumor. The WT subregion describes the whole extent of gliomas.

Magnetic Resonance Imaging (MRI) is widely used in clinical diagnosis and it is an effective method to portray the inner heterogeneity of gliomas using different radiographic phenotypes. Based on their features, distinct images and different appearance of certain subjects could be obtained easily by changing the sequence of MRI scanning, and this makes it possible to depict valuable images of subregions of gliomas with different modalities. To employ MRI scans to segment gliomas, it is a critical procedure for its

© Springer Nature Switzerland AG 2021
A. Crimi and S. Bakas (Eds.): BrainLes 2020, LNCS 12658, pp. 183–193, 2021.
https://doi.org/10.1007/978-3-030-72084-1_17

therapy. With gliomas' high heterogeneity in different tumor appearances and shapes, there are many challenging tasks in the diagnosis.

Brain Tumor Segmentation Challenge (BRATS) is organized for years, focusing on finding state-of-the-art methods handling brain tumors in multi-parametric MR scans with computer technologies [1–5]. BRATS-2020 provides training datasets consist of both LGG and HGG MR scans, containing 60 subjects and 309 subjects respectively, and a valid set consists of 125 subjects. Each subject has four 3D brain sequences data in Nifty (nii.gz) format with segmentation masks for all subregions as well. The four modalities are structural (T1) images, T1-weighted contrast-enhanced (T1ce) images, T2-weighted images and fluid-attenuated inversion recovery (Flair). In the BRATS challenge, MR scans originate from 19 institutions, and are all annotated manually. One of the main tasks in the BRATS challenge is to segment brain tumors of different subjects into sub-components using MR scans, as it plays an essential role in diagnosis treatment planning.

Popular methods of brain tumor segmentation can be classified into either generative or discriminative models [4]. With Deep Neural Network (DNN) technology developing, it seems the most popular method during the last few BraTS challenges. Based on the Convolution Neural Networks (CNN), different models are proposed and various theories are extended. Many of the CNN give the state-of-the-art performance in the semantic segmentation domain, such as VGG [5], FCN [6], DeepMedic [7], U-net [8], etc. Among all these methods, U-net is the choice of the majority to handle medical image analysis issues, due to its fitness towards medical images. Provided with sufficient data, U-net will learn to generalize to the unseen type of gliomas. Thus, there came out of lots of variety of U-net.

Section 2 involves the details about our model, including the backbone architecture, the dual pathway attention gate, the res block and the loss function we choose. Section 3 shows the result on train, valid and test set, along with the training strategy and the metrics evaluating the proposed model. Comparison between different performances of different models is also given in Sect. 3. Section 4 mainly involves the analysis and discussion of result on the proposed model.

In this work, we combine the 3D U-net with a Dual Pathway Attention (DPA) inspired by 3D Convolutional Block Attention Module (CBAM) [9] and use the residual module [10], for brain tumor segmentation from patches of MR images, as an extension of basic 3D U-net. Contribution of this work is presented as follow:

- Based on the U-shaped structure of the U-net, an additional dual residual pathway is added to its encoder layers to enhancing transmission of the high-level features during down-sampling.
- Meanwhile, a variety of attentions block is employed to skip-connection to weight the feature map conveying from down-sample block to up-sample. The attention block is based on the idea of CBAM, which is composed of spatial attention block and channel attention block, we further refined it by replacing the spatial attention block with the original attention block to conserve precious graphic memory instead of 7×7 big kernel.

2   Method

2.1   Backbone Architecture for Segmentation

Due to the high performance of the U-net in medical image segmentation, we use the 3D variant of it as the backbone of our proposed network, which takes four modalities of brain tumor MR scans along with the ground truth segmentation as input, and train them to segment subregions like WT, TC and ET simultaneously with a single network.

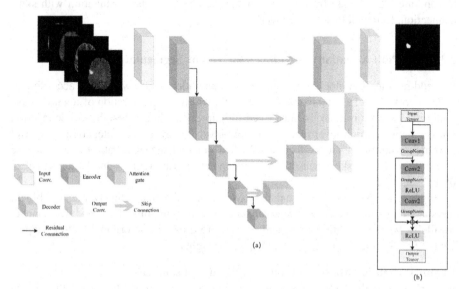

Fig. 1.   The architecture of our model. (a) the architecture of our model. (b) our modified Residual block with an additional pathway.

The proposed architecture in this work is shown in Fig. 1. As mentioned above, proposed network is a variant of 3D U-net. Based on its encoder-decoder structure with skip-connection, we improved its performance mainly by two methods: ① modified residual pathway along the down-sample layers, called dual-path residual module; ② a modified attention module. As it is shown in Fig. 1 (a), our net basically encompasses five down sample layers and four up-sample layers, namely Dual Pathway Attention (DPA) block.

2.2   Dual Residual Block

Inspired by the MFnet [11] and residual net, the residual module is modified into two pathways, each transmits a different part of the feature map along down-sampling layers. The structure of the dual-path residual module is shown in Fig. 1 (b), an extra pathway is added after the first convolution to reserve low-level features which contain more spatial information to locate the Region of Interest. Besides, this structure help solve the degradation problem that would occur to deep neural networks as well. Exerting this module to up sampling module would cause out of memory error, so the experiment goes only on down-sample layers for now. We plan to add it to up sample layers after the network is optimized and consumption of memory is reduced.

Each down-sample layer contains a max-pooling layer, a Rectified Linear Unit (ReLU) module and three basic 3D convolution layers followed by group normalization [12], additionally dual-path residual module is added to its encoder in order to ease the degradation problem for deep networks.

Each up-sample layer contains a basic trilinear up sample, two convolution layers to compute feature map from last down sample layer and its combination with skip-connection, followed by a ReLU layer.

2.3 Dual-Path Attention (DPA) Block for Tumor Segmentation

Embedding attention block into U-net is not a new idea, and it works fine according to former works [13–15]. Realizing that attention will help recalibration of feature maps, a modified attention block is introduced to skip connection. Though skip-connection bridges between high-level feature and low-level feature by concatenation, low-level information could work less effectively. To deal with that, the Dual-Path attention module is employed to help skip connections focus on the relevant regions and features instead of just sum them up.

Figure 2 shows that, there are mainly two parts of the dual pathway attention block: the spatial attention module and the channel attention module. Two modules connect parallelly, with an additional residual pathway to transmit the original feature map as the base bone of the channel-wise and spatial weights.

Channel Attention Module. Inspired by the idea of squeeze-and-Excitation (SE) network [16], a small channel-wise attention module is employed to extract the hidden features along the channel. The SE module is a bottleneck attention gate variety, comprised of three main steps as could be told from its name: squeeze, excitation, and scale. Channel attention pathway in this work is implemented as a bottleneck attention gate, using 3D max-pooling and average-pooling together to shrink the size of input feature map finally into 1, and the outputs of each pooling layer are followed by a bottleneck consist of two convolution layers to learn the relation between different channels and find out which feature maps are relevant. Different from the original SE module, two kinds of pool strategies are used at the same time instead of using a single one. In the CBAM module, the two pool layers shared the same bottleneck to take the average result of two different pool strategies, but in this work separated bottlenecks to deal with the outputs of two pool layers instead. We believe that this improvement makes results more robust than the original implementation. The channel attention outputs the computed weight of each input channel, and the channel weights are used to measure the feature information each channel contains. After getting the channel weight, the model is supposed to focus on the channels containing more features and improve the final segment result.

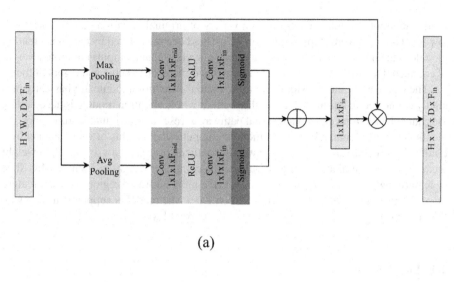

(a)

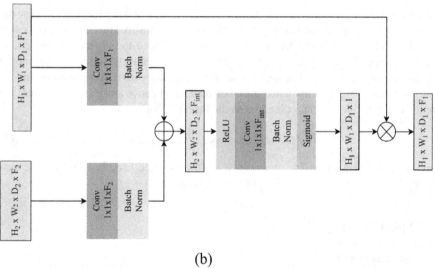

(b)

Fig. 2. Details for our attention block. Our attention module consists of two parts: (a) Spatial Attention Module and (b) Channel Attention Module. Channel extract the channel wise information from feature map from skip connection to reweight the channel.

Spatial Attention Module. As a classic module, the attention gate is popular among all computer vision domains. Employing attention module to the network could provide automatically focus mechanism towards target regions. U-net uses skip connection to fuse high-level features and low-level features. Generally, low-level features contain more spatial information that helps network locate the target area. Simply conjoin low-level features and high-level features together would not put them into fully utilization. The spatial attention module in front of concatenation will provide improvement. The spatial attention module we employ is a variety of original attention gate, it takes both low-level features from skip-connection which has been reweighted by channel attention module, and high-level features from up sample layers as input, outputs computed weight of conducted feature maps to reweight the spatial dimension to achieve the goal of focus.

The Channel Attention Module and the Spatial Attention module are basically parallelly connected in our model. First, the input of whole attention module is accepted by the channel attention gate and a residual pathway to reserve its original feature maps. As mentioned above, channel attention module computes the weights of each channel by a bottleneck structure and connects the weights with the original feature maps to reweight them. And the spatial attention gate outputs the computed weight of a single MRI slice or feature map to make sure model focus on the most informed regions of each feature map. Contacting the outputs of two attention gates, the model is supposed to focus on the most informed channels and regions of the current layer that are well trained.

2.4 Loss Function

A negative patch is a patch that centers on the non-tumor region, which tends to raise the possibility of false positive. Likewise, having too much positive patch do the opposite. Thus, both negative and positive patches would be needed to balance the model. As randomly cropped original data to $128 \times 128 \times 128$ patches as our input, a large amount of input patches is found containing no area of interest, this leads training to an imbalance issue. To moderating the imbalance issue between the negative and positive patches, we employ combined multi-class dice loss as the loss function.

$$L = \lambda_{dice} \cdot L_{dice} + \lambda_{ce} \cdot L_{Cross\ Entropy} \tag{1}$$

Here in the first equation, L_{dice} represents multi-class dice loss, L_{ce} represents cross-entropy loss, λ_{dice} and λ_{ce} their weight respectively.

3 Experiments

3.1 Datasets Detail

The datasets used in this experiment are provided by the BraTS 2020 Challenge. The training dataset contains 369 subjects in total, and each subject consists of four different modalities of MR scans, respectively T1, T2, T1-enhanced (T1C), T2 with Fluid-Attenuated Inversion Recovery (FLAIR), and the segmentation mask segmented manually. The size of each MR image is $240 \times 240 \times 150$. And the valid data include 125 subjects and share the same format with training data despite it does not include ground truth. In the experiment, training datasets is used to train and optimize our model to its best performance, and evaluate it using valid data on the BraTS 2020 official portal.

3.2 Preprocessing Methods

The model takes all four modalities as input, each modality is treated as a single channel of the model, that makes the model input channel as 4. Due to the distribution imbalance of values of different modalities, normalization such as z-score should be exerted on the original data before we feed it into networks. Also, we pack the data in sequence into the middle format in order to save data-reading time and resources.

Comparing to the 2D version, 3D U-Net consumes much more memory. In order to run 3D U-net on single 1080Ti with 11GB memory, we randomly crop the original MR images in sizes of $240 \times 240 \times 155$ into $128 \times 128 \times 128$ patches and feed it into our model.

3.3 Evaluating Metrics

Three metrics are used to evaluate the submissions of different models in the segmentation task:

Dice Coefficient Score. The dice coefficient score is employed to evaluate the similarity on the area of intersections between two graphics, here it refers to the similarity between the ground truth segment mask and the prediction segment mask. It receives popularity in medical image segmentation domains.

Hausdorff Distance. Differ with the Dice coefficient, Hausdorff distance measures the similarity by computing the distance between two different subsets. Due to its nature, Hausdorff distance is highly sensitive to the error of the graphic border. BraTS use 95% Hausdorff distance as the criteria, which is a variety of Hausdorff distance based on 95% of the distances between subsets to eliminate the influence of a small subset of outliers.

$$d_H = \max\{\max_{x \in X} \min_{y \in Y} d\{x, y\}, \max_{y \in Y} \min_{x \in X} d\{x, y\}\} \tag{2}$$

Sensitivity and Specificity. Sensitivity measures the proportion of true positives, and specificity measures the proportion of true negatives.

$$Sensitivity = \frac{TP}{TP + TN} \tag{3}$$

$$Specificity = \frac{TN}{TN + FP} \tag{4}$$

3.4 Experimental Results

Proposed network is build using PyTorch, trained on 1080Ti GPU with 11G RAM with BraTS2020 train set. As putting voxel into GPU could consume a lot of memory, batch size is set only to 2. The initial learning rate is set to 0.01, using SGD optimizer with momentum set to 0.80, batch normalization is the Group Batch Norm.

Table 1 shows the average results of different models on BraTS 2020 valid data, including mean dice accuracies and mean hausdorff_95. Our model achieves to outperformance original 3D U-net or 3D U-net with single Attention or with Residual module.

The results show that our model raises the dice coefficient score of ET and TC by 1% and 3% respectively, thus we think our model achieves a relatively good improvement on BraTS segmentation task. Trained on the same dataset with same hyper-parameters, our model still outperforms the best of its baseline.

Table 1. Comparison of Average results of different models on BraTS 2020 valid data

Model name	Mean dice			Mean Hausdorff_95		
	Enh.	Whole	Core	Enh.	Whole	Core
Baseline U-net	0.734	0.884	0.763	34.05	6.53	17.62
U-net with Attention	0.743	0.879	0.756	36.71	**5.49**	19.76
U-net with Residual	0.741	**0.888**	0.748	29.41	7.68	24.64
Proposed Model	**0.752**	0.878	**0.779**	30.65	6.30	**11.02**

Table 2 show the detailed statistics result of our model, evaluated on BraTS 2020 valid dataset on BraTS official portal. The standard deviation is 0.282, 0.112 and 0.199 for dice coefficient score of ET, WT and TC respectively. This means that our method could provide a relatively stable and reliable results on unseen brain tumors.

Meantime, the Hausdorff distance on ET subregion is high, which is considered having outliers on our segmentation results caused by inappropriate sample strategy or training methods. We plan to refine that using cascade by dividing the model into two steps: locate and segment.

Not very competitive, though, we hope our method to provide improvement to others' methods.

Table 2. Detail result of our model for the three tumor subregions evaluated on valid set

	Dice score			Specificity			Hausdorff_95		
	ET	WT	TC	ET	WT	TC	ET	WT	TC
Mean	**0.752**	0.878	**0.779**	0.999	0.9989	0.999	30.65	6.30	11.02
Std. dev	0.282	0.112	0.199	0.0005	0.0011	0.0007	96.09	10.03	34.49
Median	0.853	0.909	0.685	0.999	0.999	0.999	2	3.46	5

Trained the proposed model is used to segment the training data as well, and Table 3 shows the result. Among all 369 subjects, the results contain several outliers that dice coefficient score are zero or near zero. This leads to the fact that the Mean Hausdorff_95 is relatively high but the Median Hausdorff_95 remains a reasonable range and far lower than the mean value.

Table 3. Detail result of our model for the three tumor subregions evaluated on train set

	Dice score			Specificity			Hausdorff_95		
	ET	WT	TC	ET	WT	TC	ET	WT	TC
Mean	0.823	0.912	0.878	0.999	0.9989	0.999	13.77	4.21	4.44
Std. dev	0.188	0.053	0.108	0.0005	0.0007	0.0007	60.50	4.99	5.42
Median	0.875	0.925	0.911	0.999	0.999	0.999	1.41	3	2.82

Also, the proposed model is evaluated on testing data provided by BraTS2020. Testing data contains 166 subjects in total. Likewise, each testing data contains four modalities without ground truth. The detailed result is shown in Table 4.

Table 4. Detail result of our model for the three tumor subregions evaluated on test set

	Dice score			Specificity			Hausdorff_95		
	ET	WT	TC	ET	WT	TC	ET	WT	TC
Mean	0.773	0.861	0.790	0.999	0.9987	0.999	16.95	7.45	28.78
Std. Dev	0.229	0.133	0.270	0.0004	0.0012	0.0007	69.50	12.88	88.06
Median	0.875	0.925	0.911	0.999	0.999	0.999	1.41	4	3.39

4 Discussion

Inspired by the works of predecessors, we present a variety of 3D U-net with CBAM based attention gate and dual-path residual module, to handle the segmentation tasks of brain tumor with multi modalities with a single model and limited resources.

The attention module used here is modified to two parallel pathways contacted with extra residual. Adding a channel-wise attention would help us reweight the feature map in up sample layers, and focus on the feature maps contains more information. In order to reduce the random error, a dual pool pathway, max pool and average pool is proposed. Meanwhile, the spatial attention module is implemented by the original attention gate. Yet, we found that this modification brings some troubles to segmenting small objects. Thus, in the future, we plan to develop it to a better performance using a split attention mechanism.

According to the results of three datasets, the proposed model provides a significant improvement comparing to the base model. Also, we find that there is a common issue with the segment method. The hausdorrf_95 distance is relatively high comparing to others' methods. We find that among all subjects used to train, there are several subjects is wrongly handled, makes ET subregions of these subjects remains zero or near zero and that is the main reason makes the mean ET dice score of each data set lower and makes the mean hausdorff_95 relatively high. The fact that the mean hausdorff_95 is much higher than the median also reveals the potential cause, namely the performance on ET subregion is influenced by the outliers. The most possible reason is the inappropriate sampling strategy. Input images are cropped into 128×128 pieces randomly to save memory, that might lead to potential information lost and result in incorrect or insufficient feature. We mean to refine that issue later by use a randomly crop with center to prevent feature lost.

Meanwhile, training a 3D model with limited resources is a tough way to go. 1080Ti with 11GB RAM could work setting the batch size of 2, but it's still touching the edge of out of memory. We will improve the proposed model to prevent this problem by introducing dilated convolution to our model. After all, 3D models usually outperform 2D ones at accuracies, which makes all consumption worthy. In the future, we will perform some refinement to this model like introducing GAN [17], split attention mechanism or CRF to it to improve the final score. Cascade network is also a good choice to try.

5   Conclusion

In this work, a refined attention mechanism with dual pathway and a double pathway residual block are introduced to improve the performance on brain tumor segmentation. Dual pathway attention gate could help the network focus not only on spatial feature area but also on target-related channels. Double pathway residual block embedded in the down sample layers of the U-net to prompt the feature transmission. These two blocks are introduced to a 3D U-net. As for the training strategy, we use random crop to avoid feeding whole subjects' voxels to graphic card and to reduce the proportion of false positive samples. The proposed model evaluated on both train, valid and test datasets. The dice score of ET, WT and TC on train set are 0.823, 0.912 and 0.878 respectively. The dice score of ET, WT and TC on valid set are 0.752, 0.878 and 0.779 respectively. The dice score of ET, WT and TC on test set are 0.773, 0.861 and 0.790 respectively.

Acknowledge. The project is supported by Sichuan Science and Technology Program. It is partially funded by Grant SCITLAB-0013 of Intelligent Terminal Key Laboratory of SiChuan Province.

References

1. Bakas, S., et al.: Segmentation labels and radiomic features for the pre-operative scans of the TCGA-LGG collection. Cancer Imaging Arch. **286** (2017)
2. Bakas, S., Akbari, H., Sotiras, A., et al.: Advancing The Cancer Genome Atlas glioma MRI collections with expert segmentation labels and radiomic features. Sci. Data **4**, 170117 (2017). https://doi.org/10.1038/sdata.2017.117
3. Bakas, S., et al.: Identifying the best machine learning algorithms for brain tumor segmentation, progression assessment, and overall survival prediction in the BRATS challenge. CoRR abs/1811.02629 (2018). (1811)
4. Bakas, S., et al.: Segmentation labels and radiomic features for the pre-operative scans of the TCGA-GBM collection. Cancer Imaging Arch. Nat. Sci. Data **4**, 170117 (2017)

5. Menze, B.H., et al.: The multimodal brain tumor image segmentation benchmark (BRATS). IEEE Trans. Med. Imaging **34**(10), 1993–2024 (2014)

6. Agravat, R.R., Raval, M.S.: Deep learning for automated brain tumor segmentation in MRI images. In: Soft Computing Based Medical Image Analysis, pp. 183–201. Elsevier (2018)

7. Kamnitsas, K., et al.: DeepMedic for brain tumor segmentation. In: Crimi, A., Menze, B., Maier, O., Reyes, M., Winzeck, S., Handels, H. (eds.) BrainLes 2016. LNCS, vol. 10154, pp 138–149. Springer, Cham (2016). https://doi.org/10.1007/978-3-319-55524-9_14

8. Ronneberger, O., Fischer, P., Brox, T.: U-net: convolutional networks for biomedical image segmentation. In: Navab, N., Hornegger, J., Wells, W.M., Frangi, A.F. (eds.) MICCAI 2015. LNCS, vol. 9351, pp. 234–241. Springer, Cham (2015). https://doi.org/10.1007/978-3-319-24574-4_28

9. Woo, S., Park, J., Lee, J.-Y., Kweon, I.S.: CBAM: convolutional block attention module. In: Ferrari, V., Hebert, M., Sminchisescu, C., Weiss, Y. (eds.) ECCV 2018. LNCS, vol. 11211, pp. 3–19. Springer, Cham (2018). https://doi.org/10.1007/978-3-030-01234-2_1

10. He, K., et al.: Deep residual learning for image recognition. In: Proceedings of the IEEE Conference on Computer Vision and Pattern Recognition (2016)

11. Ha, Q., et al.: MFNet: towards real-time semantic segmentation for autonomous vehicles with multi-spectral scenes. In: 2017 IEEE/RSJ International Conference on Intelligent Robots and Systems (IROS). IEEE (2017)

12. Wu, Y., He, K.: Group normalization. In: Ferrari, V., Hebert, M., Sminchisescu, C., Weiss, Y. (eds.) ECCV 2018. LNCS, vol. 11217, pp. 3–19. Springer, Cham (2018). https://doi.org/10.1007/978-3-030-01261-8_1

13. Oktay, O., et al.: Attention U-Net: learning where to look for the pancreas. arXiv preprint arXiv:1804.03999 (2018)

14. Abraham, N., Khan, N.M.: A novel focal Tversky loss function with improved attention u-net for lesion segmentation. In: 2019 IEEE 16th International Symposium on Biomedical Imaging (ISBI 2019). IEEE (2019)

15. Li, S., et al.: Attention dense-u-net for automatic breast mass segmentation in digital mammogram. IEEE Access **7**, 59037–59047 (2019)

16. Cheng, D., et al.: SeNet: structured edge network for sea–land segmentation. IEEE Geosci. Remote Sens. Lett. **14**(2), 247–251 (2016)

17. Mondal, A.K., Dolz, J., Desrosiers, C.: Few-shot 3D multi-modal medical image segmentation using generative adversarial learning. CoRR abs/1810.12241 (2018). https://arxiv.org/abs/1810.12241

Multimodal Brain Image Analysis and Survival Prediction Using Neuromorphic Attention-Based Neural Networks

Il Song Han[✉]

Odiga, London, UK
ishan.super@gmail.com

Abstract. Accurate analysis of brain tumors from 3D Magnetic Resonance Imaging (MRI) is necessary for the diagnosis and treatment planning, and the recent development using deep neural networks becomes of great clinical importance because of its effective and accurate performance. The 3D nature of multimodal MRI demands the large scale memory and computation, while the variety of 3D U-net is widely adopted for medical image segmentation. In this study, 2D U-net is applied to the tumor segmentation and survival period prediction, inspired by the neuromorphic neural network. The new method introduces the neuromorphic saliency map for enhancing the image analysis. By mimicking the visual cortex and implementing the neuromorphic preprocessing, the map of attention and saliency is generated and applied to improve the accurate and fast medical image analysis performance. Through the BraTS 2020 challenge, the performance of the renewed neuromorphic algorithm is evaluated and an overall review is conducted on the previous neuromorphic processing and other approach. The overall survival prediction accuracy is 55.2% for the validation data, and 43% for the test data.

Keywords: Neuromorphic-attention · Brain-inspired processing · Survival prediction

1 Introduction

Brain tumor causes high mortality, and the early diagnosis is crucial for treatment planning, monitoring and analysis. The accuracy of categorical estimates ranged from 23% up to 78% for the survival prediction among the expert clinicians [1], while there are a few challenges including variations in image acquisition and lack of robust prognostic model. Automated brain image analysis would greatly aid the diagnosis, as this can alleviate the need for a clinician with a high level of training and experience and reduce the time required for image analysis [2–6].

Recent performances of deep learning methods, specifically Convolutional Neural Networks (CNNs), in several object recognition and biological image segmentation challenges increased their popularity among researches. Automatic segmentation of brain tumor has also drawn a lot of attention in the recent years due to the availability of open medical image datasets and the rapid development of CNNs [7–9].

© Springer Nature Switzerland AG 2021
A. Crimi and S. Bakas (Eds.): BrainLes 2020, LNCS 12658, pp. 194–206, 2021.
https://doi.org/10.1007/978-3-030-72084-1_18

The outstanding side of human intelligence is the presence of attention, considering the implementation of artificial intelligence. In the visual cognitive process, the attention allows for salient features to dynamically come to the forefront as needed. Attention-based learning has become one of promising approach in the research on artificial intelligence of visual recognition.

2   Methods

2.1   Neuromorphic Neural Network Inspired by Visual Cortex

Within the human brain, cognitive and perceptual processes are carried out in the regions of neocortex, with visual input being processed in the occipital lobe. Although there is no definite model of visual cortex, Hubel and Wiesel's research on feline vision demonstrated the cells in visual cortex responds to the lines of different orientations. At the same time, none of the cells responded to any specific objects like hands, face or cars, while human beings would normally recognize objects visually.

Neuromorphic neural network is based on various orientation selective features. The orientation selectivity inspired by 'simple cell' in visual cortex, provides the robustness of abstract features extraction as shown in Fig. 1 [10].

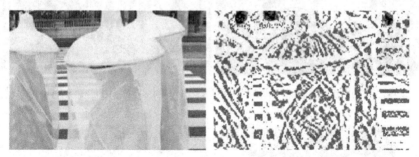

Fig. 1.   An example of neuromorphic visual information processing to provide robustness in case where the subject is shrouded in clothing

Through the down-up resizing network, we can get the saliency map from extracted abstract features. The saliency map is effective in removing noise or generating the attention, and it shows the tooth segmentation in noisy gum area in Fig. 2. The tooth saliency maps enable the clear separation of each tooth, with the collected DICOM layers of CBCT images. The segmented tooth is illustrated in Fig. 3, where a sub image on top right shows the segmented tooth in vertical cross section of sloped frontal tooth. The complex mixture of tissues and bones adds to the challenge of tooth segmentation, while a small amount of gum tissue is observed in the top right sub image [11].

Images in Fig. 4 exhibit the case of neuromorphic processing for image transformation. A new abstract image is generated by mixed networks of U-net for BraTS 2019 [12] and neuromorphic neural network, and it shows the feasibility of image pre-processing or image post-processing by neuromorphic neural network.

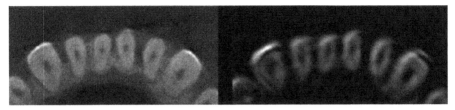

Fig. 2. Saliency map of neuromorphic neural network for segmentation of 3-dimensional teeth, (left) CBCT image of tooth roots in the gum (right) segmented tooth roots

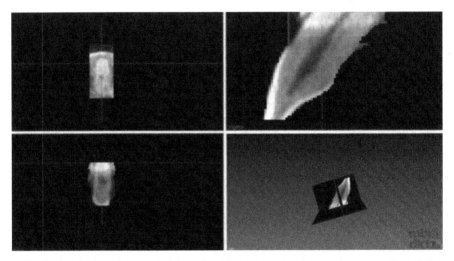

Fig. 3. 3D display of segmented frontal tooth, horizontal and vertical cross-sections [11]

Fig. 4. Abstract image generation by neuromorphic processing and U-net used for Brain Tumor segmentation at BraTS 2019, (left) original photography (right) transformed image

2.2 Neuromorphic Pre-processing for Convolutional Neural Networks with Neuromorphic Attention-Based Learner

The U-net of Fig. 5 aims to guide the learning process by an attached CNN of pre-defined kernels of neuromorphic function, in parallel. Its orientation selectivity mimicking the visual cortex provides the attention-based learning process, as the pre-determined algorithm of estimating the target object areas [12]. The detail of neuromorphic attention module was presented at BraTS 19 challenge, which demonstrated the effectiveness of training 3D objects under the constraint of 2D U-net [12]. Recently, the attention mechanism is widely applied to the deep learning for better performance, including medical imaging analytics [13].

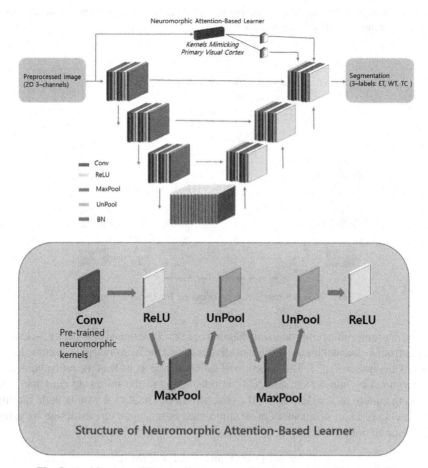

Fig. 5. Architecture of U-net with neuromorphic attention-based learner [12]

The earlier 2D U-net with neuromorphic attention module was operated on the input image data of 3 channels, instead of original 4 channels [12]. In this study, 4-channel images of MRI data are processed as 3-channel input images in Fig. 5, based on the saliency map. The saliency maps are generated by 2D neuromorphic neural network of Fig. 6, which are NM1 and NM2 used to predict the overall survival period [14].

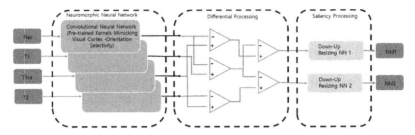

Fig. 6. 2D neuromorphic neural network for the overall survival analysis of brain tumor patient, without using the ground truth of segmentation [14]

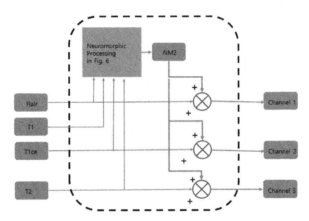

Fig. 7. Neuromorphic preprocessing for U-net in Fig. 5

The neuromorphic pre-processing diagram of Fig. 7 is designed to generate 3-channel images from 4-channel images. The objective is to add the neuromorphic features to the image data input in Fig. 5. The selection of saliency map as well as the ratio coefficient is determined by simulation, and NM2 is selected as in the following equation. The following equations are derived by the observation of simulation results with varying parameters, as a sort of manual optimization used in analogue circuit design by using SPICE simulation.

$$(\text{Channel 1}) = (\text{Flair}) + 0.2 \times (\text{NM2}) \tag{1}$$

$$(\text{Channel 2}) = (\text{T1CE}) + 0.2 \times (\text{NM2}) \tag{2}$$

$$(\text{Channel 3}) = (\text{T2}) + 0.2 \times (\text{NM2}) \tag{3}$$

2.3 Feature Selection for Survival Days Prediction

The feedforward neural network in Fig. 8 is designed with 4 inputs for the overall survival period prediction. The structure of feed-forward neural network in Fig. 8 is $4 \times 450 \times 450 \times 250 \times 150 \times 3$, where 3 outputs represent the survival period in 3 categories of short period (less than 10 months), medium period (between 10 and 15 months) and long period (more than 15 months). The segmentation result of WT is included as an input, based on the simulation of different input configurations as illustrated in Fig. 9, Fig. 10, Fig. 11 and Fig. 12. In facts, there are different views on the overall survival prediction from earlier BraTS challenges [15–19]. First, the algorithms of top-ranks differ extensively in terms of input parameters, from a single parameter of 'Age' to complicated 16 parameters. Also, it exhibits the limited example of progressive development, as seen from an example of relatively declined performance among top-rank results.

It is challenging to design the optimum parameters of inputs, outputs, and neural network structure, and the structure in Fig. 8 is determined by the performance shown in training. For an example, more categories are applied and the output of 3 categories is chosen from the performance comparison in Fig. 9.

There is rare guideline to an efficient design of input parameters for the overall survival prediction, as observed in the earlier work. With no further clinical information or expertise, the feasible evaluation can be derived by the comparison of confusion matrix using training dataset. Hence the case of 4 inputs is preferred, while it is still uncertain about its influence on post-processing.

There is an overfitting issue of feedforward neural network, which is affected by the network architecture and training process. It is quite challenging to decide how far the neural network can avoid the over-fitting, while achieving the best performance. In Fig. 11, the substantial increase in prediction accuracy is observed from 68.9% to 86.4%, with more iterations in training. Assuming the validation framework as realistic test simulation, the on-line validation of BraTS 2020 leaderboard is employed for the performance estimation. In Fig. 8, the survival days are converted from the neuron state of predicted category, though there is inherent uncertainty. Figure 12 shows another training performance of survival period prediction, based on 3 categories. With more training iterations, the confusion matrix shows the better performance, however there is no guarantee of the higher accuracy for the dataset of different characteristics.

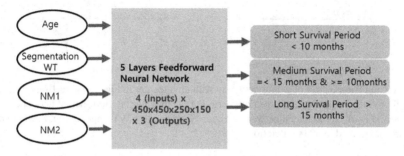

Fig. 8. Fully connected neural network for the diagnosis of overall survival prediction

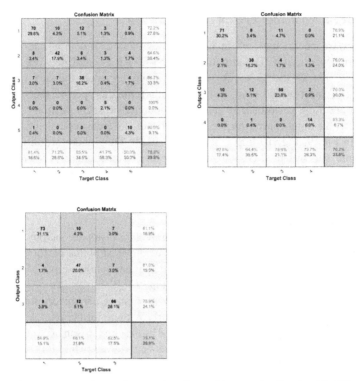

Fig. 9. Confusion matrix of neural networks with architecture in Fig. 8, with different output categories, (top left) neural network with 4 inputs and 5 output, (top right) neural network with 4 inputs and 4 outputs, (bottom left) neural network with 4 inputs and 3 outputs, where all networks are trained in 10,000 iterations.

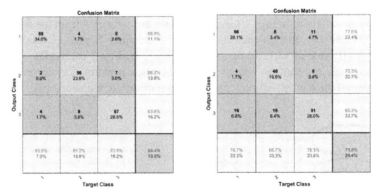

Fig. 10. Confusion matrix of neural networks with architecture in Fig. 8, where 1 represents the survival of the short period, 2 the medium period, 3 the long period, (left) neural network with 4 inputs as in Fig. 8, (right) neural network without 'Segmentation WT' input in Fig. 8, where both networks are trained in 50,000 iterations.

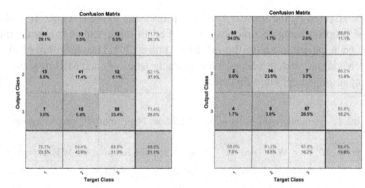

Fig. 11. Confusion matrix of neural network in Fig. 8, where 1 represents the survival of the short period, 2 the medium period, 3 the long period, (left) trained in 1000 iterations (right) trained in 50,000 iterations.

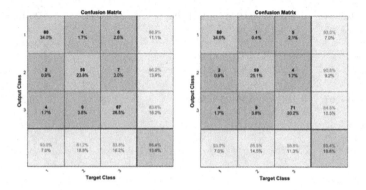

Fig. 12. Confusion matrix of neural network in Fig. 8, where 1 represents the survival of the short period, 2 the medium period, 3 the long period, (left) trained in 50,000 iterations (right) trained in 75,000 iterations.

3 Result and Discussion

3.1 Tumor Segmentation

The performance of segmentation is in Table 1, which includes the validation result of BraTS 2020, the test result of BraTS 2019 and this year's test result. The same structure of U-net with neuromorphic attention module is used, while each network is trained by it relevant data. In general, the limited performance can be observed comparing to other 3D U-net approaches. There is a similarity in DICE coefficients, with a small degradation for the case of neuromorphic pre-processing. For Hausdorff95 coefficients, there is a substantial improvement during the validation phase evaluation, with the case of neuromorphic pre-processing. The accuracy in test result decline comparing to both the validation phase and the last year's test (using the same network structure). If the assumption of equivalent training dataset can be applied to both last year and this year, the result can be affected by the sensitivity of pre-processing to the test data. It is observed that various preprocessing techniques are applied to compromise the wide intensity variation of each patient, even for the validation phase [20]. The algorithm of Fig. 6 can demand such adaptation depending on the characteristics of each MRI modality, where no such adaptation is applied to this study.

Table 1. Segmentation results with and without Neuromorphic preprocessing

	Dice_ET	Dice_WT	Dice_TC	Hausdorff95_ET	Hausdorff95_WT	Hausdorff95_TC
BraTS 2020 network using the 2020 validation dataset	0.5363	0.7717	0.6634	7.18931	8.71056	11.67522
BraTS 2019 network using the 2019 test dataset	0.5944	0.8028	0.6994	15.18193	17.24771	20.10342
BraTS 2020 network using the 2020 test dataset	0.4567	0.72837	0.56123	51.557	12.18	20.206

3.2 Overall Survival Prediction

The performance of overall survival prediction is summarized in Table 2, which includes the validation results of three different training in addition to the test result. As observed in Table 2, the best accuracy is from the network of 50,000 training iterations. With more training iterations of 75,000, the accuracy with validation data becomes worse. Also with less training iterations of 40,000, the accuracy decreases too. The accuracy using validation data is consistently lower than the accuracy using training data, which warns the possibility of over-fitting. Observing the validation result, the latest one improves the result of accuracy and error comparing the previous case without pre-processing.

The test result shows the accuracy 43% and there exist different reasons for a reduction of more than 10% in accuracy. From a neural network perspective, one reason is the overfitting and the other is the input data beyond the trained capacity. For the case of overfitting, the applied network is selected among validation results using different training process. It can avoid the overfitting issue to some degree, as there has been no further adaptation or optimization process to fit to the validation data. Assuming the dataset of similar characteristics between validation data and training data, the similar

Table 2. Performance evaluation

Feedforward Network	Accuracy	MSE	medianSE	stdSE	SpearmanR
50,000 iterations using BraTS 2020 validation data	0.552	147,898.5	52,900	162,551.7	0.333
75,000 iterations using BraTS 2020 validation data	0.414	162,916.1	75,625	171,458.1	0.231
40,000 iterations using BraTS 2020 validation data	0.517	145,296	42,849	159,594.2	0.317
BraTS 2019 network using BraTS 2019 test data	0.486	552,104.5	217,156	891,476.5	0.253
BraTS 2020 network with 5000 iterations using BraTS 2020 test data	0.43	471,979.2	102,400	1,049,491	0.051

performance is expected with a little margin. The other cause can be from the different characteristics of input data, which forces the neural network beyond its normal operation range. An appropriate processing can mitigate the issue as seen from normalization algorithms of pre-processing or post-processing [16, 20]. There is still a room to think about the cause as the unexpected variation of data characteristics, considering the case of decrease accuracy between the best results from BraTS 2018 (61%) to BraTS 2019 (57.9%).

4   Conclusion

The 2D U-net with neuromorphic attention has been revised by adding the neuromorphic pre-processing to the input image, which confirmed the improved performance on Hausdorff95 coefficients during the validation process. The accuracy of survival prediction is observed as 55.2% during the validation. Though 2D network methods are with the advantage of demanding less computing resource and computation time, the explicit advantages of 3D deep networks and ensemble architectures show the performance gap between 2D network-based approach and 3D network-based approach in terms of tumor segmentation.

This study has been motivated by the goal of developing AI technology that can diagnose brain tumors like clinicians. As a solution to the fact that multimodal MRI brain data are with a lot of intensity variation and that it takes a long period to accumulate experience as a clinician, a neuromorphic neural network that recognizes images in adverse illumination and noise has been applied. The first application of 2D neuromorphic neural network without any clinical knowledge or the ground truth of tumor segmentation achieved the overall survival prediction accuracy of 50.1% at BraTS 2018. When applying the 2D neuromorphic neural network to BraTS 2018 challenge, the accuracy was achieved through the adjustment of neuron threshold, without using 'Age' data. In BraTS 2019, the 2D neuromorphic neural network added with 'Age' data was applied to obtain the overall survival prediction accuracy of 55.2% in validation phase and 48.6% in test phase. In the test phase of BraTS 2019, the case of null neuron output was observed too excessively, resulting in the issue of intuitively changing the threshold. This year, the result of overall survival prediction is the accuracy of 55.2% from validation phase and 43% from the test phase, using a 2D neuromorphic neural network with another added input of tumor segmentation result from a 2D U-net. Adaptation according to data characteristics was not adopted in this year, and in particular, the task of test phase was done with 2D neuromorphic neural network and 2D U-net on a mobile computer during the move at the airport.

Analyzing the results of a series of studies has led to the need for a neuromorphic neural network that autonomously adjusts some of its variables to the characteristics of the multimode MRI data of given task. This can be expected from the case of other studies that have introduced various preprocessing algorithms or medical analyzes for the overall survival prediction [15–19]. However, it can be necessary to understand the dataset of test phase further, for the overall survival prediction. Considering the current development, there is a concern that there may be limitations with general deep learning interpretation methods or medical analysis models for the overall survival prediction.

The perspectives adopting 2D networks can be found from exploring the effective and compact AI algorithm for certain applications, implementing the automated diagnostics for assisting the clinical decision. Survival period prediction is one of those feasible cases, though it is still ambiguous in terms of status of art. The big gap in performance is more frequent in survival prediction using the variety of research methods, compared to the tumor segmentation.

References

1. White, N., Reid, F., Harris, A., Harries, P., Stone, P.: A systematic review of predictions of survival in palliative care: how accurate are clinicians and who are the experts? PLOS One (2016). https://doi.org/10.1371/journal.pone.0161407 of Palliat. Med. **5**(1), 22–29 (2016). https://doi.org/10.3978/j.issn.2224-5920.2015.08.04
2. Menze, B., Jakab, A., Bauer, S., Kalpathy-Cramer, J., Farahani, K., Kirby, J., et al.: The multimodal brain tumor image segmentation benchmark (BRATS). IEEE Trans. Med. Imaging **34**(10), 1993–2024 (2015). https://doi.org/10.1109/TMI.2014.2377694
3. Bakas, S., Akbari, H., Sotrias, A., Bilello, M., Rozycki, M., Kirby, J., et al.: Data descriptor: advancing the cancer genome atlas glioma MRI collections with expert segmentation labels and radiomic features. Nat. Sci. Data **117** (2017). https://doi.org/10.1038/sdata.2017.117
4. Bakas, S., Reyes, M., Jakab, A., Bauer, S., Rempfler, M., Crimi, A., et al.: Identifying the best machine learning algorithms for brain tumor segmentation, progression assessment, and overall survival prediction in the BRATS challenge, arXiv preprint: arXiv:1811.02629.2018 (2018)
5. Bakas, S., Akbari, H., Sotiras, A., Bilello, M., Rozycki, M., Kirby, J., et al.: Segmentation labels and radiomic features for the pre-operative scans of the TCGA-GBM collection. Cancer Imaging Arch. (2017). https://doi.org/10.7937/K9/TCIA.2017.KLXWJJ1Q
6. Bakas, S., Akbari, H., Sotiras, A., Bilello, M., Rozycki, M., Kirby, J., et al.: Segmentation labels and radiomic features for the pre-operative scans of the TCGA-LGG collection. Cancer Imaging Arch. https://doi.org/10.7937/K9/TCIA.2017.GJQ7R0EF (2017)

7. Myronrnko, A.: 3D MRI brain tumor segmentation using autoencoder regularization, arXiv preprint: arXiv:1810.11654v3 (2018)

8. Wang, F., Jiang, R., Zheng, L., Meng, C., Biswal, B.: 3D U-Net based brain tumor segmentation and survival days prediction, arXiv:1909.12901v2 [eess.IV], 31 March 2020

9. Wang, S., Dai, C., Mo, Y., Angelini, E., Guo, Y., Bai, W.: Automatic brain tumour segmentation and biophysics-guided survival prediction. In: Crimi, A., Bakas, S. (eds.) BrainLes 2019. LNCS, vol. 11993, pp. 61–72. Springer, Cham (2020). https://doi.org/10.1007/978-3-030-46643-5_6

10. Han, W.S., Han, I.S.: All weather human detection using neuromorphic visual processing. In: Chen, L., Kapoor, S., Bhatia, R. (eds.) Intelligent Systems for Science and Information. Studies in Computational Intelligence, vol. 542, pp. 25–44. Springer, Cham (2014). https://doi.org/10.1007/978-3-319-04702-7_2

11. Han, W.S., Han, I.S.: Object segmentation for vehicle video and dental CBCT by neuromorphic convolutional recurrent neural network. In: Bi, Y., Kapoor, S., Bhatia, R. (eds.) Intelligent Systems and Applications. Studies in Computational Intelligence, vol. 751, pp. 264–288. Springer, Cham (2018). https://doi.org/10.1007/978-3-319-69266-1_13

12. Han, W.S., Han, I.S.: Multimodal brain image segmentation and analysis with neuromorphic attention-based learning. In: Crimi, A., Bakas, S. (eds.) Brainlesion: Glioma, Multiple Sclerosis, Stroke and Traumatic Brain Injuries (Part II). LNCS, vol. 11993, pp. 14–26. Springer, Cham (2020). https://doi.org/10.1007/978-3-030-46643-5_2

13. Xu, X., Zhao, W., Zhao, J.: Brain tumor segmentation using attention based network in 3D MRI image. In: Crimi, A., Bakas, S. (eds.) Brainlesion: Glioma, Multiple Sclerosis, Stroke and Traumatic Brain Injuries (Part II). LNCS, vol. 11993, pp. 3–13. Springer, Cham (2020). https://doi.org/10.1007/978-3-030-46643-5_1

14. Han, W.-S., Han, I.: Neuromorphic neural network for multimodal brain image segmentation and overall survival analysis. In: Crimi, A., Bakas, S., Kuijf, H., Keyvan, F., Reyes, M., van Walsum, T. (eds.) BrainLes 2018. LNCS, vol. 11384, pp. 178–188. Springer, Cham (2019). https://doi.org/10.1007/978-3-030-11726-9_16

15. Feng, X., Tustison, N.J., Patel, S.H., Meyer, C.H.: Brain tumor segmentation using an ensemble of 3D U-Nets and overall survival prediction using radiomic features. Front. Comput. Neurosci. **14** (2020). Article 25

16. Agravat, R., Raval, M.: Brain tumor segmentation and survival prediction. In: Crimi, A., Bakas, S. (eds.) BrainLes 2019. LNCS, vol. 11992, pp. 338–348. Springer, Cham (2020). https://doi.org/10.1007/978-3-030-46640-4_32

17. Shboul, Z., Alam, M., Vidyaratne, L., Pei, L., Iftekharuddin, K.: Glioblastoma survival prediction. In: Crimi, Alessandro, Bakas, Spyridon, Kuijf, Hugo, Keyvan, Farahani, Reyes, Mauricio, van Walsum, Theo (eds.) BrainLes 2018. LNCS, vol. 11384, pp. 508–515. Springer, Cham (2019). https://doi.org/10.1007/978-3-030-11726-9_45

18. Puybareau, E., Tochen, G., Chazalon, J., Fabrizio, J.: Segmentation of gliomas and prediction of patient overall survival: a simple and fast procedure. In: Crimi, A., Bakas, S., Kuijf, H., Keyvan, F., Reyes, M., van Walsum, T. (eds.) Brainlesion: Glioma, Multiple Sclerosis, Stroke and Traumatic Brain Injuries (Part II). LNCS, vol. 11384, pp. 199–209. Springer, Cham (2019). https://doi.org/10.1007/978-3-030-11726-9_18

19. Weninger, L., Rippel, O., Koppers, S., Merhof, D.: Segmentation of brain tumors and patient survival prediction: methods for the BraTS 2018 challenge. In: Crimi, A., Bakas, S., Kuijf, H., Keyvan, F., Reyes, M., van Walsum, T. (eds.) Brainlesion: Glioma, Multiple Sclerosis, Stroke and Traumatic Brain Injuries (Part II). LNCS, vol. 11384, pp. 3–12. Springer, Cham (2019). https://doi.org/10.1007/978-3-030-11726-9_1
20. Jiang, Z., Ding, C., Liu, M., Tao, D.: Two-stage cascades U-Net: 1st place solution to BraTS challenge 2019 segmentation task. In: Crimi, A., Bakas, S. (eds.) Brainlesion: Glioma, Multiple Sclerosis, Stroke and Traumatic Brain Injuries (Part I). LCNS, vol. 11992, pp. 231–241. Springer, Cham (2020). https://doi.org/10.1007/978-3-030-46640-4_22

Context Aware 3D UNet for Brain Tumor Segmentation

Parvez Ahmad[1]([✉]) [ID], Saqib Qamar[2] [ID], Linlin Shen[2], and Adnan Saeed[3] [ID]

[1] National Engineering Research Center for Big Data Technology and System, Services Computing Technology and System Lab, Cluster and Grid Computing Lab, School of Computer Science and Technology, Huazhong University of Science and Technology, Wuhan 430074, China
parvezamu@hust.edu.cn

[2] Computer Vision Institute, School of Computer Science and Software Engineering, Shenzhen University, Shenzhen, China
{sqbqamar,llshen}@szu.edu.cn

[3] School of Hydropower and Information Technology, Huazhong University of Science and Technology, Wuhan 430074, China
adnansaeed@hust.edu.cn

Abstract. Deep convolutional neural network (CNN) achieves remarkable performance for medical image analysis. UNet is the primary source in the performance of 3D CNN architectures for medical imaging tasks, including brain tumor segmentation. The skip connection in the UNet architecture concatenates features from both encoder and decoder paths to extract multi-contextual information from image data. The multi-scaled features play an essential role in brain tumor segmentation. However, the limited use of features can degrade the performance of the UNet approach for segmentation. In this paper, we propose a modified UNet architecture for brain tumor segmentation. In the proposed architecture, we used densely connected blocks in both encoder and decoder paths to extract multi-contextual information from the concept of feature reusability. In addition, residual-inception blocks (RIB) are used to extract the local and global information by merging features of different kernel sizes. We validate the proposed architecture on the multi-modal brain tumor segmentation challenge (BRATS) 2020 testing dataset. The dice (DSC) scores of the whole tumor (WT), tumor core (TC), and enhancing tumor (ET) are 89.12%, 84.74%, and 79.12%, respectively.

Keywords: CNN · UNet · Contexual information · Dense connections · Residual inception blocks · Brain tumor segmentation

1 Introduction

Brain tumor is the growth of irregular cells in the central nervous system that can be life-threatening. Primary and secondary are two types of brain tumors. Primary brain tumors originate from brain cells, whereas secondary tumors metastasize into the brain from other organs. Gliomas are primary brain tumors. Gliomas

ⓒ Springer Nature Switzerland AG 2021
A. Crimi and S. Bakas (Eds.): BrainLes 2020, LNCS 12658, pp. 207–218, 2021.
https://doi.org/10.1007/978-3-030-72084-1_19

can be further sub-divided into high-grade glioblastoma (HGG) and low-grade glioblastoma (LGG). In the diagnosis and treatment planning of glioblastoma, brain tumor segmentation results can derive quantitative measurements. While radiologists have manually analyzed magnetic resonance imaging (MRI) modalities to derive information quantitatively, however segmenting 3D modalities is a time-consuming task with deviations and errors. This difficulty is further increases if organs have variation in terms of shape, size, and location. Conversely, Convolutional Neural Networks (CNNs) can apply to the MRI images to develop automatic segmentation methods. Deep CNNs have achieved remarkable performances for brain tumor segmentation [6,8,12,15,23]. A 3D UNet is a popular variation of UNet architecture for automatic brain tumor segmentation [9,13,22]. The multi-scale contextual information of the encoder-decoder paths is effective for the accurate brain tumor segmentation task. Researchers have presented variant forms of the 3D UNet to extract the enhanced contextual information from MRI [14,16]. Network's depth is a common factor to improve the performances among approaches. Residual networks [10] and dense connections [11] are effective to acquire the possible depth in the architecture. Our proposed method used dense connections and the residual-inception blocks [24] to extract the meaningful contextual information from brain MRIs. We used densely connected blocks in both encoder-decoder paths to obtain more abstract features. In the previous approach [1], we have used a low number of densely connected blocks as compared to the proposed approach. In the meantime, residual-inception block (RIB) is used to extract local and global information by merging features of different kernel sizes. Our proposed model gives scalable $3D$ UNet architecture for brain tumor segmentation in the view of these combinations. The key contributions of this study are as follows:

- We proposed a novel densely connected $3D$ encoder-decoder architecture to extract context features at each level of the network.
- We used residual-inception block (RIB) to extract local and global information by merging features of different kernel sizes.
- Our network achieves state-of-the-art performance as compared to other recent methods.

2   Proposed Method

Figure 1 is shown our proposed architecture for brain tumor segmentation. In our previous work [1], we have proposed the combined benefits of the residual and dense connections by using Atrous Spatial Pyramid Pooling (ASPP) [7]. However, insufficient contextual information in each block of the encoder-decoder paths limits the previous model's performance. Moreover, an insufficient number of higher layers degrades the scores of the model. We designed the novel densely connected 3D encoder-decoder architecture for brain tumor segmentation to address these issues. We used dense connections in our proposed architecture while enhancing the maximum features' size to 32 in the final output layer.

Therefore, the number of features is twice as compared to the previous architecture. The output features at the levels of the encoder path are 32, 64, 128, 256, and 512. The proposed work can be divided into (i) dense blocks, which are building blocks of the encoder-decoder paths, and (ii) residual-inception blocks, which are used to the first dense block of the encoder path and along with the upsampling layers of the decoder path.

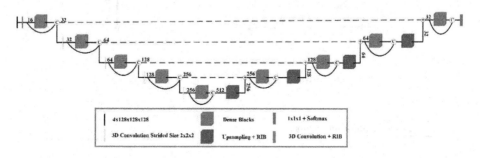

Fig. 1. Proposed densely-connected 3D UNet architecture. Each dense block (green) has three convolution layers. The first block of the encoder path has features of RIB to address the sizes of tumors. A similar approach is employed after each upsampling process (red) in the decoder part. (Color figure online)

2.1   Dense Blocks

Dense connections [11] have been exceptional in delivering high accuracy both in the medical [6,8,17] and non-medical domains [11]. Dense connections have a feature reuse property, in which output feature maps of all previous layers are the inputs to the subsequent layers. Thus, the feature reusability property of dense connection reduces the network's parameters and improve segmentation accuracy. In addition, dense connections enable multi-path flow for gradients between layers during training by back-propagation and hence does implicit deep-supervision. Therefore, inspiring by the dense networks, we use them for each dense block. Each block of the encoder-decoder paths has three convolution layers. The first dense block of the encoder path is shown in Fig. 2. Here, the output feature maps of a residual-inception block (RIB) and the first convolution layer are concatenated. The concatenated features are then passed to the first convolution layer of a dense block. In addition, we doubled the output feature maps of the first dense block. Subsequently, the output feature maps of the remaining dense blocks in the encoder path are doubled to improve the contexts. We also use a growth-rate value of 2, which aids in reducing the training parameters and enables the proposed model to fit into the GPU memory.

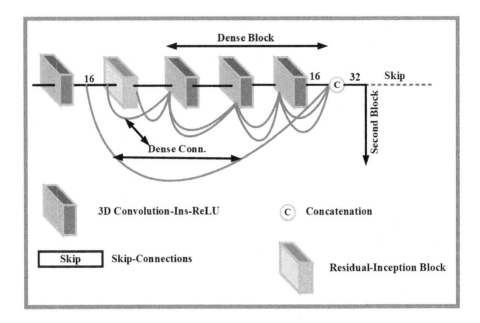

Fig. 2. Proposed densely connected structure of the first block of the encoder path. The first convolution layer and residual-inception block (RIB) (gold) generate 16 concatenated output feature maps to the first convolution layer ($3 \times 3 \times 3$ CONV-Instance Normalization-ReLU) of a dense block. In a dense block, a dropout rate of 0.2 is used after the first convolution layer. Simultaneously, all the preceding output feature maps are the inputs to the next layers of a dense block, followed by a concatenation operation. Here, the output feature maps are doubled (from 16 to 32). Finally, the resulting output feature maps are passed to the next dense blocks.

2.2 Residual-Inception Blocks

A residual-inception block (RIB) is designed with three parallel dilated convolution layers (of rates 1, 3, and 5). It is employed along with the first dense block of the encoder path. Subsequently, RIB is employed after each upsampling operation in the decoder path. The residual connections are used in the RIB to prevent the vanishing gradient problem, while the inception blocks provide multi-scale contexts to the existing residual networks. In this way, the combined RIB architecture improves the size of the features. Figure 3 is shown a RIB architecture, which is proposed to address the problem of different tumors through multiple sizes of the receptive field. Multi-scale contexual information of a RIB can reduce the number of false-positives and prevent the failed segmentation problem. Moreover, instead of concatenation operations, we used the addition operations for skip-connections, which are more computationally and memory-efficient. Furthermore, we also keep the dense connections to improve the feature's strength by concatenating feature maps of addition operations and the decoder part's dense blocks. Finally, the softmax layer is employed for the outcomes.

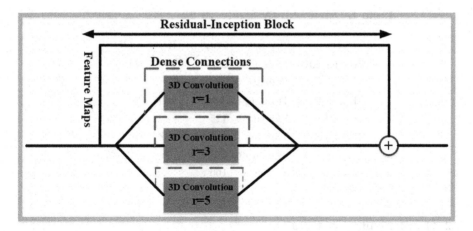

Fig. 3. Proposed RIB structure of the 3D UNet. The output feature maps of the previous layer are the inputs to the three parallel dilated layers. Output maps of every parallel layer are enhanced with dense connections (dashed red lines) to minimize the problem of multiple sizes of the tumors. (Color figure online)

3   Experimental Results

3.1   Dataset

The BRATS aims to bring the research communities together, along with their brilliant ideas for different tasks. Especially for the segmentation task, public benchmark datasets are provided by the organizers. In BRATS challenges [2–5,20] organizers provide various independent datasets for training, validation, and testing. In this paper, we use BRATS 2018, BRATS 2019, and BRATS 2020 datasets to train and evaluate our proposed work. Details of each year's BRATS dataset are shown in Table 1. Here, we can notice each training dataset's further classification into high-grade glioblastoma (HGG) and low-grade glioblastoma (LGG). Furthermore, we have only access to the training and validation datasets of BRATS 2018 and BRATS 2019. Four different types of modalities, i.e., native (T1), post-contrast T1-weighted (T1ce), T2-weighted (T2), and Fluid Attenuated Inversion Recovery (FLAIR), are related to each patient in the training, validation, and testing datasets. For each training patient, the annotated labels have the values of 1 for the necrosis and non-enhancing tumor (NCR/NET), 2 for peritumoral edema (ED), 4 for enhancing tumor (ET), and 0 for the background. The segmentation accuracy is measured by several metrics, where the predicted labels are evaluated by merging three regions, namely whole tumor (Whole Tumor or Whole: label 1, 2 and 4), tumor core (Tumor Core or Core: label 1 and 4), and enhancing tumor (Enhancing Tumor or Enhancing: label 4). The organizers performed necessary pre-processing steps for simplicity. However, the truth label is not provided for the patients of the validation and testing datasets.

Table 1. Details of BRATS 2018, 2019 and 2020 datasets.

Dataset	Type	Patients	HGG	LGG
BRATS 2018	Training	285	210	75
	Validation	66		
BRATS 2019	Training	335	259	76
	Validation	125		
BRATS 2020	Training	369	293	76
	Validation	125		
	Testing	166		

3.2 Implementation Details

Since 19 institutes are involved in data collection, these institutes used multiple scanners and imaging protocols to acquire the brain scans. Thus, normalization would be necessary to establish a similar range of intensity for all the patients and their various modalities to avoid the network's initial biases. Here, normalization of entire data may degrade the segmentation accuracy. Therefore, we normalize each MRI of each patient independently. We extract patches of size $128 \times 128 \times 128$ from 4 MRI modalities to feed them into the network. We used five-fold cross-validation, in which each time our network is trained 300 epochs. The batch size is 1. Adam is the optimizer with the initial learning rate 7×10^{-5}, which is dropped by 50% if validation loss not improved within 30 epochs. Moreover, we used augmentation techniques during the training by randomly rotating the images within a range of $[-1°, 1°]$ and random mirror flips (on the x-axis) with a probability of 0.5.

During the designing of our network, we have tuned several hyperparameters, such as the number of layers for the dense and residual-inception blocks, the initial number of channels for training, the growth rate's value, and the number of epochs, etc. Hence, to avoid any further hyperparameter tunning and inspired by a non-weighted loss function's potential, we employed the previously proposed multi-class dice loss function [21]. Thus, the earlier mentioned dice loss function can be easily adapted with our proposed model and summarized as

$$Loss = -\frac{2}{D} \sum_{d \in D} \frac{\sum_j P_{(j,d)} T_{(j,d)}}{\sum_j P_{(j,d)} + \sum_j T_{(j,d)}} \tag{1}$$

where $P_{(j,d)}$ and $T_{(j,d)}$ are the prediction obtained by softmax activation and ground truth at voxel j for class d, respectively. D is the total number of classes.

3.3 Qualitative Analysis

Figure 4 and Figure 5 are shown the segmentation results of our proposed architecture. Figure 4 shows the $T1ce$ axial slice of HGG patient from the training

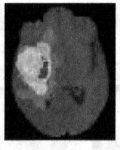   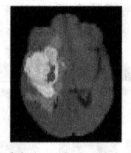

(a) Axial Truth   **(b)** Axial Segmented

Fig. 4. Segmentation results. (*a*) and (*b*) represent the truth and segmented labels overlaid on $T1ce$ axial slices. Different colors represent different parts of the tumor: red for TC, green for WT, and yellow for ET. (Color figure online)

dataset, while Fig. 5 depicts the $T1ce$ sagittal and coronal slices of four different *HGG* patients. Figure 4a and Fig. 4b depicted the truth and segmented labels overlaid on $T1ce$ axial slices. Figure 5a and Fig. 5c shows the truth labels overlaid on $T1ce$ sagittal and coronal slices, while the overlaying of the segmented labels on $T1ce$ sagittal and coronal slices is shown in Fig. 5b and Fig. 5d, respectively. Based on the visualized slices, our proposed model can accurately segment the truth labels of axial, sagittal and coronal slices.

3.4 Quantitative Analysis

We now evaluate our proposed work on BRATS datasets of 2018, 2019, and 2020. In this paper, our main contribution is to propose a variant form of 3D UNet, which can improve context information. Hence, a simple training procedure is performed to check the potential of the proposed model. For deep learning models, cross-validation is a powerful strategy, which is useful with a limited dataset in reducing the variance. Therefore, we perform a five-fold cross-validation procedure on the BRATS 2020 training dataset. After training, five models are used to evaluate the BRATS 2020 validation dataset. Furthermore, a simple post-processing step is performed to remove false-positive voxels from the training and validation predicted datasets. At a threshold value of 0.5, all enhancing tumor regions with less than 500 voxels are replaced with the necrosis [13]. Finally, an average operation is performed on five predicted training and validation sets for the final submission. The scores of the BRATS 2020 training, validation, and testing datasets are shown in Table 2 (see top three rows).

While the proposed model has obtained encouraging scores, however, GPU memory consumption is high. Therefore, we will modify our proposed work to fit into the lowest available GPU memory in the future. Nevertheless, in this paper, we evaluate the proposed model on the BRATS 2019 and 2018 datasets. Therefore, a five-fold cross-validation strategy is also performed on the BRATS 2019 and BRATS 2018 training datasets. However, the results are based only on

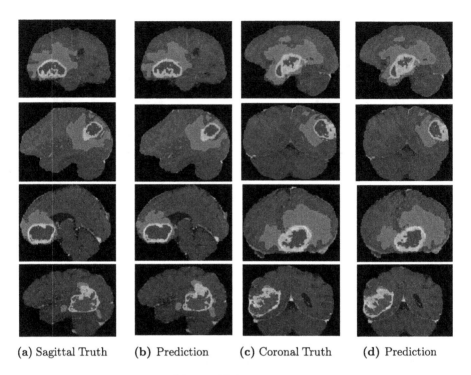

(a) Sagittal Truth   **(b)** Prediction   **(c)** Coronal Truth   **(d)** Prediction

Fig. 5. Segmentation results. (*a*) and (*c*) represent the overlaying of truth labels on *T1ce* sagittal and coronal slices. Simultaneously, *b* and *d* shows the segmented labels overlaid on *T1ce* sagittal and coronal slices. Different colors represent different parts of the tumor: red for TC, green for WT, and yellow for ET. (Color figure online)

the best models (training models of the BRATS 2019 and BRATS 2018 based on the highest mean dice scores are respectively used to evaluate the 125 cases of the BRATS 2019 validation and 66 patients of the BRATS 2018 validation datasets). The scores of the BRATS 2019 and BRATS 2018 datasets are shown in Table 2 (see the fourth row for BRATS 2019 and the fifth row for BRATS 2018). Furthermore, the Hausdorff metric scores are not included in Table 2 because the Hausdorff metric is highly susceptible to the outlier. Hence, it is not a reliable metric for medical image segmentation [25].

Table 3 and Table 4 show the comparisons between the proposed model and the state-of-the-art methods in the MICCAI BRATS 2018 and 2019 validation datasets, respectively. Table 4 shows the mean DSC value of our previous and proposed works. Context-aware $3D$ UNet approach gains state-of-the-art performances for brain tumor segmentation. We used different dilation rates (1, 3 and 5) in the residual-inception blocks to address the problem of losing information due to sparsed kernels [26]. The DSC values for WT and TC are best, and ET is lower for both BRATS 2018 and 2019 validation datasets. ET value can be improved by using some post-processing strategies (McKinley [Filtered Output] [19]). We are currently studying the influence of dilated and non-dilated convo-

Table 2. The average scores of different metrics. For BRATS 2018 and 2019, only validation scores are presented. The BRATS organizers validate all the given scores.

Dataset	Metrics	Whole	Core	Enhancing
BRATS 2020 Training	DSC	93.680	91.829	81.677
	Sensitivity	94.052	92.189	82.839
	Specificity	99.934	99.962	99.975
BRATS 2020 Validation	DSC	90.678	84.248	75.635
	Sensitivity	90.390	80.455	75.300
	Specificity	99.929	99.975	99.975
BRATS 2020 Testing	DSC	89.120	84.674	79.100
	Sensitivity	89.983	85.551	84.287
	Specificity	99.929	99.969	99.961
BRATS 2019 Validation	DSC	90.217	83.435	72.289
	Sensitivity	91.570	82.175	79.863
	Specificity	99.370	99.744	99.829
BRATS 2018 Validation	DSC	91.173	84.108	77.000
	Sensitivity	91.830	82.059	84.367
	Specificity	99.520	99.830	99.762

Table 3. Performance evaluation of different methods on the BRATS 2018 validation dataset. For comparison, only DSC scores are shown. All scores are evaluated online.

Methods	Enhancing	Whole	Core
Isensee [baseline] [13]	79.590	90.800	84.320
McKinley [Base + U + BE] [18]	79.600	90.300	84.700
Myronenko [Single Model] [22]	81.450	90.420	85.960
Proposed [Single Model]	77.000	91.173	84.108

lution layers on the DSC value of the ET. We will try to improve the mean DSC score of the ET by augmentation techniques based on the classes [27].

4   Discussion and Conclusion

We have proposed a unique $3D$ UNet model for brain tumor segmentation. The proposed architecture consists of two sub-modules: (i) dense connections at each level of the encoder-decoder paths, (ii) RIB to extract local and global contextual information by merging feature maps of different kernel's rate. This study addressed a lack of essential information by using context features at each level of encoder-decoder paths. Therefore, the mean average DSC scores of the TC and the WT are improved. In the meantime, the ET score does not improve due to various reasons: (i) zero value of the label ET in the *LGG* dataset of the

Table 4. Performance evaluation of different methods on the BRATS 2019 validation dataset. For comparison, only DSC scores are shown. All scores are evaluated online.

Methods	Enhancing	Whole	Core
Previous Work [ours [1]]	62.301	85.184	75.762
Zhao [BL+warmup+fuse] [14]	73.700	90.800	82.300
McKinley [Raw Output] [19]	75.000	91.000	81.000
McKinley [Filtered Output] [19]	77.000	91.000	81.000
Proposed [Single Model]	72.289	90.217	83.435

BRATS training and the validation datasets, respectively (ii) all the presented scores are based on unbiased corrected brain MRI volumes. (iii) the given model has 84 layers, and we are trying to build more depth to address it. At the same time, the proposed work has obtained competitive scores. However, the number of channels, which doubled at the end of each dense block in the encoder path, requires huge GPU memory. Therefore, the proposed model should be modified to fit into the low GPU memory. In the future, we will try to develop light $3D$ CNN architectures and investigate the augmentation techniques based on the classes to minimize the class imbalance problem. In summary, our proposed model has the potential to address the issue of other medical imaging tasks.

Acknowledgment. This work is supported by the National Natural Science Foundation of China under Grant No. 91959108.

References

1. Ahmad, P., Qamar, S., Hashemi, S.R., Shen, L.: Hybrid labels for brain tumor segmentation. In: Crimi, A., Bakas, S. (eds.) BrainLes 2019. LNCS, vol. 11993, pp. 158–166. Springer, Cham (2020). https://doi.org/10.1007/978-3-030-46643-5_15
2. Bakas, S., et al.: Segmentation labels and radiomic features for the pre-operative scans of the TCGA-GBM collection. The Cancer Imaging Archive (2017)
3. Bakas, S., et al.: Segmentation labels and radiomic features for the pre-operative scans of the TCGA-LGG collection. The Cancer Imaging Archive 286 (2017)
4. Bakas, S., et al.: Advancing the cancer genome atlas glioma MRI collections with expert segmentation labels and radiomic features. Sci. Data **4**, 170117 (2017). https://doi.org/10.1038/sdata.2017.117
5. Bakas, S., et al.: Identifying the best machine learning algorithms for brain tumor segmentation, progression assessment, and overall survival prediction in the BRATS challenge. CoRR abs/1811.0 (2018). arxiv:1811.02629
6. Chen, L., Bentley, P., Mori, K., Misawa, K., Fujiwara, M., Rueckert, D.: DRINet for medical image segmentation. IEEE Trans. Med. Imaging **37**(11), 2453–2462 (2018). https://doi.org/10.1109/TMI.2018.2835303
7. Chen, L.C., Papandreou, G., Schroff, F., Adam, H.: Rethinking atrous convolution for semantic image segmentation. CoRR abs/1706.0 (2017). http://arxiv.org/abs/1706.05587

8. Dolz, J., Gopinath, K., Yuan, J., Lombaert, H., Desrosiers, C., Ayed, I.B.: HyperDense-Net: a hyper-densely connected CNN for multi-modal image segmentation. CoRR abs/1804.0 (2018). http://arxiv.org/abs/1804.02967

9. Feng, X., Tustison, N., Meyer, C.: Brain tumor segmentation using an ensemble of 3D U-Nets and overall survival prediction using radiomic features. In: Crimi, A., Bakas, S., Kuijf, H., Keyvan, F., Reyes, M., van Walsum, T. (eds.) BrainLes 2018. LNCS, vol. 11384, pp. 279–288. Springer, Cham (2019). https://doi.org/10.1007/978-3-030-11726-9_25

10. He, K., Zhang, X., Ren, S., Sun, J.: Deep residual learning for image recognition. CoRR abs/1512.0 (2015). http://arxiv.org/abs/1512.03385

11. Huang, G., Liu, Z., Weinberger, K.Q.: Densely connected convolutional networks. CoRR abs/1608.0 (2016). http://arxiv.org/abs/1608.06993

12. Isensee, F., Kickingereder, P., Wick, W., Bendszus, M., Maier-Hein, K.H.: Brain tumor segmentation and radiomics survival prediction: contribution to the BRATS 2017 challenge. CoRR abs/1802.1 (2018). http://arxiv.org/abs/1802.10508

13. Isensee, F., Kickingereder, P., Wick, W., Bendszus, M., Maier-Hein, K.H.: No New-Net. In: Crimi, A., Bakas, S., Kuijf, H., Keyvan, F., Reyes, M., van Walsum, T. (eds.) BrainLes 2018. LNCS, vol. 11384, pp. 234–244. Springer, Cham (2019). https://doi.org/10.1007/978-3-030-11726-9_21

14. Jiang, Z., Ding, C., Liu, M., Tao, D.: Two-stage cascaded U-Net: 1st place solution to BraTS challenge 2019 segmentation task. In: Crimi, A., Bakas, S. (eds.) BrainLes 2019. LNCS, vol. 11992, pp. 231–241. Springer, Cham (2020). https://doi.org/10.1007/978-3-030-46640-4_22

15. Kamnitsas, K., et al.: Ensembles of multiple models and architectures for robust brain tumour segmentation. CoRR abs/1711.0 (2017). http://arxiv.org/abs/1711.01468

16. Kamnitsas, K., et al.: Efficient multi-scale 3D CNN with fully connected CRF for accurate brain lesion segmentation. Med. Image Anal. **36**, 61–78 (2017)

17. Kori, A., Soni, M., Pranjal, B., Khened, M., Alex, V., Krishnamurthi, G.: Ensemble of fully convolutional neural network for brain tumor segmentation from magnetic resonance images. In: Crimi, A., Bakas, S., Kuijf, H., Keyvan, F., Reyes, M., van Walsum, T. (eds.) BrainLes 2018. LNCS, vol. 11384, pp. 485–496. Springer, Cham (2019). https://doi.org/10.1007/978-3-030-11726-9_43

18. McKinley, R., Meier, R., Wiest, R.: Ensembles of densely-connected CNNs with label-uncertainty for brain tumor segmentation. In: Crimi, A., Bakas, S., Kuijf, H., Keyvan, F., Reyes, M., van Walsum, T. (eds.) BrainLes 2018. LNCS, vol. 11384, pp. 456–465. Springer, Cham (2019). https://doi.org/10.1007/978-3-030-11726-9_40

19. McKinley, R., Rebsamen, M., Meier, R., Wiest, R.: Triplanar ensemble of 3D-to-2D CNNs with label-uncertainty for brain tumor segmentation. In: Crimi, A., Bakas, S. (eds.) BrainLes 2019. LNCS, vol. 11992, pp. 379–387. Springer, Cham (2020). https://doi.org/10.1007/978-3-030-46640-4_36

20. Menze, B.H., et al.: The multimodal brain tumor image segmentation benchmark (BRATS). IEEE Trans. Med. Imaging **34**(10), 1993–2024 (2015). https://doi.org/10.1109/TMI.2014.2377694

21. Milletari, F., Navab, N., Ahmadi, S.A.: V-Net: fully convolutional neural networks for volumetric medical image segmentation. CoRR abs/1606.0 (2016). http://arxiv.org/abs/1606.04797

22. Myronenko, A.: 3D MRI brain tumor segmentation using autoencoder regularization. CoRR abs/1810.1 (2018). http://arxiv.org/abs/1810.11654

23. Ronneberger, O., Fischer, P., Brox, T.: U-Net: convolutional networks for biomedical image segmentation. CoRR abs/1505.0 (2015). http://arxiv.org/abs/1505.04597

24. Szegedy, C., Ioffe, S., Vanhoucke, V.: Inception-v4, Inception-ResNet and the impact of residual connections on learning. CoRR abs/1602.0 (2016). http://arxiv.org/abs/1602.07261

25. Taha, A.A., Hanbury, A.: Metrics for evaluating 3D medical image segmentation: analysis, selection, and tool. BMC Med. Imaging **15**(1), 29 (2015)

26. Wang, P., et al.: Understanding convolution for semantic segmentation. CoRR abs/1702.0 (2017). http://arxiv.org/abs/1702.08502

27. Wang, Q., Gao, J., Yuan, Y.: A joint convolutional neural networks and context transfer for street scenes labeling. IEEE Trans. Intell. Transp. Syst. **19**(5), 1457–1470 (2018). https://doi.org/10.1109/TITS.2017.2726546

Brain Tumor Segmentation Network Using Attention-Based Fusion and Spatial Relationship Constraint

Chenyu Liu[1], Wangbin Ding[1], Lei Li[2,3,4], Zhen Zhang[1], Chenhao Pei[1], Liqin Huang[1(✉)], and Xiahai Zhuang[2(✉)]

[1] College of Physics and Information Engineering, Fuzhou University, Fuzhou, China
hlq@fzu.edu.cn
[2] School of Data Science, Fudan University, Shanghai, China
zxh@fudan.edu.cn
[3] School of Biomedical Engineering, Shanghai Jiao Tong University, Shanghai, China
[4] School of Biomedical Engineering and Imaging Sciences, King's College London, London, UK

Abstract. Delineating the brain tumor from magnetic resonance (MR) images is critical for the treatment of gliomas. However, automatic delineation is challenging due to the complex appearance and ambiguous outlines of tumors. Considering that multi-modal MR images can reflect different tumor biological properties, we develop a novel multi-modal tumor segmentation network (MMTSN) to robustly segment brain tumors based on multi-modal MR images. The MMTSN is composed of three sub-branches and a main branch. Specifically, the sub-branches are used to capture different tumor features from multi-modal images, while in the main branch, we design a spatial-channel fusion block (SCFB) to effectively aggregate multi-modal features. Additionally, inspired by the fact that the spatial relationship between sub-regions of the tumor is relatively fixed, e.g., the enhancing tumor is always in the tumor core, we propose a spatial loss to constrain the relationship between different sub-regions of tumor. We evaluate our method on the test set of multi-modal brain tumor segmentation challenge 2020 (BraTs2020). The method achieves 0.8764, 0.8243 and 0.773 Dice score for the whole tumor, tumor core and enhancing tumor, respectively.

Keywords: Brain tumor · Multi-modal MRI · Segmentation

1 Introduction

Gliomas are malignant tumors that arise from the canceration of glial cells in the brain and spinal cord [16]. It is a dangerous disease with high morbidity,

L. Huang and X. Zhuang are co-senior. This work was funded by Fujian Science and Technology Project (Grant No. 2019Y9070, 2020J01472), National Natural Science Foundation of China (Grant No. 61971142), Shanghai Municipal Science and Technology Major Project (Grant No. 2017SHZDZX01).

A. Crimi and S. Bakas (Eds.): BrainLes 2020, LNCS 12658, pp. 219–229, 2021.
https://doi.org/10.1007/978-3-030-72084-1_20

recurrence and mortality. The treatment of gliomas is mainly based on resection. Therefore, accurate brain tumor segmentation plays an important role in disease diagnosis and therapy planning [4]. However, automatic tumor segmentation is still challenging, mainly due to the diverse location, appearance and shape of gliomas.

The multi-modal magnetic resonance (MR) images can provide complementary information for the anatomical structure. It has been largely used for clinical applications, such as brain, heart and intervertebral disc segmentation [11,17,20]. As reported in [13], T2 weighted (T2) and fluid attenuation inverted recovery (Flair) images highlight the peritumoral edema, while T1 weighted (T1) and T1 enhanced contrast (T1c) images visualize the necrotic and non-enhancing tumor core, and T1c further presents the region of the enhancing tumor. Therefore, the application of the multi-modal MR images for brain tumor segmentation has attracted increasing attention.

Most conventional multi-modal brain tumor segmentation approaches are based on classification algorithms, such as support vector machines [10] and random forests [12]. Recently, based on deep neural network (DNN), Havaei et al. proposed a convolutional segmentation network by using 2D multi-modal images [8], but 2D convolutions can not fully leverage the 3D contextual information. Kamnitsas et al. proposed a multi-scale 3D CNN which can perform brain tumor segmentation by processing 3D volumes directly [9]. Compared to the state-of-the-art 3D network, their model can incorporate both local and larger contextual information for segmentation. Additionally, they utilized a fully connected conditional random fields as the post-processing to refine the segmentation results. According to the hierarchical structure of the tumor regions, Wang et al. decomposed the multiple class segmentation task into three cascaded sub-segmentation tasks and each of the sub tasks is resolved by a 3D CNN [15]. Furthermore, Chen et al. proposed a end-to-end cascaded network for multi-label brain tumor segmentation [6]. However, such a cascaded method ignored the correlation among the tasks. To tackle this, Zhou et al. [18] presented a multi-task segmentation network. They jointly optimized multiple class segmentation tasks in a single model to exploit their underlying correlation.

In this work, we develop a fully automatic brain tumor segmentation method based on 3D convolution neural network, which can effectively fuse complementary tumor information from multi-modal MR images. The main contributions of our method are summarized as follows:

(1) We propose a novel multi-modal tumor segmentation network (MMTSN), and evaluate it on the multi-modal brain tumor segmentation challenge 2020 (BraTs2020) dataset [1–4,13].
(2) We propose a fusion block based on spatial and channel attention, which can effectively aggregate multi-modal features for segmentation tasks.
(3) Based on our network, we design a spatial constraint loss. The loss regularizes the spatial relationship of the sub-regions of tumor and improves the segmentation performance.

2   Method

2.1   Multi-modal Tumor Segmentation Network

Multi-modal MR images can provide different biological properties of tumor. We propose a MMTSN to fully capture this modality-specific information. Figure 1 shows the architecture of the MMTSN. It is composed of three sub segmentation branches (S_{WT}, S_{TC}, S_{ET}) and a main segmentation branch (S_{BT}).

Given a multi-modal MR image $I_{mul} = (I_{T1}, I_{T1c}, I_{T2}, I_{Flair})$, the S_{WT} is used to capture the whole tumor region (WT) by I_{T2} and I_{Flair} images; the S_{TC} aims to acquire tumor core region (TC) by I_{T1} and I_{T1c} images; and the S_{ET} is intent to extract enhanced tumor region (ET) by I_{T1c} image. Therefore, the loss functions of the three branches are defined as

$$\mathcal{L}oss_{WT} = 1 - \mathcal{D}ice(L_{WT}, \hat{L}_{WT}), \tag{1}$$

$$\mathcal{L}oss_{TC} = 1 - \mathcal{D}ice(L_{TC}, \hat{L}_{TC}), \tag{2}$$

$$\mathcal{L}oss_{ET} = 1 - \mathcal{D}ice(L_{ET}, \hat{L}_{ET}), \tag{3}$$

where $\mathcal{D}ice(A, B)$ calculates the Dice score of A and B, (L_{WT}, L_{TC}, L_{ET}) and $(\hat{L}_{WT}, \hat{L}_{TC}, \hat{L}_{ET})$ are corresponding gold standard and predicted label of regions (WT, TC, ET), respectively.

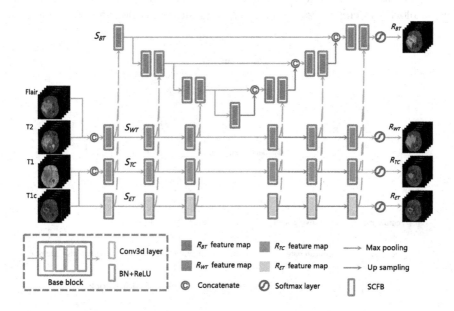

Fig. 1. Overview of the MMTSN architecture. The network contains three sub-branches to capture modality-specific information, and a main the branch to effectively fuse multi-modal features for tumor segmentation.

Having the sub-branches constructed, the multi-modal feature maps in (S_{WT}, S_{TC}, S_{ET}) can be extracted and propagated to S_{BT} for segmentation. The backbone of the S_{BT} is in U-Shape [14]. To effectively fuse complementary information, we also design a spatial-channel attention based fusion block (see Sect. 2.2 for details) for multi-modal feature aggregation. The S_{BT} jointly performs edema, enhancing and non-enhancing&necrotic regions segmentation, and the loss function is

$$\mathcal{L}oss_{BT} = 1 - \mathcal{D}ice(L_{BT}, \hat{L}_{BT}), \tag{4}$$

where L_{BT} and $\hat{L}_{BT}$ are the gold standard and predicted label of all sub-regions of the tumor, respectively. Finally, the overall loss function of the network is

$$\mathcal{L}oss_{MMTSN} = \mathcal{L}oss_{BT} + \lambda_{WT}\mathcal{L}oss_{WT} + \lambda_{TC}\mathcal{L}oss_{TC} + \lambda_{ET}\mathcal{L}oss_{ET} + \lambda_{SC}\mathcal{L}oss_{SC}, \tag{5}$$

where λ_{WT}, λ_{TC}, λ_{ET} and λ_{SC} are hyper-parameters, and the $\mathcal{L}oss_{SC}$ is the spatial constraints loss (see Sect. 2.3 for details).

2.2   Spatial-Channel Fusion Block (SCFB)

We present a spatial-channel attention based fusion block to fuse multi-modal information for segmentation. According to [5], channel attention can effectively re-calibrate channel-wise feature responses, while spatial attention highlights region of interest. Therefore, combining channel and spatial attention in our fusion block can emphasize feature maps and interest regions for the tumor.

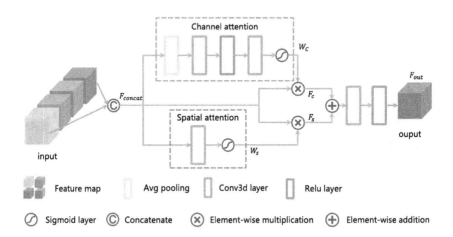

Fig. 2. The spatial-channel attention based fusion block.

The design of SCFB is shown in Fig. 2. Assume that we have three feature maps (F_{WT}, F_{TC}, F_{ET}) from (S_{WT}, S_{TC}, S_{ET}) and one previous output F_{BT} from the S_{BT}. The SCFB first concatenate $(F_{WT}, F_{TC}, F_{ET}, F_{BT})$ to obtain F_{concat}. Then, channel attention and spatial attention are applied to both emphasize informative feature maps and highlight interest regions of F_{concat}. In the SCFB, the channel attention can be defined as

$$F_c = W_c \odot F_{concat}, \tag{6}$$

$$W_c = \sigma(k^{1\times1\times1}\alpha(k^{1\times1\times1}AvgPool(F_{concat}))), \tag{7}$$

where F_c is the output feature maps of the channel attention block, W_c is the channel-wise attention weight and $\odot$ is the element-wise multiplication, $k^{a\times b\times c}$ is defined as a convolutional layer with a kernel size of $a \times b \times c$, α and σ is a ReLU layer and sigmoid activation respectively. Meanwhile, the spatial attention can be formulated as

$$F_s = W_s \odot F_{concat}, \tag{8}$$

$$W_s = \sigma(k^{1\times1\times1}F_{concat}), \tag{9}$$

where F_s is defined as output feature maps of the spatial attention block and W_c is the spatial-wise attention weight. Finally, we combine the output feature maps of channel attention block and spatial attention block by add operation. Therefore, the final output of the SCFB is

$$F_{out} = \alpha(k^{3\times3\times3}(F_c + F_s)). \tag{10}$$

2.3 Spatial Relationship Constraint

As shown in Fig. 3, there are spatial relationship between different sub-regions of tumor, i.e., TC is in WT, and the TC contains ET. Thus, we adopt these relationships as spatial constraints (SC) to regularize the segmentation results of MMTSN.

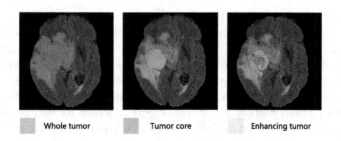

Fig. 3. Spatial relationship of different sub-regions in glioma

In Sect. 2.1, we have constructed three sub-branches (see Fig. 1) to predict the WT, TC and ET from different MR images separately. The spatial constraint can be formulated based on the prediction result of each branch,

$$Loss_{SC}^{WT,TC} = 1 - \frac{\sum\limits_{x\in\Omega}\hat{L}_{WT}(x)\cdot\hat{L}_{TC}(x)}{\sum\limits_{x\in\Omega}\hat{L}_{TC}(x)}, \tag{11}$$

$$Loss_{SC}^{TC,ET} = 1 - \frac{\sum\limits_{x \in \Omega} \hat{L}_{TC}(x) \cdot \hat{L}_{ET}(x)}{\sum\limits_{x \in \Omega} \hat{L}_{ET}(x)}, \tag{12}$$

where the Ω is the common spatial space. Ideally, the $Loss_{SC}^{WT,TC}$ (or $Loss_{SC}^{TC,ET}$) is equeal to 0 when the WT (or TC) completely contains TC (or ET). Finally, the total spatial constraint loss is

$$Loss_{SC} = Loss_{SC}^{WT,TC} + Loss_{SC}^{TC,ET}. \tag{13}$$

The auxiliary $Loss_{SC}$ enforces consistent spatial relationship between the sub-branches, so that the feature maps of each sub-branch can retain more accurate spatial information to improve the segmentation performance in the main branch.

3   Experiment

3.1   Dataset

We used the multi-modal BraTs2020 dataset to evaluate our model. The training set contains images I_{mul} from 369 patients, and the validation set contains images I_{mul} from 125 patients without the gold standard label. Each patient was scanned with four MRI sequences: T1, T1c, T2 and Flair, where each modality volume is of size $240 \times 240 \times 155$. All the images had already been skull-striped, re-sampled to an isotropic $1mm^3$ resolution, and co-registered to the same anatomical template.

3.2   Implementations

Our network was implemented in PyTorch, and trained on NVIDIA GeForce RTX 2080 Ti GPU. In order to reduce memory consumption, the network processed an image patch-wisely. For each I_{mul}, we normalized intensity values, and extracted multi-modal patches $P_{mul} = (P_{T1}, P_{T1c}, P_{T2}, P_{Flair})$ with a size of $4 \times 64 \times 64 \times 48$ from it by sliding window technique. Then the patches can be feed into the network for training and testing. Additionally, the gamma correction, random rotation and random axis mirror flip are adopted for data augmentation to prevent overfitting during model training. The hyper-parameter in λ_{WT}, λ_{ET}, λ_{TC} and λ_{SC} were set to 0.5, 0.6 , 0.6 and 0.5, respectively (see Eq. 5). Finally, the network parameters can be updated by minimizing the $Loss_{MMTSN}$ with Adam optimizer (learning rate $= 0.001$).

3.3   Results

To evaluate the performance of our framework, the Dice and 95th percentile of the Hausdorff Distance (HD95) are used as criteria. Table 1 shows the final result of our method on test set. Furthermore, To explore the advantage of our network architecture, SCFB module and the SC loss, we conducted to compare our method to four different methods on validation set:

- 3D Unet-pre: The 3D Unet which is based on input-level fusion (as shown in Fig. 4(a)) [7].
- 3D Unet-post: The 3D Unet using decision-level fusion (as shown in Fig. 4(b)) [19].
- MMTSN-WO-SCFB : Our MMTSN network but using concatenation rather than SCFB module for feature map fusion.
- MMTSN-WO-$\mathcal{L}oss_{SC}$: Our MMTSN network but without SC loss function.
- MMTSN: Our proposed multi-modal tumor segmentation network.

Table 1. Dice score and HD95 of the proposed method on the test set.

	Dice (%)			HD95 (mm)		
	ET	TC	WT	ET	TC	WT
Mean	77.31	82.43	87.64	27.17	20.23	6.45
Standard deviation	24.68	26.09	12.94	92.54	74.60	10.10
Median	85.00	92.39	91.55	1.41	2.45	3.16
25 quantile	75.95	86.08	86.49	1.00	1.41	2.00
75 quantile	90.31	95.46	94.29	2.83	4.90	6.16

Table 2. Dice score and HD95 of the proposed method and other baseline methods on the validation set.

Method	Dice (%)			HD95 (mm)		
	ET	TC	WT	ET	TC	WT
3D Unet-pre	69.79	79.05	87.67	45.64	13.48	7.04
3D Unet-post	71.98	79.27	88.22	36.31	16.30	6.28
MMTSN-WO-SCFB	73.86	79.81	**88.80**	30.67	12.60	**6.14**
MMTSN-WO-$\mathcal{L}oss_{SC}$	75.94	79.67	87.12	21.89	14.00	7.45
MMTSN	**76.37**	**80.12**	88.23	**21.39**	**6.68**	6.49

In Table 2, compared to 3D Unet-pre and 3D Unet-post, our proposed methods (MMTSN-WO-SCFB, MMTSN-WO-$\mathcal{L}oss_{SC}$ and MMTSN) performed better both in Dice and HD95. Especially in the more challenging areas (TC and ET), the MMTSN achieved the best accuracy among all compared methods. This demonstrates the effectiveness of our designed architecture (see Fig. 1).

Also in Table 2, one can be seen that the MMSTN with SCFB can achieve better results than MMTSN-WO-SCFB on both Dice score and HD95. It shows the advantage of SCFB for multi-modal feature fusion. Meanwhile, compared to MMTSN-WO-$\mathcal{L}oss_{SC}$, although MMTSN had no obvious improvement in Dice score, it greatly performed better in HD95 criterion. This reveals that SC loss can effectively achieve spatial constraints for segmentation results.

Additionally, Fig. 5 shows the visual results of three different cases. For the edema region segmentation (green), even though all of the methods obtained comparable results in the easy and median case, the MMTSN still showed potential advantages in the hard case. For enhancing tumor segmentation (yellow), one can see that the MMTSN and MMTSN-WO-$\mathcal{Loss}_{SC}$ performed better than other methods, which is consistent with the quantitative result in Table 2. For the challenging necrotic and non-enhancing segmentation (red), the figure indicates that the MMTSN can obtain relatively better visual results among all the cases.

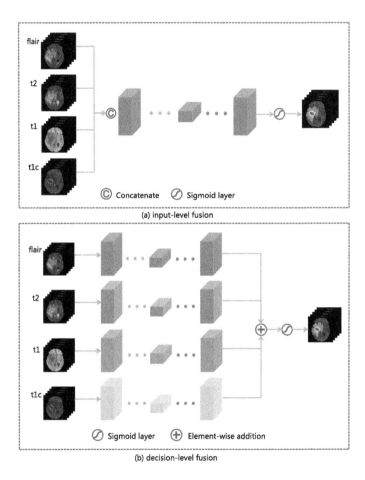

Fig. 4. The architecture of two fusion strategies. Input-level fusion directly concatenates multi-modal images as input, while decision-level fusion adds the output of each modality-specific sub-branch to get the final segmentation result. Note that skip connections are not marked, but they are actually involved in both fusion strategies.

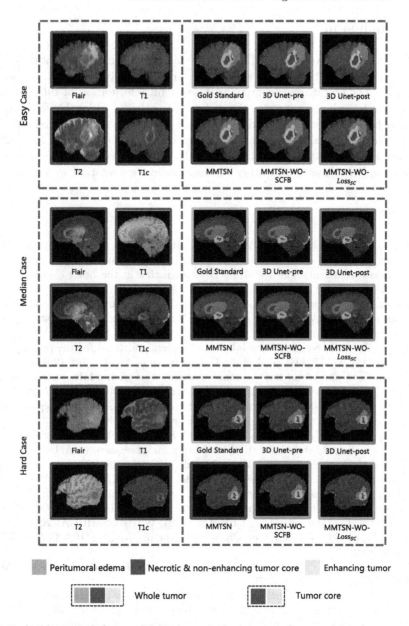

Fig. 5. Axial view of three validation cases: the easy, medium and hard case among the validation set, respectively. Our method MMTSN incorporated with SCFB and SC can achieve the best visual result. (Color figure online)

4 Conclusion

In this work, we proposed a 3D MMTSN for brain tumor segmentation. We constructed three sub-branches and a main branch to capture modality-specific and multi-modal features. In order to fuse useful information of different MR images, we introduced a spatial-channel attention based fusion block. Furthermore, a spatial loss was designed to constrain the relationship between different sub-regions of glioma. We evaluated our method on the multi-modal BraTs2020 dataset to demonstrate the effectiveness of the MMTSN framework. Future work aims to apply our method to other medical image segmentation scenarios.

References

1. Bakas, S., Reyes, M., Jakab, A., Bauer, S., Rempfler, M., Crimi, A., et al.: Identifying the best machine learning algorithms for brain tumor segmentation, progression assessment, and overall survival prediction in the brats challenge. Corr abs/1811.02629 (2018)
2. Bakas, S., et al.: Segmentation labels and radiomic features for the pre-operative scans of the TCGA-GBM collection. The Cancer Imaging Archive. Nat. Sci. Data **4**, 170117 (2017)
3. Bakas, S., et al.: Segmentation labels and radiomic features for the pre-operative scans of the TCGA-LGG collection. Cancer Imaging Archive **286** (2017)
4. Bakas, S., et al.: Advancing the cancer genome atlas glioma MRI collections with expert segmentation labels and radiomic features. Sci. Data **4**, 170117 (2017)
5. Chen, H., et al.: MMFNet: a multi-modality MRI fusion network for segmentation of nasopharyngeal carcinoma. Neurocomputing **394**, 27–40 (2020)
6. Chen, X., Liew, J.H., Xiong, W., Chui, C.-K., Ong, S.-H.: Focus, segment and erase: an efficient network for multi-label brain tumor segmentation. In: Ferrari, V., Hebert, M., Sminchisescu, C., Weiss, Y. (eds.) ECCV 2018. LNCS, vol. 11217, pp. 674–689. Springer, Cham (2018). https://doi.org/10.1007/978-3-030-01261-8_40
7. Çiçek, Ö., Abdulkadir, A., Lienkamp, S.S., Brox, T., Ronneberger, O.: 3D U-Net: learning dense volumetric segmentation from sparse annotation. In: Ourselin, S., Joskowicz, L., Sabuncu, M.R., Unal, G., Wells, W. (eds.) MICCAI 2016. LNCS, vol. 9901, pp. 424–432. Springer, Cham (2016). https://doi.org/10.1007/978-3-319-46723-8_49
8. Havaei, M., et al.: Brain tumor segmentation with deep neural networks. Med. Image Anal. **35**, 18–31 (2017)
9. Kamnitsas, K., et al.: Efficient multi-scale 3D CNN with fully connected CRF for accurate brain lesion segmentation. Med. Image Anal. **36**, 61–78 (2017)
10. Li, N., Xiong, Z.: Automated brain tumor segmentation from multi-modality MRI data based on tamura texture feature and SVM model. In: Journal of Physics Conference Series, vol. 1168 (2019)
11. Li, X., et al.: 3d multi-scale FCN with random modality voxel dropout learning for intervertebral disc localization and segmentation from multi-modality MR images. Med. Image Anal. **45**, 41–54 (2018)
12. Meier, R., Bauer, S., Slotboom, J., Wiest, R., Reyes, M.: Appearance-and context-sensitive features for brain tumor segmentation. In: MICCAI BraTS Workshop (2014)

13. Menze, B.H., et al.: The multimodal brain tumor image segmentation benchmark (BRATS). IEEE Trans. Med. Imaging **34**(10), 1993–2024 (2015)
14. Ronneberger, O., Fischer, P., Brox, T.: U-Net: convolutional networks for biomedical image segmentation. In: Navab, N., Hornegger, J., Wells, W.M., Frangi, A.F. (eds.) MICCAI 2015. LNCS, vol. 9351, pp. 234–241. Springer, Cham (2015). https://doi.org/10.1007/978-3-319-24574-4_28
15. Wang, G., Li, W., Ourselin, S., Vercauteren, T.: Automatic brain tumor segmentation using cascaded anisotropic convolutional neural networks. In: Crimi, A., Bakas, S., Kuijf, H., Menze, B., Reyes, M. (eds.) BrainLes 2017. LNCS, vol. 10670, pp. 178–190. Springer, Cham (2018). https://doi.org/10.1007/978-3-319-75238-9_16
16. Wen, P.Y., Kesari, S.: Malignant gliomas in adults. N. Engl. J. Med. **359**(5), 492–507 (2008)
17. Zhang, W., et al.: Deep convolutional neural networks for multi-modality isointense infant brain image segmentation. Neuroimage **108**, 214–224 (2015)
18. Zhou, C., Ding, C., Lu, Z., Wang, X., Tao, D.: One-pass multi-task convolutional neural networks for efficient brain tumor segmentation. In: Frangi, A.F., Schnabel, J.A., Davatzikos, C., Alberola-López, C., Fichtinger, G. (eds.) MICCAI 2018. LNCS, vol. 11072, pp. 637–645. Springer, Cham (2018). https://doi.org/10.1007/978-3-030-00931-1_73
19. Zhou, T., Ruan, S., Canu, S.: A review: deep learning for medical image segmentation using multi-modality fusion. Array **3**, 100004 (2019)
20. Zhuang, X.: Multivariate mixture model for myocardial segmentation combining multi-source images. IEEE Trans. Pattern Anal. Mach. Intell. **41**(12), 2933–2946 (2019)

Modality-Pairing Learning for Brain Tumor Segmentation

Yixin Wang[1,2,3], Yao Zhang[1,2,3], Feng Hou[1,2,3], Yang Liu[1,2,3], Jiang Tian[3], Cheng Zhong[3], Yang Zhang[4], and Zhiqiang He[1,2,4(✉)]

[1] Institute of Computing Technology, Chinese Academy of Sciences, Beijing, China
[2] University of Chinese Academy of Sciences, Beijing, China
{wangyixin19,zhangyao215,houfeng19}@mails.ucas.ac.cn
[3] AI Lab, Lenovo Research, Beijing, China
{liuyang117,tianjiang1,zhongcheng3}@lenovo.com
[4] Lenovo Corporate Research and Development, Lenovo Ltd., Beijing, China
{zhangyang20,hezq}@lenovo.com

Abstract. Automatic brain tumor segmentation from multi-modality Magnetic Resonance Images (MRI) using deep learning methods plays an important role in assisting the diagnosis and treatment of brain tumor. However, previous methods mostly ignore the latent relationship among different modalities. In this work, we propose a novel end-to-end Modality-Pairing learning method for brain tumor segmentation. Paralleled branches are designed to exploit different modality features and a series of layer connections are utilized to capture complex relationships and abundant information among modalities. We also use a consistency loss to minimize the prediction variance between two branches. Besides, learning rate warmup strategy is adopted to solve the problem of the training instability and early over-fitting. Lastly, we use average ensemble of multiple models and some post-processing techniques to get final results. Our method is tested on the BraTS 2020 online testing dataset, obtaining promising segmentation performance, with average dice scores of $0.891, 0.842, 0.816$ for the whole tumor, tumor core and enhancing tumor, respectively. We won the second place of the BraTS 2020 Challenge for the tumor segmentation task.

Keywords: Brain tumor segmentation · 3D U-Net · Multi-modality fusion

1 Introduction

Accurate diagnosis and segmentation of brain tumor are crucial to successful surgery treatment. However, manual annotation requires human experts, which is time-consuming, tedious and expensive. In recent years, motivated by the success of deep learning, researchers have attempted to apply deep learning-based approaches to segment various tumors in medical images. Fully convolutional

© Springer Nature Switzerland AG 2021
A. Crimi and S. Bakas (Eds.): BrainLes 2020, LNCS 12658, pp. 230–240, 2021.
https://doi.org/10.1007/978-3-030-72084-1_21

networks (FCN) [11], U-Net [15] and V-Net [13] are popular networks for medical image segmentation. 3D U-Net [8] quickly became the priority choice due to its ability to capture spatial context information. Furthermore, various strategies and optimization processes have also been applied to these networks to achieve higher segmentation precision.

However, designing a highly-efficient and reliable segmentation algorithm for brain tumor is much more difficult due to the variable size, shape and location of target tissues. What's more, class imbalance is another major challenge since the lesion areas are extremely small and suffer from background domination. In BraTS 2018, the winner Myronenko et al. [14] followed the encoder-decoder structure of CNN and added the variational auto-encoder(VAE) branch to reconstruct the input images jointly with segmentation in order to regularize the shared encoder. In BraTS 2019, Jiang et al. [10], who achieved the best performance on the testing dataset, proposed a two-stage cascaded U-Net to progressively refine the prediction. In addition, Zhao et al. [16] introduced a set of heuristics in data processing, model devising and result fusing process, which are combined to boost the overall accuracy of the model. These methods adopted the input-level fusion, which directly integrated the different modalities of MRI brain images. However, these various modalities are in essence different as they provide with different anatomical and functional information about brain structure and physiopathology. Specifically, BraTS datasets contain four modalities for brain tumor MRI images, which can provide complementary information due to their dependence on various acquisition. Native (T1) and post-contrast T1-weighted (T1ce) yield high contrast between gray and white matter tissues, which highlight the tumor without peritumoral. T2-weighted (T2) and T2 Fluid Attenuated Inversion Recovery (Flair) enhance the image contrast for the whole peritumoral edema.

Inspired by [5,6,17], we propose a Modality-Pairing network to segment brain tumor substructures. The proposed network consists of paralleled branches, using different modalities as input. The first branch uses Flair and T2 to extract features of the whole tumor, while the second takes T1 and T1ce to learn other tumor representations. These two branches are densely-connected to learn the complementary information effectively. Furthermore, a Modality-Pairing loss is utilized to encourage the consistency between the two sets of high-level feature representations. In addition, a learning rate warmup strategy and an ensemble strategy of multiple models are adopted to improve the segmentation performance. Finally, a post-processing stage is implemented to remove spurious or incoherent segmentation objects. We validate the proposed methods on the BraTS2020 training and validation dataset through qualitative and quantitative analyses. Experimental results show that our method can boost the overall segmentation accuracy.

2   Methods

Modality-Pairing Network. Figure 1 shows the architecture of our proposed
network. Inspired by the recent works on multi-phase tasks, we present a better way to integrate information from different modalities effectively. Instead of
merging the four modalities (T1, T1ce, T2, Flair) at the input of the network,
we consider a Modality-Pairing Network which consists of paralleled branches.
Each branch focuses on specific modalities and all the branches are properly connected. The purpose of the pairing paths is to derive the features most relevant
to each modality and obtain more abundant information among different modalities. We divide the four modalities into two groups and combine two modalities
in each group. Experiments show a better choice to combine T1 and T1ce, T2
and Flair. These two groups are fed into the two branches separately.

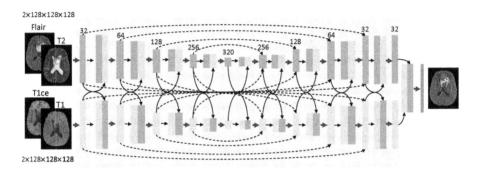

Fig. 1. Overview of proposed Modality-Pairing architecture. Each gray region represents a convolutional block. The numbers in the blocks denote channel numbers. Red
arrows correspond to convolutions and dotted arrows indicate skip connections between
feature maps. Blue and orange boxes are used to distinguish feature maps from different
branches. (Color figure online)

All the branches share the same 3D U-Net architecture. It has a U-shape like
structure with an encoding and decoding signal path. The encoder includes a
series of encoding blocks to contract features, while the decoder is a symmetric
path to recover spatial information. The two paths are connected using skip
connections to recombine essential high-resolution features. Each block contains
two $3 \times 3 \times 3$ convolutions, each followed by instance normalization and leaky
ReLU instead of popular batch normalization and original ReLU. In addition,
deep supervision loss is aggregated using multi-level deep supervision output.
The output of each deep level is the combination of both branches and computed
with corresponding downsampled ground-truth segmentation.

For better fusion of features from different branches, we design a series of
connection operation between layers across branches. Patches from Flair and
T2, T1ce and T1 are concatenated separately to generate two feature maps as
input to each branch A and B. Let x_i^m be the output of ith layer of branch m.

This vector can be obtained from the output of the previous layer x_{i-1}^m by a mapping $\mathcal{H}(\cdot)$:

$$x_i^m = \mathcal{H}(x_i^m) = W * \sigma\left(\mathcal{I}\left(x_{i-1}^m\right)\right), i \in \{0, \ldots, l\}. \tag{1}$$

The encoder layers can be defined as:

$$x_{i+1}^A = \mathcal{H}\left(\left[x_i^A, x_i^B\right]\right), x_{i+1}^B = \mathcal{H}\left(\left[x_i^B, x_i^A\right]\right), \tag{2}$$

where W represents the weight matrix, $*$ denotes convolution operation, $\mathcal{I}(\cdot)$ and $\sigma(\cdot)$ represent instance normalization and leaky ReLU, respectively.

Similar to the encoder, the decoder comprises connections across branches. Besides, the original skip-connections are extended to multi-modality as follows:

$$\begin{aligned} x_i^A &= \mathcal{H}\left(\left[x_{i-1}^A, x_{l-i+1}^A, x_{l-i+1}^B\right]\right), \\ x_i^B &= \mathcal{H}\left(\left[x_{i-1}^B, x_{l-i+1}^B, x_{l-i+1}^A\right]\right). \end{aligned} \tag{3}$$

Before the final classification layer, feature maps X^A, X^B from two branches are fused to generate a final feature map and then fed into a 4-class softmax classifier(i.e. background, ED, NET, ET). In the multi-modality connection settings, our model can yield more powerful feature representations within and in-between different modalities through different branches.

Loss. Dice Loss and Cross Entropy Loss are effective for medical segmentation tasks. In our model, however, different modalities are separately utilized to extract feature maps, which may lead to prediction variance between two branches. In order to better handle this problem, following [17], we adopt a Modality Pairing loss to minimize the distance between the two sets of high-level semantic features. Given two feature maps A, B from two branches, we aim to exploit the consistency between them:

$$\mathcal{L}_{MP}\left(X_i^A, X_i^B\right) = -\frac{\sum_{j=1}^N \left(X_{ij}^A - \overline{X_i^A}\right)\left(X_{ij}^B - \overline{X_i^B}\right)}{\sqrt{\sum_{j=1}^N \left(X_{ij}^A - \overline{X_i^A}\right)^2 \sum_{j=1}^N \left(X_{ij}^B - \overline{X_i^B}\right)^2}}, \tag{4}$$

where N is the total number of voxels and X_{ij} represent j^{th} voxel in i^{th} sample. Based on the above statement, the total loss can be given as follows:

$$L = \lambda_1 \cdot L_{Dice} + \lambda_2 \cdot L_{CE} + \lambda_3 \cdot L_{MP}, \tag{5}$$

where L_{Dice}, L_{CE}, L_{MP} denote the average Dice Loss, Cross Entropy Loss and Modality-Pairing Loss, respectively. $\lambda_1, \lambda_2, \lambda_3$ are three loss weights which are set as $1, 1, 0.5$ based on our Dice results of 5-fold cross validation on the training dataset. Meanwhile, deep supervisions are introduced to add auxiliary outputs from decoder layers for better gradient propagation. This auxiliary deep supervision loss only uses L_{Dice} and L_{CE}, with $\lambda_1 = \lambda_2 = 1$. As a result, the whole network training is optimized by minimizing the loss from both main branches and the auxiliary loss functions.

Learning Rate Strategy. Learning rate warmup [7] is adopted in our training strategy, where we start training with a much smaller learning rate and then increase it over a few epochs until it reaches our set 'initial' learning rate. The effect of learning rate warmup is to prevent deeper layers from training instability and 'early over-fitting'. In detail, we start from a learning rate η_{min}, with $\eta_{min} = 0.0005$, and then increase it by a constant amount 0.0005 at every epoch until it reaches $\eta = \eta_{max}$ with $\eta_{max} = 0.01$. This warmup strategy is followed up by a poly learning rate policy $(1 - \text{epoch} / \text{epoch}_{max})^{0.9}$.

Post-processing. In order to remove spurious or incoherent object, we make connect component analysis and delete any small component (voxels<10). What's more, it is noted that there exists no enhancing tumor region in some LGG cases. Therefore, we empirically replace the enhancing tumor regions with less than a threshold (500 voxels) by necrosis.

Ensemble of Multiple Models. Due to the high variance of single deep learning models, we adopt model ensemble strategy to combine the segmentation predictions from trained single models. For simple models, the average of models has greater capacity than single models, which can reduce bias substantially. Specifically, at the end of the inference on the validation dataset, the predicted probability distributions from each single model are averaged as $y_{en}^{M}(x)$:

$$y_{en}^{M}(x) = \frac{1}{M} \sum_{m=1}^{M} y_m(x), \tag{6}$$

where $y_m(x)$ denotes the output probability of model $m \in \{1, ..., M\}$ at voxel x. Then, each voxel is assigned by the label with the highest probability.

3 Experiments

3.1 Dataset

BraTS2020 training dataset [1–4,12] consists of 369 multi-contrast MRI scans, out of which 293 have been acquired from glioblastoma (GBM/HGG) and 76 from lower grade glioma (LGG). All the multi-modalilty scans contain four modalities: a) native (T1), b) post-contrast T1-weighted (T1Gd), c) T2-weighted (T2), and d) T2 Fluid Attenuated Inversion Recovery (T2-FLAIR) volumes, which are acquired with different clinical protocols and various scanners from multiple (n=19) institutions. Each of these modalities captures different properties of brain tumor subregions: GD-enhancing tumor (ET — label 4), the peritumoral edema (ED — label 2), and the necrotic and non-enhancing tumor core (NCR/NET — label 1).

3.2 Experimental Settings

All the experiments are implemented in Pytorch and trained on NVIDIA Tesla V100 32GB GPU. 5-fold cross validation is adopted while training models on the training dataset. We do data pre-processing following nnU-Net [9]. Due to the large size of input image, the input patch size is set as $128 \times 128 \times 128$ and batch size as 2. Stochastic gradient descent optimizer (SGD) with an initial learning rate of 0.01, a nesterov momentum of 0.99 and weight decay are used. The maximum number of training iterations is set to 1000 epochs with 20 epochs of linear warmup. The whole model is trained in an end-to-end manner.

3.3 Evaluation Metrics

Consistent with the BraTS challenge, we adopt four evaluation metrics. 'Dice' measures volumetric overlap between segmentation results and annotations, while 'Hausdorff distance (HD95)' measures the 95th percentile of values in the set of closet distances between two surfaces. The diagnostic test accuracy 'Sensitivity' and 'Specificity' are also considered to determine potential over- or under-segmentations. 'Sensitivity' shows the percentage of positive instances correctly identified positive and 'Specificity' calculates the proportion of actual negatives that are correctly identified.

3.4 Experimental Results

The prediction of four subregions is aggregated to generate the whole tumor(WT), tumor core(TC) and enhancing tumor(ET). We first train and test our proposed Modality-Pairing model on BraTS2020 training datasets. Then, the best models and an ensemble model are chosen on BraTS2020 validation dataset.

BraTS2020 Training Dataset. For BraTS2020 training dataset, we use 5-fold cross validation based on a random split manner. We report the result of best single model in Table 1. Compared with Vanilla U-Net, proposed Modality-Pairing method improves the segmentation results of enhancing tumor greatly. Additionally, Fig. 2 also shows some examples of segmentation results of the proposed Modality-Pairing method. As can be seen, the segmentation results are sensibly similar to ground-truth with accurate boundaries and some minor tumor areas identified.

Table 1. The segmentation results of best single model on the training dataset.

Method	Tumor	Dice	Sensitivity	Specificity	Hausdorff95
Vanilla U-Net	Enhancing Tumor	0.848	0.863	1.000	12.145
	Whole Tumor	0.923	0.909	0.999	4.508
	Tumor Core	0.900	0.880	1.000	3.368
Modality-Pairing learning	Enhancing Tumor	0.863	0.875	1.000	7.179
	Whole Tumor	0.924	0.910	0.999	4.131
	Tumor Core	0.898	0.877	1.000	3.448

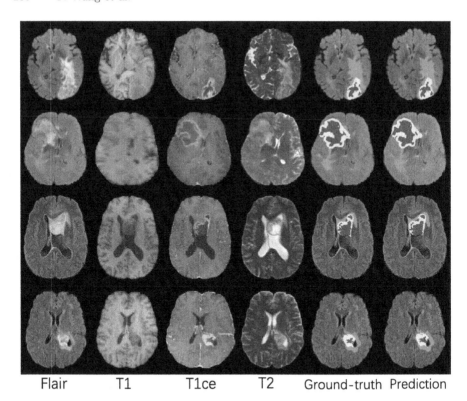

Flair   T1   T1ce   T2   Ground-truth Prediction

Fig. 2. Visual segmentation results of proposed Modality-Pairing method. From left to right, show the axial slice of MRI images in four modalities, ground-truth and predicted results. The labels include enhancing tumor (yellow), edema (green) and necrotic and non-enhancing tumor (red). (Color figure online)

Table 2. The segmentation results of single Modality-Pairing model on the validation dataset.

	Dice			Sensitivity			Specificity			Hausdorff95		
	ET	WT	TC	ET	WT	TC	ET	WT	TC	ET	WT	TC
Mean	0.785	0.907	0.837	0.783	0.901	0.804	1.000	0.999	1.000	32.25	4.39	8.34
StdDev	0.272	0.072	0.178	0.286	0.103	0.211	0.000	0.001	0.000	101.03	5.97	33.64
Median	0.880	0.931	0.902	0.886	0.931	0.891	1.000	1.000	1.000	1.73	2.83	3.16
25quantile	0.790	0.890	0.808	0.795	0.885	0.733	1.000	1.000	1.000	1.00	1.73	1.41
75quantile	0.923	0.949	0.944	0.951	0.964	0.943	1.000	1.000	1.000	3.00	4.24	5.83

BraTS2020 Validation Dataset. We report the results of our proposed model on the validation dataset, consisting of 125 cases with no ground-truth segmentation mask. All the results are evaluated by the official competition platform

(CBICA IPP[1]). We first evaluate the performance of each single model and present the best results in Table 2. Figure 3 also shows the box plots of segmentation accuracy. According to the Dice and Hausdorff95 metrics, among the three subregions, WT achieves the best performance compared to TC and ET. ET is much more difficult to be detected in some cases with a much higher standard deviation. In the ensemble, we choose the top three Modality-Pairing single models from the 5 folds based on the performance on the validation dataset and three Vanilla U-Net single models with proposed learning rate strategy. Table 3 shows the segmentation results after applying our ensemble method. It can be clearly seen that the ensemble strategy improves the Dice results of enhancing tumor (ET) from 0.785 to 0.793, tumor core (TC) from 0.837 to 0.850, performing better than single best models, which shows the effectiveness of the average ensemble strategy.

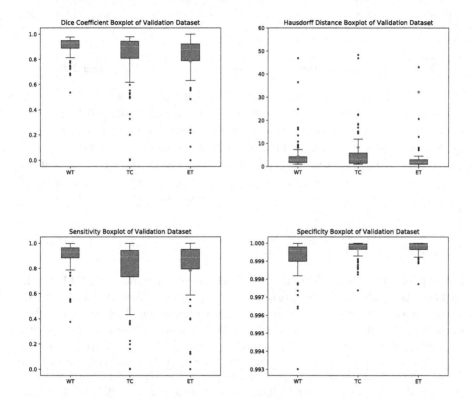

Fig. 3. Box plots of the Dice, Hausdorff95, Sensitivity and Specificity metrics for single best model on the validation dataset evaluated on the Whole Tumor (WT), Enhancing Tumor(ET), and Tumor Core (TC) regions.

---

[1] https://ipp.cbica.upenn.edu.

Table 3. The segmentation results of ensemble model on the validation dataset

	Dice			Sensitivity			Specificity			Hausdorff95		
	ET	WT	TC	ET	WT	TC	ET	WT	TC	ET	WT	TC
Mean	0.787	0.908	0.856	0.786	0.905	0.822	1.000	0.999	1.000	35.01	4.71	5.70
StdDev	0.276	0.078	0.130	0.289	0.103	0.175	0.000	0.001	0.000	105.54	7.62	10.17
Median	0.885	0.931	0.907	0.887	0.934	0.895	1.000	1.000	1.000	1.73	2.83	3.00
25quantile	0.809	0.892	0.827	0.801	0.887	0.768	1.000	0.999	1.000	1.00	1.73	1.41
75quantile	0.925	0.950	0.944	0.960	0.967	0.947	1.000	1.000	1.000	2.83	4.47	5.39

BraTS2020 Testing Dataset. The ensemble strategy was used as the final method to BraTS2020 challenge, with the same ensemble model in the validation phase. Table 4 shows our final results on the testing dataset(166 cases), which is provided by the challenge organizer. Our team achieved the second place out of all 78 participating teams.

Table 4. The segmentation results of ensemble model on the testing dataset

	Dice			Sensitivity			Specificity			Hausdorff95		
	ET	WT	TC	ET	WT	TC	ET	WT	TC	ET	WT	TC
Mean	0.816	0.891	0.842	0.847	0.911	0.853	1.000	0.999	1.000	17.79	6.24	19.54
StdDev	0.197	0.112	0.244	0.211	0.118	0.225	0.000	0.001	0.001	74.87	28.98	74.78
Median	0.857	0.925	0.925	0.916	0.941	0.931	1.000	0.999	1.000	1.41	2.83	2.00
25quantile	0.788	0.884	0.865	0.823	0.905	0.854	1.000	0.999	1.000	1.00	1.41	1.41
75quantile	0.921	0.950	0.959	0.963	0.968	0.966	1.000	1.000	1.000	2.24	4.66	3.74

4   Discussion and Conclusion

In this paper, we propose a Modality-Pairing learning method using 3D U-Net as backbone network, which exploits a better way to fuse the four modalities of MRI brain images to get compromise for a precise segmentation. This method utilizes paralleled branches to separately extract feature from different modalities and combines them via effective layer connections. On the BraTS2020 online testing dataset, our method achieves average Dice scores of 0.891, 0.842, 0.816 for the whole tumor, tumor core and enhancing tumor, respectively. The approach won the second place in the BraTS 2020 challenge segmentation task, with 78 teams participating in the challenge.

References

1. Bakas, S., et al.: Advancing the cancer genome atlas glioma MRI collections with expert segmentation labels and radiomic features. Sci. Data **4** (2017). https://doi.org/10.1038/sdata.2017.117

2. Bakas, S., et al.: Segmentation labels and radiomic features for the pre-operative scans of the TCGA-GBM collection (2017). https://doi.org/10.7937/K9/TCIA.2017.KLXWJJ1Q

3. Bakas, S., et al.: Segmentation labels and radiomic features for the pre-operative scans of the TCGA-LGG collection (2017). https://doi.org/10.7937/K9/TCIA.2017.GJQ7R0EF

4. Bakas, S., Reyes, M., Jakab, A., Bauer, S., Rempfler, M., Crimi, A., et al.: Identifying the best machine learning algorithms for brain tumor segmentation, progression assessment, and overall survival prediction in the brats challenge (2018)

5. Dolz, J., Gopinath, K., Yuan, J., Lombaert, H., Desrosiers, C., Ben Ayed, I.: Hyperdense-net: a hyper-densely connected CNN for multi-modal image segmentation. IEEE Trans. Med. Imaging **38**, 1116–1126 (2018). https://doi.org/10.1109/TMI.2018.2878669

6. Fidon, L., et al.: Scalable multimodal convolutional networks for brain tumour segmentation. In: Descoteaux, M., Maier-Hein, L., Franz, A., Jannin, P., Collins, D.L., Duchesne, S. (eds.) MICCAI 2017. LNCS, vol. 10435, pp. 285–293. Springer, Cham (2017). https://doi.org/10.1007/978-3-319-66179-7_33

7. Goyal, P., et al.: Accurate, large minibatch SGD: training imagenet in 1 hour (2017)

8. Çiçek, Ö., Abdulkadir, A., Lienkamp, S.S., Brox, T., Ronneberger, O.: 3D U-Net: learning dense volumetric segmentation from sparse annotation. In: Ourselin, S., Joskowicz, L., Sabuncu, M.R., Unal, G., Wells, W. (eds.) MICCAI 2016. LNCS, vol. 9901, pp. 424–432. Springer, Cham (2016). https://doi.org/10.1007/978-3-319-46723-8_49

9. Isensee, F., et al.: Abstract: nnU-NET: self-adapting framework for U-NET-based medical image segmentation. In: Bildverarbeitung für die Medizin 2019 - Algorithmen - Systeme - Anwendungen. Proceedings des Workshops, vom 17, bis 19. März 2019 in Lübeck, p. 22 (2019). https://doi.org/10.1007/978-3-658-25326-4_7

10. Jiang, Z., Ding, C., Liu, M., Tao, D.: Two-stage cascaded U-Net: 1st place solution to BraTS challenge 2019 segmentation task. In: Crimi, A., Bakas, S. (eds.) BrainLes 2019. LNCS, vol. 11992, pp. 231–241. Springer, Cham (2020). https://doi.org/10.1007/978-3-030-46640-4_22

11. Long, J., Shelhamer, E., Darrell, T.: Fully convolutional networks for semantic segmentation (2014)

12. Menze, B.H., Jakab, A., Bauer, S., Kalpathy-Cramer, J., Farahani, K., Kirby, J., et al.: The multimodal brain tumor image segmentation benchmark (brats). IEEE Trans. Med. Imaging **34**(10), 1993–2024 (2015)

13. Milletari, F., Navab, N., Ahmadi, S.: V-net: fully convolutional neural networks for volumetric medical image segmentation. In: Fourth International Conference on 3D Vision, 3DV 2016, Stanford, CA, USA, 25–28 October 2016, pp. 565–571 (2016). https://doi.org/10.1109/3DV.2016.79

14. Myronenko, A.: 3D MRI brain tumor segmentation using autoencoder regularization. In: Crimi, A., Bakas, S., Kuijf, H., Keyvan, F., Reyes, M., van Walsum, T. (eds.) BrainLes 2018. LNCS, vol. 11384, pp. 311–320. Springer, Cham (2019). https://doi.org/10.1007/978-3-030-11726-9_28

15. Ronneberger, O., Fischer, P., Brox, T.: U-Net: convolutional networks for biomedical image segmentation. In: Navab, N., Hornegger, J., Wells, W.M., Frangi, A.F. (eds.) MICCAI 2015. LNCS, vol. 9351, pp. 234–241. Springer, Cham (2015). https://doi.org/10.1007/978-3-319-24574-4_28

16. Zhao, Y.-X., Zhang, Y.-M., Liu, C.-L.: Bag of tricks for 3D MRI brain tumor segmentation. In: Crimi, A., Bakas, S. (eds.) BrainLes 2019. LNCS, vol. 11992, pp. 210–220. Springer, Cham (2020). https://doi.org/10.1007/978-3-030-46640-4_20
17. Zhou, Y., et al.: Hyper-pairing network for multi-phase pancreatic ductal adenocarcinoma segmentation. In: Shen, D., et al. (eds.) MICCAI 2019. LNCS, vol. 11765, pp. 155–163. Springer, Cham (2019). https://doi.org/10.1007/978-3-030-32245-8_18

Transfer Learning for Brain Tumor Segmentation

Jonas Wacker[1]([✉]), Marcelo Ladeira[2], and Jose Eduardo Vaz Nascimento[2,3]

[1] EURECOM, Biot, France
[2] University of Brasília, Brasília, DF, Brazil
[3] Syrian-Lebanese Hospital, Brasília, DF, Brazil

Abstract. Gliomas are the most common malignant brain tumors that are treated with chemoradiotherapy and surgery. Magnetic Resonance Imaging (MRI) is used by radiotherapists to manually segment brain lesions and to observe their development throughout the therapy. The manual image segmentation process is time-consuming and results tend to vary among different human raters. Therefore, there is a substantial demand for automatic image segmentation algorithms that produce a reliable and accurate segmentation of various brain tissue types. Recent advances in deep learning have led to convolutional neural network architectures that excel at various visual recognition tasks. They have been successfully applied to the medical context including medical image segmentation. In particular, fully convolutional networks (FCNs) such as the U-Net produce state-of-the-art results in the automatic segmentation of brain tumors. MRI brain scans are volumetric and exist in various co-registered modalities that serve as input channels for these FCN architectures. Training algorithms for brain tumor segmentation on this complex input requires large amounts of computational resources and is prone to overfitting. In this work, we construct FCNs with pretrained convolutional encoders. We show that we can stabilize the training process this way and achieve an improvement with respect to dice scores and Hausdorff distances. We also test our method on a privately obtained clinical dataset.

Keywords: Brain tumor segmentation · Transfer learning

1 Introduction

According to the American Brain Tumor Association[1], in the United States alone, each year 68,470 people are diagnosed with a primary brain tumor and more than twice that number is diagnosed with a metastatic tumor. Gliomas

---

[1] https://www.abta.org.

J. Wacker—Before with University of Brasília.

A. Crimi and S. Bakas (Eds.): BrainLes 2020, LNCS 12658, pp. 241–251, 2021.
https://doi.org/10.1007/978-3-030-72084-1_22

belong to the group of primary brain tumors and represent around 28% of all brain tumors. About 80% of all the malignant (cancerous) tumors are gliomas, which makes them the most common malignant kind. They can be divided into low-grade (WHO grade II) and high-grade (WHO grades III-IV) gliomas [10], conforming to the World Health Organization (WHO) classification.

Patients with the more aggressive high-grade gliomas (anaplastic astrocytomas and glioblastoma multiforme) have a median survival-rate of two years or less – even with aggressive chemoradiotherapy and surgery. The more slowly-growing low-grade (astrocytomas or oligodendrogliomas) variant comes with a life expectancy of several years [11].

In any case, Magnetic Resonance Imaging (MRI) modalities are used by radiotherapists before and during the treatment process. Brain tumor regions can be manually segmented into heterogeneous sub-regions (i.e., edema, enhancing and non-enhancing core) that become visible when comparing MRI modalities with different contrast levels [1–3]. Commonly employed MRI modalities include T1 (spin-lattice relaxation), T1c (contrast-enhanced), T2 (spin-spin relaxation), Fluid-Attenuated Inversion Recovery (FLAIR) and many others. Each modality corresponds to gray-scale images that highlight different kinds of tissue.

The Brain Tumor Segmentation (BraTS) benchmark [4,11] revealed that there is a high disagreement among medical specialists when delineating the boundaries of various tumor subregions. Furthermore, the appearance of gliomas varies strongly among patients regarding shape, location and size.

Therefore, the segmentation task is considered quite challenging, especially when intensity gradients between lesions and healthy tissue are smooth. This is often the case since gliomas are infiltrative tumors.

Due to the high level of difficulty and the time-consuming nature of the task, there has been an emerging need for automatic segmentation methods over the last years. The goal of the BraTS benchmark is to compare these methods on a publicly available dataset. The data contains pre-operative multimodal MRI scans of high-grade (glioblastoma) and low-grade glioma patients acquired from different institutions.

Figure 1 shows the four MRI modalities used in BraTS of an example patient along with the ground-truth annotations.

State-of-the-art methods for automatic brain tumor segmentation use fully-convolutional networks (FCNs) such as the U-Net [13] and its volumetric extensions (e.g. V-Net [12]). A major challenge in this regard is the volumetric multimodal input that leads to high memory requirements and long training times despite the use of expensive GPUs.

In this work, we apply FCNs with encoders that are pretrained on the well-known ImageNet large-scale image dataset [14] in order to stabilize the training process and to improve prediction performance. This approach has led to outstanding results in two-dimensional image segmentation benchmarks such as the Carvana Image Masking Challenge[2]. The resulting U-Net architecture is called TernausNet [6]. We show that despite the difference between the Ima-

---

[2] https://www.kaggle.com/c/carvana-image-masking-challenge.

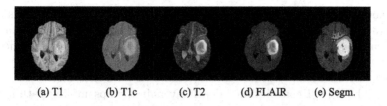

(a) T1   (b) T1c   (c) T2   (d) FLAIR   (e) Segm.

Fig. 1. Example of a high-grade glioma case. The four MRI modalities (T1, T1c, T2, and FLAIR) of a single patient are shown (a-d). The rightmost image (e) shows the tumor segmentation. Orange corresponds to edema, white to the tumor core and red to the active tumor region. (Color figure online)

geNet dataset and the MRI images used in this work, a performance gain can be achieved while stabilizing training convergence. The PyTorch implementation of our method is publicly available on GitHub[3].

In order to further evaluate our approach in a practical context, we have acquired MRI scans of five glioma patients from the Syrian-Lebanese hospital in Brasilia (Brazil) that we consider compatible with the BraTS data. We aligned this data with the scans of the BraTS challenge, which allows us to test our method on data that comes from an institution that did not provide any training or validation data.

2   Method

2.1   Extending the AlbuNet Architecture

We make use of a ResNet34 [5] encoder that results in the AlbuNet variant [15] of TernausNet. Figure 2 depicts our U-Net based architecture. The ResNet34 downsampling layers are complemented with an upsampling path that uses transpose convolutional layers and receives intermediate inputs from the downsampling path. Both paths are symmetric in the dimensionality of intermediate outputs that they produce. However, no residual connections are used in the upsampling path that also contains less convolutional layers than the downsampling path.

In order to match the expected RGB three-channel input of the ResNet34 architecture, we select only the T1c, T2 and FLAIR modalities and discard T1 scans. We then normalize voxel intensities with respect to each scan. The resulting voxel intensities are directly treated as RGB channels, where R corresponds to FLAIR, G to T1c and B to T2.

We choose to discard the T1 modality without enhanced contrast instead of any other because this modality is rarely used by radiotherapists to delineate tumor boundaries. Furthermore, the privately acquired dataset presented in Sect. 3 does not contain T1 scans. However, including the T1 modality in our model may still improve predictions. We leave the extension of the architecture to 4-channel inputs for future research.

---

[3] https://github.com/joneswack/brats-pretraining.

A novelty of our architecture is that we extend the 3×3 convolutions inside ResNet34 with $1 \times 3 \times 3$ convolutions. Since the number of parameters stays the same, pretrained ResNet34 weights can still be loaded without modification. However, the effect is that the architecture is able to treat volumes of any depth now. The encoder simply processes the volumetric data slice-wise.

We add $3 \times 1 \times 1$ convolutional depth layers in the upsampling path in order to exploit segmentation correlations in stacks of slices. The depth dimension can be changed to values different from 3 without further adaptation since the depth of the input is not reduced by this convolutional layer thanks to input padding. The architecture can be easily reduced to process slices instead of volumes by disabling the depth layers and choosing an input volume depth of one.

The last layer is a softmax over four channels where the channels correspond to the segmentation labels (edema, tumor core, enhancing tumor and background). Therefore, a single forward pass yields segmentations for all segmentation classes.

2.2   Loss Function

We use the Multiple Dice Loss as in [7] as a training objective:

$$\mathcal{L}_{DSC} = 1 - \frac{1}{K} \sum_{l=1}^{K} DSC \quad \text{with} \quad DSC = \frac{2 \sum_n r_{ln} p_{ln}}{\sum_n r_{ln} + \sum_n p_{ln}}, \tag{1}$$

where r_{ln} is the reference segmentation and p_{ln} the predicted segmentation for voxel n and class l. DSC is the dice score coefficient that measures the similarity over two sets and can take on values between 0 (minimum) and 1 (maximum). Since there are three segmentation classes without background, we have $K = 3$. The Multiple Dice Loss guarantees equal importance to each class irrespective of their proportion inside the scan. Using multiplications instead of set intersections and additions instead of set unions makes the dice loss function differentiable.

2.3   Choice of Hyperparameters

The batch size as well as the input patch size are determined by our hardware setup where we occupy the maximum amount of memory possible in order to speed up the training process and to exploit the trade-off between fast and well-directed weight updates.

For the case of 2D inputs, we use randomly sampled patches of $1 \times 128 \times 128$ voxels, which leads to a batch size of 64. For the 3D case, we use a patch size of $24 \times 128 \times 128$ voxels and a batch size of 24. The patch width and height are chosen to be roughly half of the input dimension of the data provided by the BraTS benchmark. However, it has to be noted that the brain of the patient only occupies a part of the space so that 128×128 patches cover a great part of it.

We use the Adam optimizer with a learning rate of 10^{-3} which yielded the best results in a preliminary evaluation. Training is carried out over 50 epochs where each epoch contains 100 input batch samples drawn uniformly at random from the non-zero area of the MRI input volumes.

2.4 Preprocessing and Data Augmentation

We normalize each voxel inside an input channel for each patient. Moreover, we crop the MRI input values to their non-zero regions in order to sample more efficiently and to reduce the memory footprint.

Finally, we apply a set of spatial transforms (data augmentation) to the input patches in order to improve the generalization error. These include elastic deformations, reflections and noise as well as blur. These transformations help the network to deal with unseen low-resolution scans inside the test dataset that occur frequently. The dataloader that we use for data augmentation and preprocessing is publicly available[4].

2.5 Prediction

We use the same patch sizes as the ones for training when predicting entire patient volumes. In order to improve prediction performance, we use a sliding window approach where predictions are carried out on overlapping input patches. The step sizes of the sliding window are 32 along the width and the height dimension and 24 along the depth dimension of the input. Overlapping predictions are averaged out yielding a slight gain in prediction performance compared to non-overlapping predictions.

3   Evaluation

We evaluate *AlbuNet* (from now on denoted as *AlbuNet2D*) as well as our proposed extension (*AlbuNet3D*) with and without pretraining. Moreover, we compare these models on a privately acquired clinical dataset. For the private clinical dataset we only have access to the tumor core labels.

3.1   Extended AlbuNet with and Without Pretraining

We evaluate the effect of ImageNet pretraining on the brain tumor segmentation results for *AlbuNet2D* and *AlbuNet3D* as presented in Sect. 2.

Figure 3 shows validation dice scores during training of AlbuNet3D on the BraTS '20 training data with a held out validation set over 50 epochs. For each of the tumor regions the pretrained version keeps a small margin above the version without pretraining throughout the entire process. Initializing the model with pretrained weights also has a stabilizing effect on the training process as the

---

[4] https://github.com/MIC-DKFZ/batchgenerators.

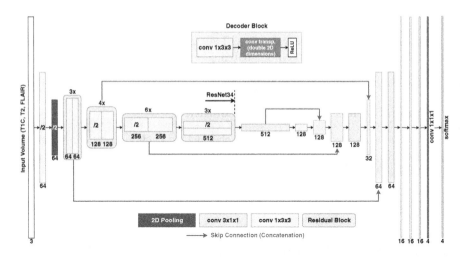

Fig. 2. The extended AlbuNet architecture. All 33 downsampling convolutional layers are extracted from the ResNet34 architecture that is pretrained on the ImageNet dataset. Until this point, the architecture is unmodified except for the initial pooling layer for which the kernel size is reduced from 3×3 to 2×2. Residual connections are maintained and only the final fully-connected layer is removed. The decoding blocks complement this downsampling path to retrieve equally dimensioned segmentation results. $3 \times 1 \times 1$ depth layers are a novelty in our architecture.

standard deviations around the mean scores shrink. This holds in particular for the tumor core and enhancing tumor cases. Convergence happens more or less at the same speed and good results can be achieved after roughly 2000 iterations where each iteration processes one training batch of size 24 with patch size $(24 \times 128 \times 128)$ voxels.

We used a single 16 GB GPU of the NVIDIA Tesla P100 graphics card for training. Training for 5000 iterations took 14 h for *AlbuNet3D* and 5.5 h for *AlbuNet2D* that follows a similar convergence pattern as the 3D architecture.

Using the prediction method presented in Sect. 2, the prediction for a single patient takes 45 s for the *AlbuNet2D* and 15 s for the *AlbuNet3D* architecture respectively.

3.2   Clinical Dataset of the Syrian-Lebanese Hospital

We acquired a private dataset of 25 glioma patients from the Syrian-Lebanese hospital in Brasilia (Brazil). This dataset is not used for training and only serves evaluation purposes. The types of available MRI contrasts as well as their resolutions vary among patients. We selected 5 of these patients by hand for which T1c, T2 and FLAIR MRI modalities were provided. These were the most compatible ones with the BraTS data among the 25 patients. For none of the patients, the T1 modality used in the BraTS benchmark was available. Two of the five

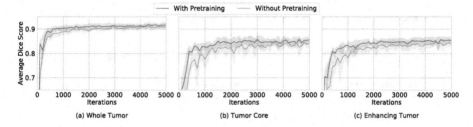

Fig. 3. Validation Dice Scores during training of the AlbuNet3D architecture. We carried out five training runs for the version with and without pretraining, respectively. Thick lines denote the mean dice score over five runs and the shaded error corresponds to the standard deviation from the mean.

patients have two different T1c modalities, a volumetric one with a high depth resolution and a shallow one with a high axial resolution.

A great part of this work was to reverse-engineer the BraTS data preprocessing pipeline and to apply it to the 5 aforementioned patients.

This pipeline includes the following steps:

(a) Conversion of MRI data as well as annotations from DICOM to NIFTI format using 3D Slicer[5].
(b) Conversion of each volume to an axial orientation using fslswapdim of the FMRIB Software Library (FSL) [8].
(c) Application of medical brain mask annotations to each scan in order to remove the skull of the patient.
(d) Rigid Co-Registration of each brain volume to a reference scan that had the highest resolution (volumetric T1c in all cases) using nearest-neighbor interpolation. This was done using the FSL FLIRT linear co-registration tool [9].
(e) Resampling of all volumes and segmentations to an isotropic resolution ($1\,\text{mm} \times 1\,\text{mm} \times 1\,\text{mm}$) using nearest-neighbour interpolation.

For three of our five patients, only one T1c modality was available. This did not change the preprocessing in any way. The only difference is that there are only three modalities (volumetric T1c, T2 and FLAIR) instead of four (volumetric T1c, T1c, T2 and FLAIR). In the three cases where only one T1c modality was available we used this modality for both the T1 and the T1c input channel.

3.3   Final Results

Figure 4 visualizes the *enhancing tumor* dice score distribution for a single run of each method in a boxplot. It becomes apparent that our 3D extension of AlbuNet outperforms *AlbuNet2D* (improved quartiles and medians). The use of pretrained weights has a positive effect for both the 2D and the 3D architectures.

---

[5] https://www.slicer.org/.

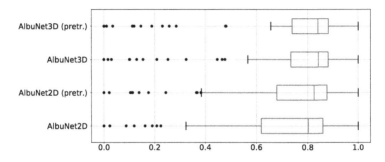

Fig. 4. Boxplot of the enhancing tumor dice scores obtained from the evaluation on the BraTS'20 validation data.

Nonetheless, there are a large number of outliers for each method that suggest a need for further improvements in robustness.

Table 1 compares mean/median dice scores and Hausdorff distances for all segmentation labels and methods taken over 5 runs for the validation data and a single test run of the best-performing method. Except for the mean tumor core dice/Hausdorff scores, the pretrained AlbuNet3D architecture obtains the best validation results. The test score means are generally improved except for the whole tumor dice score while the median dice scores remain almost the same (except tumor core). Therefore, we can assume that the test set produced less outliers than the validation set. It also confirms that our method generalizes reasonably well to unseen data.

Table 1. Mean (Median) dice scores and Hausdorff distances for the Brats'20 validation and test data.

Method	Dice			Hausdorff		
	ET	WT	TC	ET	WT	TC
Validation Data (Average over 5 runs)						
AlbuNet3D - pretrained	**0.7187**	**0.8851**	0.7822	**31.5322**	**5.9929**	15.6479
	(0.8382)	**(0.9127)**	**(0.8830)**	**(2.2361)**	**(3.4308)**	**(4.0196)**
AlbuNet3D	0.7048	0.8808	**0.7926**	39.0381	7.5833	**14.0151**
	(0.8321)	(0.9079)	(0.8775)	(2.3214)	(3.5406)	(4.5917)
AlbuNet2D - pretrained	0.6849	0.8786	0.7527	40.3850	7.9363	17.6747
	(0.8222)	(0.9043)	(0.8700)	(2.3972)	(3.9187)	(4.6775)
AlbuNet2D	0.6842	0.8733	0.7447	38.6672	8.7802	19.4597
	(0.8182)	(0.9041)	(0.8622)	(2.4742)	(3.8365)	(5.1250)
Test Data (Single final run)						
AlbuNet3D - pretrained	0.7819	0.8744	0.8271	18.3230	5.4160	20.3297
	(0.8373)	(0.9117)	(0.9158)	(1.7321)	(3.0000)	(2.4495)

Table 2 shows tumor core dice scores of five selected patients from our privately acquired clinical dataset whose scans were the most compatible with the BraTS challenge. The overall average performance (AVG) drops for all methods compared to the BraTS'20 evaluation in Table 1, which shows that it is not straightforward to apply existing methods to new clinical data.

The effect of pretraining is weaker on the tumor core labels as seen before. Since the clinical dataset is slightly different from the training data, these patients could be treated as outlier cases which is particularly true for patient 3 (lowest scores). A special treatment beyond pretraining is needed for outliers that is subject to further research.

Table 2. Average tumor core dice scores and standard deviations over 5 runs for five patients (P1-P5) of the Syrian-Lebanese hospital.

Method	P1	P2	P3	P4	P5	AVG
AlbuNet3D - pretrained	0.8227	0.7918	0.4410	0.8112	**0.7838**	0.7301
	(±0.06)	(±0.02)	(±0.10)	(±0.01)	(±0.04)	(±0.16)
AlbuNet3D	**0.8235**	**0.8046**	**0.4618**	**0.8190**	0.7809	**0.7380**
	(±0.02)	(±0.02)	(±0.09)	(±0.01)	(± 0.03)	(±0.16)
AlbuNet2D - pretrained	0.8042	0.7805	0.4026	0.8010	0.5855	0.6748
	(±0.04)	(±0.02)	(±0.06)	(±0.01)	(±0.28)	(±0.18)
AlbuNet2D	0.7939	0.7798	0.3631	0.8001	0.6217	0.6717
	(±0.04)	(±0.03)	(±0.07)	(±0.02)	(±0.24)	(±0.19)

4 Conclusion

We successfully extended *AlbuNet2D* to process volumetric input patches and showed that *AlbuNet3D* yields further performance improvements.

In order to outperform the state-of-the-art on the BraTS benchmark, future research should focus on the question of how to extend our method to all four input channels (T1, T1c, T2 and FLAIR). In this regard it would be possible to train an ensemble where each network uses a three-channel subset of the four provided channels. Another option would be to pretrain a model on four input channels, preferably on a large MRI dataset. This may improve the robustness of our method with respect to outliers.

We have shown that encoders pretrained on ImageNet improve the segmentation results of U-Net based architectures for the task of brain tumor segmentation and that they lead to a more robust training process. Unfortunately, this does not hold for our privately acquired clinical dataset where future research for robustness improvement is needed. The available MRI data in a practical clinical context is much more heterogeneous than in the BraTS benchmark. Only a subset of the MRI modalities may be available for a patient and scans vary strongly in resolution, contrast and orientation. Therefore, it is crucial to continue our line of research to develop methods that can flexibly process changing input modalities.

Acknowledgement. We would like to thank the Syrian-Lebanese hospital in Brasilia for supplying us with MRI data for 25 glioma patients. We also appreciate the guidance that we received from their experts in the oncology department.

References

1. Bakas, S., et al.: Advancing the cancer genome atlas glioma MRI collections with expert segmentation labels and radiomic features. Scientific Data **4**(July), 1–13 (2017). https://doi.org/10.1038/sdata.2017.117
2. Bakas, S., et al.: Segmentation labels and radiomic features for the pre-operative scans of the TCGA-GBM collection (2017). https://doi.org/10.7937/K9/TCIA.2017.KLXWJJ1Q
3. Bakas, S., et al.: Segmentation labels and radiomic features for the pre-operative scans of the TCGA-LGG collection (2017). https://doi.org/10.7937/K9/TCIA.2017.GJQ7R0EF
4. Bakas, S., et al.: Identifying the best machine learning algorithms for brain tumor segmentation, progression assessment, and overall survival prediction in the BRATS challenge. CoRR (2018). http://arxiv.org/abs/1811.02629
5. He, K., Zhang, X., Ren, S., Sun, J.: Deep residual learning for image recognition. In: ICCV (2015). https://doi.org/10.3389/fpsyg.2013.00124
6. Iglovikov, V., Shvets, A.: TernausNet: U-Net with VGG11 encoder pre-trained on imagenet for image segmentation (2018). http://arxiv.org/abs/1801.05746
7. Isensee, F., Kickingereder, P., Wick, W., Bendszus, M., Maier-Hein, K.H.: Brain tumor segmentation and radiomics survival prediction: contribution to the BRATS 2017 challenge. In: Crimi, A., Bakas, S., Kuijf, H., Menze, B., Reyes, M. (eds.) BrainLes 2017. LNCS, vol. 10670, pp. 287–297. Springer, Cham (2018). https://doi.org/10.1007/978-3-319-75238-9_25
8. Jenkinson, M., Beckmann, C.F., Behrens, T.E., Woolrich, M.W., Smith, S.M.: Fsl. NeuroImage **62**(2), 782–790 (2012). https://doi.org/10.1016/j.neuroimage.2011.09.015
9. Jenkinson, M., Smith, S.: Global optimisation method for robust affine registration of brain images. Med. Image Anal. **5**, 143–56 (2001). https://doi.org/10.1016/S1361-8415(01)00036-6
10. Louis, D.N., et al.: The 2016 world health organization classification of tumors of the central nervous system: a summary. Acta Neuropathol. **131**(6), 803–820 (2016). https://doi.org/10.1007/s00401-016-1545-1
11. Menze, B.H., et al.: The multimodal brain tumor image segmentation benchmark (BRATS). IEEE Trans. Med. Imaging **34**(10), 1993–2024 (2015). https://doi.org/10.1109/TMI.2014.2377694
12. Milletari, F., Navab, N., Ahmadi, S.A.: V-Net: fully convolutional neural networks for volumetric medical image segmentation, pp. 1–11 (2016). https://doi.org/10.1109/3DV.2016.79. http://arxiv.org/abs/1606.04797

13. Ronneberger, O., Fischer, P., Brox, T.: U-Net: convolutional networks for biomedical image segmentation. In: Navab, N., Hornegger, J., Wells, W.M., Frangi, A.F. (eds.) MICCAI 2015. LNCS, vol. 9351, pp. 234–241. Springer, Cham (2015). https://doi.org/10.1007/978-3-319-24574-4_28

14. Russakovsky, O., Deng, J., et al.: ImageNet large scale visual recognition challenge. Int. J. Comput. Vis. **115**(3), 211–252 (2015). https://doi.org/10.1007/s11263-015-0816-y

15. Shvets, A., Iglovikov, V., Rakhlin, A., Kalinin, A.A.: Angiodysplasia detection and localization using deep convolutional neural networks. CoRR (2018). http://arxiv.org/abs/1804.08024

Efficient Embedding Network for 3D Brain Tumor Segmentation

Hicham Messaoudi[1], Ahror Belaid[1(✉)], Mohamed Lamine Allaoui[1],
Ahcene Zetout[1], Mohand Said Allili[2], Souhil Tliba[1,3], Douraied Ben Salem[4,5],
and Pierre-Henri Conze[4,6]

[1] Medical Computing Laboratory (LIMED), University of Abderrahmane Mira,
06000 Bejaia, Algeria
ahror.belaid@univ-bejaia.dz
http://www.univ-bejaia.dz/limed/
[2] Université du Québec en Outaouais, Gatineau, QC J8X 3X7, Canada
[3] Neurosurgery Department, University Hospital Center,
Biological Engineering of Cancers, 06000 Bejaia, Algeria
[4] Laboratory of Medical Information Processing (LaTIM), UMR 1101, Inserm,
Univ. Brest, 22 avenue Camille Desmoulins, 29238 Brest, France
[5] Neuroradiology Department, CHRU Brest,
Boulevard Tanguy-Prigent, 29609 Brest, France
[6] IMT Atlantique, Technopôle Brest Iroise, 29238 Brest, France

Abstract. 3D medical image processing with deep learning greatly suffers from a lack of data. Thus, studies carried out in this field are limited compared to works related to 2D natural image analysis, where very large datasets exist. As a result, powerful and efficient 2D convolutional neural networks have been developed and trained. In this paper, we investigate a way to transfer the performance of a two-dimensional classification network for the purpose of three-dimensional semantic segmentation of brain tumors. We propose an asymmetric U-Net network by incorporating the EfficientNet model as part of the encoding branch. As the input data is in 3D, the first layers of the encoder are devoted to the reduction of the third dimension in order to fit the input of the EfficientNet network. Experimental results on validation and test data from the BraTS 2020 challenge demonstrate that the proposed method achieve promising performance.

Keywords: Convolutional encoder-decoders · Embedding networks · Transfer learning · 3D image segmentation · EfficientNet

1 Introduction

Gliomas are the most common type of primary brain tumors of the central nervous system. They can be of low-grade or high-grade type. High-Grade Gliomas (HGG) are an aggressive type of malignant brain tumors that grow rapidly. Furthermore, Low-Grade Gliomas (LGG) are classified into grade I and grade

© Springer Nature Switzerland AG 2021
A. Crimi and S. Bakas (Eds.): BrainLes 2020, LNCS 12658, pp. 252–262, 2021.
https://doi.org/10.1007/978-3-030-72084-1_23

II. These tumors represent less than 50% of glial tumors and are considered as isolated tumor cells within the nervous parenchyma, with slow initial growth. Gliomas are also characterized by their infiltrating character resulting in ambiguous and fuzzy boundaries.

However, the treatment of LGG remains difficult due to the variability of tumor size, location, histology and biological behavior. Furthermore, because of the pressure that the tumor exhibits, normal tissues get deformed, making it even harder to distinguish normal tissues from tumoral areas. The diagnosis of these tumors followed by early treatments are critical for patient survival. Indeed, in most cases, patients who are suffering from LGG die in the next ten years of the initial diagnosis. For all these reasons, accurate and reproducible segmentation of gliomas is a pre-requisite step for investigating brain MRI data.

Magnetic Resonance Imaging (MRI) has been quickly imposed as being an essential medical imaging modality for disease diagnosis. MRI is particularly useful for brain tumor diagnosis, patient follow-up, therapy evaluation and human brain mapping [5,11,15]. The main advantage related to the use of MRI is its ability of acquiring non-invasive and non-irradiant medical images. It is also very sensitive to the contrast and provides an excellent spatial resolution which is entirely appropriate for the exploration of the brain tissues nature. In addition, the imaging easily derives 3D volumes according to brain tissues.

The multimodal Brain Tumor Segmentation (BraTS) challenge [1–4,8] aims at encouraging the development of state-of-the-art methods for the segmentation of brain tumors by providing a large 3D MRI dataset of annotated LGG and HGG. The BraTS 2020 training dataset include 369 cases (293 HGG and 76 LGG), each with 4 modalities describing: native (T1), post-contrast T1-weighted (T1Gd), T2-weighted (T2), and T2 Fluid Attenuated Inversion Recovery (T2-FLAIR) volumes, which were acquired with different clinical protocols and various MRI scanners from multiple (n = 19) institutions. Each tumor was segmented into edema, necrosis and non-enhancing tumor, and active/enhancing tumor. Annotations were combined into 3 nested sub-regions: Whole Tumor (WT), Tumor Core (TC) and Enhancing Tumor (ET).

In the last decade, Convolutional Neural Networks (CNNs) have outperformed all others traditional methodologies in biomedical image segmentation. Particularly, the U-Net architecture [10] is currently experiencing a huge success, as most winning contributions to recent medical image segmentation challenges were exclusively build around U-Net.

The aim of this work is to investigate how to re-use powerful deep convolutional networks that exist in the literature for 2D image analysis purposes. Indeed, a lot of powerful 2D classification networks are well trained on very large datasets. Through transfer learning, these pre-trained networks can be easily re-used for other classification problems. However, transferring both learning and feature detection power of classification networks to another types of problem is not obvious. Especially, in problems using convolutional encoder-decoder architectures such as U-Net, this consists in integrating the pre-trained model as part of the encoder branch. When the dimensions of the classifier and the processed network are the same (i.e. when they processed data having the same dimensions), the integration of the classifier is almost immediate. Otherwise, an adaptation process is necessary to be able to adapt the classifier to deal with the dimensions of the images under study.

Related works can be found in the literature where the segmentation of 3D medical images are based on 2D networks whose encoder has been pre-trained on ImageNet [6,7,13]. Although taking advantage of a pre-trained encoder, these networks do not integrate the spatial coherence in 3D, since they process data slice by slice. In our case, we aim at studying how to transfer the skills of a powerful and pre-trained network on 2D images to a convolutional encoder-decoder processing 3D images without loosing the consistency across the third dimension. It is known that the field of 2D image classification is well studied and well developed, generally because of the availability of large annotated 2D natural image datasets. The public availability of such large 3D databases is non-existent, and even less so in the medical field. Taking advantage of such powerful and well-studied networks by re-using them in a 3D problem is an idea worth investigating. In summary, the main proposed contributions are:

- Proposal of an efficient way to transfer any 2D classification architecture to 3D segmentation purposes without losing the 3D consistency.
- The proposed idea is generalizable in order to integrate any low-dimensional classification architecture into another high-dimensional architecture without losing spatial coherence.

In particular, we are interested in the recently published EfficientNet network [12] which achieves state-of-the-art top-1 and top-5 accuracy on ImageNet, while being widely more smaller and faster on inference than the best existing deep architectures. This paper is organized as follow: the proposed architecture is detailed in the next Sect. 2, the results are summarized in Sect. 3, followed by a conclusion in Sect. 4.

2   Method

The proposed segmentation approach follows a convolutional encoder-decoder architecture. It is built from an asymmetrically large encoder to extract image features and a smaller decoder to reconstruct segmentation masks. We embed as part of the encoder branch the recently proposed network called EfficientNet [12].

2.1   Data Pre-processing

Because of the limitations in GPU memory and time-consuming computation, we were forced to take some precautions. We process each modality separately, and resize the images dimensions by reducing the background using the largest crop size of $192 \times 160 \times 108$, and compromise the batch size to be 1. We do not use any additional training data and employ the provided training set only. We normalize all input images to have zero mean and unit variance.

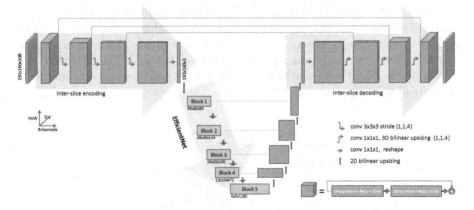

Fig. 1. Schematic illustration of the proposed network architecture. Input is a one-channel cropped 3D MRI. The inter-slice encoder as well as the decoder consist of a succession of residual blocks with GroupNorm normalization. The output of the decoder has three channels with the same spatial size as the input. Under each EfficientNet block are shown corresponding output feature dimensions.

Since the EfficientNet models range from 0 to 7, we restricted our experimentations in this preliminary work only by testing the baseline EfficientNet-B0 [12]. Indeed, this model presents a compromise between performance and complexity. Using EfficientNet-B7 at this stage requires more resources, a choice which may strongly complexify our model. In any case, if the proposed architecture works well with this baseline model, its generalization towards EfficientNet-B7, will be straightforward and will definitely improve the performances.

2.2   Encoder Branch

The encoding process goes through two steps. First, we encode three-dimensional data into two-dimensional data, while keeping the height and width at their original size and compressing only the depth to 3 channels. Second, the data is now ready to start the second encoding step, which is none other than the EfficientNet network without its fully connected layers.

As shown in Fig. 1, EfficientNet is represented as blocks as in its original version. However, only the blocks involved in skip connection layers are represented. The inter-slices encoding part uses convolutional blocks which consists of two convolutional layers with normalization and ReLU, followed by skip connection. Following the works of [9], we choose to use Group Normalization [14], which divides the channels into groups before normalizing them by using mean and variance of each group. It seems to perform better than traditional batch normalization, especially when the batch size is small.

Let us assume that the input volume is of width W, height H, depth D, with C channels. The data passes through a 3D-2D shrinking step. Thus, the depth is reduced by factors 3, 3 and 4 to reach a final depth size of 3, which corresponds to the required number of channels of EfficientNet. In this shrinking procedure, the width and height of the single 3D batch are not modified. The different depth reduction factors can be changed and adapted according to the third dimension of the data. In our study, dimensions change as shown in Table 1. After this process, the reduced data reached through the 2D shrinking process are given as inputs of the EfficientNet model.

Table 1. Dimension shrinking.

Layer	Dimension
Input	$W \times H \times D \times C$
Block 1	$W \times H \times \frac{D}{f_1} \times C_1$
Block 2	$W \times H \times \frac{D}{f_2} \times C_2$
$\vdots$	$\vdots$
Block n	$W \times H \times 3 \times C_n$
Block n+1	$W \times H \times 3 \times 1$
Output	$W \times H \times 3$

2.3 Decoder Branch

Asymmetrically to the encoding part, the decoder is composed entirely of homogeneous blocks as shown in Fig. 1. Obviously, the decoding part linked to the EfficientNet is a 2D decoder whereas the inter-slice decoding is a 3D decoder. Each decoder level begins by upsampling the spatial dimension, doubling the number of features by a factor of 2 followed by skip connections. A sigmoid function is used as an activation for the output of the decoder which has three channels corresponding to the number of classes with the same spatial size as inputs.

2.4 Loss

Many networks are trained with a cross-entropy loss function, however the resulting delineations may not be ideal in terms of Dice score. As an alternative, one can employ a soft Dice loss function to train the proposed network. While several formulations of the dice loss exist in the literature, we prefer to use a soft Dice loss which has given good results in segmentation challenges in the past [9]. The soft Dice loss function is differentiable and is given by:

$$\mathcal{L}_{Dice} = \frac{2 \sum P_{true} P_{pred}}{\sum P_{true}^2 + \sum P_{pred}^2 + \epsilon} , \tag{1}$$

were P_{true} and P_{pred} represent respectively the ground truth and the predicted labels. Brain MRI segmentation is a challenging task partly due to severe class imbalance. Tackling this issue by only using a fixed loss function, cross entropy or Dice, for the entire training process is not an optimal strategy. Therefore, a linear combination of the two loss functions is often considered as the best practice, and leads to more robust and optimal segmentation models. In practice, the final loss function is as follows:

$$\mathcal{L} = \mathcal{L}_{Cross} - \mathcal{L}_{Dice}. \qquad (2)$$

2.5 Training

The proposed network architecture is trained with centered cropped data of size $192 \times 160 \times 108$ voxels, ensuring that the useful content of each slice remains within the boundaries of the cropped area, training was made on BraTS2020 [8] dataset. Constrained by the poor performance of the material, we set the batch size to 1. Training has been done using the Adam optimizer known for little memory requirements, with an initial learning rate of 10^{-4} reduced by a factor of 10 whenever the loss has not improved for 50 epochs.

3 Results

We designed the proposed network on Tensorflow and run it on the 368 training cases of BraTS 2020. We process the four modalities separately and average the sigmoid outputs. The validation dataset was provided to test the performance of the models on unseen data. It consists of 125 cases with 4 modalities and without their corresponding segmentation. The results of the segmentation of the Whole Tumor (WT), Tumor Core (TC) and Enhancing Tumor (ET) are summarized in Table 2. All reported values were computed by the online evaluation platform (https://ipp.cbica.upenn.edu/). Figure 2 shows a typical segmentation results extracted from the validation dataset.

Table 2. Results on BraTS 2020 validation data. Metrics were computed by the online evaluation platform.

	Dice		
	ET	WT	TC
Mean	65.37	84.13	68.04
StdDev	31.93	10.67	31.29
Median	81.23	87.08	78.30

With an average Dice score of 84.13 for the WT class, on the validation set, the proposed model seems efficient and accurate enough to handle the training

Table 3. Results on BraTS 2020 test data. Metrics were computed by the online evaluation platform.

	Dice		
	ET	WT	TC
Mean	69.59	80.68	75.20
StdDev	26.10	14.93	28.94
Median	78.81	86.38	87.28

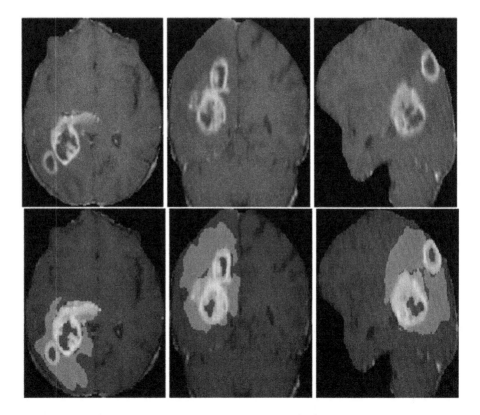

Fig. 2. Typical result on the BraTS validation set (2020). From left to right: axial, coronal end sagittal views in T1ce. Enhancing tumor is shown in yellow, necrosis in red and edema in green. (Color figure online)

and inference over a 3D dataset. On the other hand, results on enhanced and core tumors are less efficient. This may be due to the fact that we process each MRI modality separately.

After the compression on the depth, the network retains the shape of the brain and its structure, even if they appear slightly blurred and degraded. The network learns to extract three axial sections at more or less regular levels. We

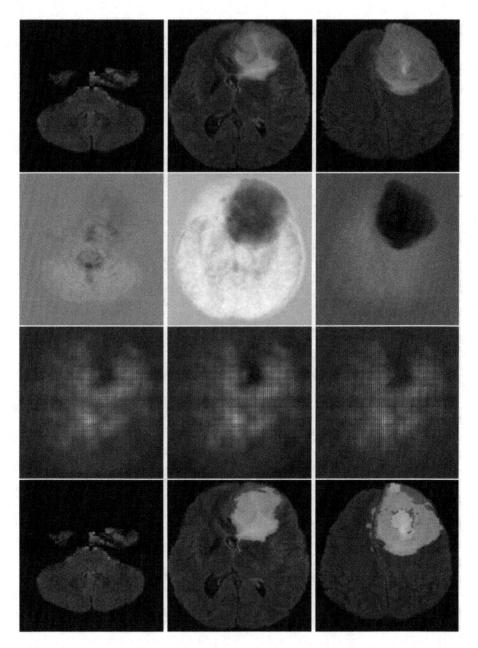

Fig. 3. Result visualization from the BraTS training set (2020). From up to down: three axial slices of an unique sample in Flair, corresponding channels output of the 3D encoder, corresponding channels input of the 3D decoder and corresponding labels. With Necrotic and Non-Enhancing Tumor core in red, Gadolinium-enhancing tumor in blue and peritumoral edema in green (Color figure online)

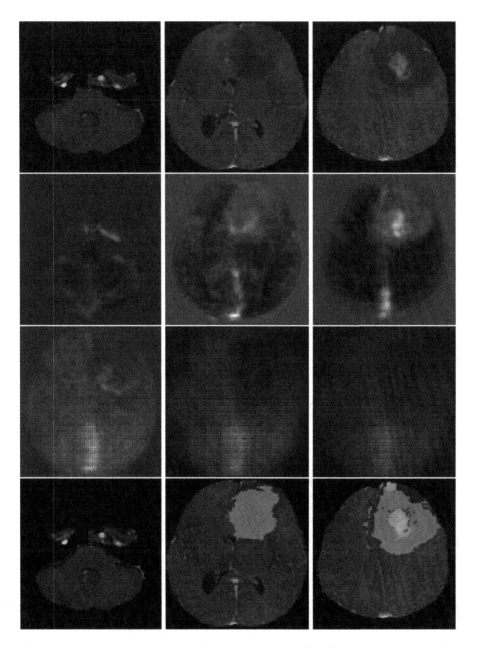

Fig. 4. Result visualization from the BraTS training set (2020). Three axial slices of an unique sample in T1ce, corresponding channels output of the 3D encoder, corresponding channels input of the 3D decoder and corresponding labels. With Necrotic and Non-Enhancing Tumor core in red, Gadolinium-enhancing tumor in blue and peritumoral edema in green (Color figure online)

notice the presence of the tumor on the three learned sections, the axial sections generally involved in 3-channel compression are the sections that denote the presence of tumor information. The high intensity on the tumor parts indicates that the features of the network are focused on the detection of the tumor. These details are reported in Fig. 3 and Fig. 4.

The results on the test set are reported on Table 3. We can notice a large improvement on ET and TC scores and a slight decrease on WT compared to the validation set. From Fig. 3, we can see that the images resulting from the 3D compression of the axial slices are well oriented on the learning of the characteristics of the whole tumor, thus the visible region on FLAIR. In Fig. 4 we find that the compressed data are much more oriented in learning the structural characteristics of enhanced and necrotic tumor components, which are the visible regions in T1CE.

4   Conclusion

In this paper, we introduced a generic 3D U-Net architecture that allows performance transfer by re-using and embedding any 2D classifier network. The encoder as well as the decoder are composed of two stages. The 3D input data goes through a process of depth shrinking in order to transform the 3D data into 2D data. This process is a succession of blocks of 3D convolutions and max-pooling reducing the third dimension only. The transformed output data can be then encoded by any 2D classification network. Moreover, decoding also goes through a 2D decoding phase followed by a 3D decoding procedure. Because of the limited computational resources, we resized the images and trained separately the fourth modalities using four modality-specific networks. Nevertheless, the preliminary results seem to be promising.

Our goal was not to surpass all the BraTS sophisticated segmentation techniques, but to provide a functional way to re-use 2D classification architectures for 3D medical image segmentation purposes. As can be seen, the learning transfer from weights trained on 2D natural images can be exploited for processing 3D medical images. We are convinced that we can significantly improve the results by robustifying the learning technique, keeping the original size of the data and stacking the 4 modalities together.

Acknowledgements. This work has been sponsored by the General Directorate for Scientific Research and Technological Development, Ministry of Higher Education and Scientific Research (DGRSDT), Algeria.

References

1. Bakas, S., Akbari, H., Sotiras, A., Bilello, M., Rozycki, M., Kirby, J., et al.: Segmentation labels and radiomic features for the pre-operative scans of the TCGA-GBM collection. Cancer Imaging Arch. (2017). https://doi.org/10.7937/K9/TCIA.2017. KLXWJJ1Q

2. Bakas, S., Akbari, H., Sotiras, A., Bilello, M., Rozycki, M., Kirby, J., et al.: Segmentation labels and radiomic features for the pre-operative scans of the TCGA-GBM collection. Cancer Imaging Arch. (2017). https://doi.org/10.7937/K9/TCIA.2017.GJQ7R0EF

3. Bakas, S., Akbari, H., Sotiras, A., Bilello, M., Rozycki, M., Kirby, J.S., et al.: Advancing the cancer genome atlas glioma MRI collections with expert segmentation labels and radiomic features. Nat. Sci. Data 4, 170117 (2017). https://doi.org/10.1038/sdata.2017.117

4. Bakas, S., Reyes, M., Jakab, A., Bauer, S., Rempfler, M., Crimi, A., et al.: Identifying the best machine learning algorithms for brain tumor segmentation, progression assessment, and overall survival prediction in the brats challenge. arXiv preprint arXiv:1811.02629 (2018)

5. Bauer, S., Wiest, R., Nolte, L.-P., Reyes, M.: A survey of MRI-based medical image analysis for brain tumor studies. Phys. Med. Biol. 58(13), R97–R129 (2013). https://doi.org/10.1088/0031-9155/58/13/R97

6. Conze, P.-H., Brochard, S., Burdin, V., Sheehan, F.T., Pons, C.: Healthy versus pathological learning transferability in shoulder muscle MRI segmentation using deep convolutional encoder-decoders. Comput. Med. Imaging Graph. 83, 101733 (2020). https://doi.org/10.1016/j.compmedimag.2020.101733

7. Conze, P.-H., et al.: Abdominal multi-organ segmentation with cascaded convolutional and adversarial deep networks. arXiv preprint arXiv:2001.09521 (2020)

8. Menze, B.H., Jakab, A., Bauer, S., Kalpathy-Cramer, J., Farahani, K., Kirby, J., et al.: The multimodal brain tumor image segmentation benchmark (BRATS). IEEE Trans. Med. Imaging 34(10), 1993–2024 (2015). https://doi.org/10.1109/TMI.2014.2377694

9. Myronenko, A.: 3D MRI brain tumor segmentation using autoencoder regularization. In: Crimi, A., Bakas, S., Kuijf, H., Keyvan, F., Reyes, M., van Walsum, T. (eds.) BrainLes 2018. LNCS, vol. 11384, pp. 311–320. Springer, Cham (2019). https://doi.org/10.1007/978-3-030-11726-9_28

10. Ronneberger, O., Fischer, P., Brox, T.: U-Net: convolutional networks for biomedical image segmentation. In: Navab, N., Hornegger, J., Wells, W.M., Frangi, A.F. (eds.) MICCAI 2015. LNCS, vol. 9351, pp. 234–241. Springer, Cham (2015). https://doi.org/10.1007/978-3-319-24574-4_28

11. Souadih, K., Belaid, A., Ben Salem, D., Conze, P.-H.: Automatic forensic identification using 3D sphenoid sinus segmentation and deep characterization. Med. Biol. Eng. Comput. 58(2), 291–306 (2019). https://doi.org/10.1007/s11517-019-02050-6

12. Tan, M., Le Q.V.E.: EfficientNet: rethinking model scaling for convolutional neural networks. In: Proceedings of Machine Learning Research, 36th International Conference on Machine Learning (ICML), Long Beach, California, USA, vol. 97, pp. 10691–10700 (2019)

13. Vu, M.H., Grimbergen, G., Nyholm, T., Löfstedt, T.: Evaluation of multi-slice inputs to convolutional neural networks for medical image segmentation. arXiv preprint arXiv:1912.09287 (2019)

14. Wu, Y., He, K.: Group normalization. Int. J. Comput. Vis. 128(3), 742–755 (2019). https://doi.org/10.1007/s11263-019-01198-w

15. Zaouche, R., et al.: Semi-automatic method for low-grade gliomas segmentation in magnetic resonance imaging. IRBM 39(2), 116–128 (2018). https://doi.org/10.1016/j.irbm.2018.01.004

Segmentation of the Multimodal Brain Tumor Images Used Res-U-Net

Jindong Sun[1]🄳, Yanjun Peng[1,2(✉)], Dapeng Li[1], and Yanfei Guo[1]

[1] College of Computer Science and Engineering, Shandong University of Science and Technology, Qingdao, China
pengyanjuncn@163.com
[2] Shandong Province Key Laboratory of Wisdom Mining Information Technology, Qingdao, China

Abstract. Gliomas are the most common brain tumors, which have a high mortality. Magnetic resonance imaging (MRI) is useful to assess gliomas, in which segmentation of multimodal brain tissues in 3D medical images is of great significance for brain diagnosis. Due to manual job for segmentation is time-consuming, an automated and accurate segmentation method is required. How to segment multimodal brain accurately is still a challenging task. To address this problem, we employ residual neural blocks and a U-Net architecture to build a novel network. We have evaluated the performances of different primary residual neural blocks in building U-Net. Our proposed method was evaluated on the validation set of BraTS 2020, in which our model makes an effective segmentation for the complete, core and enhancing tumor regions in Dice Similarity Coefficient (DSC) metric (0.89, 0.78, 0.72). And in testing set, our model got the DSC results of 0.87, 0.82, 0.80. Residual convolutional block is especially useful to improve performance in building model. Our proposed method is inherently general and is a powerful tool to studies of medical images of brain tumors.

Keywords: Brain tumor segmentation · Deep learning · Magnetic resonance images

1 Introduction

Gliomas are the most frequent primary brain tumors, which have the highest mortality rate [1,3,4,18]. They can be categorized to low-grade gliomas (LGG) and high-grade gliomas (HGG). HGG is more aggressive form of the disease, which has a median survival rate of two years or less. The slower growing low-grade variants, such as low-grade astrocytomas and oligodendrogliomas, usually makes life expectancy of several years [15]. MRI is a basic modality commonly used in brain structure analysis, which provides images with high contrast for soft tissues and high spatial resolution and can be useful to evaluate unknown health risk [2,6,15].

© Springer Nature Switzerland AG 2021
A. Crimi and S. Bakas (Eds.): BrainLes 2020, LNCS 12658, pp. 263–273, 2021.
https://doi.org/10.1007/978-3-030-72084-1_24

In recent years, lots of automatic approaches have been proposed for accurate segmentation in brain tumors, and these works can be roughly categorized into machine learning methods, deep learning methods and both-combined methods. Machine learning method is based on probabilistic models, which can learn from brain tumor patterns that do not follow a specific model, such as Conditional Random Field (CRF), Random Forrest (RF) and Support Vector Machine (SVM). Deep learning method learns the feature representation in a data-driven way [7], such as convolutional neural network (CNN), parallelized long short-term memory network (LSTM) and fully convolutional network (FCN). In addition, some authors combined probabilistic model (CRF, RF or SVM) and deep learning method to develop a novel method [5, 10, 12].

The fully convolutional neural networks (FCN), a new variant of CNN, gained the great interest in the segmentation competition of PASCAL VOC 2012. The deep convolutional learning model with substantially enlarged depth advanced the state-of-art performance on segmentation tasks that it alleviated the optimization degradation issue by approximating the objective function with residual functions instead of simply stacking layers, and residual block are skip connections between layers of the network. FCN based approaches are the pioneering work of deep learning in medical image segmentation, although the segmented result is not good enough.

In the end-to-end methods, with the combination of encoding layers or decoding layers, they achieved the success of image segmentation in pixel level. Compared to primary convolutional neural network, the end-to-end method can avoid a lot of duplicate calculations. U-Net architecture, based on fully convolution, had been successfully applied to medical image segmentation [9, 17, 19, 21] . This model is a popular and efficient network for segmentation in brain tumors. Naser and Deen [16] proposed a new approach to achieve segmentation in gliomas. They combined U-Net model for convolutional segmentation and pre-trained VGG16 model for transferring learning and a fully connected classifier for tumor grading. For clinical usage, the challenge is how to pursue the best accuracy for segmentation within limited computational budgets. Li et al. [13] proposed a multi-modality aggregation network (MMAN), which was able to extract multi-scale features of brain tissues and harness complementary information from multi-modality MRI images for fast and accurate segmentation. They applied dilated convolutional layers with different kernel size to obtain large-scale features without increasing too many parameters and computational costs. Ding et al. [8] developed a novel multi-path adaptive fusion network. In this model, they applied the idea of skip-connection in ResNets to the dense block so as to effectively reserve and propagate more low-level visual features. Liu et al. [14] investigated the performance of U-Net model in brain tumor, stroke, white matter hyperintensities (WMHs), eye, cardiac, liver, musculoskeletal, skin cancer, and neuronal pathology. They reported the different extended U-shaped networks and analyzed their pros and cons.

In this work, inspired on the groundbreaking proposal on U-Net, we focus on building the U-Net architecture by using residual convolutional blocks. We evaluated performances of different residual blocks. In addition, it is a key element to keep gradients independent and distributed identically. We aim to get better segmentation score in BraTS 2020 challenge.

2   Method

2.1   Pre-processing

In this work, we applied cropping and random-slicing methods. As for cropping, due to the GPU memory limitation, we cropped the zero-pixel region which in MRI images before training. The zero-pixel area of image boundary does not help to improve the segmentation accuracy. The original size of MRI images is array size of $155 \times 240 \times 240$. In model, we employed max-pooling function four times that every dimension size must be divided by 16 (2^4). Therefore, considering factors above, we set the size of 3D MRI images as $144 \times 192 \times 192$. As for multimodal 3D images, it is $4 \times 144 \times 192 \times 192$.

For each MRI images, we cropped to nine slices randomly. This step can effectively prevent overfitting during training stage. We randomly take 9 consecutive 3D sequences with length of 16 in the first dimension of the MRI images into training. After randomly cropping, the array size of MRI images is $9 \times 16 \times 192 \times 192$. As for multimodal 3D images, it is $4 \times 9 \times 16 \times 192 \times 192$. In addition, we do the same operation for each epoch during training stage. So the sequences that input to the neural network are generally different for per image and per epoch. This randomization makes the neural network model powerful generalization, especially in limited training data sets. We ensure that all pixels of the brain are trained in training step.

In addition, we employed z-score normalization in medical images [11]. It is accomplished by linearly transforming the original intensities between mean and standard deviation into the corresponding learned landmarks, which defined as:

$$z = \frac{x - \mu}{\sigma} \tag{1}$$

where μ is the mean of the MRI sequence in pixel level and σ is the standard deviation of the MRI sequence in pixel level.

2.2   Architecture

We build the architecture of deep learning referring to Fig. 1. It is an end-to-end method of deep learning, which is also a pixel-to-pixel method. Each layer in this model is five-dimensional array size of bs $\times$ c $\times$ h $\times$ w $\times$ d, where bs is batch size dimension, c is the channel or multimodal (Flair, T1, T1c and T2) sequences and h, w, d are spatial dimensions. Each convolutional layer and de-convolutional layer contains batch normalization and activation function.

In building this architecture, we refined three primary residual blocks and employed these blocks into encoding stage, in which there are res-block-1, res-block-2 and res-block-3. The residual block is a kind of skip-connect architecture,

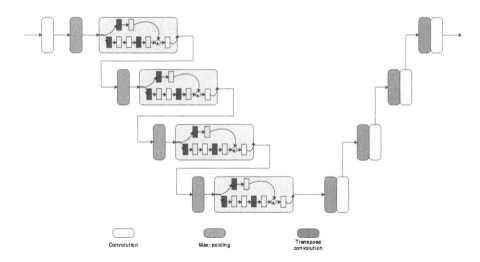

Fig. 1. Our proposed model using res-block-1 in encoding stage. Res-block-1 is shown in Fig. 2

avoiding gradient vanished with increasing depth of network, in which the gradient is effectively transferred to the shallow layer during training network. We apply the randomized leaky rectified liner unit (RReLU) as activated function for neural network.

We designed the res-block-1 block by using a dual-path convolution, an addition operation and a RReLU function. Referring to Fig. 2, we employed two convolutional layers with batch normalization in main path, where RReLU was adopted after the first convolutional layer. And in the skip-path, we employed a convolutional layer and a batch normalization layer. These two path are added by weighted, after that the output data feature activated by a RReLU function.

Referring to Fig. 3, we designed the res-block-2 block by using the same dual-path architecture like res-block-1. In res-block-2, we putted RReLU function to the first position, so that the output feature which computed at the dual-path added each other, and then it putted fused feature to next neural unit.

Referring to Fig. 4, the res-block-3 is the main single convolutional block with a primary skip-connect weights, in which the last convolutional layer is connected after weighted addition operation.

In this model, we apply RReLU as activated function for neural network [20], which defined as:

$$RReLU(x) = \begin{cases} x & x > 0 \\ ax & otherwise \end{cases} \qquad (2)$$

where a is randomly sampled from uniform distribution U(L, R). L is lower bound of the uniform distribution and R is upper bound of the uniform distribution. We set L of 1/8 and set R of 1/3.

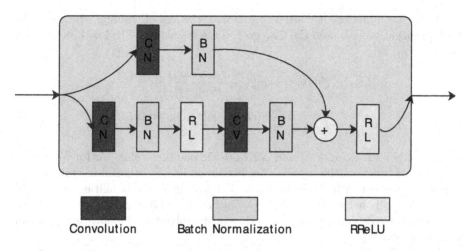

Fig. 2. Res-block-1.

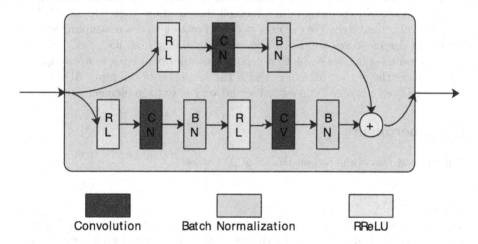

Fig. 3. Res-block-2.

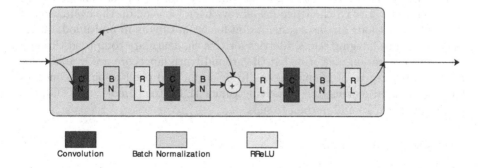

Fig. 4. Res-block-3.

As for loss function, it is used to calculate the loss of training which used in back propagation. We used the Categorical Cross-entropy. The loss function can be described as:

$$loss(x, class) = -\log(\frac{\exp(x(class))}{\sum_j \exp(x(j))})$$
$$= -x(class) + \log(\sum_j \exp(x(j))) \qquad (3)$$

which combines $LogSoftmax$ and $NLLLoss$ in one single class. As for $NLLLoss$ $(-x(class))$ function, the negative log likelihood loss, it is useful to train a classification problem with $class$ classes, and obtaining log-probabilities in a neural network is easily achieved by adding a $LogSoftmax$ layer in the last layer. The Categorical Cross-entropy function is useful to solve the classification problem with multi-classes.

In encoding stage, we employ lots of convolutional layers to extract features from MRI images. And we set parameters of convolutional function with kernel size of 3, stride of 1, padding of 1. Channels, in encoding stage, are 32, 64, 128, 256 and 512 respectively. We use max-pooling function to down-sampling so that model get deep features and learn segmentation ability from its.

In decoding stage, we employ transposed convolutional layer to up-sampling, which makes the output 3D images with the same size of the input 3D images. The transposed convolution is effective and very easy to implement.

3   Experiments

Our method was evaluated on BraTS 2020 dataset.

3.1   Dataset

The BraTS 2020 dataset contains four modes for every patient: Flair, T1, T1c and T2. We trained our model in BraTS 2020 training set, which contains 369 MRI scans including high-grade and low-grade brain tumor. In addition, the validation set contains 125 scans of glioblastoma and testing set contains 166 scans of glioblastoma. BraTS challenge has always been focusing on the evaluation of state-of-art methods for the segmentation for brain tumors in multimodal magnetic resonance imaging scans. Metrics for this challenge are computed through the online evaluation platform that the ground truth labels are not available for public. Every region of gliomas needs to be segmented pixel-to-pixel sequences for 4 meaningful regions: the enhancing tumor (ET), the tumor core (TC), the whole tumor (WT) and normal tissues.

3.2   Setup

Some of the hyper-parameters of the architectures were shown in Table 1. We approached brain tumor segmentation as a multi-class classification problem, segmented normal tissue, necrosis, edema, non-enhancing, and enhancing tumor from MR images respectively. However, the given MR images are not suitable for pouring into neural network directly that the redundant data will cost large GPU memory. So we cropped images of effective parts in three dimensions. Similarly, the same process was used on label set. Additionally, in brain tumor segmentation, the number of samples of necrosis and enhancing tumor is small in training set. To deal with that, we normalized all pixel-level image using z-score (zero-mean) normalization, which made the input data follow a normal distribution and speeded up training. The learning rate was linearly decreased each epoch during the training stage. Our model was developed using PyTorch. We train the model using four GPUs of Nvidia RTX 2080 TI with 40 h.

3.3 Evaluation

The evaluation metrics of brain tumor segmentations consist of three types of measures: Dice similarity coefficient (DSC), Sensitivity and Specificity. The DSC measures the spatial overlap between the automatic segmentation and the label. It is defined as:

$$DSC = \frac{2TP}{FP + 2TP + FN} \tag{4}$$

where FP, FN and TP are false positive, false negative detections and true position, respectively. Sensitivity, also called the true positive rate or probability of detection, measures the proportion of positives that are correctly identified as such:

$$Sensitivity = \frac{TP}{TP + FN} \tag{5}$$

A larger value of Sensitivity denotes a higher proximity of abnormal tissue between label and prediction of segmentation. Finally, specificity, also called the true negative rate, measures the proportion of negatives. It is defined as:

$$Specificity = \frac{TN}{TN + FP} \tag{6}$$

where TN is true negative detections. A larger value of Specificity denotes a higher proximity of normal tissue between label and prediction of segmentation.

3.4 Result

We evaluate our proposed model on validation with three different residual block, and compared with other state-of-art methods. Lastly, we report the result of segmentation on BraTS 2020 testing dataset. The performance of our model is presented on Fig. 5.

Referring to Table 2, the Res-Block-1 get better performance of segmentation than others. In Dice metric of WT region, the Res-Block-2 gain a little advantage. In addition, all the res-block get good score in segmentation of WT region.

Table 1. Hyper-parameters of our proposed model. The weights and bias in initialization are each convolutional layers' setting. We set the randomized leaky ReLU with default parameter setting.

Stage	Hyper-parameters	Value
Initialization	Weights	1.0
	Bias	0.0
Training	Max epoch	100
	Batch size	4
	Learning rate	0.002
	Learning rate decay	0.95
	GPU RTX 2080Ti	4
Randomized leaky ReLU	Lower	1/8
	Upper	1/3
	Inplace	True

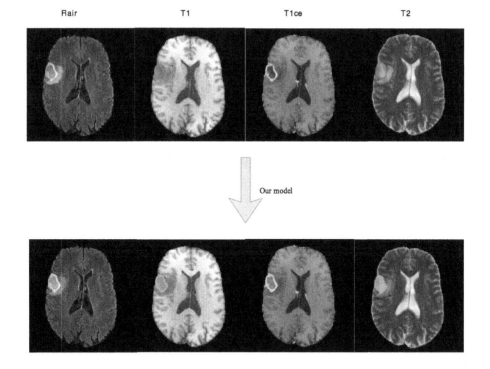

Fig. 5. Our predict MRI images using res-block-1 block.

Table 2. Segmentation result of Res-Block-1, Res-Block-2 and Res-Block-3 on BraTS 2020 Validation dataset

Method	Dice			Sensitivity			Specificity		
	ET	WT	TC	ET	WT	TC	ET	WT	TC
Res-Block-1	0.723	0.892	0.788	0.732	0.908	0.776	0.999	0.998	0.999
Res-Block-2	0.690	0.897	0.748	0.717	0.920	0.750	0.999	0.998	0.999
Res-Block-3	0.680	0.891	0.740	0.686	0.885	0.693	0.999	0.999	0.999

Table 3. Segmentation result of our proposed model, DeepLab and U-Net model on BraTS 2020 Validation dataset

Method	Dice			Sensitivity			Specificity		
	ET	WT	TC	ET	WT	TC	ET	WT	TC
Our method	0.723	0.892	0.788	0.732	0.908	0.776	0.999	0.998	0.999
DeepLab	0.707	0.884	0.774	0.742	0.876	0.754	0.999	0.999	0.999
U-Net	0.688	0.871	0.703	0.720	0.835	0.687	0.999	0.999	0.999

However, it is diverse to design res-block in U-shaped like model. Further experimental investigations are needed to estimate the performance of these decoding method in segmentation of medical image.

We compared our model with DeepLab and U-Net model by applying the same pre-processing methods for quantitative study. Our study is focus on performance of deep learning neural model. The results are reported on Table 3. The most difficult tasks in this brain tumor segmentation is marking the tumor core region for LGG and the enhancing tissues for HGG. To compare with two classical end-to-end model and referring to Table 3, our proposed model outperformed these models in Dice metrics.

Table 4. Segmentation result of our proposed model on BraTS 2020 testing dataset

Method	Dice			Sensitivity			Specificity		
	ET	WT	TC	ET	WT	TC	ET	WT	TC
Mean	0.803	0.872	0.823	0.808	0.896	0.819	0.999	0.999	0.999
StdDev	0.202	0.132	0.250	0.220	0.138	0.246	0.0003	0.001	0.0006
Median	0.854	0.913	0.910	0.886	0.936	0.920	0.999	0.999	0.999
25quantile	0.774	0.850	0.832	0.792	0.879	0.796	0.999	0.998	0.999
75quantile	0.916	0.944	0.951	0.935	0.965	0.956	0.999	0.999	0.999

The BraTS challenge testing result is reported on Table 4. In segmentation of the core and the enhancing tumor, our proposed method have better performance on testing set.

4   Conclusion

In this work, we propose a U-shaped architecture using residual block. We evaluated performance of different residual block in this U-shaped architecture. Residual block is an effective block to build deep neural network in feature extraction stage. In brain tumor segmentation, there are lots of deep-learning models including 2D and 3D model that the architecture becomes more and more complex as the development of computer hardware and the result of segmentation becomes more and more precise. Our research approach is a powerful tool to studies of 3D medical images of brain tumors and our proposed model is an effective deep-learning model, especially in 3D brain tumor segmentation.

Acknowledgement. This work was supported in part by National Natural Science Foundation of China under Grant No. 61976126, Shandong Natural Science Foundation under Grant No. ZR2019MF003, No. ZR2017MF054.

References

1. Bakas, S., et al.: Segmentation labels and radiomic features for the pre-operative scans of the TCGA-LGG collection. Cancer Imaging Arch. **286** (2017)
2. Bakas, S., et al.: Segmentation labels and radiomic features for the pre-operative scans of the TCGA-GBM collection, July 2017. https://doi.org/10.7937/K9/TCIA.2017.KLXWJJ1Q
3. Bakas, S., et al.: Advancing the cancer genome atlas glioma MRI collections with expert segmentation labels and radiomic features. Sci. Data **4**, 170117 (2017)
4. Bakas, S., et al.: Identifying the best machine learning algorithms for brain tumor segmentation, progression assessment, and overall survival prediction in the brats challenge. arXiv preprint arXiv:1811.02629 (2018)
5. Bauer, S., Nolte, L.-P., Reyes, M.: Fully automatic segmentation of brain tumor images using support vector machine classification in combination with hierarchical conditional random field regularization. In: Fichtinger, G., Martel, A., Peters, T. (eds.) MICCAI 2011. LNCS, vol. 6893, pp. 354–361. Springer, Heidelberg (2011). https://doi.org/10.1007/978-3-642-23626-6_44
6. Bauer, S., Wiest, R., Nolte, L.P., Reyes, M.: A survey of MRI-based medical image analysis for brain tumor studies. Phys. Med. Biol. **58**(13), R97 (2013)
7. Çiçek, Ö., Abdulkadir, A., Lienkamp, S.S., Brox, T., Ronneberger, O.: 3D U-Net: learning dense volumetric segmentation from sparse annotation. In: Ourselin, S., Joskowicz, L., Sabuncu, M.R., Unal, G., Wells, W. (eds.) MICCAI 2016. LNCS, vol. 9901, pp. 424–432. Springer, Cham (2016). https://doi.org/10.1007/978-3-319-46723-8_49
8. Ding, Y., Gong, L., Zhang, M., Li, C., Qin, Z.: A multi-path adaptive fusion network for multimodal brain tumor segmentation. Neurocomputing **412**, 19–30 (2020)
9. Dong, H., Yang, G., Liu, F., Mo, Y., Guo, Y.: Automatic brain tumor detection and segmentation using U-Net based fully convolutional networks. In: Valdés Hernández, M., González-Castro, V. (eds.) MIUA 2017. CCIS, vol. 723, pp. 506–517. Springer, Cham (2017). https://doi.org/10.1007/978-3-319-60964-5_44

10. Havaei, M., Jodoin, P.M., Larochelle, H.: Efficient interactive brain tumor segmentation as within-brain KNN classification. In: 2014 22nd International Conference on Pattern Recognition, pp. 556–561. IEEE (2014)

11. Jain, A., Nandakumar, K., Ross, A.: Score normalization in multimodal biometric systems. Pattern Recogn. **38**(12), 2270–2285 (2005)

12. Lee, C.-H., Wang, S., Murtha, A., Brown, M.R.G., Greiner, R.: Segmenting brain tumors using pseudo–conditional random fields. In: Metaxas, D., Axel, L., Fichtinger, G., Székely, G. (eds.) MICCAI 2008. LNCS, vol. 5241, pp. 359–366. Springer, Heidelberg (2008). https://doi.org/10.1007/978-3-540-85988-8_43

13. Li, J., Yu, Z.L., Gu, Z., Liu, H., Li, Y.: MMAN: multi-modality aggregation network for brain segmentation from MR images. Neurocomputing **358**, 10–19 (2019)

14. Liu, L., Cheng, J., Quan, Q., Wu, F.X., Wang, Y.P., Wang, J.: A survey on U-shaped networks in medical image segmentations. Neurocomputing **409**, 244–258 (2020)

15. Menze, B.H., et al.: The multimodal brain tumor image segmentation benchmark (BRATS). IEEE Trans. Med. Imaging **34**(10), 1993–2024 (2014)

16. Naser, M.A., Deen, M.J.: Brain tumor segmentation and grading of lower-grade glioma using deep learning in MRI images. Comput. Biol. Med. **121**, 103758 (2020)

17. Noori, M., Bahri, A., Mohammadi, K.: Attention-guided version of 2D UNet for automatic brain tumor segmentation. In: 2019 9th International Conference on Computer and Knowledge Engineering (ICCKE), pp. 269–275. IEEE (2019)

18. Pereira, S., Pinto, A., Alves, V., Silva, C.A.: Brain tumor segmentation using convolutional neural networks in MRI images. IEEE Trans. Med. Imaging **35**(5), 1240–1251 (2016)

19. Ronneberger, O., Fischer, P., Brox, T.: U-Net: convolutional networks for biomedical image segmentation. In: Navab, N., Hornegger, J., Wells, W.M., Frangi, A.F. (eds.) MICCAI 2015. LNCS, vol. 9351, pp. 234–241. Springer, Cham (2015). https://doi.org/10.1007/978-3-319-24574-4_28

20. Xu, B., Wang, N., Chen, T., Li, M.: Empirical evaluation of rectified activations in convolutional network. arXiv preprint arXiv:1505.00853 (2015)

21. Xu, F., Ma, H., Sun, J., Wu, R., Liu, X., Kong, Y.: LSTM multi-modal UNet for brain tumor segmentation. In: 2019 IEEE 4th International Conference on Image, Vision and Computing (ICIVC), pp. 236–240. IEEE (2019)

Vox2Vox: 3D-GAN for Brain Tumour Segmentation

Marco Domenico Cirillo[1,2(✉)], David Abramian[1,2], and Anders Eklund[1,2,3]

[1] Department of Biomedical Engineering, Linköping University, Linköping, Sweden
{marco.domenico.cirillo,david.abramian,anders.eklund}@liu.se
[2] Center for Medical Image Science and Visualization, Linköping University, Linköping, Sweden
[3] Division of Statistics and Machine learning, Department of Computer and Information Science, Linköping University, Linköping, Sweden

Abstract. Gliomas are the most common primary brain malignancies, with different degrees of aggressiveness, variable prognosis and various heterogeneous histological sub-regions, i.e., peritumoral edema, necrotic core, enhancing and non-enhancing tumour core. Although brain tumours can easily be detected using multi-modal MRI, accurate tumor segmentation is a challenging task. Hence, using the data provided by the BraTS Challenge 2020, we propose a 3D volume-to-volume Generative Adversarial Network for segmentation of brain tumours. The model, called Vox2Vox, generates realistic segmentation outputs from multi-channel 3D MR images, segmenting the whole, core and enhancing tumor with mean values of 87.20%, 81.14%, and 78.67% as dice scores and 6.44 mm, 24.36 mm, and 18.95 mm for Hausdorff distance 95 percentile for the BraTS testing set after ensembling 10 Vox2Vox models obtained with a 10-fold cross-validation. The code is available at https://github.com/mdciri/Vox2Vox.

Keywords: MRI · Vox2Vox · Generative adversarial networks · Deep learning · Artificial intelligence · 3D image segmentation

1 Introduction

Gliomas are the most frequent intrinsic tumours of the central nervous system. Based on the presence or absence of marked mitotic activity, necrosis and florid microvascular proliferation, a malignancy grade, WHO grade II, III or IV [33], is assigned. Gliomas with WHO grade II are also called low grade gliomas (LGG), whereas gliomas with higher WHO grade are called high grade gliomas (HGG). Although both these brain tumour types can easily be detected, they have a diffuse, infiltrative way of growing in the brain, and they exhibit peritumoural edema, such as an increase in water content in the area surrounding the tumour. This makes it arduous to define the tumour border by visual assessment, both in analysis and also during surgery [5].

A. Crimi and S. Bakas (Eds.): BrainLes 2020, LNCS 12658, pp. 274–284, 2021.
https://doi.org/10.1007/978-3-030-72084-1_25

For this reason, researchers recently started resorting to powerful techniques, able to segment complex objects and, in this way, guide the surgeons during the operation with a suitable accuracy. Indeed, machine learning [27] and especially deep learning [14,16,19,22,25,31] can provide state-of-the-art segmentation results.

1.1 Related Works

Nowadays generative adversarial networks (GANs) [10] are gaining popularity in computer vision, since they can learn to synthesise virtually any type of image. Specifically, GANs can be used for style transfer [9], image synthesis from noise [17], image to image translation [15], and also image segmentation [29]. GANs have become especially popular in medical imaging [34] since medical imaging datasets are much smaller compared to general computer vision datasets such as ImageNet. Additionally, in medical imaging it is common to collect several image modalities for each subject before proceeding with the analysis, and, when this is not possible, CycleGAN introduced in [35] can be used to synthesize the missing modalities.

GANs have also been used for medical image segmentation. Indeed, Han *et al.* in [11] proposed a GAN to segment multiple spinal structures in MRIs; Li *et al.* in [21] developed a novel transfer-learning framework using a GAN for robust segmentation of different human epithelial type 2 (HEp-2) cells; Dong *et al.* in [7] implemented a U-Net style GAN for accurate and timely organs-at-risk (OARs) segmentation; Nema *et al.* in [26] designed a 2D GAN, called RescueNet, to segment brain tumours from MR images; etc. Anyhow, Yi *et al.* [34] provide a complete and recent review of GANs applied in medicine.

Hence, inspired by these works and especially by the Pix2Pix GAN [15], which can generate an image of type A from a paired image of type B, the aim of this project is to do 3D image segmentation using 3D Pix2Pix GAN, named Vox2Vox, to segment brain gliomas. While a normal convolutional neural network, such as U-Net [28], performs the segmentation pixel by pixel, or voxel by voxel, through maximizing a segmentation metric or metrics (i.e. dice score, intersection over union, etc.), a GAN will also punish segmentation results that do not look realistic. Our hypothesis is that this can result in better segmentations.

2 Method

2.1 Data

The MR images used for this project are the Multimodal Brain tumour Segmentation Challenge (BraTS) 2020 training ones [1–4,23]. The BraTS 2020 training dataset contains MR volumes of shape $240 \times 240 \times 155$ from 369 patients, and for each patient four types of MR images were collected: native (T1), post-contrast T1-weighted (T1Gd), T2-weighted (T2), and T2 Fluid Attenuated Inversion Recovery (FLAIR). The BraTS 2020 validation dataset contains MR volumes from 125 patients. The images were acquired from 19 different institutions with different clinical protocols. The training set was segmented manually, by one to four raters, following the same annotation protocol, and their annotations were approved by experienced neuro-radiologists, whereas no segmentation was provided for the validation set. Moreover, all data were co-registered to the same anatomical template, interpolated to the same resolution $(1\,\text{mm}^3)$ and skull-stripped. Figure 1 shows an example of one training T1 MR image overlapped with its true segmentation in the three different planes.

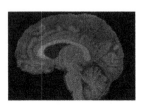 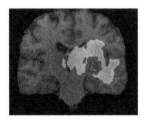 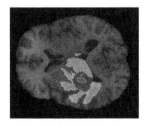

Fig. 1. From left to right there is a T1 MR image in the sagittal, coronal and transverse plane overlapped with its true segmentation. Peritumoural edema (ED), necrotic and non-enhancing tumour core (NCR/NET), and GD-enhancing tumour (ET) are highlighted in yellow, red and cyan respectively. (Color figure online)

2.2 Image Pre-processing and Augmentation

For each MR image intensity normalization is done per channel, whereas the background voxels are fixed to 0. On the other hand, the grey-scale ground-truths are transformed into categorical, so each target has four channels, as the number of the classes to segment: background, peritumoural edema (ED), necrotic and non-enhancing tumour core (NCR/NET), and GD-enhancing tumour (ET) labeled with 0, 1, 2, 4 respectively. BraTS, moreover, released 125 and 166 additional volumes as validation and test sets, respectively.

Since these volumes are memory demanding, patch augmentation is applied to extract one sub-volume of $128 \times 128 \times 128$ from each original volume. In this way, only 23.5% of the whole training set is used in every training epoch. Moreover, in order to prevent the networks from overfitting and memorizing the exact details of the training images, random 3D flipping, random 3D rotations between $0°$ and $30°$, power-law gamma intensity transformation (gain and gamma randomly chosen between [0.8–1.2]), elastic deformation with square deformation grid with displacements sampled from a normal distribution with standard deviation 5 voxels [28], or a combination of these with probability 0.5 are applied as image augmentation techniques.

2.3 Model Architecture

The Vox2Vox model, as the Pix2Pix one [15], consists of a generator and a discriminator. The generator, illustrated by Fig. 2, is built with U-Net and

Res-Net [13] architecture style, see Fig. 2; whereas the discriminator is build with PatchGAN [15] architecture style, see Fig. 3. Following are the details of both model's architecture[1]. The generator consists of:

I: a 3D image with 4 channels: T1, T2, T1Gd, and T2 FLAIR;

E: four down-sampling blocks, each of them made by 3D convolutions using kernel size 4, stride 2 and same padding, followed by instance normalization [32] and Leaky ReLU activation function. The number of filters used at the first 3D convolution is 64 and at each down-sampling the number is doubled;

B: four residual blocks, each made by 3D convolutions using kernel size 4, stride 1 and same padding, followed by instance normalization and Leaky ReLU activation function. Every convolution-normalization-activation output is concatenated with the previous one;

D: three up-sampling blocks, each of them made by 3D transpose convolutions using kernel size 4 and stride 2, followed by instance normalization and ReLU activation. Each 3D convolution input is concatenated with the respective encoder output layer;

O: a 3D transpose convolution using 4 filters (as the number of the classes to segment), kernel size 4 and stride 2, followed by softmax activation function. The output has shape $128 \times 128 \times 128 \times 4$ and constitutes the segmentation prediction for each class.

On the other hand, the discriminator consists of:

I: the 3D image with 4 channels and its segmentation ground-truth or the generator's segmentation prediction;

E: same as the generator;

O: one 3D convolution using 1 filter, kernel size 4, stride 1 and same padding. The output has shape $8 \times 8 \times 8 \times 1$ and constitutes the quality of the segmentation prediction created by the generator.

All 3D convolution and 3D transpose convolution layers use kernel size 4 (as in [15]), the He *et al.* weight initialization method [12], and same padding. Moreover, all Leaky ReLU layers have slope coefficient 0.3.

In order to reduce GPU memory consumption, we decided to used a single convolution layer in the encoder and decoder blocks, while a standard U-Net uses two convolutions layers per block. As our convolutions use stride 2, it means that the encoder/decoder will directly downsample/upsample the input volume a factor 2. In future work we will investigate if using two convolutions layers per blocks improves segmentation performance.

2.4 Losses

Since Vox2Vox contains two models, the generator and the discriminator, two loss functions are used. The discriminator loss, L_D, is the sum of the L_2 error of

---

[1] In the model descriptions we use *I, E, B, D,* and *O* to refer to Input(s), Encoder, Bottleneck, Decoder, and Output respectively.

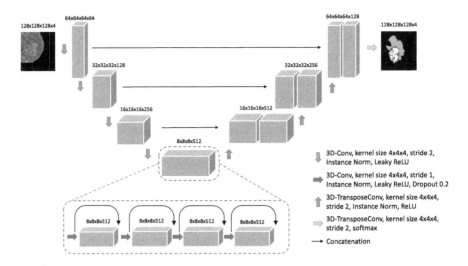

Fig. 2. The generator model.

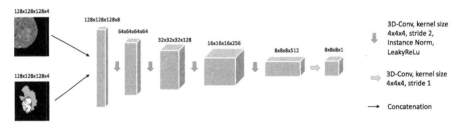

Fig. 3. The discriminator model.

the discriminator output, $D(\cdot, \cdot)$, between the original image x and the respective ground-truth y with a tensor of ones, and the L_2 error of the discriminator output between the original image and the respective segmentation prediction $\hat{y}$ given by the generator with a tensor of zeros, i.e.:

$$L_D = L_2 \left[D(x, y), \mathbf{1} \right] + L_2 \left[D(x, \hat{y}), \mathbf{0} \right], \tag{1}$$

whereas, the generator loss, L_G, is the sum of the L_2 error of the discriminator output between the original image and the respective segmentation prediction given by the generator with a tensor of ones, and the generalized dice loss [24, 30], $GDL(\cdot, \cdot)$, between the ground-truth and the generator's output multiplied by the scalar weight coefficient $\alpha \geq 0$, i.e.:

$$L_G = L_2 \left[D(x, \hat{y}), \mathbf{1} \right] + \alpha \, GDL \left(y, \hat{y} \right). \tag{2}$$

By looking at Eq. 2, it is easy to conclude that if $\alpha = 0$: Vox2Vox is a pure GAN and it minimizes only the unsupervised loss given by the discriminator; whereas if $\alpha \to \infty$: Vox2Vox ignores the discriminator, and behaves as a 3D U-Net with the architecture shown in Fig. 2.

2.5   Optimization and Regularization

Both the generator and the discriminator are trained using the Adam optimizer [18] with the parameters: $\lambda = 2 \cdot 10^{-4}$, $\beta_1 = 0.5$, and $\beta_2 = 0.999$. Dropout regularization with a dropout probability of 0.2 is used after each 3D convolutional operation in the generator's bottleneck (see Fig. 2). Moreover, as Yi *et al.* reported in [34], the discriminator loss helps the generator to guarantee the spatial consistency in the final results, behaving as a shape regularizer. In other words, the discriminator takes care that the generated brain segmentation looks realistic (i.e. like manual segmentations). In the end, we expect that Vox2Vox performs better with a trade-off α which does not disregard completely the discriminator loss and, at the same time, does not disregard the generator either.

2.6   Model Ensembling and Post-processing

Model ensembling is a technique that combines the outputs of several models in order to obtain more robust predictions [16]. Hence, once the Vox2Vox model is built, M-fold cross-validation over the training data can be done, which results in M models. Instead of performing the ensembling independently for each voxel, we propose a neighborhood averaging ensembling, by training a CNN that combines the outputs of the M Vox2Vox models. This CNN, called Ensembler, is simply made by just one 3D convolution, which takes as input a tensor with shape $(128 \times 128 \times 128 \times 4M)$, and returns a tensor with shape $(128 \times 128 \times 128 \times 4)$, using stride 1, kernel size 3 and softmax activation function. Since the input of our Ensembler are softmax outputs from each Vox2Vox model, the probabilities were zero-centered by subtracting 0.5 prior to training. In the end, the Ensembler is trained over 100 epochs, minimizing the generalized dice loss, and early stopping with a patience of 10 epoch (for the validation loss) is used.

Furthermore, since the M models are trained to detect 4 classes, it is normal that they sometimes detect a class which is not present for a subject and the ensembling follows such mis-segmentation. Therefore, post-processing can be useful in order to reduce mis-segmentation. Since the mis-segmentation normally results in small false positives, a cluster size threshold can be used to remove clusters smaller than a specific volume V.

3   Results

The Vox2Vox model is implemented using Python 3.7, Tensorflow 2.1 and its Keras library. The model is trained and validated on sub-volumes of size $128 \times 128 \times 128$ from 369 and 125 subjects respectively, using batch size 4, over 200 epochs on a computer equipped with 128 GB RAM and an Nvidia GeForce RTX 2080 Ti graphics card with 11 GB of memory. Once the training is completed, the 166 test volumes are cropped in order to have shape $160 \times 192 \times 128$. In this way, the testing set can be given as input to the fully convolutional Vox2Vox, because each axis is now divisible by $2^4 = 16$, where 4 is the generator's and discriminator's depth.

Table 1 reports the dice and the Hausdorff distance 95 percentile scores for all the classes of interest varying the α parameter in Eq. 2 for the training dataset. Note that the classes are reset as: whole tumour (WT = ET $\cup$ ED $\cup$ NCR/NET), tumour core (TC = ET $\cup$ NCR/NET) and enhancing tumour (ET). It also clearly shows that the best trade-off for the α parameter is 5, but at the same time that it cannot detect the enhancing tumor (ET) class properly over the whole training dataset. Anyway, it also shows that the discriminator helps to achieve good results, because the metrics obtained with $\alpha = 5$ are better than when the model only considers the generator loss (high values of α) and also when the model is a pure GAN ($\alpha = 0$).

Anyway, it has to be pointed out that the Vox2Vox GAN is trained to detected the classes presented in the data and not the combination of them.

Table 1. Mean dice score and Hausdorff distance 95 percentile for the different brain tumour areas over the training set. The metrics here are obtained training the model with sub-volumes of $128 \times 128 \times 128$ voxels, for different values of $\alpha = 0, 1, 3, 5, 10, 25, 50, 100, 200, 250$.

α	Dice score [%]			Hausdorff distance 95 [mm]		
	WT	TC	ET	WT	TC	ET
0	58.96	23.57	37.63	77.74	104.35	73.84
1	86.36	72.48	62.50	13.12	17.26	40.50
3	92.98	90.82	79.12	4.49	4.06	31.32
5	**93.21**	**91.70**	**79.31**	**3.80**	**3.18**	**30.91**
10	87.50	80.66	70.55	12.09	9.92	34.40
25	84.35	83.01	71.04	9.62	8.75	32.53
50	87.87	76.79	61.54	9.55	8.33	31.57
100	91.81	89.22	77.31	5.28	5.03	31.63
200	88.72	90.16	77.62	6.33	3.38	31.19
250	86.65	85.67	78.35	9.32	8.90	31.87

Moreover, it is evident that the ET class is the most problematic one to detect. Indeed, there are just 27 subjects (7.3%) in the training dataset that do not contain any ET voxels. So, the post-processing should focus on that class in the end, see Sect. 2.6.

With the alpha parameter set to 5, 10-fold cross-validation ($M = 10$) over the training set is applied, using all the image augmentation techniques listed in Sect. 2.2 are applied to avoid overfitting. Every epoch takes approximately 20 min to complete using a Keras.utils.Sequence generator. Successively, once the 10 Vox2Vox models are trained, their outputs are combined by the Ensembler CNN explained previously in Sect. 2.6.

As post-processing, a threshold $th = 1000$ voxels ($V = 1\text{cm}^3$) is set for the ET class: if the final segmentation has a number of ET voxels fewer than th ones, the ET voxels are converted into the NCR/NET class. Table 2 reports the metrics calculated on the training, validation and test sets for each class by the CBICA Image Processing Portal[2] with our proposed ensembling and post-processing.

Table 2. Mean, standard deviation and median dice score and Hausdorff distance 95 percentile for the three different brain tumour classes over the training, validation, and test set. The predictions are calculated ensembling 10 models trained after a 10-fold cross-validation and post-processed with a threshold $th = 1000$ voxels for the ET class. The values reported here were calculated by the CBICA Image Processing Portal.

Dataset		Dice score [%]			Hausdorff distance 95 [mm]		
		WT	TC	ET	WT	TC	ET
Training	Mean	91.63	89.25	79.56	3.66	3.52	30.04
	StdDev	6.36	11.49	24.74	3.87	4.57	96.67
	Median	93.39	92.50	87.16	2.44	2.23	1.73
Validation	Mean	89.26	79.19	75.04	6.39	14.07	36.00
	StdDev	8.28	22.30	28.68	11.44	47.56	105.28
	Median	91.75	88.13	85.87	3.0	3.74	2.23
Test	Mean	87.20	81.14	78.67	6.44	24.36	18.95
	StdDev	14.07	25.47	20.17	11.09	79.78	74.91
	Median	91.42	90.59	83.65	3.16	3.0	2.23

The training metrics for the WT and TC class decreased compared to those reported in Table 1, probably due to the image augmentation techniques introduced during the training, but just slightly; on the other hand, the ET ones slightly increased, probably thanks to the ensembling and the post-processing. The metrics obtained for both sets are high, but there are still some bad predictions that compromise the mean values, which also explains the large standard deviations. For this reason, the median values for each class reported in Table 2 may be more representative of the typical segmentation performance of the model.

4   Conclusions

Table 2 establishes that ensembling multiple Vox2Vox models generates high quality segmentation outputs that looks realistic thanks to their discriminators' support, achieving median values of: 93.40%, 92.49%, and 86.48% dice scores and 2.44 mm, 2.23 mm, and 1.73 mm Hausdorff distance 95 percentile over the training dataset; 91.75%, 88.13%, and 83.14% and 3.0 mm, 3.74 mm, and 2.44 mm over

---

[2] https://ipp.cbica.upenn.edu, the name of this group for the challenge is IMT_AE.

the validation dataset; and 91.42%, 90.59%, and 83.65% and 3.16 mm, 3.0 mm and 2.23 mm over the testing dataset for whole tumour, core tumour and enhancing tumour respectively.

Moreover, as future work, the Vox2Vox can be improved for the following BraTS challenges in many ways, i.e.: training the model to optimize the BraTS metrics and not those provided in the data, optimizing the PatchGAN architecture, and ensembling Vox2Vox trained with different augmentation techniques, following the suggestions reported in [6].

In the end, the Vox2Vox model can be used not only for image segmentation but also for further image augmentation. Indeed, Vox2Vox could be combined with a 3D noise to image GAN [8,20] which can synthesize realistic segmentation outputs, that are then translated to realistic MR volumes. The combination of these two GANs might result in a really fast batch generation of MR images with their targets.

Acknowledgement. This study was supported by LiU Cancer, VINNOVA Analytic Imaging Diagnostics Arena (AIDA), and the ITEA3/VINNOVA funded project Intelligence based iMprovement of Personalized treatment And Clinical workflow supporT (IMPACT). Funding was also provided by the Center for Industrial Information Technology (CENIIT) at Linköping University.

References

1. Bakas, S., et al.: Segmentation labels and radiomic features for the pre-operative scans of the TCGA-GBM collection. Cancer Imaging Arch. (2017)
2. Bakas, S., et al.: Segmentation labels and radiomic features for the pre-operative scans of the TCGA-LGG collection. Cancer Imaging Arch. (2017)
3. Bakas, S., et al.: Advancing the cancer genome atlas glioma MRI collections with expert segmentation labels and radiomic features. Sci. Data **4**, 170117 (2017)
4. Bakas, S., et al.: Identifying the best machine learning algorithms for brain tumor segmentation, progression assessment, and overall survival prediction in the BRATS challenge. arXiv preprint arXiv:1811.02629 (2018)
5. Blystad, I.: Clinical Applications of Synthetic MRI of the Brain, vol. 1600. Linköping University Electronic Press, Linköping (2017)
6. Cirillo, M.D., Abramian, D., Eklund, A.: What is the best data augmentation for 3D brain tumor segmentation? arXiv preprint arXiv:2010.13372 (2020)
7. Dong, X., et al.: Automatic multiorgan segmentation in thorax CT images using U-Net-GAN. Med. Phys. **46**(5), 2157–2168 (2019)
8. Eklund, A.: Feeding the zombies: synthesizing brain volumes using a 3D progressive growing GAN. arXiv preprint arXiv:1912.05357 (2019)
9. Gatys, L.A., Ecker, A.S., Bethge, M.: Image style transfer using convolutional neural networks. In: Proceedings of the IEEE Conference on Computer Vision and Pattern Recognition, pp. 2414–2423 (2016)
10. Goodfellow, I., et al.: Generative adversarial nets. In: Advances in Neural Information Processing Systems, pp. 2672–2680 (2014)
11. Han, Z., Wei, B., Mercado, A., Leung, S., Li, S.: Spine-GAN: semantic segmentation of multiple spinal structures. Med. Image Anal. **50**, 23–35 (2018)

12. He, K., Zhang, X., Ren, S., Sun, J.: Delving deep into rectifiers: surpassing human-level performance on ImageNet classification. In: Proceedings of the IEEE International Conference on Computer Vision, pp. 1026–1034 (2015)

13. He, K., Zhang, X., Ren, S., Sun, J.: Deep residual learning for image recognition. In: Proceedings of the IEEE Conference on Computer Vision and Pattern Recognition, pp. 770–778 (2016)

14. Isensee, F., Kickingereder, P., Wick, W., Bendszus, M., Maier-Hein, K.H.: No new-net. In: Crimi, A., Bakas, S., Kuijf, H., Keyvan, F., Reyes, M., van Walsum, T. (eds.) BrainLes 2018. LNCS, vol. 11384, pp. 234–244. Springer, Cham (2019). https://doi.org/10.1007/978-3-030-11726-9_21

15. Isola, P., Zhu, J.Y., Zhou, T., Efros, A.A.: Image-to-image translation with conditional adversarial networks. In: Proceedings of the IEEE Conference on Computer Vision and Pattern Recognition, pp. 1125–1134 (2017)

16. Kamnitsas, K., et al.: Ensembles of multiple models and architectures for robust brain tumour segmentation. In: Crimi, A., Bakas, S., Kuijf, H., Menze, B., Reyes, M. (eds.) BrainLes 2017. LNCS, vol. 10670, pp. 450–462. Springer, Cham (2018). https://doi.org/10.1007/978-3-319-75238-9_38

17. Karras, T., Aila, T., Laine, S., Lehtinen, J.: Progressive growing of GANs for improved quality, stability, and variation. ICLR (2018)

18. Kingma, D.P., Ba, J.: Adam: a method for stochastic optimization. arXiv preprint arXiv:1412.6980 (2014)

19. Kong, X., Sun, G., Wu, Q., Liu, J., Lin, F.: Hybrid pyramid U-Net model for brain tumor segmentation. In: Shi, Z., Mercier-Laurent, E., Li, J. (eds.) IIP 2018. IAICT, vol. 538, pp. 346–355. Springer, Cham (2018). https://doi.org/10.1007/978-3-030-00828-4_35

20. Kwon, G., Han, C., Kim, D.: Generation of 3D brain MRI using auto-encoding generative adversarial networks. In: Shen, D., et al. (eds.) MICCAI 2019. LNCS, vol. 11766, pp. 118–126. Springer, Cham (2019). https://doi.org/10.1007/978-3-030-32248-9_14

21. Li, Y., Shen, L.: cC-GAN: a robust transfer-learning framework for HEp-2 specimen image segmentation. IEEE Access **6**, 14048–14058 (2018)

22. McKinley, R., Meier, R., Wiest, R.: Ensembles of densely-connected CNNs with label-uncertainty for brain tumor segmentation. In: Crimi, A., Bakas, S., Kuijf, H., Keyvan, F., Reyes, M., van Walsum, T. (eds.) BrainLes 2018. LNCS, vol. 11384, pp. 456–465. Springer, Cham (2019). https://doi.org/10.1007/978-3-030-11726-9_40

23. Menze, B.H., et al.: The multimodal brain tumor image segmentation benchmark (BRATS). IEEE Trans. Med. Imaging **34**(10), 1993–2024 (2014)

24. Milletari, F., Navab, N., Ahmadi, S.A.: V-net: fully convolutional neural networks for volumetric medical image segmentation. In: 2016 Fourth International Conference on 3D Vision (3DV), pp. 565–571. IEEE (2016)

25. Myronenko, A.: 3D MRI brain tumor segmentation using autoencoder regularization. In: Crimi, A., Bakas, S., Kuijf, H., Keyvan, F., Reyes, M., van Walsum, T. (eds.) BrainLes 2018. LNCS, vol. 11384, pp. 311–320. Springer, Cham (2019). https://doi.org/10.1007/978-3-030-11726-9_28

26. Nema, S., Dudhane, A., Murala, S., Naidu, S.: RescueNet: an unpaired GAN for brain tumor segmentation. Biomed. Sig. Process. Control **55**, 101641 (2020)

27. Polly, F., Shil, S., Hossain, M., Ayman, A., Jang, Y.: Detection and classification of HGG and LGG brain tumor using machine learning. In: 2018 International Conference on Information Networking (ICOIN), pp. 813–817. IEEE (2018)

28. Ronneberger, O., Fischer, P., Brox, T.: U-Net: convolutional networks for biomedical image segmentation. In: Navab, N., Hornegger, J., Wells, W.M., Frangi, A.F. (eds.) MICCAI 2015. LNCS, vol. 9351, pp. 234–241. Springer, Cham (2015). https://doi.org/10.1007/978-3-319-24574-4_28

29. Sato, M., Hotta, K., Imanishi, A., Matsuda, M., Terai, K.: Segmentation of cell membrane and nucleus by improving Pix2pix. In: BIOSIGNALS, pp. 216–220 (2018)

30. Sudre, C.H., Li, W., Vercauteren, T., Ourselin, S., Jorge Cardoso, M.: Generalised dice overlap as a deep learning loss function for highly unbalanced segmentations. In: Cardoso, M.J., et al. (eds.) DLMIA/ML-CDS -2017. LNCS, vol. 10553, pp. 240–248. Springer, Cham (2017). https://doi.org/10.1007/978-3-319-67558-9_28

31. Topol, E.J.: High-performance medicine: the convergence of human and artificial intelligence. Nat. Med. **25**(1), 44–56 (2019)

32. Ulyanov, D., Vedaldi, A., Lempitsky, V.: Instance normalization: the missing ingredient for fast stylization. arXiv preprint arXiv:1607.08022 (2016)

33. Wesseling, P., Capper, D.: WHO 2016 classification of gliomas. Neuropathol. Appl. Neurobiol. **44**(2), 139–150 (2018)

34. Yi, X., Walia, E., Babyn, P.: Generative adversarial network in medical imaging: a review. Med. Image Anal. **58**, 101552 (2019)

35. Zhu, J.Y., Park, T., Isola, P., Efros, A.A.: Unpaired image-to-image translation using cycle-consistent adversarial networks. In: Proceedings of the IEEE International Conference on Computer Vision, pp. 2223–2232 (2017)

Automatic Brain Tumor Segmentation with Scale Attention Network

Yading Yuan[✉]

Department of Radiation Oncology, Icahn School of Medicine
at Mount Sinai, New York, NY, USA
yading.yuan@mssm.edu

Abstract. Automatic segmentation of brain tumors is an essential but challenging step for extracting quantitative imaging biomarkers for accurate tumor detection, diagnosis, prognosis, treatment planning and assessment. Multimodal Brain Tumor Segmentation Challenge 2020 (BraTS 2020) provides a common platform for comparing different automatic algorithms on multi-parametric Magnetic Resonance Imaging (mpMRI) in tasks of 1) Brain tumor segmentation MRI scans; 2) Prediction of patient overall survival (OS) from pre-operative MRI scans; 3) Distinction of true tumor recurrence from treatment related effects and 4) Evaluation of uncertainty measures in segmentation. We participate the image segmentation challenge by developing a fully automatic segmentation network based on encoder-decoder architecture. In order to better integrate information across different scales, we propose a dynamic scale attention mechanism that incorporates low-level details with high-level semantics from feature maps at different scales. Our framework was trained using the 369 challenge training cases provided by BraTS 2020, and achieved an average Dice Similarity Coefficient (DSC) of 0.8828, 0.8433 and 0.8177, as well as 95% Hausdorff distance (in millimeter) of 5.2176, 17.9697 and 13.4298 on 166 testing cases for whole tumor, tumor core and enhanced tumor, respectively, which ranked itself as the 3rd place among 693 registrations in the BraTS 2020 challenge.

1   Introduction

Gliomas are the most common primary brain malignancies and quantitative assessment of gliomas constitutes an essential step of tumor detection, diagnosis, prognosis, treatment planning and outcome evaluation. As the primary imaging modality for brain tumor management, multi-parametric Magnetic Resonance Imaging (mpMRI) provides various different tissue properties and tumor spreads. However, proper interpretation of mpMRI images is a challenging task not only because of the large amount of three-dimensional (3D) or four-dimensional (4D) image data generated from mpMRI sequences, but also because of the intrinsic heterogeneity of brain tumor. As a result, computerized analysis have been of great demand to assist clinicians for better interpretation of mpMRI images for brain tumor. In particular, the automatic segmentation of brain tumor and its sub-regions is an essential step in quantitative image analysis of mpMRI images.

© Springer Nature Switzerland AG 2021
A. Crimi and S. Bakas (Eds.): BrainLes 2020, LNCS 12658, pp. 285–294, 2021.
https://doi.org/10.1007/978-3-030-72084-1_26

The brain tumor segmentation challenge (BraTS) [1–5] aims to accelerate the research and development of reliable methods for automatic brain tumor segmentation by providing a large 3D mpMRI dataset with ground truth annotated by multiple physicians. This year, BraTS 2020 provides 369 cases for model training and 125 cases for model validation. The MRI scans were collected from 19 institutions and acquired with different protocols, magnetic field strengths and manufacturers. For each patient, a native T1-weighted, a post-contrast T1-weighted, a T2-weighted and a T2 Fluid-Attenuated Inversion Recovery (FLAIR) were provided. These images were rigidly registered, skull-stripped and resampled to $1 \times 1 \times 1$ mm isotropic resolution with image size of $240 \times 240 \times 155$. Three tumor subregions, including the enhancing tumor, the peritumoral edema and the necrotic and other non-enhancing tumor core, were manually annotated by one to four raters following the same annotation protocol and finally approved by experienced neuro-radiologists (Fig. 1).

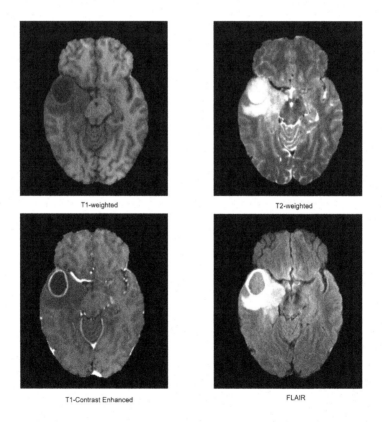

Fig. 1. An example of MRI modalities used in BraTS 2020 challenge

2   Related Work

With the success of convolutional neural networks (CNNs) in biomedical image segmentation, all the top performing teams in recent BraTS challenges exclusively built their solutions around CNNs. In BraTS 2017, Kamnitsas et al. [6] combined three different network architectures, namely 3D FCN [7], 3D U-Net [9], and DeepMedic [8] and trained them with different loss functions and different normalization strategies. Wang et al. [10] employed a FCN architecture enhanced by dilated convolutions [11] and residual connections [12]. In BraTS 2018, Myronenko [13] utilized an asymmetrical U-Net with a large encoder to extract image features, and a smaller decoder to recover the label. A variational autoencoder (VAE) branch was added to reconstruct the input image itself in order to regularize the shared encoder and impose additional constraints on its layers. Isensee et al. [14] introduced various training strategies to improve the segmentation performance of U-Net. In BratTS 2019, Jiang et al. [15] proposed a two-stage cascaded U-Net, which was trained in an end-to-end fashion, to segment the subregions of brain tumor from coarse to fine, and Zhao et al. [16] investigated different kinds of training heuristics and combined them to boost the overall performance of their segmentation model.

The success of U-Net and its variants in automatic brain tumor segmentation is largely contributed to the skip connection design that allows high resolution features in the encoding pathway be used as additional inputs to the convolutional layers in the decoding pathway, and thus recovers fine details for image segmentation. While intuitive, the current U-Net architecture restricts the feature fusion at the same scale when multiple scale feature maps are available in the encoding pathway. Studies [17,18] have shown feature maps in different scales usually carry distinctive information in that low-level features represent detailed spatial information while high-level features capture semantic information such as target position, therefore, the full-scale information may not be fully employed with the scale-wise feature fusion in the current U-Net architecture.

To make full use of the multi-scale information, we propose a novel encoder-decoder network architecture named scale attention network (SA-Net), where we re-design the inter-connections between the encoding and decoding pathways by replacing the scale-wise skip connections in U-Net with full-scale skip connections. This allows SA-Net to incorporate low-level fine details with the high-level semantic information into a unified framework. In order to highlight the important scales, we introduce the attention mechanism [19,25] into SA-Net such that when the model learns, the weight on each scale for each feature channel will be adaptively tuned to emphasize the important scales while suppressing the less important ones. Figure 2 shows the overall architecture of SA-Net.

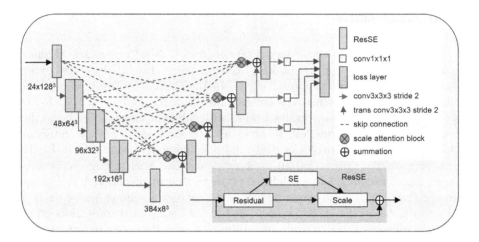

Fig. 2. Architecture of SA-Net. Input is a $4 \times 128 \times 128 \times 128$ tensor followed by one ResSE block with 24 features. Here ResSE stands for a squeeze-and-excitation block embedded in a residual module [19]. By progressively halving the feature map dimension while doubling the feature width at each scale, the endpoint of the encoding pathway has a dimension of $384 \times 8 \times 8 \times 8$. The output of the decoding pathway has three channels with the same spatial size as the input, i.e., $3 \times 128 \times 128 \times 128$.

3   Methods

3.1   Overall Network Structure

SA-Net follows a typical encoding-decoding architecture with an asymmetrically larger encoding pathway to learn representative features and a smaller decoding pathway to recover the segmentation mask in the original resolution. The outputs of encoding blocks at different scales are merged to the scale attention blocks (SA-block) to learn and select features with full-scale information. Due to the limit of GPU memory, we randomly crop the input image from $240 \times 240 \times 155$ to $128 \times 128 \times 128$, and concatenate the four MRI modalities of each patient into a four channel tensor to yield an input to SA-Net with the dimension of $4 \times 128 \times 128 \times 128$. The network output includes three channels, each of which presents the probability that the corresponding voxel belongs to WT, TC, and ET, respectively.

3.2   Encoding Pathway

The encoding pathway is built upon ResNet [12] blocks, where each block consists of two Convolution-Normalization-ReLU layers followed by additive skip connection. We keep the batch size to 1 in our study to allocate more GPU memory resource to the depth and width of the model, therefore, we use instance normalization [24] that has been demonstrated with better performance than batch normalization when batch size is small. In order to further improve the representative capability of the model, we add a squeeze-and-excitation module [19] into each residual block with reduction ratio $r = 4$ to form a ResSE block. The initial scale includes one ResSE block with the initial number of features (width) of 24. We then progressively halve the feature map dimension while doubling the feature width using a strided (stride $= 2$) convolution at the first convolution layer of the first ResSE block in the adjacent scale level. All the remaining scales include two ResSE blocks except the endpoint of the encoding pathway, which has a dimension of $384 \times 8 \times 8 \times 8$. We only use one ResSE block in the endpoint due to its limited spatial dimension.

3.3 Decoding Pathway

The decoding pathway follows the reverse pattern as the encoding one, but with a single ResSE block in each spatial scale. At the beginning of each scale, we use a transpose convolution with stride of 2 to double the feature map dimension and reduce the feature width by 2. The upsampled feature maps are then added to the output of SA-block. Here we use summation instead of concatenation for information fusion between the encoding and decoding pathways to reduce GPU memory consumption and facilitate the information flowing. The endpoint of the decoding pathway has the same spatial dimension as the original input tensor and its feature width is reduced to 3 after a $1 \times 1 \times 1$ convolution and a sigmoid function.

In order to regularize the model training and enforce the low- and middle-level blocks to learn discriminative features, we introduce deep supervision at each intermediate scale level of the decoding pathway. Each deep supervision subnet employs a $1 \times 1 \times 1$ convolution for feature width reduction, followed by a trilinear upsampling layer such that they have the same spatial dimension as the output, then applies a sigmoid function to obtain extra dense predictions. These deep supervision subnets are directly connected to the loss function in order to further improve gradient flow propagation.

3.4 Scale Attention Block

The proposed scale attention block consists of full-scale skip connections from the encoding pathway to the decoding pathway, where each decoding layer incorporates the output feature maps from all the encoding layers to capture fine-grained details and coarse-grained semantics simultaneously in full scales. As an example illustrated in Fig. 3, the first stage of the SA-block is to transform the input feature maps at different scales in the encoding pathway, represented as $\{S_e, e = 1, ..., N\}$ where N is the number of total scales in the encoding pathway except the last block ($N = 4$ in this work), to a same dimension, i.e., $\bar{S}_{ed} = f_{ed}(S_e)$. Here e and d are the scale level at the encoding and decoding pathways, respectively. The transform function $f_{ed}(S_e)$ is determined as follows. If $e < d$, $f_{ed}(S_e)$ downsamples S_e by $2^{(d-e)}$ times by maxpooling followed by a Conv-Norm-ReLU block; if $e = d$, $f_{ed}(S_e) = S_e$; and if $e > d$, $f_{ed}(S_e)$

upsamples S_e through tri-linear upsampling after a Conv-Norm-ReLU block for channel number adjustment. After summing these transformed feature maps as $P_d = \sum_e \bar{S}_{ed}$, a spatial pooling is used to average each feature to form a information embedding tensor $G_d \in R^{C_d}$, where C_d is the number of feature channels in scale d. Then a $1-to-N$ Squeeze-Excitation is performed in which the global feature embedding G_d is squeezed to a compact feature $g_d \in R^{C_d/r}$ by passing through a fully connected layer with a reduction ratio of r, then another N fully connected layers with sigmoid function are applied for each scale excitation to recalibrate the feature channels on that scale. Finally, the contribution of each scale in each feature channel is normalized with a softmax function, yielding a scale-specific weight vector for each channel as $w_e \in R^{C_d}$, and the final output of the scale attention block is $\widetilde{S}_d = \sum_e w_e \cdot \bar{S}_{ed}$.

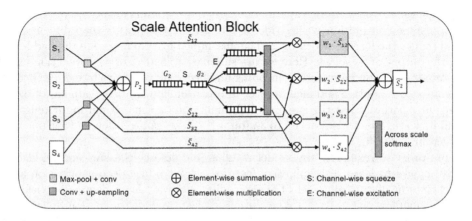

Fig. 3. Scale attention block. Here $S1, S2, S3$ and $S4$ represent the input feature maps at different scales from the encoding pathway. $d = 2$ in this example.

3.5   Implementation

Our framework was implemented with Python using Pytorch package. As for pre-processing, since MRI images are non-standarized, we simply normalized each modality from each patient independently by subtracting the mean and dividing by the standard deviation of the entire image. The model was trained with randomly sampled patches of size $128 \times 128 \times 128$ voxels and batch size of 1. Training the entire network took 300 epochs from scratch using Adam stochastic optimization method. The initial learning rate was set as 0.003, and learning rate decay and early stopping strategies were utilized when validation loss stopped decreasing. In particular, we kept monitoring both the validation loss ($L^{(valid)}$) and the exponential moving average of the validation loss ($\widetilde{L}^{(valid)}$) in each epoch. We kept the learning rate unchanged at the first 150 epochs, but dropped the learning rate by a factor of 0.3 when neither $L^{(valid)}$ nor $\widetilde{L}^{(valid)}$

improved within the last 30 epochs. The models that yielded the best $L^{(valid)}$ and $\widetilde{L}^{(valid)}$ were recorded for model inference.

Our loss function used for model training includes two terms:

$$L = L_{jaccard} + L_{focal} \tag{1}$$

$L_{jaccard}$ is a generalized Jaccard distance loss [20–23], which we developed in our previous work for single object segmentation, to multiple objects, and L_{focal} is the voxel-wise focal loss function that focuses more on the difficult voxels. Since the network output has three channels corresponding to the whole tumor, tumor core and enhanced tumor, respectively, we simply added the three loss functions together.

In order to reduce overfitting, we randomly flipped the input volume in left/right, superior/inferior, and anterior/posterior directions on the fly with a probability of 0.5 for data augmentation. We also adjusted the contrast in each image input channel by a factor randomly selected from [0.9, 1.1]. We used 5-fold cross validation to evaluate the performance of our model on the training dataset, in which a few hyper-parameters were also experimentally determined. All the experiments were conducted on Nvidia GTX 1080 TI GPU with 11 GB memory.

During testing, we applied the sliding window around the brain region and extracted 8 patches with a size $128 \times 128 \times 128$ (2 windows in each dimension), and averaged the model outputs in the overlapping regions before applying a threshold of 0.5 to obtain a binary mask of each tumor region.

4   Results

We trained SA-Net with the training set (369 cases) provided by the BraTS 2020 challenge, and evaluated its performance on the training set via 5-fold cross validation, as well as on the validation set, which includes 125 cases with unknown segmentation. Table 1 shows the segmentation results in terms of Dice similarity coefficient (DSC) for each region. As compared to the results that were obtained from a model using the vanilla U-Net structure with scale-wise skip connection and feature concatenation, the proposed SA-Net consistently improved segmentation performance for each target, yielding an average 1.47% improvement.

When applying the trained models on the challenge validation dataset, a bagging-type ensemble strategy was implemented to combine the outputs of eleven models obtained through 5-fold cross validation to further improve the segmentation performance. In particular, our model ensemble included the models that yielded the best $L^{(valid)}$ (validation loss) and $\widetilde{L}^{(valid)}$ (moving average of the validation loss) respectively in each fold, plus the model that was trained with all the 369 cases. We uploaded our segmentation results to the BraTS 2020 server for performance evaluation in terms of DSC, sensitivity, specificity and Hausdorff distance for each tumor region, as shown in Table 2.

Table 1. Comparison between SA-Net and U-Net segmentation results (DSC) in 5-fold cross validation using 369 training image sets. WT: whole tumor; TC: tumor core; ET: enhanced tumor.

		fold-0	fold-1	fold-2	fold-3	fold-4	ALL
SA-Net	WT	0.9256	0.9002	0.9217	0.9085	0.9195	0.9151
	TC	0.8902	0.8713	0.8758	0.8538	0.8953	0.8773
	ET	0.8220	0.7832	0.8198	0.8107	0.8268	0.8125
	AVG	0.8793	0.8516	0.8724	0.8577	0.8805	0.8683
U-Net	WT	0.9218	0.9003	0.9104	0.9021	0.9113	0.9092
	TC	0.8842	0.8735	0.8772	0.8307	0.8700	0.8671
	ET	0.7982	0.7672	0.7922	0.7955	0.8007	0.7908
	AVG	0.8680	0.8470	0.8599	0.8428	0.8607	0.8557

Table 2. Segmentation results of SA-Net on the BraTS 2020 validation sets in terms of Mean DSC and 95% Hausdorff distance (mm). WT: whole tumor; TC: tumor core; ET: enhanced tumor.

	DSC			HD95		
	WT	TC	ET	WT	TC	ET
The best single model	0.9044	0.8422	0.7853	5.4912	8.3442	20.3507
Ensemble of 11 models	0.9108	0.8529	0.7927	4.0975	5.8879	18.1957

During the testing phase, only one submission was allowed. Table 3 summarizes our final results, which ranked our method as the 3rd place among 693 registrations in Brats 2020 challenge.

Table 3. Segmentation results of SA-Net on the BraTS 2020 testing sets in terms of Mean DSC and 95% Hausdorff distance (mm). WT: whole tumor; TC: tumor core; ET: enhanced tumor.

	DSC			HD95		
	WT	TC	ET	WT	TC	ET
Ensemble of 11 models	0.8828	0.8433	0.8177	5.2176	17.9697	13.4298

5   Summary

In this work, we presented a fully automated segmentation model for brain tumor segmentation from multimodality 3D MRI images. Our SA-Net replaces the long-range skip connections between the same scale in the vanilla U-Net with full-scale skip connections in order to make maximum use of feature maps in full scales for accurate segmentation. Attention mechanism is introduced to adaptively adjust the weights of each scale feature to emphasize the important scales while suppressing the less important ones. As compared to the vanilla U-Net structure with scale-wise skip connection and feature concatenation, the proposed scale attention block not only improved the segmentation performance by 1.47%, but also reduced the number of trainable parameters from 17.8M (U-Net) to 16.5M (SA-Net), which allowed it to achieve a top performance with limited GPU resource in this challenge.

Acknowledgment. This work is partially supported by a research grant from Varian Medical Systems (Palo Alto, CA, USA) and grant UL1TR001433 from the National Center for Advancing Translational Sciences, National Institutes of Health, USA.

References

1. Menze, B.H., et al.: The multimodal brain tumor image segmentation benchmark (BRATS). IEEE Trans. Med. Imaging **34**(10), 1993–2024 (2015)
2. Bakas, S., et al.: Advancing the cancer genome atlas glioma MRI collections with expert segmentation labels and radiomic features. Nat. Sci. Data **4**, 170117 (2017)
3. Bakas, S., et al.: Identifying the best machine learning algorithms for brain tumor segmentation, progression assessment, and overall survival prediction in the BRATS challenge. arXiv preprint arXiv:1811.02629 (2018)
4. Bakas, S., et al. Segmentation labels and radiomic features for the pre-operative scans of the TCGA-GBM collection. Cancer Imaging Arch. (2017)
5. Bakas, S., et al.: Segmentation labels and radiomic features for the pre-operative scans of the TCGA-LGG collection. Cancer Imaging Arch. (2017)
6. Kamnitsas, K., et al.: Ensembles of multiple models and architectures for robust brain tumour segmentation. In: Crimi, A., Bakas, S., Kuijf, H., Menze, B., Reyes, M. (eds.) BrainLes 2017. LNCS, vol. 10670, pp. 450–462. Springer, Cham (2018). https://doi.org/10.1007/978-3-319-75238-9_38
7. Long, J., et al.: Fully convolutional networks for semantic segmentation. In: CVPR, pp. 3431–3440 (2015)
8. Kamnitsas, K., et al.: Efficient multi-scale 3D CNN with fully connected CRF for accurate brain lesion segmentation. Med. Image Anal. **36**, 61–78 (2017)
9. Ronneberger, O., Fischer, P., Brox, T.: U-Net: convolutional networks for biomedical image segmentation. In: Navab, N., Hornegger, J., Wells, W.M., Frangi, A.F. (eds.) MICCAI 2015. LNCS, vol. 9351, pp. 234–241. Springer, Cham (2015). https://doi.org/10.1007/978-3-319-24574-4_28
10. Wang, G., Li, W., Ourselin, S., Vercauteren, T.: Automatic brain tumor segmentation using cascaded anisotropic convolutional neural networks. In: Crimi, A., Bakas, S., Kuijf, H., Menze, B., Reyes, M. (eds.) BrainLes 2017. LNCS, vol. 10670, pp. 178–190. Springer, Cham (2018). https://doi.org/10.1007/978-3-319-75238-9_16
11. Chen, L.-C., et al.: DeepLab: semantic image segmentation with deep convolutional nets, atrous convolution, and fully connected CRFs. IEEE Trans. Pattern Anal. Mach. Intell. **40**(4), 834–848 (2018)
12. He, K., et al.: Deep residual learning for image recognition. Proc. CVPR **770–778**, 2016 (2016)
13. Myronenko, A.: 3D MRI brain tumor segmentation using autoencoder regularization. In: Crimi, A., Bakas, S., Kuijf, H., Keyvan, F., Reyes, M., van Walsum, T. (eds.) BrainLes 2018. LNCS, vol. 11384, pp. 311–320. Springer, Cham (2019). https://doi.org/10.1007/978-3-030-11726-9_28

14. Isensee, F., Kickingereder, P., Wick, W., Bendszus, M., Maier-Hein, K.H.: No new-net. In: Crimi, A., Bakas, S., Kuijf, H., Keyvan, F., Reyes, M., van Walsum, T. (eds.) BrainLes 2018. LNCS, vol. 11384, pp. 234–244. Springer, Cham (2019). https://doi.org/10.1007/978-3-030-11726-9_21

15. Jiang, Z., Ding, C., Liu, M., Tao, D.: Two-stage cascaded U-Net: 1st place solution to BraTS challenge 2019 segmentation task. In: Crimi, A., Bakas, S. (eds.) BrainLes 2019. LNCS, vol. 11992, pp. 231–241. Springer, Cham (2020). https://doi.org/10.1007/978-3-030-46640-4_22

16. Zhao, Y.-X., Zhang, Y.-M., Liu, C.-L.: Bag of tricks for 3D MRI brain tumor segmentation. In: Crimi, A., Bakas, S. (eds.) BrainLes 2019. LNCS, vol. 11992, pp. 210–220. Springer, Cham (2020). https://doi.org/10.1007/978-3-030-46640-4_20

17. Zhou, Z., et al.: UNet++: redesigning skip connections to exploit multiscale features in image segmentation. IEEE Trans. Med. Imaging **39**(6), 1856–1867 (2019)

18. Roth, H., et al.: Spatial aggregation of holistically-nested convolutional neural networks for automated pancreas localization and segmentation. Med. Imaging Anal. **45**, 94–107 (2018)

19. Hu, J., et al.: Squeeze-and-excitation networks. In: Proceedings of the CVPR 2018, pp. 7132–7141 (2018)

20. Yuan, Y., et al.: Automatic skin lesion segmentation using deep fully convolutional networks with Jaccard distance. IEEE Trans. Med. Imaging **36**(9), 1876–1886 (2017)

21. Yuan, Y.: Hierarchical convolutional-deconvolutional neural networks for automatic liver and tumor segmentation. arXiv preprint arXiv:1710.04540 (2017)

22. Yuan, Y.: Automatic skin lesion segmentation with fully convolutional-deconvolutional networks. arXiv preprint arXiv:1703.05154 (2017)

23. Yuan, Y., et al.: Improving dermoscopic image segmentation with enhanced convolutional-deconvolutional networks. IEEE J. Biomed. Health Inform. **23**(2), 519–526 (2019)

24. Wu, Y., He, K.: Group normalization. In: Ferrari, V., Hebert, M., Sminchisescu, C., Weiss, Y. (eds.) ECCV 2018. LNCS, vol. 11217, pp. 3–19. Springer, Cham (2018). https://doi.org/10.1007/978-3-030-01261-8_1

25. Li, X., et al.: Selective kernel networks. In: Proceedings of the CVPR 2019, pp. 510–519 (2019)

Impact of Spherical Coordinates Transformation Pre-processing in Deep Convolution Neural Networks for Brain Tumor Segmentation and Survival Prediction

Carlo Russo[1]([✉]) [iD], Sidong Liu[1,2] [iD], and Antonio Di Ieva[1] [iD]

[1] Computational NeuroSurgery (CNS) Lab, Macquarie University, Sydney, Australia
carlo.russo@mq.edu.au
[2] Centre for Health Informatics, Australian Institute of Health Innovation,
Macquarie University, Sydney, Australia

Abstract. Pre-processing and Data Augmentation play an important role in Deep Convolutional Neural Networks (DCNN). Whereby several methods aim for standardization and augmentation of the dataset, we here propose a novel method aimed to feed DCNN with spherical space transformed input data that could better facilitate feature learning compared to standard Cartesian space images and volumes. In this work, the spherical coordinates transformation has been applied as a preprocessing method that, used in conjunction with normal MRI volumes, improves the accuracy of brain tumor segmentation and patient overall survival (OS) prediction on Brain Tumor Segmentation (BraTS) Challenge 2020 dataset. The LesionEncoder framework has been then applied to automatically extract features from DCNN models, achieving 0.586 accuracy of OS prediction on the validation data set, which is one of the best results according to BraTS 2020 leaderboard.

Keywords: Deep Convolutional Neural Network · Polar transformation · Spherical coordinates · BraTS

1 Introduction

Magnetic Resonance Imaging (MRI) is used in everyday clinical practice to assess brain tumors. However, the manual segmentation of each volume representing the extension of the tumor is time-demanding and operator-dependent, as it is often non-reproducible and depends upon neuroradiologists' expertise. Several automatic or semi-automatic segmentation algorithms have been introduced to help segment brain tumors, and Deep Convolutional Neural Networks (DCNN) have recently shown very promising results. To further improve the accuracy of automatic methods, the Multimodal Brain Tumor Segmentation (BraTS) challenge [1–3] is organized annually within the International

---

C. Russo and S. Liu—Authors contributed equally.

© Springer Nature Switzerland AG 2021
A. Crimi and S. Bakas (Eds.): BrainLes 2020, LNCS 12658, pp. 295–306, 2021.
https://doi.org/10.1007/978-3-030-72084-1_27

Conference on Medical Image Computing and Computer Assisted Intervention (MIC-CAI). The BraTS 2020 challenge includes a task for automatic segmentation of the total area containing the tumor (Whole Tumor - WT), as well as the Necrosis and Active tumor cells area (Tumor Core - TC, Enhancing Tumor - ET, Necrosis and ET are contained in TC).

Furthermore, glioma patients often have a dire survival prognosis following surgical resection and radiochemotherapy [4]. Thus, a further task to predict patient overall survival (OS) has been added into the challenge, aimed at improving the prediction of patient survival outcome in order to add information that are relevant to the decision-making process.

DCNNs are data driven algorithms. They require huge amount of data to obtain good results. In medical imaging, such big datasets are not often available, thus pre-processing and data augmentation plays an important role. While pre-processing methods are usually used to standardize input data, they can also be used to enhance meaningful data inside the original input images: an example is cropping the region of interest when the input data includes lots of redundant and misleading information.

Therefore, we propose a novel spherical space transformation method to enhance information on specific points of the tumor as well as enable the DCNN learning process to be invariant to rotation and scaling of the input images. Furthermore, we extended the use of lesion features extracted from the latent space of the segmentation models using the LesionEncoder framework, which replaces the classic imaging/radiomic features, such as volumetric parameters, intensity, morphologic, histogram-based and textural features, which showed high predictive power in patient OS prediction.

2 Brain Tumor Segmentation

2.1 Background

Dataset. The dataset consists of four MRI sequences used to determine the segmentation and extract survival features, namely T1-weighted, post-contrast T1, T2-weighted and FLAIR images. Training dataset have 336 4-channel volumes with ground truth segmentation. Validation dataset is composed by data from 125 patients [5, 6]. Testing dataset is composed by additional 166 patients.

Baseline Model. The DCNN that we chose to use as baseline for our method is derived from Myronenko [7], which is based on Variational Auto Encoder (VAE) U-Net with adjusted input shape and loss function according to the type of transformation used into the preprocessing phase. The VAE proposed by Myronenko is composed by a U-Net with two decoder branches: a segmentation decoder branch, used to obtain the final segmentation, and an additional decoder branch to reconstruct the original volumes, used to regularize the shared encoder. The loss function is given by the formula:

$$L = L_{dice} + 0.1 * L_{L2} + 0.1 * L_{KL}$$

where L_{L2} is the L2 loss on the VAE branch and L_{KL} is the KL divergence penalty term.

Trained Models. We trained different models by changing pre-processing method (Cartesian and spherical) and some layers hyperparameters. Although the models share the same VAE structure proposed by Myronenko, there are a few differences. More specifically, Cartesian_v1 includes standard Dropout with rate 0.2, Kernel size filters 3 × 3 × 3 in Convolution layers outside each ResNet-like block of the VAE and an additional 3 × 3 × 3 convolution layer before the final output block. While Cartesian_v2 uses SpatialDropout3D with the same ratio, 1 × 1 × 1 convolution filters and no additional layers. The spherical model has the same structure of Cartesian_v1 but with Spherical transformation pre-processing on inputs. The spherical transformed model (using Cartesian_v2 structure) has not been trained yet by the 2020 challenge deadline. A bis-Cartesian model has been also trained using the structure of the Cartesian v1 model but using a lower coefficient of the KL loss, set to 0.0001. The model has not given better segmentation results, although it is showing improved results on OS task.

2.2 Spherical Coordinates Transformation

Our team previously presented the spherical coordinate transformation pre-processing as a method to improve segmentation results [8]. The Spherical transformed volume is shown in Fig. 1.

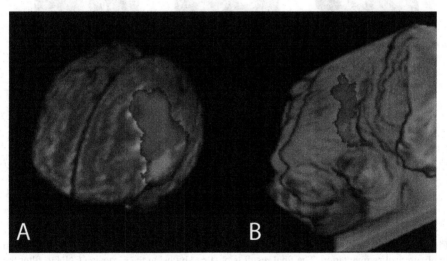

Fig. 1. Example of the representation of a radiologic volume in spherical coordinate system. A) A brain MRI volume with its 3D segmentation of the tumor, and B) the same volume transformed into a spherical coordinate system using the center of the volume as the origin.

Each pre-processed volume uses an origin point. Thus, to achieve good performance on the training, it is important to correctly select origin points, included within the tumor. For this reason, we used a cascade of three DCNNs, the first one predicting a coarse segmentation, and then refining the segmentation by using origin points included in the previous model. The first pass model of the cascade could also be a model trained on non-transformed input (a Cartesian model), but the use of the Spherical model already in the first cascade's pass enabled pre-training weights to be used for the next training steps.

The Spherical coordinate transformation also adds an extreme augmentation. This is a beneficial step as it adds invariance to the rotation and scaling to the DCNN model. However, such an invariance also has a drawback especially when dealing with WT segmentation: apparently, WT segmentation works better with a Cartesian model, whereas using Spherical pre-processing adds many false positive regions to the WT. Thus, we used a Cartesian model to filter out the false positive regions found by the spherical pre-processing as shown in Fig. 2.

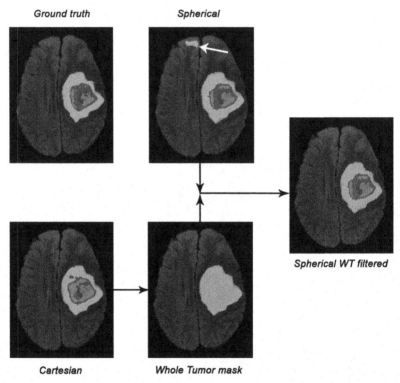

Fig. 2. An example of the application of the Cartesian filter to the spherical model's segmentation to obtain final segmentation. The final result is obtained by using the Cartesian CNN to filter the WT mask obtained on spherical segmentation. Such a step helps to erase false positive voxel regions wrongly selected by the spherical CNNs (see segmented object indicated by the yellow arrow).

We used this proposed method for the first time in a BraTS challenge, achieving similar results obtained in our original paper regarding the improvement of the accuracy of the model trained on transformed input compared to the baseline model [8]. We also tested the intersection of the segmentation on the three different classes (Spherical -

Cartesian intersection 3CH) instead of filtering only the WT class. Finally, we ensembled the best segmentations from Cartesian_v2 and the intersection method to improve the results further.

2.3 Post-processing

After filtering the Spherical segmentation with the Cartesian filter, we used a post-processing method to improve ET segmentation. We noticed that many false positive ET segmentations are due to isolated voxels. For this reason, we applied a binary opening operator to isolate thin branches over ET spots and then filter out the spots having less than 30 voxels. When the ET segmentation is still present after these filters, the original ET segmentation is restored and used as the final one. Otherwise, the ET segmentation is completely erased, meaning that no ET is present in the current volume.

2.4 Segmentation Results

Table 1 shows the summary of segmentation results on the validation dataset. The most promising methods tested so far were the Cartesian_v2 and the spherical models: used alone without post-processing, the Cartesian_v2 gave the best results in WT and TC segmentation, while the spherical model worked better on ET segmentation.

Table 1. Evaluation of segmentation performance on BraTS 2020 official validation dataset

Model names	Dice			Sensitivity			Specificity			Hausdorff95_ET		
	ET	WT	TC	ET	WT	TC	ET	WT	TC	ET	WT	TC
Cartesian only (v1)	0.7046	0.8904	0.7852	0.7204	0.9193	0.7829	0.9997	0.9988	0.9996	39.8783	7.6745	15.8352
Cartesian only (v2)	0.7072	**0.8999**	**0.8250**	0.7187	0.8947	0.8324	0.9997	0.9992	0.9995	46.4816	6.0702	8.1284
Cartesian only (v1 bis)	0.6926	0.8954	0.7905	0.6953	0.9064	0.7860	0.9997	0.9991	0.9996	44.5790	6.7375	8.2056
Spherical only	0.7428	0.8555	0.7783	**0.7882**	**0.9419**	**0.8455**	0.9995	0.9975	0.9991	23.3944	9.6847	8.4968
Spherical with Cartesian v2 WT filter	0.7533	0.8983	0.7846	0.7867	0.8807	0.8302	0.9995	**0.9993**	0.9992	**22.4573**	5.9040	8.1578
Cartesian v2 with postprocessing	0.7604	**0.8999**	**0.8250**	0.7747	0.8947	0.8324	0.9997	0.9992	0.9995	33.8563	6.0702	8.1284
Spherical -Cartesian intersection 3CH and postprocessing	**0.7662**	0.8983	0.8060	0.7471	0.8807	0.7803	**0.9998**	**0.9993**	**0.9997**	27.3789	5.9040	7.2708
Ensemble of methods	**0.7662**	**0.8999**	**0.8250**	0.7471	0.8947	0.8324	**0.9998**	0.9992	0.9995	27.3789	6.0702	8.1284

Merging Spherical model labels with Cartesian model WT label filter gave a further improvement in the segmentation of ET class, while the best result for the WT and TC classes segmentation remained using only the Cartesian_v2 model.

The best overall improvement on ET has been shown by post-processing the segmentation of the intersected Cartesian and Spherical models on the three channels (3CH), even if the TC class dice score decreased and WT class did not improve further. The final model used on the testing dataset is an Ensemble of methods taking the ET label from the 3CH model and WT and TC labels from the Cartesian v2 model.

Results of the final model on testing dataset shown in Table 2 seems to confirm a good accuracy, above all on the ET segmentation, although is not possible to make a comparison with the other models since the Challenge only allows to test one method on the dataset.

Table 2. Evaluation of segmentation performance on BraTS 2020 official testing dataset

	Dice			Sensitivity			Specificity			Hausdorff95_ET		
	ET	WT	TC	ET	WT	TC	ET	WT	TC	ET	WT	TC
Mean	0.7898	0.8687	0.8066	0.7864	0.8970	0.8515	0.9997	0.9990	0.9995	17.9747	6.7349	22.2474
StdDev	0.1951	0.1373	0.2609	0.2095	0.1283	0.2291	0.0003	0.0011	0.0012	74.7789	10.4801	74.6658
Median	0.8338	0.9128	0.9068	0.8480	0.9322	0.9285	0.9998	0.9993	0.9997	1.7321	3.6056	3.0000
25quantile	0.7508	0.8581	0.8310	0.7442	0.8825	0.8438	0.9996	0.9986	0.9995	1.4142	2.0590	1.4937
75quantile	0.8897	0.9412	0.9463	0.9051	0.9677	0.9748	1.0000	0.9996	0.9999	3.0000	6.5312	7.2111

3   Prediction of Patient OS

Our team also participated in Task 2 of the BraTS Challenge: prediction of patient overall OS from pre-operative MRI scans. Instead of using the pre-defined imaging / radiomic features, such as volumetric parameters, intensity, morphologic, histogram-based and textural features, we used the features automatically extracted from MRI scans using the novel LesionEncoder (LE) framework [9]. The LE features were further processed using Principle Component Analysis (PCA) to reduce dimensionality, and then used as input to a generalized linear model (GLM) [10] to predict patient OS.

3.1   LesionEncoder Framework

The LE framework was proposed in a recent work for COVID-19 severity assessment and progression prediction [9]. The original LE adopted the U-Net structure [11], which consists of an encoder and a decoder based on the EfficientNet [12]. While the encoder learns and captures the lesion features in the input images, the decoder maps the lesion features back to the original image space and generates the segmentation maps. The features learnt by the encoder in the latent space encapsulate rich information of the lesions, therefore, can be used for lesion segmentation, as well as other tasks such as classification and prediction.

In this study, we used the VAE as backbone to build the LE. As described in the previous section, three different configurations have been applied to the VAE model, resulting in three different lesion encoders: i.e., $LE_{Cartesian}$ (Cartesian_v2), $LE_{Spherical}$ (Spherical) and $LE_{bisCartesian}$ (Cartesian_v1_bis). The latent variables of the input images/MRI scans extracted by individual lesion encoders were then used as the features to predict patient OS. For each MRI scan, a high-dimensional feature vector ($d = 256$) was derived. As the high-dimensional feature space tended to lead to overfitting, we therefore used PCA to control the feature dimensionality by setting different numbers of principle components ($\hat{d} = [2, 60]$ in this study) for further analysis.

3.2 Tweedie Regressor

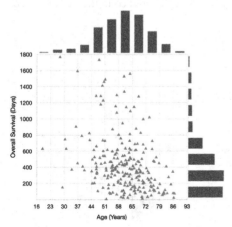

Fig. 3. Joint age and OS distribution of the training samples

Figure 3 shows the joint age and OS distribution of the patients in the training cohort. The age distribution, as shown at the top of the figure, seems to be a normal distribution. The OS distribution, on the right side of the figure, seems to be heavily skewed, with the majority of cases having OS less than 400 days. To model the tailed distribution of the OS values, we therefore used a Tweedie distribution [13], a special case of exponential dispersion models whose skewness can be controlled by a power parameter ($r = [1.1, 1.9]$ in this study).

A GLM model [10] based on the Tweedie distribution, i.e., Tweedie Regressor [13], was built to predict OS values. The Tweedie Regressor was implemented using scikit-learn (v0.23.2). As the resection status and age are essential predictors of OS, both of them were merged with the LesionEncoder features as input to the Tweedie Regressor for OS prediction.

3.3 Performance Evaluation

Two evaluation schemes were used to assess the prediction performance. The results were first evaluated based on accuracy of the classification of subjects as long-survivors (>15 months/450 days), short-survivors (<10 months / 300 days), and mid-survivors (survival rate between 10 and 15 months/300 – 450 days). In addition, a pairwise error analysis between the predicted and actual OS (in days) was performed, evaluated using the following metrics: mean square error (MSE), median square error (median SE), standard deviation of the square errors (std SE), and the Spearman correlation coefficient (Spearman R). Amongst the 235 patients in the training set, 118 underwent a surgical gross total resection (GTR) and 10 underwent a subtotal resection (STR); in 107 cases, no information about the resection status are available. All of the 29 subjects in the validation set had a GTR resection status. The extent of resection was considered in the model as it has been shown to correlate to post-surgical outcome [14].

3.4 OS Prediction Results

Cross-Validation on the Training Set. We used fivefold cross-validation to train and validate the proposed method. An internal validation set (20%) was split from the dataset in each fold with the remaining 80% as the training set. For each of the three lesion encoders, i.e., LE_Cartesian, LE_Spherical and LE_bisCartesian, this process was repeated 5 times, leading to 5 different sub-models. Figure 4 illustrates the projected feature space of the features extracted using LE_Spherical (a), and the scatter plots (b, c) of the predicted OS vs. actual OS of the training samples. There was high variance in the performance of the sub-models (from 0.362 to 0.574).

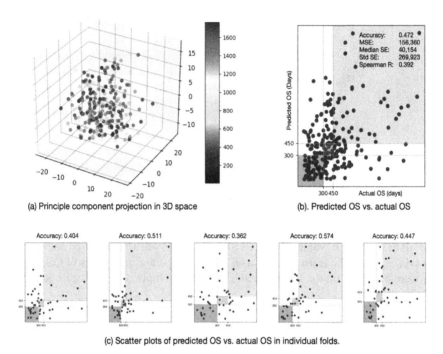

(a) Principle component projection in 3D space

(b). Predicted OS vs. actual OS

(c) Scatter plots of predicted OS vs. actual OS in individual folds.

Fig. 4. Five-fold cross-validation on the training set using the lesion encoder with spherical configuration (LE_Spherical). (a) 3D projection of the top 3 principle components derived from PCA. The color table indicates the patients' OS (days). (b) Scatter plot of the predicted OS vs. actual OS of the aggregate of holdout samples in individual folds. Gray, green and red boxes indicate the correct predictions of the short-, mid- and long-survivors, respectively. (c) Scatter plots of the holdout samples in each cross-validation fold. (Color figure online)

The prediction results of the 5 sub-models were further aggregated, and the results were summarized in Table 3. LE_Cartesian achieved the highest accuracy (0.494) and Spearman R (0.429), while LE_Spherical had the lowest MSE, median SE and std SE. These two models outperformed LE_bisCartesian; however, the differences were not substantial (<0.034 in accuracy).

Prediction Performance on the Validation Set. The 5 different sub-models with the same configuration were then applied to the official validation set (n = 29) to predict the OS of each validation case. The mean value of the 5 predictions of each case were then averaged to derive the final prediction. Results of the three models with different configurations are summarized in Table 4. The LE_bisCartesian model achieved the highest

Table 3. Summaries of OS prediction performance on the training set

Algorithm	Accuracy	MSE	median SE	std SE	Spearman R
LE_Cartesian	0.494	164,806.753	43,224.50	402,510.04	0.429
LE_Spherical	0.472	156,359.5	40,153.61	269,922.79	0.392
LE_bisCartesian	0.460	174,067.83	49,956.34	318,133.99	0.329

accuracy (0.552); however, its MSE and std SE were higher, and Spearman R was lower than the other models. The LE_Cartesian model had the lowest MSE and the highest Spearman R, showing a better representation of the overall distribution of the OS values.

Table 4. Summaries of OS prediction results of the validation dataset

Algorithm	Accuracy	MSE	median SE	std SE	Spearman R
LE_Cartesian (M1)	0.483	**75,908.78**	32,322.27	**90,074.92**	**0.498**
LE_Spherical (M2)	0.517	125,308.38	**19,226.91**	216,570.80	0.374
LE_bisCartesian (M3)	0.552	130,941.15	27,762.22	232,271.86	0.261
M1 & M2	**0.586**	88,311.58	27,114.54	142,969.74	0.412
M1 & M3	**0.586**	93,565.43	24,579.67	145,669.08	0.360
M2 & M3	0.552	125,659.88	32,131.13	223,333.91	0.300
M1 & M2 & M3	**0.586**	99,776.58	24,539.13	168,419.90	0.338

These findings showed a complementary nature of different models; therefore, we combined the outputs of these models to test whether the prediction performance could be improved further. Four combinations were tested, which consistently showed equal or better accuracy (between 0.552 and 0.586) compared to using individual models.

Our final submission for OS prediction on the validation dataset, which was based on the **M1&M2** model, was ranked the 4th place in accuracy among the 42 participating teams. In the meanwhile, it achieved the 5th place in both MSE and Spearman R, the 8th place in median SE, and the 10th place in std SE (checked on 23 October 2020).

We further applied the **M1&M2&M3** model on the official test dataset (n = 107). The model's performance, as shown in Table 5, was lower on the test dataset compared to the validation dataset, implying a marked difference between the two datasets and overfitting of the model. However, without knowing the results of other models, either of our own or from other participating teams, it is difficult to confirm whether such performance drop is caused by a less representative training dataset or a less generalizable model, or both.

Table 5. Summaries of OS prediction results of the testing dataset.

Algorithm	Accuracy	MSE	median SE	std SE	Spearman R
M1 & M2 & M3	0.495	469,253.36	67,626.86	1,281,569.74	0.379

4 Discussion

Spherical Coordinate Transformation Pre-processing. Spherical coordinate transformation pre-processing of the input dataset contribute to explore data in a different way, thus changing the learning process and achieving different features compared to the classical DCNN model learning process. Those different features can help to improve the segmentation process as well as contributing to deep feature extraction to be used in patients' OS prediction. Even if the spherical pre-processing method contributes to improving baseline model results, simple post-processing methods also have a strong impact on segmentation accuracy. However, overall segmentation results obtained by this method are not amongst the best ones compared to other teams in BraTS 2020 leaderboard, and additional efforts should be done to fine tune both Cartesian and spherical training phase.

LesionEncoder Framework. The LesionEncoder framework extends the use of lesion features beyond conventional lesion segmentation. There is a wealth of information in the brain tumors including shape, texture, location, extent and distribution of involvement of the abnormality, that can be extracted by the lesion encoder. While it has been demonstrated in COVID-19 progression prediction [9] and severity assessment [15], here we demonstrated a new application of LE in patient OS prediction. It may have strong potential in a wide range of other clinical and research applications, e.g., brain tumor pseudo-progression detection [16] and ophthalmic disease screening [17].

Model Optimization. Various dimension reduction methods have been tested in this study, including PCA, Independent Component Analysis (ICA), t-distributed stochastic neighbor embedding (T-SNE). In the training phase, PCA was found to have lower variability in accuracy than the other methods; as a result, it was chosen to process the high-dimensional features. We used a linear search strategy to optimize the two most important parameters of the OS prediction model, including the number of principle components in PCA ($\hat{d} = [2, 60]$) and the power of Tweedie distribution ($r = [1.1, 1.9]$). The optimal parameters for LE$_{\text{Cartesian}}$ were ($\hat{d} = 10, r = 1.6$), and ($\hat{d} = 3, r = 1.6$) for both LE$_{\text{Spherical}}$ and LE$_{\text{bisCartesian}}$. In addition, it will be important to demonstrate the scale invariance in the Tweedie regressor within different datasets in our future work.

5 Conclusion

In conclusion, we have introduced a novel and very promising method to pre-process brain tumors' MR images by means of a spherical coordinates transformation to be used in DCNN models for brain tumor segmentation. The LesionEncoder framework has been applied to automatically extract imaging features from DCNN models, demonstrating good performance for the survival prediction task.

References

1. Menze, B.H., Jakab, A., Bauer, S., Kalpathy-Cramer, J., Farahani, K., Kirby, J., et al.: The multimodal brain tumor image segmentation benchmark (BRATS). IEEE Trans. Med. Imaging **34**(10), 1993–2024 (2015). https://doi.org/10.1109/TMI.2014.2377694

2. Bakas, S., Akbari, H., Sotiras, A., Bilello, M., Rozycki, M., Kirby, J.S., et al.: Advancing The Cancer Genome Atlas glioma MRI collections with expert segmentation labels and radiomic features. Nat. Sci. Data **4**, 170117 (2017). https://doi.org/10.1038/sdata.2017.117

3. Bakas, S., Reyes, M., Jakab, A., Bauer, S., Rempfler, M., Crimi, A., et al.: Identifying the best machine learning algorithms for brain tumor segmentation, progression assessment, and overall survival prediction in the BRATS challenge, arXiv preprint arXiv:1811.02629 (2018)

4. Louis, D.N., et al.: The 2016 World Health Organization Classification of Tumors of the Central Nervous System: a summary. Acta Neuropathol. **131**(6), 803–820 (2016). https://doi.org/10.1007/s00401-016-1545-1

5. Bakas, S., Akbari, H., Sotiras, A., Bilello, M., Rozycki, M., Kirby, J., et al.: Segmentation labels and radiomic features for the pre-operative scans of the TCGA-GBM collection. Cancer Imaging Arch. (2017).https://doi.org/10.7937/K9/TCIA.2017.KLXWJJ1Q

6. Bakas, S., Akbari, H., Sotiras, A., Bilello, M., Rozycki, M., Kirby, J., et al.: Segmentation labels and radiomic features for the pre-operative scans of the TCGA-LGG collection. Cancer Imaging Arch. (2017). https://doi.org/10.7937/K9/TCIA.2017.GJQ7R0EF

7. Myronenko, A.: 3D MRI brain tumor segmentation using autoencoder regularization. In: Crimi, A., Bakas, S., Kuijf, H., Keyvan, F., Reyes, M., van Walsum, T. (eds.) BrainLes 2018. LNCS, vol. 11384, pp. 311–320. Springer, Cham (2019). https://doi.org/10.1007/978-3-030-11726-9_28

8. Russo, C., Liu, S., Di Ieva, A.: Spherical coordinates transform pre-processing in deep convolution neural networks for brain tumor segmentation in MRI. arXiv preprint, arXiv:2008.07090 (2020)

9. Feng, Y.Z., Liu, S., Cheng, Z.Y., Quiroz, J.C., et al.: Severity assessment and progression prediction of COVID-19 patients based on the LesionEncoder framework and chest CT. medRxiv (2020). https://doi.org/10.1101/2020.08.03.20167007

10. McCullagh, P., Nelder, J.: Generalized Linear Models, 2nd edn. Chapman and Hall/CRC, Boca Raton (1989).ISBN 0-412-31760-5

11. Ronneberger, O., Fischer, P., Brox, T.: U-Net: convolutional networks for biomedical image segmentation. arXiv:1505.04597 (2015)

12. Tan, M., Le, Q.V.: EfficientNet: rethinking model scaling for convolutional neural networks. arXiv:1905.11946v3 (2019)

13. Jørgensen, B.: The theory of exponential dispersion models and analysis of deviance. Monografias de matemática, no. 51 (1992)

14. Hervey-Jumper, S.L., Berger, M.S.: Evidence for improving outcome through extent of resection. Neurosurg. Clin. N. Am. **30**(1), 85–93 (2019). https://doi.org/10.1016/j.nec.2018.08.005

15. Quiroz, J.C., Feng, Y.Z., Cheng, Z.Y., Rezazadegan, D., et al.: Severity assessment of COVID-19 based on clinical and imaging data. medRxiv (2020). https://doi.org/10.1101/2020.08.12.20173872

16. Gao, Y., Xiao, X., Han, B., Li, G., et al.: A deep learning methodology for differentiating glioma from radiation necrosis using multimodal MRI: algorithm development and validation. JMIR Medical Informatics, preprint. preprints.jmir.org/preprint/19805 (2020)
17. Liu, S., Graham, S., Schulz, A., Yiannikas, C., et al.: A deep learning based algorithm identifies glaucomatous discs using monoscopic fundus photos. Ophthalmol. Glaucoma 1(1), 15–22 (2018)

Overall Survival Prediction
for Glioblastoma on Pre-treatment MRI
Using Robust Radiomics and Priors

Yannick Suter[1,2(✉)], Urspeter Knecht[2,3], Roland Wiest[4], and Mauricio Reyes[1,2]

[1] Insel Data Science Center, Inselspital, Bern University Hospital, Bern, Switzerland
[2] ARTORG Center for Biomedical Engineering Research, University of Bern,
Bern, Switzerland
yannick.suter@artorg.unibe.ch
[3] Radiology Department, Spital Emmental, Burgdorf, Switzerland
[4] Support Center for Advanced Neuroimaging, Inselspital, Bern University Hospital,
Bern, Switzerland

Abstract. Patients with Glioblastoma multiforme (GBM) have a very
low overall survival (OS) time, due to the rapid growth an invasiveness
of this brain tumor. As a contribution to the overall survival (OS) prediction task within the Brain Tumor Segmentation Challenge (BraTS),
we classify the OS of GBM patients into overall survival classes based
on information derived from pre-treatment Magnetic Resonance Imaging
(MRI). The top-ranked methods from the past years almost exclusively
used shape and position features. This is a remarkable contrast to the
current advances in GBM radiomics showing a benefit of intensity-based
features. This discrepancy may be caused by the inconsistent acquisition
parameters in a multi-center setting. In this contribution, we test if normalizing the images based on the healthy tissue intensities enables the
robust use of intensity features in this challenge. Based on these normalized images, we test the performance of 176 combinations of feature
selection techniques and classifiers. Additionally, we test the incorporation of a sequence and robustness prior to limit the performance drop
when models are applied to unseen data. The most robust performance
on the training data (accuracy: 0.52 ± 0.09) was achieved with random
forest regression, but this accuracy could not be maintained on the test
set.

Keywords: Glioblastoma · Overall survival · Radiomics · Priors ·
MRI · Normalization

1 Introduction

Glioblastoma multiforme (GBM) is a very infiltrative and fast-growing brain
tumor. GBM patients typically have a very short overall survival time of only
around 16 months [23]. Currently, no curative treatments are available. The

A. Crimi and S. Bakas (Eds.): BrainLes 2020, LNCS 12658, pp. 307–317, 2021.
https://doi.org/10.1007/978-3-030-72084-1_28

standard-of-care consists of maximum safe resection followed by temozolomide-based chemoradiation [22,26]. Due to the low median OS, the patients are closely monitored with follow-up MRI. MRI-based biomarkers may some day enable clinicians to detect progression earlier and analyze patterns in the disease paths. We consider investigating pre-treatment MRI features related to OS as an intermediate step towards the discovery of such biomarkers.

The Brain Tumor Segmentation Challenge (BraTS) includes a task for the overall survival estimation from pre-treatment MRI [3–6,17]. This study is a contribution to this challenge (team ubern-mia).

Analyzing the highly-ranked challenge entries from the past two years, we observed that (a) intensity-based features are rarely used, (b) the classification performance varies considerably between the training, validation and test set, and (c), the maximum reachable accuracy did not improve substantially [6]. Approaches using the patients age, basic tumor shape features, and position information were able to reach high ranks and avoid over-fitting to the training data. This is disparate to the currently large volume of radiomics-based studies being published (e.g. [2,13,15]). At the same time, the robustness of radiomic features to image quality, inter-rater variability on the regions-of-interest, and multi-center and -vendor acquisition was shown to be a challenge (e.g., [21,24, 27]).

For intensity-based radiomic features, consistent grey-value binning across patients and centers is key. The basic z-score normalization and subsequent intensity scaling for the dataset used in this challenge leads to considerably different histogram bin counts across patients. We address this issue by image normalization based on the mean intensities of the grey and white matter. To further address the robustness of radiomic features, we follow [24], in testing the use of features previously determined robust (robustness prior), and limiting the features used to the two MR sequences the clinicians predominantly consider when examining pre-treatment data (sequence prior). Since simple models only considering the age of the patients, with shape and position features were highly ranked in the previous editions of the challenge, we include it in our experiments as a benchmark.

2   Materials and Methods

Figure 1 gives an overview of the pipeline and experiments on the training set. We follow the good practice recommendations outlined in [18].

2.1   Data

The training set consists of 236 patients with age, extent of resection, and the ground truth OS available. For each patient, four MRI sequences are available: pre- and post-contrast T1-weighted (T1, T1c), T2-weighted (T2), and T2-weighed-fluid attenuated inversion recovery (FLAIR). These sequences are provided already co-registered, resampled to isotropic $1\,mm^3$ voxels, and skull-stripped [17].

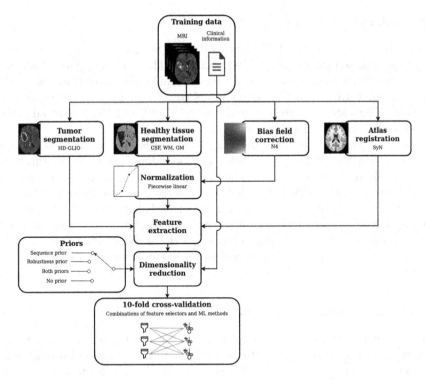

Fig. 1. Overview of the pipeline used for this study. The best-performing model is subsequently re-trained on the whole training dataset, and applied to the validation and test data.

2.2 Segmentation and Normalization

We use HD-GLIO to segment the contrast-enhancing tumor and the T2/FLAIR-abnormality [11,14].

To normalize the images across patients, we matched the intensities of the healthy grey and white matter to the mean of the training set for all MRI sequences using a piece-wise linear transform. To identify these healthy regions, we trained a U-Net with the nnUNet-framework [11] with silver-standard ground truth labels for cerebrospinal fluid (CSF), white and grey matter (WM, GM) obtained with FSL [12] on BraTS 2016 data. Based on these healthy tissue labels, a MR sequence specific piece-wise linear intensity transform was applied to all images to match the healthy tissue intensity and map to a fixed intensity range for consistent grey-value binning. Only the WM and GM labels were used, the CSF segmentation was often contaminated by incomplete skull-stripping.

All images were bias-field corrected with the N4 algorithm [25].

2.3 Radiomic Features

All available features in the *PyRadiomics* library (version 2.0) [9] were extracted for all four MR sequences and two segmentation labels. Intensity-based features were extracted for the original, wavelet image, and Laplacian of Gaussian (LoG) filtered image with $\sigma = 2$ and 3 to possibly account for some interpolation effects and noise. The features were extracted for 40, 70, 100, and 130 bins by adjusting the bin width on the normalized images.

For tumor position features, the images were registered to the MNI152 atlas [16], using `mri_robust_register` [20]. The bounding box limits in this atlas space for all three axes were used as a feature. Additionally, we introduced the distance to the ventricle as a feature, since ventricle-touching tumors may spread fast through the CSF. Since the position and shape of the ventricles were often heavily affected by the mass effect of the tumor, we applied the symmetric normalization method (SyN) [1], to properly deform the affected areas.

All features were z-score normalized to have zero mean and a standard deviation of one.

2.4 Dimensionality Reduction

The number of features has to be adjusted to the number of training samples. Following [18], we used nine features as input to the machine learning (ML) methods. As a first step, we excluded all features with a concordance index (c-index) lower than 0.55. Since we extracted the same features for different bin widths, we keep one version of a given feature across bin widths by selecting the feature with the highest c-index. Since feature selection methods have issues with highly correlated inputs, we further iteratively exclude the feature with the lower c-index for all feature pairs with a correlation coefficient of 0.95 or higher.

2.5 Priors

We test the introduction of two priors. When using the sequence prior, we limit the features to the T1c and FLAIR MRI images, which are most commonly considered for pre-treatment evaluation by neuroradiologists.

The second prior uses the results from robustness testing against expected variability in the image quality (voxel size, slice spacing, k-space sub-sampling, noise), and inter-rater agreement [24]. For this robustness prior, features with an intraclass correlation coefficient ICC(2,1) of 0.85 or higher were considered.

These priors were tested separately and jointly.

2.6   Survival Time Transform

Many ML methods require or profit from normally distributed target variables. Therefore we tested processing the survival time in days with a quantile, power, and robust scaler implemented in *scikit-learn* [19].

Feature Selection and Classifier Evaluation. We tested all combinations of 16 feature selection and eleven ML methods from the literature. The feature selection methods included ReliefF (RELF), Fischer Score (FSCR), Chi-square score (CHSQ), joint mutual information (JMI), conditional infomax feature extraction (CIFE), double input symmetric relevance (DISR), mutual information maximization (MIM), conditional mutual information maximization (CMIM), interaction capping (ICAP), minimum redundancy maximum relevance (MRMR), and mutual information feature selection (MIFS). The ML methods Automatic Relevance Determination Regression (ARD), AdaBoost, Decision Tree, Extra Tree(s), Gaussian processes, Nearest Neighbors, radius neighbors, passive aggressive, RANSAC, stochastic gradient descent, Support Vector regression (SVR), random forests, multilayer perceptrons (MLP), linear, and Theil-Sen regression were applied.

The performance for all combinations was tested under ten-fold stratified cross-validation on the training set based on the balanced accuracy to account for possible class imbalance.

The code used for this contribution is available on https://github.com/ysuter/brats20-survivalprediction.git.

3   Results

The mean accuracy and balanced accuracy for the experiment without priors for all feature selection and ML methods is shown in Fig. 2. For this setup, the MIFS feature selector with an Extra Tree model performed best.

3.1   Age, Shape and Position Model

The best model using only the age, tumor shape and position features achieved a cross-validated accuracy of 0.436.

3.2   Influence of Priors

Using a prior widened the interquartile range across splits for all tested models and lead to lower mean performance.

3.3 Influence of Target Variable Transform

Of all tested overall survival time transforms, the Yeo-Johnson power transform [28] achieved the best results. For all models including intensity-based features, this transform could raise the performance when compared to using the raw survival time in days. Additionally, the power transform lead to a lower interquartile range (i.e. a more consistent performance) across splits for the models using the robustness prior.

3.4 Final Model Selection and Challenge Outcome

The models using no prior with power transformed OS time showed the most consistent performance during the cross-validation and highest accuracy and balanced accuracy. Since only the classification accuracy was considered for the ranking, we decided to use this approach.

This best performing models used a Mutual Information feature selector followed by a Random Forest Regressor [7]. The nine features used are listed in Table 1.

The consistent performance on the training set (Fig. 3 and Table 2) could not be maintained on the testing set, where the accuracy dropped to 0.364.

Table 1. Features used for the submitted model, listed with decreasing importance according to the mutual information feature selector. LoG: Laplacian of Gaussian, GLSZM: Grey level size zone matrix

Category	Name	Underlying image	MR sequence	Label	Bin count
GLSZM	Grey level non-uniformity	LoG, $\sigma = 2$	T1c	Edema	40
GLSZM	Grey level non-uniformity	LoG, $\sigma = 2$	T1c	Enhancing	40
1st order	90th percentile	Original	T1	Edema	40
Position	Bounding box, vertical lower end	–	–	Enhancing	–
1st order	Median	Original	T1	Edema	40
1st order	10th percentile	Original	T2	Edema	40
GLSZM	Grey level non-uniformity	Wavelet HHH	T1	Enhancing	40
Shape	Maximum 2D columns diameter	–	–	Enhancing	–
1st order	Range	original	T2	Enhancing	40

Table 2. Performance comparison for the tested feature sets, and combinations of feature selectors and ML methods. For each model type, only the best performing combination is shown.

Model	Power transform	Balanced acc. mean ± std	Accuracy mean ± std	MSE/days^2 mean ± std	Spearman's rho mean ± std
Age, shape, position	–	0.422 ± 0.132	0.436 ± 0.126	163606 ± 68909	0.159 ± 0.1829
Age, shape, position	✓	0.422 ± 0.132	0.436 ± 0.126	163606 ± 68909	0.159 ± 0.182
No prior	–	0.505 ± 0.042	0.502 ± 0.046	104823 ± 41600	0.462 ± 0.120
No prior	✓	**0.512 ± 0.091**	**0.515 ± 0.092**	112653 ± 57885	0.438 ± 0.138
Robustness prior	–	0.481 ± 0.108	0.467 ± 0.111	193950 ± 65143	0.279 ± 0.174
Robustness prior	✓	0.460 ± 0.124	0.460 ± 0.128	127121 ± 77249	0.330 ± 0.162
Sequence prior	–	0.456 ± 0.118	0.454 ± 0.107	138117 ± 75640	0.288 ± 0.229
Sequence prior	✓	0.465 ± 0.115	0.462 ± 0.121	122523 ± 73261	0.377 ± 0.147
Both priors	–	0.457 ± 0.115	0.451 ± 0.110	132059 ± 56725	0.301 ± 0.211
Both priors	✓	0.470 ± 0.116	0.465 ± 0.118	123631 ± 71540	0.361 ± 0.147

Performance of feature selection and machine learning technique combinations no priors, 10-fold cross-validation

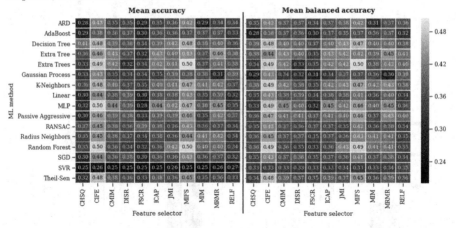

Fig. 2. Performance comparison for the experiment using no prior or target variable transform. Mean accuracy and balanced accuracy for all tested combinations of feature selectors and ML techniques under ten-fold cross-validation.

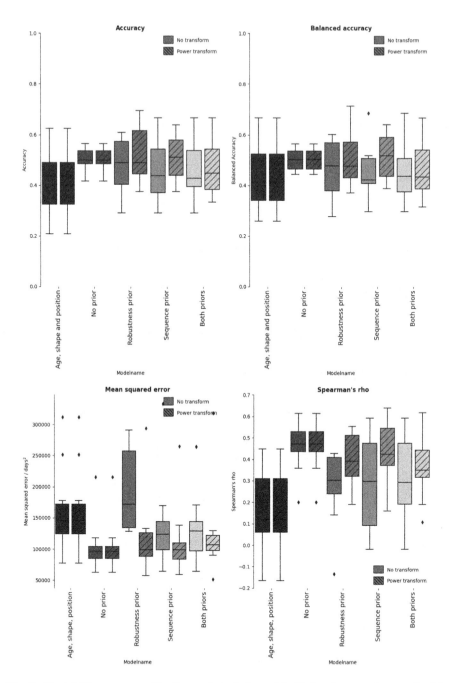

Fig. 3. Variability across splits for each tested model. Not using a prior resulted in the most consistent performance in the cross-validation. Using a prior increased the variability, while using a power transform on the OS time increased the accuracy in the sequence prior, both prior and no prior experiments.

4   Conclusion

When comparing the tested priors, no apparent benefit could be seen, since they were introduced to improve the transfer from single- to multi-center data. It would be interesting to see if there is a benefit if they were tuned to be specific for the given BraTS data.

We observed that the normalization based on a healthy tissue segmentation decreases the performance variability when not using a prior. Unfortunately, this type of normalization did not lead to a consistent performance on the training and testing set.

Considering the top ranked methods of this years challenge, simple models using the patient age, together with position, shape and lesion count features remain hard to beat. To close the gap between the current GBM radiomics literature outside of BraTS promoting intensity-based features, further investigations are needed regarding: (a) Normalization methods to profit from the added intensity information in a multi-center setting; and (b) the information content in pre-treatment MRI regarding the OS. Since the treatment and other important parameters such as the MGMT methylation status [10], we expect to reach a hard limit of the performance achievable with this limited data.

A part of the data in the BraTS dataset was acquired prior to the current imaging recommendations [8] and treatment practice [22]. An analysis on how this impacts the applicability of models developed with BraTS data on current data would be of high relevance.

Acknowledgements. We gratefully acknowledge the funding received from the Swiss Cancer League (Krebsliga Schweiz), grant KFS-3979-08-2016 and the NVIDIA Corporation for donating a Titan Xp GPU. Computations were partly performed on Ubelix, the HCP cluster at the University of Bern.

References

1. Avants, B.B., Epstein, C.L., Grossman, M., Gee, J.C.: Symmetric diffeomorphic image registration with cross-correlation: evaluating automated labeling of elderly and neurodegenerative brain. Med. Image Anal. **12**(1), 26–41 (2008)
2. Bae, S., et al.: Radiomic MRI phenotyping of glioblastoma: improving survival prediction. Radiology **289**(3), 797–806 (2018). https://doi.org/10.1148/radiol.2018180200. http://pubs.rsna.org/doi/10.1148/radiol.2018180200
3. Bakas, S., et al.: Segmentation labels and radiomic features for the pre-operative scans of the TCGA-GBM collection. Cancer Imaging Arch. (2017). https://doi.org/10.1038/sdata.2017.117
4. Bakas, S., et al.: Segmentation labels and radiomic features for the pre-operative scans of the TCGA-LGG collection. Cancer Imaging Arch. (2017). https://doi.org/10.1038/sdata.2017.117
5. Bakas, S., et al.: Advancing the cancer genome atlas glioma MRI collections with expert segmentation labels and radiomic features. Sci. Data **4**, 170117 (2017)
6. Bakas, S., et al.: Identifying the best machine learning algorithms for brain tumor segmentation, progression assessment, and overall survival prediction in the brats challenge. arXiv preprint arXiv:1811.02629 (2018)

7. Breiman, L.: Random forests. Mach. Learn. **45**(1), 5–32 (2001). https://doi.org/10.1023/A:1010933404324

8. Ellingson, B.M., Wen, P.Y., Cloughesy, T.F.: Modified criteria for radiographic response assessment in glioblastoma clinical trials. Neurotherapeutics **14**(2), 307–320 (2017). https://doi.org/10.1007/s13311-016-0507-6

9. van Griethuysen, J.J., et al.: Computational radiomics system to decode the radiographic phenotype. Cancer Res. **77**(21), e104–e107 (2017). https://doi.org/10.1158/0008-5472.CAN-17-0339. https://cancerres.aacrjournals.org/content/77/21/e104

10. Hegi, M.E., et al.: MGMT gene silencing and benefit from temozolomide in glioblastoma. New Engl. J. Med. **352**(10), 997–1003 (2005)

11. Isensee, F., Petersen, J., Kohl, S.A., Jäger, P.F., Maier-Hein, K.H.: nnU-Net: breaking the spell on successful medical image segmentation, vol. 1, pp. 1–8. arXiv preprint arXiv:1904.08128 (2019)

12. Jenkinson, M., Beckmann, C.F., Behrens, T.E., Woolrich, M.W., Smith, S.M.: FSL. Neuroimage **62**(2), 782–790 (2012)

13. Kickingereder, P., et al.: Radiomic profiling of glioblastoma: identifying an imaging predictor of patient survival with improved performance over established clinical and radiologic risk models. Radiology **280**(3), 880–889 (2016). https://doi.org/10.1148/radiol.2016160845. http://fsl.fmrib

14. Kickingereder, P., et al.: Automated quantitative tumour response assessment of MRI in neuro-oncology with artificial neural networks: a multicentre, retrospective study. Lancet Oncol. **20**(5), 728–740 (2019). https://doi.org/10.1016/S1470-2045(19)30098-1

15. Lao, J., et al.: A deep learning-based radiomics model for prediction of survival in glioblastoma multiforme. Sci. Rep. **7**(1), 1–8 (2017). https://doi.org/10.1038/s41598-017-10649-8

16. Mazziotta, J., et al.: A probabilistic atlas and reference system for the human brain: international consortium for brain mapping (ICBM). Philos. Trans. Roy. Soc. Lond. Ser. B: Biol. Sci. **356**(1412), 1293–1322 (2001)

17. Menze, B.H., Jakab, A., Van Leemput, K.: The multimodal brain tumor image segmentation benchmark (BRATS). IEEE Trans. Med. Imaging **34**(10), 1993–2024 (2015). https://doi.org/10.1109/TMI.2014.2377694

18. Papanikolaou, N., Matos, C., Koh, D.M.: How to develop a meaningful radiomic signature for clinical use in oncologic patients. Cancer Imaging **20**, 1–10 (2020). https://doi.org/10.1186/s40644-020-00311-4

19. Pedregosa, F., et al.: Scikit-learn: machine learning in Python. J. Mach. Learn. Res. **12**, 2825–2830 (2011)

20. Reuter, M., Rosas, H.D., Fischl, B.: Highly accurate inverse consistent registration: a robust approach. NeuroImage **53**(4), 1181–1196 (2010). https://doi.org/10.1016/J.NEUROIMAGE.2010.07.020

21. Saha, A., Yu, X., Sahoo, D., Mazurowski, M.A.: Effects of MRI scanner parameters on breast cancer radiomics. Expert Syst. Appl. **87**, 384–391 (2017)

22. Stupp, R., et al.: Radiotherapy plus concomitant and adjuvant temozolomide for glioblastoma. New Engl. J. Med. **352**(10), 987–996 (2005). https://doi.org/10.1056/NEJMoa043330

23. Suchorska, B., et al.: Complete resection of contrast-enhancing tumor volume is associated with improved survival in recurrent glioblastoma-results from the DIRECTOR trial. Neuro-Oncol. **18**(4), 549–556 (2016). https://doi.org/10.1093/neuonc/nov326https://doi.org/10.1093/neuonc/nov326

24. Suter, Y., et al.: Radiomics for glioblastoma survival analysis in pre-operative MRI: exploring feature robustness, class boundaries, and machine learning techniques. Cancer Imaging **20**(1), 1–13 (2020)

25. Tustison, N.J., et al.: N4ITK: improved N3 bias correction. IEEE Trans. Med. Imaging **29**(6), 1310–1320 (2010)

26. Weller, M., et al.: European Association for Neuro-Oncology (EANO) guideline on the diagnosis and treatment of adult astrocytic and oligodendroglial gliomas (2017). https://doi.org/10.1016/S1470-2045(17)30194-8

27. Weninger, L., Rippel, O., Koppers, S., Merhof, D.: Segmentation of brain tumors and patient survival prediction: methods for the BraTS 2018 challenge. In: Crimi, A., Bakas, S., Kuijf, H., Keyvan, F., Reyes, M., van Walsum, T. (eds.) BrainLes 2018. LNCS, vol. 11384, pp. 3–12. Springer, Cham (2019). https://doi.org/10.1007/978-3-030-11726-9_1

28. Yeo, I., Johnson, R.A.: A new family of power transformations to improve normality or symmetry. Biometrika **87**(4), 954–959 (2000). https://doi.org/10.1093/biomet/87.4.954

Glioma Segmentation Using Encoder-Decoder Network and Survival Prediction Based on Cox Analysis

Enshuai Pang[1,2], Wei Shi[1,2], Xuan Li[1,2], and Qiang Wu[1,2(✉)]

[1] School of Information Science and Engineering, Shandong University, Jinan, China
wuqiang@sdu.edu.cn
[2] Institute of Brain and Brain-Inspired Science, Shandong University, Jinan, China

Abstract. Glioma imaging analysis is a challenging task. In this paper, we used the encoder-decoder structure to complete the task of glioma segmentation. The most important characteristic of the presented segmentation structure is that it can extract more abundant features, and at the same time, it greatly reduces the amount of network parameters and the consumption of computing resources. Different textures, first order statistics and shape-based features were extracted from the BraTS 2020 dataset. Then, we use cox survival analysis to perform feature selection on the extracted features. Finally, we use randomforest regression model to predict the survival time of the patients. The result of survival prediction with five-fold cross-validation on the training dataset is better than the baseline system.

Keywords: Gliomas segmentation · Abundant features · Cox analysis · Randomforest · Survival prediction

1 Introduction

Glioma is the most common form of brain tumor as a collection or tissue of abnormal cells [7]. It may has different degrees of invasiveness, prognosis and histological subareas and can be described by various intensity distributions of MRI patters, reflecting different biological characteristics of tumors [1]. The growth rate and location of glioma determine its influence on nervous system and glioma usually be classified as benign tumors and malignant tumors. Glioma especially the malignant tumor has seriously endangered people's lives, therefore early detection and treatment is extremely important [8].

Recently, convolutional neural networks have achieved good results in various fields of images, such as image classification, target detection, and image segmentation, and they have gradually been applied to medical image segmentation and classification. In this competition, we used the encoder-decoder structure to complete the segmentation task of glioma and employ cox analysis to extract essential pattern for survival prediction task.

© Springer Nature Switzerland AG 2021
A. Crimi and S. Bakas (Eds.): BrainLes 2020, LNCS 12658, pp. 318–326, 2021.
https://doi.org/10.1007/978-3-030-72084-1_29

With the increasing incidence of brain tumors in the population and the gradual maturity of magnetic resonance imaging in disease analysis, survial prediction based on MRI can supply essential support to the clinical prognosis. Now, the development of artificial intelligence technologies can provide an effective way to predict the survival time of patients based on MRI. In this paper, we propose an integrated prediction model based on cox analysis to predict the survival time of patients with glioma. By performing feature extraction and feature selection on multimodal MRI samples, a survival prediction framework based on cox analysis and randomforest regression is established.

2   Segmentation Method

Considering the availability of actual 2D data, we use the traditional 2D segmentation network as the backbone network [9]. At the same time, due to the results of image segmentation are affected by the features extracted from the network, we use well-designed feature mining units which can extract rich features as the basic network building units [14,15]. The segmentation framework is shown in Fig. 1. The number of features extracted by feature mining unit is shown in Table 1.

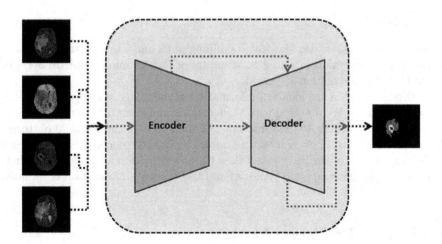

Fig. 1. Brain tumor segmentation framework.

Table 1. The number of features extracted by feature mining unit.

	FMU1	FMU2	FMU3	FMU4	FMU5
Encode	32	64	128	256	512
Decode	32	64	128	256	

2.1   Encoder

In the encoder part, we use five feature mining units (FMU) and four down-sample units to encode the input data. This structure can remove the noise and retain the effective features of the input data which are conducive to segmentation.

The feature mining unit is composed of the following parts: firstly, the input features are expanded by a 1×1 convolution unit, which can increase the number of feature channels, extract more coding information of input data, and facilitate the use of subsequent residual units; at the same time, in order to improve the diversity of convolution operations, we use convolution operations with different convolution kernels that interact with each other through skip connection, and the feature channel information extracted by these operations will be spliced together; finally, we use 1×1 convolution operation to reduce the number of feature channels, so as to remove the invalid coding information and retain the effective coding information [14–16]. Due to the complexity of the above operations, in order to prevent the gradient from disappearing, we use the bottleneck structure [12]. The steps of all convolution operations are 1×1, and each convolution operation is followed by a batch normalization operation and relu activation function [13].

2.2   Decoder

In the decoder part, we use four feature mining units and four up-sample units to decode the coding information. The structure of feature mining unit in decoder is exactly the same as that in encoder.

In order to obtain the location information of the image, each feature mining unit of the decoder is connected with the feature mining unit of the corresponding position of the encoder. At the same time, in order to keep the essential patterns, the output information of each feature mining unit in the decoder is up-sampled to the original size of the image, and then these information are spliced together, and the invalid information is removed and the effective information is retained [10, 11].

3   Survival Prediction Method

We explore four modalities MRI data to obtain different glioma regions, and then we use the pyradiomics [17] toolkit to extract the corresponding features. The obtained features are selected by cox analysis, and finally we use randomforest regression to predict the survival time of patients.

3.1   Feature Extraction

We use the pyradiomics[17] toolkit to extract features from multiple MRI modalities of the patients. In the specific operation, we used the peritumoral edema area, the enhanced area and the necrotic and non-enhanced area as the mask on MRI data to extract features. A total of 303 features were extracted. The primary features are shown in Table 2.

Table 2. The extracted feature type and dimension

Feature type	Dimension
First order features	54
Shape features	42
Gray level cooccurrence matrix features	72
Gray level size zone matrix features	48
Gray level run length matrix features	39
Gray level dependence matrix features	36
Neighbouring gray tone difference matrix features	12

3.2   Feature Selection

Because too many features are extracted by above step, we need to select effective features. The significance of feature selection is to reserve the relevant features and eliminate the repetitive features. In this paper, we use the cox survival analysis method for feature selection to ensure the validity of features and improve the accuracy of regression model. The specific feature selection steps are as follows:

(1) First, we use cox survival analysis for univariate screening. Cox univariate analysis is performed on the extracted 303 features separately and those features with a ph value of less than 0.05 are reserved.

(2) Cox multivariate analysis is performed on the extracted features after screening in step (1). We need to conduct a ph hypothesis test on the features to ensure that the partial residuals of the feature variables and the rank significance of the survival time cannot be less than 0.05. We use SPSS software to draw the scatter diagram of the partial residuals of the feature variables and the rank significance of the survival time to show the process of ph hypothesis test. From the scatter plot, we can see that if the characteristic variables meet the ph hypothesis test, there is no proportional relationship between the partial residuals of the characteristic variables and the rank of the survival time. The ph hypothesis test on the feature is shown in Fig. 2.

(3) We perform cox regression multivariate analysis on the feature variables that satisfy the ph hypothesis in step (2), and the final selected features have a ph value less than 0.1. We selected three types of features, namely: minorAxisLength, sphericity extracted from the segmentation result with the peritumoral edema area as the mask, and busyness extracted from the segmentation result with the enhanced area as the mask.

(4) In addition, we add the age of the patients to the final selected feature set. We used SPSS software to perform a bivariate analysis on the patient's age and survival time. Through bivariate analysis, we found that the patient's age and survival time were significantly correlated at the 0.01 level, pearson correlation was negative and the correlation coefficient was -0.353, indicating that the older the patient, the shorter the survival time. The bivariate analysis is shown in Table 3.

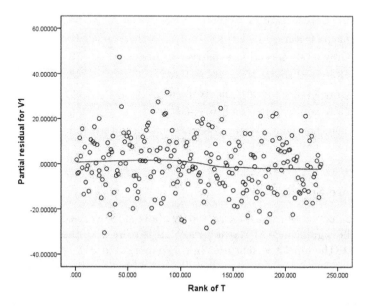

Fig. 2. The ph hypothesis test on the feature.

Table 3. Bivariate analysis of patient's age and survival time.(Correlation is significant at the 0.01 level.)

	Age	T
Age Person Correlation Sig.(2-tailed) N	1	−.353
		.000
	234	234
T Person Correlation Sig.(2-tailed) N	−.353	1
	.000	
	234	234

3.3   Regression Framework

In this paper, we use cox analysis and randomforest regression to complete the prediction task and achieve the final survival prediction results of glioma patients. The prediction of overall survival regression framework is shown in Fig. 3.

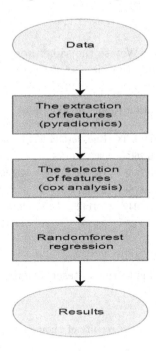

Fig. 3. Regression framework.

4   Experiments

4.1   Data

The dataset used in the experiment are all from the Brain Tumor Segmentation (BraTS) 2020 competition. These data are all manually segmented by raters following the same annotation principle, and finally reviewed by neuro-radiologists. Each patient includes four modalities data, namely T1, T2, T1ce, Flair. The segmented areas of glioma include background (label 0), necrotic and non-enhanced area (label 1), peritumoral edema area (label 2) and enhanced area (label 4) [1–5]. The training dataset includes 76 lower glioma patients and 293 higher glioma patients. The validation dataset includes 125 patients and the testing dataset includes 166 patients. The number of higher and lower glioma patients in the validation dataset and testing dataset is unclear.

All the input data are standardized. In addition, data processing operations mainly include:(1) if there is no tumor area in ground truth, we delete the slices of ground truth and the corresponding four modalities, (2) we use data clipping to reduce the proportion of the background area, (3) data rotation increases the amount of data.

4.2   Results

Segmentation of Glioma. We use the same processing validation dataset to verify. The results of the validation dataset are shown in Table 4.

Table 4. The segmentation results of gliomas on the validation dataset.

	Dice ET	Dice WT	Dice TC	Hausdorff95 ET	Hausdorff95 WT	Hausdorff95 TC
Mean	0.7538	0.8811	0.7605	34.2391	18.0901	29.0570
StdDev	0.2770	0.0988	0.2658	100.8688	28.2531	85.1708
Median	0.8495	0.9102	0.8768	2.2361	4.3589	4.4721
25quantile	0.7558	0.8800	0.7249	1.4142	2.4495	2.0000
75quantile	0.9068	0.9346	0.9328	3.6056	13.4907	10.8190

We use the same processing testing dataset to verify. The results of the testing dataset are shown in Table 5.

Table 5. The segmentation results of gliomas on the testing dataset.

	Dice ET	Dice WT	Dice TC	Hausdorff95 ET	Hausdorff95 WT	Hausdorff95 TC
Mean	0.7944	0.8652	0.8166	19.2869	14.5314	28.7860
StdDev	0.2018	0.1288	0.2528	75.1032	20.5876	88.4671
Median	0.8445	0.9100	0.9106	1.7321	4.3008	2.8284
25quantile	0.7659	0.8436	0.8407	1.0000	2.2361	1.7321
75quantile	0.9082	0.9396	0.9509	2.4495	16.9329	6.4807

Prediction of Patient Survival Time. In the training phase, we use five fold cross validation to train the model. We divide the data into five parts, four parts as the training dataset, and the remain part is using as the validation dataset. We use the standards provided by the contestants to evaluate the performance of our model. We set a month to be 30 days. The data is divided into three categories: long-term data(>450 days), medium-term data ($\geq$300 days and $\leq$ 450 days) and short-term data (<300 days) [6]. The results of patient survival time task on the competition training dataset are shown in Table 6.

In order to find a best regression method, in addition to randomforest regression, we also evaluated other regression method including: linear regression, GBRT and Decision tree. In contrast, we can see that the model using cox combined with randomforest regression has provided the better performance.

The results of patient survival prediction task based on our proposed framework on validation dataset and testing dataset are shown in Table 7 and Table 8 respectively.

Table 6. The results of five-fold cross-validation on training dataset.

Methods	Accuracy
linear regression	0.435 ± 0.086
Decision tree regression	0.495 ± 0.077
GBDT	0.508 ± 0.047
Randomforest	**0.538 ± 0.028**

Table 7. The results of prediction of patient survival time on validation dataset.

CasesExpected	CasesEvaluated	Accuracy	MSE	medianSE	stdSE	SpearmanR
29	29	0.517	151013.386	91990.89	185785.149	0.155

Table 8. The results of prediction of patient survival time on testing dataset.

CasesExpected	CasesEvaluated	Accuracy	MSE	medianSE	stdSE	SpearmanR
107	107	0.523	463328.788	53620.034	1435798.081	0.32

5   Conclusion

In this paper,we used the encoder-decoder structure to complete the segmentation task of glioma. The most important feature of our proposed segmentation structure is that it can extract more abundant features, and at the same time, it greatly reduces the amount of network parameters and the consumption of computing resources. We also propose a survival prediction framework for glioma patients. The pyradiomics [17] toolkit is employed to extract features from the segmentation results of the above segmentation model and multimodal MRI data. Cox survival analysis is used to select features. Finally, we use randomfores regression to predict patients survival time. Our segmentation and prediction model has shown good performance on the validation dataset compared with the baseline methods. At the same time, our segmentation and prediction model has also shown good performance on the testing dataset.

Acknowledgements. This work was supported by the Clinical Research Center of Shandong University. (No. 2020SDUCRCB002)

References

1. Menze, B.H., et al.: The multimodal brain tumor image segmentation benchmark (BRATS). IEEE Trans. Med. Imaging **34**(10), 1993–2024 (2015). https://doi.org/10.1109/TMI.2014.2377694
2. Bakas, S., et al.: Advancing The Cancer Genome Atlas glioma MRI collections with expert segmentation labels and radiomic features. Nat. Sci. Data **4**, 170117 (2017). https://doi.org/10.1038/sdata.2017.117

3. Bakas, S., et al.: Identifying the best machine learning algorithms for brain tumor segmentation, progression assessment, and overall survival prediction in the BRATS Challenge, arXiv preprint arXiv:1811.02629 (2018)

4. Bakas, S., et al.: Segmentation labels and radiomic features for the pre-operative scans of the tCGA-GBM collection. Cancer Imaging Arch. (2017). https://doi.org/10.7937/K9/TCIA.2017.KLXWJJ1Q

5. Bakas, S., et al.: Segmentation labels and radiomic features for the pre-operative scans of the TCGA-LGG collection. Cancer Imaging Arch. (2017). https://doi.org/10.7937/K9/TCIA.2017.GJQ7R0EF

6. Multimodal Brain Tumor Segmentation Challenge 2019, Evaluation Framework. https://www.med.upenn.edu/cbica/brats2019/evaluation.html

7. DeAngelis, L.M.: Brain tumors. N. Engl. J. Med. **344**, 114–123 (2001). https://doi.org/10.1056/NEJM200101113440207

8. Onishi, M., Ichikawa, T., Kurozumi, K., et al.: Angiogenesis and invasion in glioma. Brain Tumor Pathol. **28**(1), 13–24 (2011). https://doi.org/10.1007/s10014-010-0007-z

9. Ronneberger, O., Fischer, P., Brox, T.: U-Net: convolutional networks for biomedical image segmentation. In: Navab, N., Hornegger, J., Wells, W.M., Frangi, A.F. (eds.) MICCAI 2015. LNCS, vol. 9351, pp. 234–241. Springer, Cham (2015). https://doi.org/10.1007/978-3-319-24574-4_28

10. Lin, T.-Y., et al.: Feature pyramid networks for object detection. In: Proceedings of the IEEE Conference on Computer Vision and Pattern Recognition, pp. 2117–2125 (2017). https://doi.org/10.1109/CVPR.2017.106

11. Kong, X., Sun, G., Wu, Q., Liu, J., Lin, F.: Hybrid pyramid U-Net model for brain tumor segmentation. In: Shi, Z., Mercier-Laurent, E., Li, J. (eds.) IIP 2018. IAICT, vol. 538, pp. 346–355. Springer, Cham (2018). https://doi.org/10.1007/978-3-030-00828-4_35

12. He, K., et al.: Deep residual learning for image recognition. In: Proceedings of the IEEE Conference on Computer Vision and Pattern Recognition, pp. 770–778 (2016). https://doi.org/10.1109/CVPR.2016.90

13. Ioffe, S., Szegedy, C.: Batch normalization: accelerating deep network training by reducing internal covariate shift, arXiv preprint arXiv:1502.03167 (2015)

14. Szegedy, C., et al.: Going deeper with convolutions. In: Proceedings of the IEEE Conference on Computer Vision and Pattern Recognition, pp. 1–9 (2015). https://doi.org/10.1109/CVPR.2015.7298594

15. Szegedy, C., et al.: Rethinking the inception architecture for computer vision. In: Proceedings of the IEEE Conference on Computer Vision and Pattern Recognition, pp. 2818–2826 (2016). https://doi.org/10.1109/CVPR.2014.347

16. Iandola, F.N., et al.: SqueezeNet: AlexNet-level accuracy with 50x fewer parameters and ¡0.5 MB model size, arXiv preprint arXiv:1602.07360 (2016)

17. Van Griethuysen, J.J.M., Fedorov, A., Parmar, C., et al.: Computational radiomics system to decode the radiographic phenotype. Cancer Res. **77**(21), e104–e107. https://doi.org/10.1158/0008-5472.CAN-17-0339

Brain Tumor Segmentation with Self-ensembled, Deeply-Supervised 3D U-Net Neural Networks: A BraTS 2020 Challenge Solution

Théophraste Henry[1(✉)], Alexandre Carré[1], Marvin Lerousseau[1,2],
Théo Estienne[1,2], Charlotte Robert[1,3], Nikos Paragios[4], and Eric Deutsch[1,3]

[1] Université Paris-Saclay, Institut Gustave Roussy, Inserm, Radiothérapie
Moléculaire et Innovation Thérapeutique, 94805 Villejuif, France
theophraste.henry@gustaveroussy.fr
[2] Université Paris-Saclay, CentraleSuplec, 91190 Gif-sur-Yvette, France
[3] Gustave Roussy, Département d'oncologie-radiothérapie, 94805 Villejuif, France
[4] Therapanacea, Paris, France

Abstract. Brain tumor segmentation is a critical task for patient's disease management. In order to automate and standardize this task, we trained multiple U-net like neural networks, mainly with deep supervision and stochastic weight averaging, on the Multimodal Brain Tumor Segmentation Challenge (BraTS) 2020 training dataset. Two independent ensembles of models from two different training pipelines were trained, and each produced a brain tumor segmentation map. These two labelmaps per patient were then merged, taking into account the performance of each ensemble for specific tumor subregions. Our performance on the online validation dataset with test time augmentation were as follows: Dice of 0.81, 0.91 and 0.85; Hausdorff (95%) of 20.6, 4, 3, 5.7 mm for the enhancing tumor, whole tumor and tumor core, respectively. Similarly, our solution achieved a Dice of 0.79, 0.89 and 0.84, as well as Hausdorff (95%) of 20.4, 6.7 and 19.5 mm on the final test dataset, ranking us among the top ten teams. More complicated training schemes and neural network architectures were investigated without significant performance gain at the cost of greatly increased training time. Overall, our approach yielded good and balanced performance for each tumor subregion. Our solution is open sourced at https://github.com/lescientifik/open_brats2020.

Keywords: Deep learning · Brain tumor · Semantic segmentation

---

T. Henry and A. Carré—Equally contributing authors.

© Springer Nature Switzerland AG 2021
A. Crimi and S. Bakas (Eds.): BrainLes 2020, LNCS 12658, pp. 327–339, 2021.
https://doi.org/10.1007/978-3-030-72084-1_30

1   Introduction

1.1   Clinical Overview

Gliomas are the most frequent primitive brain tumors in adult patients and exhibit various degrees of aggressiveness and prognosis. Magnetic Resonance Imaging (MRI) is required to fully assess tumor heterogeneity, and the following sequences are conventionally used: T1 weighted sequence (T1), T1-weighted contrast enhanced sequence using gadolinium contrast agents (T1Gd), T2 weighted sequence (T2), and fluid attenuated inversion recovery (FLAIR) sequence.

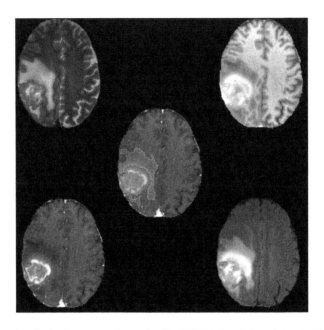

Fig. 1. Example of a brain tumor from the BraTS 2020 training dataset. **Red:** enhancing tumor (ET), **Green:** non enhancing tumor/ necrotic tumor (NET/NCR), **Yellow:** peritumoral edema (ED). Upper Left: T2 weighted sequence, Upper Right: T1 weighted sequence, Lower Left: T1-weighted contrast enhanced sequence, Lower Right: FLAIR sequence Middle: T1-weighted contrast enhanced sequence with labelmap overlay (Color figure online)

Four distinct tumoral subregions can be defined from MRI: the "enhancing tumor" (ET) which corresponds to area of relative hyperintensity in the T1Gd with respect to the T1 sequence; the "non enhancing tumor" (NET) and the "necrotic tumor" (NCR) which are both hypo-intense in T1-Gd when compared to T1; and finally the "peritumoral edema" (ED) which is hyper-intense in FLAIR sequence. These almost homogeneous subregions can be clustered together to compose three "semantically" meaningful tumor subparts: ET is the first cluster, addition of ET, NET and NCR represents the "tumor core" (TC) region, and addition of ED to TC represents the "whole tumor" (WT). Example of each sequence and tumor subvolumes is provided in Fig. 1 using 3D Slicer [10].

Accurate delineation of each tumor subregion is critical to patient's disease management, especially in a post-surgical context. Indeed, the radiation oncologist is required to segment the tumor, including the surgical resection cavity, the residual enhancing tumor and surrounding edema according to the Radiation Therapy Oncology Group (RTOG) [22]. Correct segmentation could also unveil prognostic factors through the use of radiomics or deep-learning based approach [9].

1.2 Multimodal Brain Tumor Segmentation Challenge 2020

The Multimodal Brain Tumor Segmentation Challenge 2020 [3–6,19] was split in three different tasks: segmentation of the different tumor sub-regions, prediction of patient overall survival (OS) from pre-operative MRI scans, and evaluation of uncertainty measures in segmentation. The Segmentation challenge consisted in accurately delineating the ET, TC and WT part of the tumor. The main evaluation metrics were an overlap measure and a distance metric. The commonly used Dice Similarity Coefficient (DSC) measures the overlap between two sets. In the context of ground truth comparison, it can be defined as follows:

$$DSC = \frac{2TP}{2TP + FP + FN} \tag{1}$$

with TP the true positives (number of correctly classified voxels), FP the false positives and FN the false negatives. It is interesting to note that this metric is insensitive to the extent of the background in the image. The Hausdorff distance [15] is complementary to the Dice metric, as it measures the maximal distance between the margin of the two contours. It greatly penalizes outliers: a prediction could exhibits almost voxel-perfect overlap, but if a single voxel is far away from the reference segmentation, the Hausdorff distance will be high. As such, this metric can seem noisier than the Dice index, but is very handy to evaluate the clinical relevance of a segmentation. As an example, if a tumor segmentation encompasses distant healthy brain tissue, it would require manual correction from the radiation oncologist to prevent disastrous consequences for the patient, even if the overall overlap as measured by the Dice metric is good enough.

2   Methods

Two independent training pipelines were designed, with a common neural network architecture based on the 3D U-Net with minor variations (described below). These two different training approaches were kept separate in order to promote network predictions' diversity. The specific details of each pipeline will be described below, and referred to as pipeline A and pipeline B.

2.1 Neural Network Architecture

After neural network architecture exploration, the chosen network used an encoder decoder architecture, heavily inspired by the 3D U-Net architecture from Çiçek et al. [35]. The architecture used is displayed in Fig. 2.

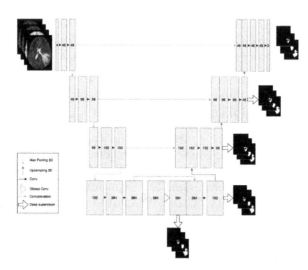

Fig. 2. Neural Network Architecture: 3D U-Net [35] with minor modifications

In the following description, a stage is defined as an arbitrary number of convolutions that does not change the spatial dimensions of the feature maps. All convolutions were followed by a normalization layer and a nonlinear activation (ReLU layer [21]). Group normalization [32] (A) and Instance normalization [29] (B) were used as a replacement for Batch Normalization [16] due to a small batch size during training and good theoretical performance on non-medical datasets.

The encoder had four stages. Each stage consisted of two $3 \times 3 \times 3$ convolutions. The first convolution increased the number of filters to the predefined value for the stage (48 for stage 1), while the second one kept the number of output channels unchanged. Between each stage, spatial downsampling was performed by a MaxPool layer with a kernel size of $2 \times 2 \times 2$ with stride 2. After each spatial downsampling, the number of filters was doubled. After the last stage, two $3 \times 3 \times 3$ dilated convolutions with a dilation rate of 2 were performed, and then concatenated with the last stage output.

The decoder part of the network was almost symmetrical to the encoder. Between each stage, spatial upsampling was performed using a trilinear interpolation. Shortcut connections between encoder and decoder stages that shared the same spatial sizes were performed by concatenation. The decoder stage performing at the lowest spatial resolution was made up of only one $3 \times 3 \times 3$ convolution. Last convolutional layer used a $1 \times 1 \times 1$ kernel with 3 output channels and a sigmoid activation.

The previous winner of the Brats challenge [1] limited their downsampling steps to 3. We hypothesized that further downsampling of the features maps, given the limited size of the input ($128 \times 128 \times 128$), would lead to irreversible loss of spatial information. As the last stage of the encoder takes much less GPU memory than the first, the dilation trick [8] was used to perform a pseudo fifth stage at the same spatial resolution as the fourth stage.

3D attention U-Nets were also trained, using the Convolutional Block Attention Module [31] added at the end of each encoder stage.

2.2   Loss Function

Inspired by the conciseness of the 2019 winning solution [1], the neural network was trained using only the Dice Loss [20] (A). The loss L is computed batch-wise and channel-wise, without weighting:

$$DSC = 1 - \frac{1}{N} \sum_n \frac{S_n * R_n + \varepsilon}{S_n^2 + R_n^2 + \varepsilon} \tag{2}$$

with n the number of output channels, S the output of the neural network after sigmoid activation, R the ground truth label and ϵ a smoothing factor (set to 1 in our experiment). For diversity, the pipeline B used a slightly different formulation of the Dice Loss, without squaring the terms of the denominator. Similarly, optimization was made directly on the final tumor regions to predict (ET, TC and WT) and not on their components (ET, NET-NCR, ED). The neural network output was a 3-channel volume, each channel representing the probability map for each tumor region.

Deep supervision [30] was performed after the dilated convolutions, and after each stage of the decoder (except the last) as in [23]. Deep supervision was achieved by adding an extra $1 \times 1 \times 1$ convolution with sigmoid activation and trilinear upsampling. Like the main output, each of this additional convolution resulted in a 3-channel volume, each channel representing the probability map for each tumor region (ET, TC and WT). The final loss was the unweighted sum of the main output loss, and the four auxiliary losses.

2.3   Image Pre-processing

Since MRI intensities vary depending on manufacturers, acquisition parameters, and sequences, input images needed to be standardized. Min-max scaling of each MRI sequence was performed separately, after clipping all intensity values to the 1 and 99 percentiles of the non-zero voxels distribution of the volume (A). Pipeline B performed a z-score normalization of the non-zero voxels of each IRM sequence independently.

Images were then cropped to a variable size using the smallest bounding box containing the whole brain, and randomly re-cropped to a fixed patch size of $128 \times 128 \times 128$. This allowed to remove most of the useless background that was present in the original volume, and to learn from an almost complete view of each brain tumor.

2.4 Data Augmentation Techniques

To prevent overfitting, on-the-fly data augmentation techniques were applied in both pipelines, according to a predefined probability. The augmentations and their respective probability of application were:

- input channel rescaling: multiplying each voxel by a factor uniformly sampled between 0.9 and 1.1 (A: 80% probability, B: 20%).
- input channel intensity shift: Adding each voxel a constant uniformly sampled between -0.1 and 0.1 (A: not performed, B: 20% probability).
- additive gaussian noise, using a centered normal distribution with a standard deviation of 0.1.
- input channel dropping: all voxel values of one of the input channels were randomly set to zero (A: 16% probability, B: not performed).
- random flip along each spatial axis (A: 80% probability, B: 50%).

2.5 Training Details

Models were produced by a five-fold cross-validation. The validation set was only used to monitor the network performance during training, and to benchmark its performance at the end of the training procedure.

Pipeline A: For each fold, the neural network was trained for 200 epochs with an initial learning rate of 1e−4, progressively reduced by a cosine decay after 100 epochs [12]. A batch size of 1 and the Ranger optimizer [18,33,34] were used. After 200 epochs, we performed a training scheme inspired from the fast stochastic weight averaging procedure [2]. The initial learning rate was restored to half of its initial value (5e-5), and training was done for another 30 epochs with cosine decay. Every 3 epochs, the model weights were saved. This procedure was repeated 5 times for a total of 150 additional epochs. At the end, the saved weights were averaged, effectively creating a new "self-ensembled" model. The Adam optimizer [17] was used without weight decay for the stochastic weight averaging procedure.

Pipeline B: The maximum number of training iterations was set to 400. The best model kept was the one with the lowest loss value on the validation set. A batch size of 3 and Adam optimizer with an initial learning of 1e−4 and no weight decay. Cosine annealing scheduler was used.

Common: In order to train a bigger neural network, float 16 precision (FP16) was used, which reduced memory consumption, accelerated the training procedure, and may lead to extra performance [12].

The neural network was built and trained using Pytorch v1.6 (which has native FP16 training capability) on Python 3.7. The model could fit on one graphic card (GPU).

2.6 Inference

Inference was performed in a two-steps fashion. First, models available from each pipeline were ensembled separately, by simple predictions averaging. Consequently, two labelmaps per case, one for each pipeline, were created. Three different models per fold (except one fold due to time constraint) were available for pipeline A: a 3D attention U-net version, a U-net version trained on an unfiltered version of the training dataset, and a U-net version trained on a filtered subset of the training dataset. The filtering process was based on previous training runs: cases with high training loss at the end of the training procedure were flagged as potentially wrong and removed from the complete training set, thus creating a "cleaned" version of the training dataset. The top two performing models per fold were chosen for ensembling (A). For Pipeline B, the five cross-validated models (one per fold) were ensembled. Then, the two labelmaps are merged based on the individual performance of each ensemble on the online validation set, as described below.

First Step. For each pipeline, the initial volume was preprocessed like the training data, then cropped to the minimal brain extent, and finally zero-padded to have each of the spatial dimensions divisible by 8. Test time augmentation (TTA) was done using 16 different augmentations for each of the models generated by the cross-validation, for a total of 80 predictions per sample. We used flips, and 90-180-270 rotations only in the axial plane, as rotation in other planes led to worse performance on the local validation set. Final prediction was made by averaging the predictions, using a threshold of 0.5 to binarize the prediction. Labelmap reconstruction was then performed in a straightforward manner: ET prediction was left untouched, the NET-NEC region of the tumor was deduced from a boolean operation between the ET label and the TC label, and similarly for the edema between the TC and the WT label ($NonEnhanching = TC - ET$; $edema = WT - TC$).

Second Step. The first step gave two labelmaps per case. Based on the online validation dataset, the mean whole tumour dice metric of the pipeline B's ensemble was consistently higher than that of the pipeline A's ensemble. We hypothesized that models from pipeline B were better for predicting edema. To keep the score intact on ET and TC from models A, ET and NET/NCR predicted labels had to be left untouched. If A predicted background or edema and B predicted edema or background respectively, B predicted labels were kept. The merging procedure is shown in Table 1.

2.7 Ablation Study for Pipeline A

Experiments with and without dataset filtering and attention block were produced for pipeline A. Cross-validated results can be found in Table 2. There was no clear benefit of either strategy, hence we decided to keep the two best available models for each fold for this pipeline.

Table 1. Merging procedure of the two labelmaps. 0: background, 1: necrotic and non-enhancing tumor core (NET), 2: peri-tumoral edema (ED), 4: enhancing tumor (ET)

		Model A			
		0	*1*	*2*	*4*
Model B	*0*	0	1	0	4
	1	0	1	2	4
	2	2	1	2	4
	4	0	1	2	4

Table 2. Ablation study: results from cross-validation on the training set.

Dice: mean (std)	ET	WT	TC
U-Net like	0.8077 (0.011)	0.9070 (0.006)	0.8705 (0.013)
+ Patients removal	0.8126 (0.019)	0.9043 (0.005)	0.8686 (0.012)
+ Attention block	0.8144 (0.022)	0.9037 (0.008)	0.8701 (0.018)

3 Results

3.1 Online Validation Dataset

Table 3 displays the results for the online validation data. Our models produced a Dice metric greater than 0.8. for each tumor region. Our two-pass merging strategy had no impact on the ET and TC segmentation performance of the pipeline A's ensemble, while greatly improving WT segmentation. Single pass strategy already yielded good performance for all three tumor regions. Larger value of Hausdorff distance for ET compared to other tumor subregions is explained by the absence of the ET label for some cases. Consequently, predicting even one voxel of ET would lead to a major penalty for this metric. Example of segmented tumor from the online validation set is displayed in Fig. 3. It is hard to visually

Table 3. Performance on the complete BraTs'20 Online Validation Data for the merging strategy, unless otherwise specified.

Metric (mean)	ET	WT	TC
Dice (Pipeline A alone)	0.80585	0.89518	0.85415
Dice (Pipeline B alone)	0.72738	0.91123	0.84921
Dice	0.80585	0.91148	0.85416
Sensitivity	0.81488	0.91938	0.84485
Specificity	0.99970	0.99915	0.99963
Hausdorff (95%)	20.55756	4.30103	5.69298

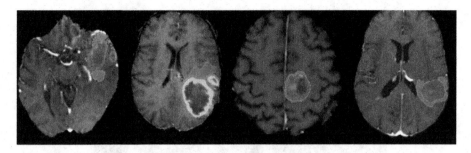

Fig. 3. From left to right: ground truth example from the training set, and generated segmentations from our solution for three patients among the online validation set; respectively: best mean dice score (ET:0.95, WT:0.96, TC:0.98), average mean dice score (ET:0.73, WT:0.92, TC:0.93), and worst mean dice score (ET:0.23, WT:0.95, TC:0.13). **Red**: enhancing tumor (ET), **Green**: non enhancing tumor/ necrotic tumor (NET/NCR), **Yellow**: peritumoral edema (ED) (Color figure online)

discriminate best from the average result, based on the mean dice score per patient (average across the three tumor sub-regions). However, our worst generated mask showed obvious error: contrast enhanced arteries were mislabeled as enhancing tumor.

3.2 Testing Dataset

Our final results on the testing dataset are displayed in Table 4. These results ranked us among the top 10 teams for the segmentation challenge. A significant discrepancy between validation and testing datasets for the TC Hausdorff distance was visible, while all other metrics showed small but limited overfit.

Table 4. Performance on the BraTs'20 Testing Data.

Metric (mean)	ET	WT	TC
Dice	0.78507	0.88595	0.84273
Sensitivity	0.81308	0.91690	0.85934
Specificity	0.99967	0.99905	0.99964
Hausdorff (95%)	20.36071	6.66665	19.54915

4   Discussion

Our solution to the BraTS'20 challenge is based on standard approaches carefully crafted together: we used U-net 3D neural networks, trained with on-the-fly data

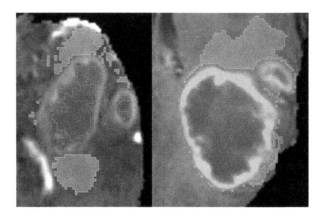

Fig. 4. Zoomed version of the first two vignettes of Fig. 3 Left: ground truth example from the training set. Right: generated segmentations from our solution for the best mean dice score patient on the validation set. **Red**: enhancing tumor (ET), **Green**: non enhancing tumor/ necrotic tumor (NET/NCR), **Yellow**: peritumoral edema (ED). It is interesting to note that both exhibit the same pattern: central non enhancing tumor core with surrounding enhancing ring and diffuse peritumoral edema. (Color figure online)

augmentations using the Dice Loss and deep supervision, and inferred using test time augmentation and models predictions ensembling.

Many modern "bells and whistles" were tried: short additive residual connections [11], dense blocks [14], more recent neural networks backbone based on inverted residual bottleneck [13], newer decoder structure like biFPN layer [26], or semi-supervised setting using brain dataset from the Medical Decathlon [24]. None of these refinements led to significant improvement on the local validation set. We hypothesize that this was probably due to GPU memory constraints. Indeed, while these layers improve the model accuracy at a relatively small parameter cost, it increases significantly the size of the activation maps of the model, forcing us to use smaller networks (reduction of the number of output channels per convolutional layer). Reducing the crop size of the patch was not an option as this would have most probably reduced the network performance due to the lack of context. Moreover, all of these additions led to a significant increase of the training time, reducing the searchable space in the limited timeframe of the challenge.

Stochastic weight averaging at the end of the training was the most notable refinement we used. This training scheme was a remnant from the mean teacher semi-supervised training [28]. We did not benchmark its real potential but expect it to produce a more generalizable model, to prevent from overfitting on the training set and to remember the noisy labels. Indeed, it has been shown that a high learning rate could prevent such behavior, and we expect that our training benefits from the multiple learning rate restarts [27].

Notably, while our results were not state of the art for the BraTS 2020 challenge, the segmentation performance of our method is in the usual range of inter-rater agreement for lesion segmentation [7,25] and could already be valuable for clinical use. As an example, Fig. 4 zooms in the tumor segmentation of the first two annotations of Fig. 3 (respectively manual ground truth annotations and best validation case).

5   Conclusion

The task of brain tumor segmentation, while challenging, can be solved with good accuracy using 3D U-Net like neural network architecture, with a carefully crafted pre-processing, training and inference procedure. We open-sourced our training pipeline at https://github.com/lescientifik/open_brats2020, allowing future researchers to build upon our findings, and improve our segmentation performance .

References

1. Jiang, Z., Ding, C., Liu, M., Tao, D.: Two-stage cascaded U-Net: 1st place solution to BraTS challenge 2019 segmentation task. In: Crimi, A., Bakas, S. (eds.) BrainLes 2019. LNCS, vol. 11992, pp. 231–241. Springer, Cham (2020). https://doi.org/10.1007/978-3-030-46640-4_22
2. Athiwaratkun, B., Finzi, M., Izmailov, P., Wilson, A.G.: There are many consistent explanations of unlabeled data: why you should average, p. 22 (2019)
3. Bakas, S., et al.: Advancing the cancer genome atlas glioma MRI collections with expert segmentation labels and radiomic features. Sci. Data **4**, 170117 (2017). https://doi.org/10.1038/sdata.2017.117
4. Bakas, S., et al.: Identifying the best machine learning algorithms for brain tumor segmentation, progression assessment, and overall survival prediction in the BRATS challenge. arXiv:1811.02629 [cs, stat], April 2019
5. Bakas, S., et al.: Segmentation labels and radiomic features for the pre-operative scans of the TCGA-GBM collection. Cancer Imaging Arch. (2017). https://doi.org/10.7937/K9/TCIA.2017.GJQ7R0EF
6. Bakas, S., et al.: Segmentation labels and radiomic features for the pre-operative scans of the TCGA-LGG collection. Cancer Imaging Arch. (2017). https://doi.org/10.7937/K9/TCIA.2017.KLXWJJ1Q
7. Chassagnon, G., et al.: AI-Driven quantification, staging and outcome prediction of COVID-19 pneumonia. Med. Image Anal. 101860 (2020). https://doi.org/10.1016/j.media.2020.101860, http://www.sciencedirect.com/science/article/pii/S1361841520302243
8. Chen, L.C., Papandreou, G., Kokkinos, I., Murphy, K., Yuille, A.L.: DeepLab: semantic image segmentation with deep convolutional nets, atrous convolution, and fully connected CRFs. arXiv:1606.00915 [cs], May 2017
9. Dercle, L., Henry, T., Carré, A., Paragios, N., Deutsch, E., Robert, C.: Reinventing radiation therapy with machine learning and imaging bio-markers (radiomics): state-of-the-art, challenges and perspectives. Methods (2020). https://doi.org/10.1016/j.ymeth.2020.07.003, http://www.sciencedirect.com/science/article/pii/S1046202319303184

10. Fedorov, A., et al.: 3D slicer as an image computing platform for the quantitative imaging network. Magn. Reson. Imaging **30**(9), 1323–1341 (2012). https://doi.org/10.1016/j.mri.2012.05.001, https://www.ncbi.nlm.nih.gov/pmc/articles/PMC3466397/

11. He, K., Zhang, X., Ren, S., Sun, J.: Deep residual learning for image recognition. arXiv:1512.03385 [cs], December 2015

12. He, T., Zhang, Z., Zhang, H., Zhang, Z., Xie, J., Li, M.: Bag of tricks for image classification with convolutional neural networks. arXiv:1812.01187 [cs], December 2018

13. Howard, A.G., et al.: MobileNets: efficient convolutional neural networks for mobile vision applications. arXiv:1704.04861 [cs], April 2017

14. Huang, G., Liu, Z., van der Maaten, L., Weinberger, K.Q.: Densely connected convolutional networks. arXiv:1608.06993 [cs], January 2018

15. Huttenlocher, D., Klanderman, G., Rucklidge, W.: Comparing images using the Hausdorff distance. IEEE Trans. Pattern Anal. Mach. Intell. **15**(9), 850–863 (1993). https://doi.org/10.1109/34.232073. Conference Name: IEEE Transactions on Pattern Analysis and Machine Intelligence

16. Ioffe, S., Szegedy, C.: Batch normalization: accelerating deep network training by reducing internal covariate shift, February 2015. https://arxiv.org/abs/1502.03167v3

17. Kingma, D.P., Ba, J.: Adam: a method for stochastic optimization. arXiv:1412.6980 [cs], January 2017

18. Liu, L., et al.: On the variance of the adaptive learning rate and beyond. arXiv:1908.03265 [cs, stat], April 2020

19. Menze, B.H., Jakab, A., Bauer, S., Kalpathy-Cramer, J., Farahani, K., Kirby, J., et al.: The multimodal brain tumor image segmentation benchmark (BRATS). IEEE Trans. Med. Imaging **34**(10), 1993–2024 (2015). https://doi.org/10.1109/TMI.2014.2377694

20. Milletari, F., Navab, N., Ahmadi, S.A.: V-Net: fully convolutional neural networks for volumetric medical image segmentation. arXiv:1606.04797 [cs], June 2016

21. Nair, V., Hinton, G.E.: Rectified linear units improve restricted Boltzmann machines. In: ICML 2010, Haifa, Israel, no. 8, pp. 807–814. Omnipress, Madison (2010). ISBN 9781605589077

22. Niyazi, M., et al.: ESTRO-ACROP guideline "target delineation of glioblastomas". Radiother. Oncol. **118**(1), 35–42 (2016)

23. Qin, X., Zhang, Z., Huang, C., Gao, C., Dehghan, M., Jagersand, M.: BASNet: boundary-aware salient object detection. In: 2019 IEEE/CVF Conference on Computer Vision and Pattern Recognition (CVPR), Long Beach, CA, USA, pp. 7471–7481. IEEE, June 2019. https://doi.org/10.1109/CVPR.2019.00766, https://ieeexplore.ieee.org/document/8953756/

24. Simpson, A.L., et al.: A large annotated medical image dataset for the development and evaluation of segmentation algorithms. arXiv:1902.09063 [cs, eess], February 2019

25. Tacher, V., et al.: Semiautomatic volumetric tumor segmentation for hepatocellular carcinoma: comparison between C-arm cone beam computed tomography and MRI. Acad. Radiol. **20**(4), 446–452 (2013). https://doi.org/10.1016/j.acra.2012.11.009. http://www.sciencedirect.com/science/article/pii/S107663321200606X

26. Tan, M., Pang, R., Le, Q.V.: EfficientDet: scalable and efficient object detection. arXiv:1911.09070 [cs, eess], July 2020

27. Tanaka, D., Ikami, D., Yamasaki, T., Aizawa, K.: Joint optimization framework for learning with noisy labels. arXiv:1803.11364 [cs, stat], March 2018

28. Tarvainen, A., Valpola, H.: Mean teachers are better role models: weight-averaged consistency targets improve semi-supervised deep learning results. arXiv:1703.01780 [cs, stat], April 2018

29. Ulyanov, D., Vedaldi, A., Lempitsky, V.: Instance normalization: the missing ingredient for fast stylization. arXiv:1607.08022 [cs], November 2017

30. Wang, L., Lee, C.Y., Tu, Z., Lazebnik, S.: Training deeper convolutional networks with deep supervision. arXiv:1505.02496 [cs], May 2015

31. Woo, S., Park, J., Lee, J.Y., Kweon, I.S.: CBAM: convolutional block attention module. arXiv:1807.06521 [cs], July 2018

32. Wu, Y., He, K.: Group normalization. arXiv:1803.08494 [cs], June 2018

33. Yong, H., Huang, J., Hua, X., Zhang, L.: Gradient centralization: a new optimization technique for deep neural networks. arXiv:2004.01461 [cs], April 2020. Version: 2

34. Zhang, M.R., Lucas, J., Hinton, G., Ba, J.: Lookahead optimizer: k steps forward, 1 step back. arXiv:1907.08610 [cs, stat], December 2019

35. Çiçek, O., Abdulkadir, A., Lienkamp, S.S., Brox, T., Ronneberger, O.: 3D U-Net: learning dense volumetric segmentation from sparse annotation. arXiv:1606.06650 [cs], June 2016

Brain Tumour Segmentation Using a Triplanar Ensemble of U-Nets on MR Images

Vaanathi Sundaresan[1(✉)] [ID], Ludovica Griffanti[1,2] [ID], and Mark Jenkinson[1,3,4] [ID]

[1] Wellcome Centre for Integrative Neuroimaging, Oxford Centre for Functional MRI of the Brain, Nuffield Department of Clinical Neurosciences, University of Oxford, Oxford, UK
vaanathi.sundaresan@ndcn.ox.ac.uk
[2] Wellcome Centre for Integrative Neuroimaging, Oxford Centre for Human Brain Activity, Department of Psychiatry, University of Oxford, Oxford, UK
[3] Australian Institute for Machine Learning (AIML), School of Computer Science, The University of Adelaide, Adelaide, Australia
[4] South Australian Health and Medical Research Institute (SAHMRI), Adelaide, Australia
https://www.ndcn.ox.ac.uk/team/vaanathi-sundaresan

Abstract. Gliomas appear with wide variation in their characteristics both in terms of their appearance and location on brain MR images, which makes robust tumour segmentation highly challenging, and leads to high inter-rater variability even in manual segmentations. In this work, we propose a triplanar ensemble network, with an independent tumour core prediction module, for accurate segmentation of these tumours and their sub-regions. On evaluating our method on the MICCAI Brain Tumor Segmentation (BraTS) challenge validation dataset, for tumour sub-regions, we achieved a Dice similarity coefficient of 0.77 for both enhancing tumour (ET) and tumour core (TC). In the case of the whole tumour (WT) region, we achieved a Dice value of 0.89, which is on par with the top-ranking methods from BraTS'17-19. Our method achieved an evaluation score that was the equal 5th highest value (with our method ranking in 10th place) in the BraTS'20 challenge, with mean Dice values of 0.81, 0.89 and 0.84 on ET, WT and TC regions respectively on the BraTS'20 unseen test dataset.

Keywords: Tumour segmentation · Triplanar ensemble · U-Net · Brain MRI

1 Introduction

Gliomas, the most common class of brain tumours, occur with different levels of aggressiveness with highly heterogeneous sub-regions including invaded

---

L. Griffanti and M. Jenkinson—Contributed equally to this work.

© Springer Nature Switzerland AG 2021
A. Crimi and S. Bakas (Eds.): BrainLes 2020, LNCS 12658, pp. 340–353, 2021.
https://doi.org/10.1007/978-3-030-72084-1_31

edematous tissue or peritumoral edema (ED) and tumour core region including necrotic core (NCR), non-enhancing tumour (NET), and enhancing tumour
(ET) [1,2]. Accurate and reproducible automated detection of gliomas would aid
in the timely diagnosis and staging of tumours in a clinical setting, and reliable
analysis in large population studies. However, the intrinsic histological variations
of gliomas are further complicated by the heterogeneity in characteristics (e.g.
intensity) of tumours on MRI scans. Various sub-regions of gliomas occur with
wide variations in their appearance and shape depending on their biological conditions, making their segmentation highly challenging and often leading to high
inter-rater variability even in expert clinicians' segmentations across different
datasets [1].

The MICCAI Brain Tumor Segmentation (BraTS) challenges aim to provide accurate segmentation of brain tumours on multimodal MR images [1–
5]. Several methods, including recent deep learning methods, have been proposed in BraTS challenges. The top ranking methods used convolutional neural network (CNN) architectures [6–9] mostly using ensemble networks [10–12]
and/or encoder-decoder frameworks [13,14]. U-Nets [15], one of the most popular encoder-decoder networks, were used successfully with accurate results for
tumour segmentation [16–19]. Regarding the model dimensions, both 2D [6] and
3D networks [7] were used with additional post-processing steps (e.g. conditional
random fields used in [7], [9]) and occasionally within multi-scale architectures
[7,8] and multi-step cascaded frameworks [12,19]. Some methods aimed to leverage the advantages of both 2D and 3D architecture by using triplanar ensembles
of CNNs [12], providing accurate segmentation with fewer parameters than 3D
networks. Further, modifications to the loss functions have been proposed [9,12]
for overcoming class imbalance, reliable tumour core detection and accurate segmentation of tumour boundaries.

We propose a fully automated deep learning method for brain tumour segmentation using a triplanar ensemble architecture consisting of a 2D U-Net in
each plane (axial, sagittal and coronal) of MR images. Our method uses a combination of loss functions in order to overcome the class imbalance and includes an
independent tumour core prediction module to refine the segmentation of tumour
core sub-regions. We study the effect of various components of our architecture
on the segmentation results by performing an ablation study. We evaluate our
method on BraTS'20 training and validation datasets, which exhibit wide heterogeneity in tumour characteristics and provides a benchmark to assess the
robustness of our segmentation method. Finally, the results of our method on
BraTS'20 test dataset shows that our method provides accurate segmentation
of tumour regions, ranking among the top 10 best performing methods of the
challenge.

2   Materials and Methods

2.1   Data

We evaluated the performance of the proposed method on the publicly available BraTS'20 dataset, consisting of pre-operative multimodal MRI scans, with

369 training cases. For each subject, the given input modalities include FLAIR, T1-weighted (T1), post-contrast T1-weighted (T1-CE) and T2-weighted (T2) images. The manual segmentations for the training dataset consists of 3 labels [1], [2]: NCR/NET, ED and ET. The input modalities (FLAIR, T1, T1-CE, T2) were already co-registered to the same anatomical template of dimension 240 × 240 × 155, interpolated to the same resolution (1 mm^3 isotropic) and skull-stripped. In addition to the training data, 125 validation cases were provided without manual segmentations (referred from now on as "unlabelled validation dataset") to enable an initial validation of the method via the challenge's online evaluation platform (using the ground truth at their end). Additionally, 166 test cases without manual segmentations (referred from now on as "unseen test dataset") were released for a duration of 48 h for the final testing via the online evaluation platform (again, using their local ground truth).

2.2   Preprocessing

We cropped the images to a standard size of 192 × 192 × 160 voxels so that field of view (FOV) is close to the brain, and applied Gaussian normalisation to the intensity values. We then extracted 2D slices from the volumes from axial, sagittal and coronal planes with dimensions of 192 × 192, 192 × 160 and 192 × 160 voxels respectively.

2.3   CNN Architecture

We used the triplanar architecture proposed in [20]. Briefly, as shown in Fig. 1, the triplanar architecture consists of three 2D U-Nets, one for each plane, taking FLAIR, T1, T1-CE and T2 slices as input channels. On the training dataset, we observed that while the manual segmentation for the cumulative tumour core (TC, which is ET + NCR/NET) (Fig. 1a) was quite consistent, there were wide variations in those of individual sub-regions (NCR/NET and ET), due to their underlying histological heterogeneity. Therefore, in order to reduce the inconsistencies in the boundaries of ET and NCR/NET labels, we modified the ground truth labels to include TC in addition to the provided labels, obtaining 4 labels: ET, ED, NCR/NET and TC. During training, we used the ET, NCR/NET and ED labels to train the U-Nets in the triplanar architecture, while we used the TC label to train an independent axial U-Net as shown in Fig. 1b. We later used the TC regions in the post-processing step (refer Sect. 2.4) to further refine the final output labels (ET, NCR/NET and ED).

We trimmed the depth of the classic U-Net [15] in each plane to a depth of 3-layers (Fig. 1b), to reduce the computational load. While axial U-Nets use 3 × 3 convolutional kernels in the initial layer, the other two U-Nets use 5 × 5 kernels. This helped to learn more generic lesion patterns, thus avoiding any discontinuities in segmentation along the z-dimension. In the ensemble model, we trained the U-Nets in each plane independently using 2D slices extracted in each plane. We used a combination of cross-entropy (CE) and Dice loss (DL) functions in order to overcome the effect of class imbalance between tumour/edema and

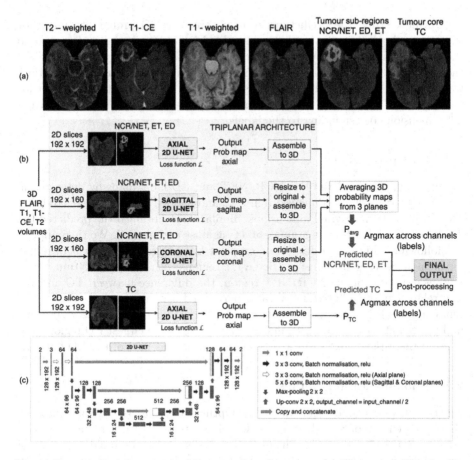

Fig. 1. Proposed triplanar ensemble network architecture. (a) Input modalities in the axial plane along with manual segmentations for NCR/NET (blue), ET (yellow), ED (red) and TC (magenta), (b) the proposed network and (c) 3-layer deep U-Net blocks used in (b). Slices with 4 channels (input modalities) were provided to all U-nets. (Color figure online)

healthy tissue. The loss function was computed batch-wise as shown below in Eq. 1.

$$L = CE + DL = - \sum_{c=1}^{C} y_c(x) \log(p_c(x)) - \frac{1}{C} \sum_{c=1}^{C} \frac{2 \times \sum_{x=1}^{N} M_c(x) \cdot PL_c(x)}{\sum_{x=1}^{N} M_c(x) + \sum_{x=1}^{N} PL_c(x)} \tag{1}$$

where $p_c(x)$ denotes the output of the soft-max layer, C is the number of classes in labels, $y_c(x) \in \{0, 1\}$ indicates the binary value at voxel x for each class, M_c indicates the manual segmentation and PL_c indicates the predicted label map obtained by determining the argmax of labels from the soft-max output.

During testing, the predictions were obtained as 2D probability maps for slices in each plane and were later assembled into 3D volumes and resized to the original dimensions. We then averaged the 3D probability volumes to get the final probability volume (P_{avg}) for the triplanar architecture. In addition, we obtained a 3D probability map (P_{TC}) from the independent axial U-Net for predicting TC label. Note that 3D probability maps P_{avg} and P_{TC} still have a 4th dimension corresponding to the labels.

2.4   Post-processing

We obtained the predicted ET, NCR/NET and ED label maps by determining the *argmax* of labels (4th dimension) in P_{avg} and padded them with zeros to bring them back to their original dimensions. Similarly, we obtained the predicted the TC label map from P_{TC} as argmax of TC against background. We then applied the following additional rules, based on prior knowledge and patterns observed in the manual segmentations: predicted ET regions with volume $<200\,mm^3$ were relabeled as part of the NCR/NET region; the difference between TC and ET was relabelled as NCR/NET. We also performed a morphological clean-up by removing small isolated noisy stray regions in ED output (volume $<200\,mm^3$ and located at a distance $>75\,mm$ from the centre of the largest ED region) and filled in the missed voxels in the TC - ED interfaces as part of the ED output.

2.5   Implementation Details

The models were trained using the Adam Optimiser with $\epsilon=10^{-4}$. We empirically chose a batch size of 8, with an initial learning rate of 1×10^{-3}, reducing it by a factor of 1×10^{-1} every 2 epochs (set empirically due to the rapid reduction of loss values at these early epochs), until it reached a fixed value of 1×10^{-5}. Every individual U-Net took ≈ 50 epochs for convergence. The networks were trained on an NVIDIA Tesla V100, taking ≈ 12 mins (for 3 planes + TC network) per epoch training/validation split of 90%/10%.

2.6   Data Augmentation

Data augmentation was applied in an online manner by randomly selecting from following transformations: translation (x/y-offset $\in$ [-10, 10]), rotation ($\theta \in$ [-10, 10]) and random noise injection (Gaussian, $\mu = 0$, $\sigma^2 \in$ [0.01, 0.09]), increasing the dataset by a factor of 2 (chosen empirically) for all planes. The hyperparameters for the transformations were randomly sampled from the above specified closed intervals using a uniform distribution.

2.7   Performance Evaluation Metrics

Metrics computed by the online evaluation platform in BraTS'20 are (i) Dice Similarity Coefficient measured as $2 \times TP/(2 \times TP + FP + FN)$, (ii) sensitivity

measured as TP/(TP + FN), (iii) specificity measured as TN/(TN + FP) and
(iv) the 95th percentile of the Hausdorff Distance (H95), where TP, FP, FN and
TN are number of true positive, false positive, false negative and true negative
voxels respectively. Regarding the ground truth labels, while the manual segmentations consist of ET, ED and NCR/NET classes, evaluation performance
metrics were determined by the evaluation platform for the following sub-regions
of tumour: (1) ET, (2) TC (NCR/NET + ET) and (3) whole tumour (WT, which
is TC + ED).

3 Experiments and Results

Cross-Validation on the Labelled Training Data: We used 369 labelled
subjects from BraTS'20 training data to perform 5-fold cross-validation with
a training-validation-testing split ratio of 255-37-73 subjects (255-37-77 for the
last fold). The results on the test splits (evaluated using the challenge online
platform) are shown in Table 1. Among the three sub-regions, we achieved the
best segmentation performance for the whole tumour (WT) with a mean Dice
value of 0.93. A few sample outputs of our method are shown along with manual
segmentations in Fig. 2.

Table 1. Results of 5-fold cross-validation on BraTS'20 Training data.

	Dice			Sensitivity			Specificity			H95 (mm)		
	ET	WT	TC	ET	WT	TC	ET	WT	TC	ET	WT	TC
Mean	0.83	0.93	0.87	0.83	0.90	0.85	0.99	0.99	0.99	20.3	3.4	6.3
Std.	0.22	0.08	0.17	0.23	0.11	0.19	.0005	.0005	.0005	80.3	5.1	28.0
Median	0.90	0.95	0.93	0.91	0.93	0.93	0.99	0.99	0.99	1.0	1.7	2.2
25 quantile	0.83	0.91	0.86	0.83	0.87	0.83	0.99	0.99	0.99	1.0	1.0	1.4
75 quantile	0.95	0.97	0.96	0.96	0.97	0.97	0.99	0.99	0.99	2.2	3.2	4.2

ET - enhancing tumour, WT - whole tumour, TC - tumour core.

Ablation Study on the Labelled Training Data: In order to determine the
effect of individual components of our architecture on the segmentation performance, we evaluated the segmentation results with the following components:
(i) axial 2D U-Net only, (ii) axial + sagittal 2D U-Nets, (iii) triplanar network
(axial + sagittal + coronal U-Nets) and (iv) triplanar network + axial U-Net for
TC label detection. We used a cross-validation strategy (with the same training-
validation-test split mentioned above) to evaluate the performance metrics. The
values of performance metrics for the ablation study are shown in Table 2 and
the corresponding boxplots are shown in Fig. 3. The segmentation of sub-regions
improved with the addition of the TC network, with significant increase in Dice
and sensitivity values ($p < 0.01$), especially for the ET and TC sub-regions. We

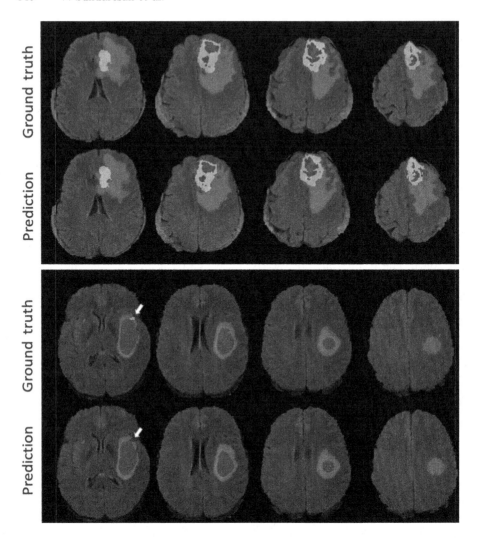

Fig. 2. Results on BraTS'20 training data from two sample subjects (top and bottom panels). Manual segmentations and predicted outputs on axial slices from two sample subjects (NCR/NET - blue, ET - yellow and ED - red). Subject in the top panel: Dice (ET/WT/TC) - 0.96/0.98/0.97, sensitivity (ET/WT/TC) - 0.95/0.98/0.99, specificity (all) - 0.99, H95 (all) - 1 mm; subject in the bottom panel: Dice (ET/WT/TC) - 0/0.98/0.94, sensitivity (ET/WT/TC) - 0/0.97/0.89, specificity (ET/WT/TC) - 1/0.99/1, H95 (ET/WT/TC) - 373.1/1/1.4. In the bottom panel, the white arrows indicate the false prediction of the ET region, leading to the dice value of 0.00. (Color figure online)

also observed significant improvement in the specificity of the WT segmentation (p < 0.001) using the triplanar architecture when compared to individual U-Nets.

Table 2. Ablation study results for BraTS'20 Training data. Mean and standard deviation (in brackets) values for various components of the method reported for tumour sub-regions. In the lower part p-values of paired two-tailed t-test results between individual pairs of components are shown, with significant values underlined.

	Dice			Sensitivity			Specificity			H95 (mm)		
	ET	WT	TC	ET	WT	TC	ET	WT	TC	ET	WT	TC
A	0.78	0.89	0.80	0.78	0.86	0.76	0.9997	0.9994	0.9997	26.6	5.7	7.3
	(0.25)	(0.10)	(0.24)	(0.26)	(0.13)	(0.26)	(.0006)	(.0007)	(.0005)	(89.7)	(7.9)	(20.9)
A+S	0.79	0.90	0.82	0.79	0.87	0.83	0.9997	0.9995	0.9995	25.9	4.9	7.4
	(0.24)	(0.10)	(0.21)	(0.26)	(0.13)	(0.23)	(.0006)	(.0006)	(.0008)	(89.7)	(8.4)	(21.4)
TP	0.79	0.90	0.83	0.79	0.86	0.82	0.9997	0.9996	0.9997	25.8	4.6	7.0
	(0.24)	(0.09)	(0.21)	(0.26)	(0.14)	(0.23)	(.0006)	(.0006)	(.0006)	(89.6)	(6.1)	(28.0)
TP+TC	0.82	0.92	0.87	0.83	0.90	0.85	0.9998	0.9997	0.9997	20.3	3.4	6.3
	(0.22)	(0.07)	(0.17)	(0.23)	(0.11)	(0.20)	(.0005)	(.0005)	(.0005)	(80.1)	(5.3)	(28.0)
p-values												
A vs A+S	0.36	0.20	0.10	0.48	0.60	≤0.001	0.43	0.006	≤0.001	0.90	0.22	0.94
A vs TP	0.37	0.17	0.03	0.47	0.77	0.004	0.42	≤0.001	0.19	0.90	0.03	0.92
A vs TP+TC	0.005	≤0.001	≤0.001	0.007	≤0.001	≤0.001	0.14	≤0.001	0.77	0.31	≤0.001	0.60
A+S vs TP	0.99	0.93	0.56	0.99	0.82	0.60	0.99	0.40	0.01	0.99	0.48	0.87
A+S vs TP+TC	0.06	≤0.001	0.002	0.04	≤0.001	0.08	0.50	0.001	≤0.001	0.38	0.002	0.55
TP vs TP+TC	0.06	≤0.001	0.01	0.04	≤0.001	0.02	0.50	0.02	0.30	0.38	0.004	0.70

Tumour sub-regions: ET - enhancing tumour, WT - whole tumour, TC - tumour core.
Methods: A - axial, A+S - axial + sagittal, TP - triplanar, TP+TC - triplanar + TC network.

Evaluation on the Unlabelled Validation Data: The models trained from the 5-fold cross-validation of the training data were applied to the validation data and the final predictions were obtained using majority voting (from 5 models). The results obtained from the online evaluation platform are shown in Table 3. A few sample validation data outputs of our method are shown along with input modalities in Fig. 4. The results followed a trend similar to the training data obtaining the best results for WT segmentation with mean Dice and H95 values of 0.89 and 4.4 respectively.

Table 3. Results for BraTS'20 Validation data.

	Dice			Sensitivity			Specificity			H95 (mm)		
	ET	WT	TC	ET	WT	TC	ET	WT	TC	ET	WT	TC
Mean	0.77	0.89	0.77	0.76	0.87	0.73	0.99	0.99	0.99	29.4	4.4	15.3
Std.	0.27	0.09	0.27	0.29	0.13	0.29	.0005	.0009	.0003	96.2	5.4	57.3
Median	0.87	0.93	0.90	0.88	0.91	0.86	0.99	0.99	0.99	2.0	3.0	3.0
25 quantile	0.77	0.89	0.73	0.74	0.85	0.61	0.99	0.99	0.99	1.0	2.0	1.7
75 quantile	0.91	0.95	0.94	0.94	0.95	0.93	0.99	0.99	0.99	3.6	4.6	8.5

ET - enhancing tumour, WT - whole tumour, TC - tumour core.

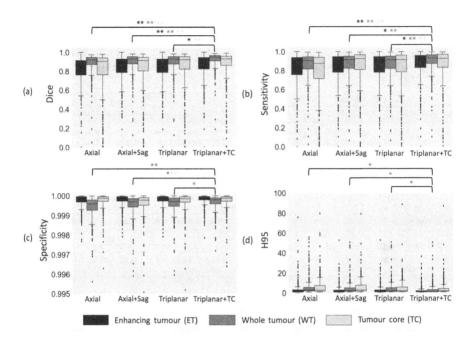

Fig. 3. Boxplots of results from the ablation study, showing (a) Dice, (b) sensitivity, (c) specificity and (d) H95 for axial, axial+sagittal, triplanar and triplanar+TC cases. Significant differences in performance metrics are indicated by asterisks (* - $p < 0.01$, ** - $p < 0.001$) in the corresponding colours for tumour sub-regions. Note that only the significant differences in performance metrics between triplanar+TC and other cases are shown (for all combinations of individual pairs, refer to Table 2).

Results on the Unseen Test Data: Finally, the results obtained from the online evaluation platform on the BraTS'20 test data are shown in table 4. Similar to the validation stage, majority voting on the results of 5 models (from 5-fold cross-validation) was used to predict the tumour region labels. As seen from the table, the Dice values for the WT region was higher than for the ET and TC regions as in the case of the validation stage. However, the Dice and sensitivity values for the ET and TC regions were considerably higher than those on the validation data, and were almost on par with those obtained with the fold-validation on the training dataset.

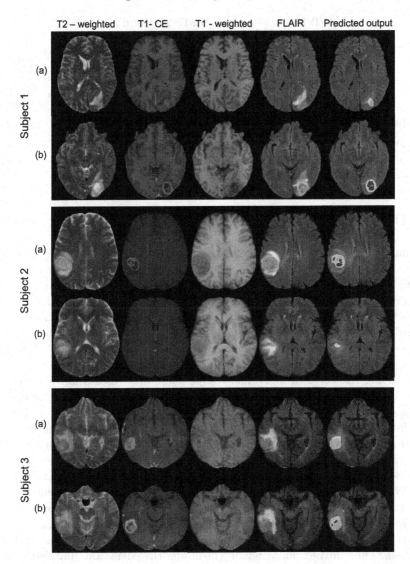

Fig. 4. Results on BraTS'20 validation data from three sample subjects. Predicted outputs on two axial slices (a) and (b) from from three sample subjects (NCR/NET - blue, ET - yellow and ED - red), along with the provided input modalities: T2-weighted, T1-CE, T1 and FLAIR. Subject1: Dice (ET/WT/TC) - 0.86/0.97/0.96, sensitivity (ET/WT/TC) - 0.81/0.96/0.98, specificity (all) - 0.99, H95 (ET/WT/TC) - 1.4/1/1 mm; Subject2: Dice (ET/WT/TC) - 0.93/0.97/0.96, sensitivity (ET/WT/TC) - 0.93/0.99/0.97, specificity (all) - 0.99, H95 (all) - 1 mm; Subject3: Dice (ET/WT/TC) - 0.95/0.93/0.95, sensitivity (ET/WT/TC) - 0.97/0.99/0.98, specificity (all) - 0.99, H95 (ET/WT/TC) - 1/2/1 mm. (Color figure online)

Table 4. Results for BraTS'20 Test data.

	Dice			Sensitivity			Specificity			H95 (mm)		
	ET	WT	TC	ET	WT	TC	ET	WT	TC	ET	WT	TC
Mean	0.81	0.89	0.84	0.84	0.88	0.83	0.99	0.99	0.99	15.3	6.3	15.2
Std.	0.20	0.11	0.24	0.21	0.13	0.25	.0004	.0007	.0007	69.5	28.9	63.9
Median	0.85	0.92	0.92	0.92	0.92	0.94	0.99	0.99	0.99	1.4	2.8	2.2
25 quantile	0.78	0.87	0.87	0.83	0.86	0.82	0.99	0.99	0.99	1.0	1.7	1.4
75 quantile	0.93	0.95	0.96	0.96	0.96	0.97	0.99	0.99	0.99	2.2	4.8	3.8

ET - enhancing tumour, WT - whole tumour, TC - tumour core.

4   Discussion and Conclusions

In this work, we proposed an end-to-end automated tumour segmentation method using a triplanar ensemble architecture of 2D U-Nets. Our method segmented ET, WT and TC sub-regions of the tumour with dice values of 0.83, 0.93 and 0.87 on the training dataset. On an independent unlabelled validation dataset, our method achieved Dice values of 0.77, 0.89 and 0.77 for ET, WT and TC sub-regions respectively. On the BraTS'20 unseen test dataset, our method achieved the Dice values of 0.81, 0.89 and 0.84 for the ET, WT and TC regions respectively.

Studying the effect of individual components on segmentation performance aided in better understanding the proposed method. For all the tissue classes, a single axial network performs the worst, probably due to the lack of contextual information from the contiguous slices. The triplanar architecture provides better performance than the individual 2D networks with higher Dice and sensitivity values. Moreover, we achieved significant improvement in the performance metrics with the addition of a specific TC network, especially for the TC class. The WT segmentation improves with the addition of each component and the significantly lower H95 values indicate a more precise tumour segmentation.

From the results on the training and validation data, performance for the WT is higher than for the other two classes, indicating that the method segments the cumulative tumour region (including the edematous/invaded region) with higher accuracy than the differentiation between core and edema. However, a few misclassified tumour core regions were due to cases where the ET class is very small (as indicated by white arrows in Fig. 2). This results in ET either being incorrectly predicted (by the network) or falsely relabeled (in the post-processing step) as a part of the NCR/NET class. This substantially affects the Dice score for the ET class for the subject. In general, we observed that applying generalisable/consistent prior information or post-processing operations for TC and ET classes was not possible due to the wide range of tumour characteristics and variation in ground truth, which presented a major challenge for the segmentation task. Interestingly, for the TC region, our method performed better on the training and the unseen test datasets when compared to the validation dataset, as observed from the higher values of the metrics in Tables 1 and 4 as compared to those in Table 3. The higher values on the training dataset could be due to the fact that the cross-validation results are generally more prone to over-fitting, and hence are less reliable when compared to the results on the unseen validation data. However, our method also provided consistently good performance on the unseen test dataset (Table 4). Since this was an unseen dataset, we cannot exclude the possibility that the tumour characteristics and image intensity profiles of the test dataset could be more similar to those of the training dataset.

On the BraTS'20 unseen test dataset, our method achieved higher Dice values compared to the validation Dice values, obtaining an evaluation score that was equal 5th highest and 10th place in the overall ranking in the challenge. As an indirect comparison with existing methods using different validation datasets, our method achieved a Dice value of 0.89 (on the BraTS'20 validation dataset) for the WT class, comparable to the top-ranking methods of BraTS'17-19 ($\sim$0.90) [7,14,21], [18,19] and a Dice value of 0.77 for the ET class, on par with top-ranking methods of BraTS'17 ($\sim$0.78) [7,21]. It is worth noting that even though the previous BraTS challenges used data from different subjects, this indirect comparison could be quite useful in determining the potential of our method, since the comparison involves the same task and type of data.

Summarising, our proposed triplanar ensemble method achieves accurate segmentation of whole tumours and their sub-regions on brain MR images from multimodal data of the BraTS'20 challenge. After the challenge, we will make our method publicly available as a Docker container that could be used as an independent tumour segmentation tool. Future directions include further improvement of tumour sub-region segmentation by leveraging the salient features (e.g. using attention networks).

Acknowledgements. The authors of this paper declare that their method for the BraTS'20 challenge has not used any pre-trained models nor additional datasets other than those provided by the organizers. This work was supported by Wellcome Centre for Integrative Neuroimaging, which has core funding from the Wellcome Trust (203139/Z/16/Z). VS is supported by Wellcome Centre for Integrative Neuroimaging (203139/Z/16/Z). LG is supported by the Oxford Parkinson's Disease Centre (Parkinson's UK Monument Discovery Award, J-1403), the MRC Dementias Platform UK (MR/L023784/2), and the National Institute for Health Research (NIHR) Oxford Health Biomedical Research Centre (BRC). MJ is supported by the National Institute for Health Research (NIHR), Oxford Biomedical Research Centre (BRC) and Wellcome Trust (215573/Z/19/Z). The computational aspects of this research were supported by the Wellcome Trust Core Award (203141/Z/16/Z) and the NIHR Oxford BRC. The views expressed are those of the authors and not necessarily those of the NHS, the NIHR or the Department of Health.

References

1. Menze, B.H., Jakab, A., Bauer, S., Kalpathy-Cramer, J., Farahani, K., Kirby, J., et al.: The multimodal brain tumor image segmentation benchmark (BRATS). IEEE Trans. Med. Imaging **34**(10), 1993–2024 (2015). https://doi.org/10.1109/TMI.2014.2377694
2. Bakas, S., Reyes, M., Jakab, A., Bauer, S., Rempfler, M., Crimi, A., et al.: Identifying the best machine learning algorithms for brain tumor segmentation, progression assessment, and overall survival prediction in the BRATS challenge, arXiv preprint arXiv:1811.02629 (2018)
3. Bakas, S., Akbari, H., Sotiras, A., Bilello, M., Rozycki, M., Kirby, J.S., et al.: Advancing The Cancer Genome Atlas glioma MRI collections with expert segmentation labels and radiomic features. Nat. Sci. Data **4**, 170117 (2017). https://doi.org/10.1038/sdata.2017.117
4. Bakas, S., Akbari, H., Sotiras, A., Bilello, M., Rozycki, M., Kirby, J., et al.: Segmentation labels and radiomic features for the pre-operative scans of the TCGA-GBM collection. Cancer Imaging Arch. (2017). https://doi.org/10.7937/K9/TCIA.2017.KLXWJJ1Q
5. Bakas, S., Akbari, H., Sotiras, A., Bilello, M., Rozycki, M., Kirby, J., et al.: Segmentation labels and radiomic features for the pre-operative scans of the TCGA-LGG collection. Cancer Imaging Arch. (2017). https://doi.org/10.7937/K9/TCIA.2017.GJQ7R0EF
6. Pereira, S., Pinto, A., Alves, V., Silva, C.A.: Brain tumor segmentation using convolutional neural networks in MRI images. IEEE Trans. Med. Imaging **35**(5), 1240–1251 (2016)
7. Kamnitsas, K., Ledig, C., Newcombe, V.F., Simpson, J.P., Kane, A.D., Menon, D.K., et al.: Efficient multi-scale 3D CNN with fully connected CRF for accurate brain lesion segmentation. Med. Image Anal. **36**, 61–78 (2017)
8. Havaei, M., Davy, A., Warde-Farley, D., Biard, A., Courville, A., Bengio, Y., et al.: Brain tumor segmentation with deep neural networks. Med. Image Anal. **35**, 18–31 (2017)
9. Shen, H., Wang, R., Zhang, J., McKenna, S.J.: Boundary-aware fully convolutional network for brain tumor segmentation. In: Descoteaux, M., Maier-Hein, L., Franz, A., Jannin, P., Collins, D.L., Duchesne, S. (eds.) MICCAI 2017. LNCS, vol. 10434, pp. 433–441. Springer, Cham (2017). https://doi.org/10.1007/978-3-319-66185-8_49
10. Kamnitsas, K., et al.: Ensembles of multiple models and architectures for robust brain tumour segmentation. In: Crimi, A., Bakas, S., Kuijf, H., Menze, B., Reyes, M. (eds.) BrainLes 2017. LNCS, vol. 10670, pp. 450–462. Springer, Cham (2018). https://doi.org/10.1007/978-3-319-75238-9_38
11. McKinley, R., Meier, R., Wiest, R.: Ensembles of densely-connected CNNs with label-uncertainty for brain tumor segmentation. In: Crimi, A., Bakas, S., Kuijf, H., Keyvan, F., Reyes, M., van Walsum, T. (eds.) BrainLes 2018. LNCS, vol. 11384, pp. 456–465. Springer, Cham (2019). https://doi.org/10.1007/978-3-030-11726-9_40
12. McKinley, R., Rebsamen, M., Meier, R., Wiest, R.: Triplanar ensemble of 3D-to-2D CNNs with label-uncertainty for brain tumor segmentation. In: Crimi, A., Bakas, S. (eds.) BrainLes 2019. LNCS, vol. 11992, pp. 379–387. Springer, Cham (2020). https://doi.org/10.1007/978-3-030-46640-4_36
13. Yang, T., Ou, Y., Huang, T.: Automatic segmentation of brain tumor from MR images using SegNet: selection of training data sets. In: Proceedings of the 6th MICCAI BraTS Challenge, pp. 309–312 (2017)

14. Myronenko, A.: 3D MRI brain tumor segmentation using autoencoder regularization. In: Crimi, A., Bakas, S., Kuijf, H., Keyvan, F., Reyes, M., van Walsum, T. (eds.) BrainLes 2018. LNCS, vol. 11384, pp. 311–320. Springer, Cham (2019). https://doi.org/10.1007/978-3-030-11726-9_28

15. Ronneberger, O., Fischer, P., Brox, T.: U-Net: convolutional networks for biomedical image segmentation. In: Navab, N., Hornegger, J., Wells, W.M., Frangi, A.F. (eds.) MICCAI 2015. LNCS, vol. 9351, pp. 234–241. Springer, Cham (2015). https://doi.org/10.1007/978-3-319-24574-4_28

16. Kim, G.: Brain tumor segmentation using deep fully convolutional neural networks. In: Crimi, A., Bakas, S., Kuijf, H., Menze, B., Reyes, M. (eds.) BrainLes 2017. LNCS, vol. 10670, pp. 344–357. Springer, Cham (2018). https://doi.org/10.1007/978-3-319-75238-9_30

17. Isensee, F., Kickingereder, P., Wick, W., Bendszus, M., Maier-Hein, K.H.: Brain tumor segmentation and radiomics survival prediction: contribution to the BRATS 2017 challenge. In: Crimi, A., Bakas, S., Kuijf, H., Menze, B., Reyes, M. (eds.) BrainLes 2017. LNCS, vol. 10670, pp. 287–297. Springer, Cham (2018). https://doi.org/10.1007/978-3-319-75238-9_25

18. Isensee, F., Kickingereder, P., Wick, W., Bendszus, M., Maier-Hein, K.H.: No new-net. In: Crimi, A., Bakas, S., Kuijf, H., Keyvan, F., Reyes, M., van Walsum, T. (eds.) BrainLes 2018. LNCS, vol. 11384, pp. 234–244. Springer, Cham (2019). https://doi.org/10.1007/978-3-030-11726-9_21

19. Jiang, Z., Ding, C., Liu, M., Tao, D.: Two-stage cascaded U-Net: 1st place solution to BraTS challenge 2019 segmentation task. In: Crimi, A., Bakas, S. (eds.) BrainLes 2019. LNCS, vol. 11992, pp. 231–241. Springer, Cham (2020). https://doi.org/10.1007/978-3-030-46640-4_22

20. Sundaresan, V., Zamboni, G., Rothwell, P.M., Jenkinson, M., Griffanti, L.: Triplanar ensemble U-Net model for white matter hyperintensities segmentation on MR images. BioRxiv (2020). https://doi.org/10.1101/2020.07.24.219485

21. Wang, G., Li, W., Ourselin, S., Vercauteren, T.: Automatic brain tumor segmentation using cascaded anisotropic convolutional neural networks. In: Crimi, A., Bakas, S., Kuijf, H., Menze, B., Reyes, M. (eds.) BrainLes 2017. LNCS, vol. 10670, pp. 178–190. Springer, Cham (2018). https://doi.org/10.1007/978-3-319-75238-9_16

MRI Brain Tumor Segmentation Using a 2D-3D U-Net Ensemble

Jaime Marti Asenjo[1]([✉]) and Alfonso Martinez-Larraz Solís[2]

[1] Medical Physics Department, HM Sanchinarro Hospital, c\ Oña 10, 28050 Madrid, Spain
jmartiasenjo@hmhospitales.com
[2] Data Science, Agroviz Inc., Madrid, Spain
alfonso@agroviz.inc

Abstract. Three 2D networks, one for each patient-plane (axial, sagittal and coronal) plus a 3-D network were ensemble for tumor segmentation over MRI images, with final Dice scores of 0.75 for the enhancing tumor (ET), 0.81 whole tumor (WT) and 0.78 for tumor core (TC). A survival prediction model was design on Matlab, based on features extracted from the automatic segmentation. Gross tumor size and location seem to play a major role on survival prediction. A final accuracy of 0.617 was achieved.

1 Task 1: Brain Tumor Segmentation in MRI Scans

1.1 Materials and Methods

The goal of this task is the segmentation of the glioma tumor volume and differentiate among its three parts: NCR/NET (necrotic and non-enhancing tumor), ED (peritumoral edema) and ET (enhancing tumor). For that purpose, scans of 369 patients are provided [1–5]. For each patient there are four different MRI sequences (T1, T1GD: Post-contrast T1-weighted, T2-weighted, T2-Flair: Fluid Attenuated Inversion Recovery) and a segmentation.

Convolutional Neural Networks (CNNs) are the selected model for the task, as they have proven the best performance in segmentation tasks and especially in medical image segmentation [6, 7]. Our approach is based on a combination of 2D and 3D networks.

1.2 2D Networks

For each patient three different dataset are generated, one for each 2D anatomical plane: axial, sagittal and coronal. For each one of the three datasets we have applied a UNet based architecture because it has been proven to show some of the best performance at medical images and organ segmentation [6]. A fully convolutional neural network based in two parts defines this network: the encoder, which will extract most of the important features from the image, and the decoder, that will expand the featured map until the output can be compared to the ground truth.

U-Nets also have links between these two parts to avoid gradient vanishing [8]. Our network is based on a U-Net architecture with major changes, both in the encoder and the decoder.

A. Crimi and S. Bakas (Eds.): BrainLes 2020, LNCS 12658, pp. 354–366, 2021.
https://doi.org/10.1007/978-3-030-72084-1_32

Encoder. As it has been previously reported [9], using a pre-trained network on ImageNet [10] as the U-net encoder, improves the detection of low-level features. We have used efficientNet, designed by Mingxing et al. [11], as a pre-trained encoder. The thought behind this was to increase both the number of layers (deep) and the number of channels for every layer (wide). Using compound scaling, Mingxing et al. built 8 models with different sizes and named EfficientNet-b (0–7) (Fig. 1). This network presents multiple advantages: The number of parameters is small compared to other networks, it also has shown great performance at classifying images and therefore, detecting image features. Moreover, overfitting and underfitting might be avoided as there are different size networks.

After running some tests, the EfficientNet-b5, b6 and b7 showed the best performance for this task. These were the models chosen as the encoder of the final networks and named effUNet-b5, effUNet-b6 and effUNet-b7.

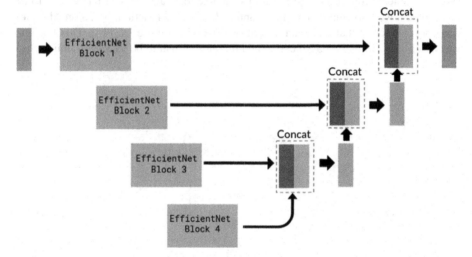

Fig. 1. EffUnet architecture

Decoder. As a decoder we implemented an expansion of the feature map in five steps, doubling the size each time. Each of the expansions is contained within a decoder block as shown in Eq. 1.

$$Conv3x3 -> ELU -> Upsample2x2 -> Conv3x3 \qquad (1)$$

Finally, there is a convolutional layer with a size-1 kernel, that returns a HxWxC matrix, being H and W the size of the original image and C the number of classes (different organs). We used a softmax activation function to calculate the probability of each class.

Input. As it has been pre-trained on ImageNet, decoder Input layer was a 3-channel layer. It has been modified to a 4-channel in order to accept all MRI sequences for each patient. The 4th layer was initialized with random weights from a Gaussian distribution, with the same mean and standard deviation as the other three pre-trained layers.

Training. Data set has been divided into a train dataset (85%) and a validation dataset (15%) in order to adjust the learning rate and control the underfitting and the overfitting.

Data augmentation was used in the training set and is based on elastic transformations, rotations, horizontal and vertical flips and zoom-in and zoom-out. For this task the Albumentations library [12] was used.

The final loss function was formed by the combination of three different loss functions (Eq. 2): Distribution based loss: First, a Cross Entropy Loss like (log likelihood loss) that has shown good performance on multiclass segmentation, second, a region-based function: DICE loss (Lovasz loss) that improves the convergence and finally a boundary based loss function: Hausdorff Distance (HD) loss.

However, the problem about Hausdorff Distance as loss function is that the Stocastic Grandient Algorithm requires a diferentiable loss function and HD is not. To fix this problem we implemented a similar and differentiable function, based on "distance transform" function that has been described Davood Karimi et al. [13]. This function was described for binary classification and therefore It had to be adapted to multi-class purposes.

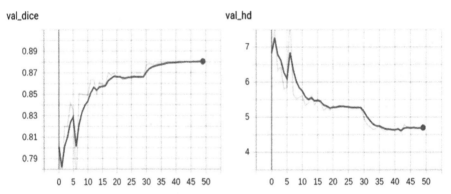

Fig. 2. Example of DICE and Hausdorff distance in validation set using the Soft Dice Loss in the first 30 epochs and the Soft Dice HD Loss in the last 20 epochs.

$$SoftDiceHDLoss = \alpha_1 \cdot CrossEntropy + \alpha_2 \cdot (1 - DICE) + \alpha_3 \cdot HD \quad (2)$$

Where α_1, α_2, and α_3 are coefficients to balance components so they have the same importance in the loss calculation. α_1 was assigned a fixed value, α_2 was equal to 1-α_1 and α_3 was equal to $\alpha_1 \cdot CrossEntropy + \alpha_2 \cdot (1 - DICE)$ divided by HD. A final value for α_1 was 0.5. This function has a high cost of computation time, but showed better HD and DICE results as seen in Fig. 2. This loss function was calculated as a single function, and the backpropagation was done from this value. The optimizer was Adam [14] with an initial learning rate of (LR) = 1E−04. During all the training there was a LR-schedule that divided the LR by 2 every five epochs if the validation loss had not decreased to at least 0.001. The batch size was 20 and all models were trained during 100 epochs each one with a softmax activation at the end of the model to calculate the class probabilities.

Inference. Segmentation was obtained for all three anatomical planes. In each plane predictions are calculated for each one of the three networks. These three different inferences in each plane give us one volume per class contained in each voxel, the probability that each voxel would belong to one of the different classes in which the model was being trained.

1.3 3D Network

3D models required large GPU memory specifications. To avoid this limitation, every image set has been divided into 12 parts, size $80 \times 80 \times 77$ pixels. All patches are used as train and validation sets. We did not discard any of the patches even if they did not contain any region of interest. These smaller image datasets allowed us to work with a standard UNet-3D (Fig. 3), with a 1024 central feature map and a batch size of 2.

Design. The 3D network is a standard UNet-3D based on the model proposed by MedicalZoo [15] but modified to a 1024 channels central feature map.

Input. To avoid memory issues all data has been divided as it was mentioned. A channel for each MRI sequence is added, so the final input data is a $4 \times 80 \times 80 \times 77$ pixel volume.

Training. Dataset was divided as for the 2D models (85% train, 15% validation) and the same data augmentation was performed. A similar loss function to the 2D model was designed, however the Haussdorff distance has a too high computational cost and

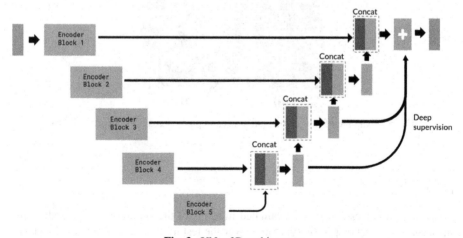

Fig. 3. UNet-3D architecture

was finally not included. Optimizer was Adam and a learning rate policy of dividing it by 2 if the loss had not decreased after 5 epochs.

Batch size was only one, due to the GPU memory limitations. The accumulative gradient technique was used in order to avoid an excess on gradient variability [16]. This technique simulates a larger batch size by accumulating several gradients before back propagating. Model was trained for 200 Epochs.

1.4   2D-3D Ensembling

A Random forest algorithm [17] was trained to ensemble all networks. The input data was the voxel probability for every class given by every model. This ensembling model was also trained with the training dataset. A 5 buckets cross validation method was used.

1.5   Post-processing

We identified the main tumor volume as the largest connected (solid) volume. Distance from every smaller volume (secondary tumor volumes), not connected to the main tumor, were calculated over the training dataset. After segmentation, all secondary volumes located at a distance larger than the average distance plus 1.5 times the standard deviation, were considered false positive, and therefore removed and classified as background (Fig. 4).

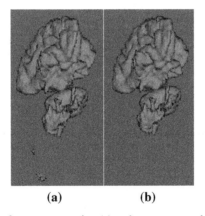

(a)   **(b)**

Fig. 4. Example of raw segmentation (a) and post-processed segmentation (b)

1.6   Results

Segmentation performance was evaluated in terms of dice coefficient, Haussdorf distance, sensibility and specificity, results over the testing dataset appear in Table 1 and Table 2. Some examples of tumor segmentation are shown in Fig. 5.

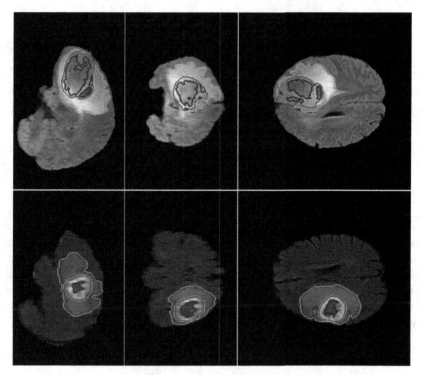

Fig. 5. Network segmentation examples

Table 1. Testing data results: Dice and Haussdorf distance

	Dice ET	Dice WT	Dice TC	$^{95\%}\text{H}_\text{D}$ ET	$^{95\%}\text{H}_\text{D}$ WT	$^{95\%}\text{H}_\text{D}$ TC
Mean	0.75174	0.8085	0.78179	38.08898	33.85716	41.86391
STD	0.28135	0.2625	0.61504	109.79924	99.34806	112.4294
Median	0.84113	0.90625	0.91697	1.73205	3.6055	2.44949
25-quartile	0.76507	0.85348	0.8421	1.0	2.23607	1.41421
75 -quartile	0.90385	0.9344	0.95529	2.44949	6.16441	5.28887

2   Task 2: Survival Task

2.1   Materials and Methods

Feature extraction was performed with software Matlab (© 1994–2020 The MathWorks, Inc.) with the Image processing toolbox. A large number of features were obtained from the segmentation matrix and the four MRI sequences. Machine learning models contained in the Matlab Machine Learning toolbox were used for the survival prediction.

Table 2. Testing data results: Sensitivity and Specificity

	Sens. ET	Sens. WT	Sens. TC	Spec. ET	Spec. WT	Spec. TC
Mean	0.80076	0.84012	0.79194	0.9996	0.99876	0.99967
STD	0.305	0.28579	0.32761	0.0004	0.00114	0.00065
Median	0.93247	0.96076	0.94873	0.99971	0.99905	0.99981
25-quartile	0.82263	0.87783	0.821	0.99946	0.9982	0.99961
75 -quartile	0.97077	0.98422	0.98239	0.99991	0.99957	0.99994

2.2 Data Sources

Original data for each patient was a set of four MRI studies (T1, T1ce, T2, and flair sequences), manual segmentation and patient age. According to glioma tumor growth models [18, 19], laplacian and gradient of the tumor play a major role. Mathematical transformations were applied to the MRI and segmentation matrices. Laplacian (Eq. 3) and the vector module of the 3D-gradient (Eq. 4) were obtained for all MRI sequences and the segmentation matrix. Also, module of 3D gradient multiplied by the distance to the whole tumor centroid (Eq. 5).

$$\Delta F = \frac{\partial^2 F}{\partial x^2} + \frac{\partial^2 F}{\partial y^2} + \frac{\partial^2 F}{\partial z^2} \tag{3}$$

$$|\nabla F| = \sqrt{\left(\frac{\partial F}{\partial x}\right)^2 + \left(\frac{\partial F}{\partial y}\right)^2 + \left(\frac{\partial F}{\partial z}\right)^2} \tag{4}$$

$$|\nabla F| \cdot |r| = |\nabla F| \cdot \left(\sqrt{(x - x_o)^2 + (y - y_o)^2 + (z - z_o)^2}\right) \tag{5}$$

Transformations by Eqs. 3, 4 and 5 were calculated over MRI images (pixel value) and segmentation masks for a specific region, being a binary matrix, value 1 inside and value 0 outside the specific region.

2.3 Regions of Interest (ROIs)

Training data contained three different classes in the segmentation file: necrotic and non-enhancing tumor (class 1), GD-enhancing tumor (class 4) and peritumoral edema (class 2). In addition, some other volumes were constructed:

A complete tumor, containing the sum of all three classes. Gross tumor volume (GTV), containing necrotic and both (enhancing and non-enhancing) tumor volumes (classes 1 and 4). Also, a three ring shaped structures, containing the boundaries of the NCR/NET (necrotic and non-enhancing tumor) volume, GTV, and whole tumor. Rings contain only 2 pixel inwards plus 2 pixel outwards the boundary of each volume.

2.4 Feature Extraction

The possible number of radiomic features that can be extracted is virtually unlimited [20]. Most common features for radiomics [21–23] were selected. After extraction, a dimensionality reduction had to be performed in order to filter those features that have a stronger correlation to patient survival. Features were extracted attending to different characteristics:

Geometrical. Average value, sum, and standard deviation within every class region for this features were collected. Matlab function *regionprops3d* was used to obtain the sphere equivalent diameter, extent, surface, volume and principal axis length for every class and the whole tumor.

Statistical. Pixel value for every MRI sequence was also considered. Average, entropy, standard deviation, variance, mode, median and kurtosis were collected for every class region including the complete tumor and the boundaries structure.

Location. Tumor location have shown a significant relation to patient survival time [24]. Image was cropped to its boundaries in all dimensions (superior-inferior, anterior-posterior and left-right). Then it, was divided into three main parts, superior, middle and inferior, with the same number of images alongside the craniocaudal direction (longitudinal axis) (Fig. 6a). Also, every axial image, was divided different areas. Images within the middle volume were divided into 5 parts as the volume around and within the brainstem might be significant. This central square contains half-length of both diagonals (Fig. 6c). Superior and inferior images were divided only into 4 parts (Fig. 6b, Fig. 6d). Location was collected as 13 different categorical and binary features, being 1 the value of the region containing the centroid of the complete tumor structure.

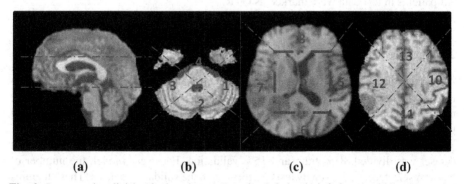

(a)　　　　　(b)　　　　　(c)　　　　　(d)

Fig. 6. Image region division for location: a) Superior-Inferior, b) inferior, c) middle, d) superior

As a different variable, distances to skull were obtained for the whole tumor centroid and the GTV. Right, left, anterior, posterior and superior distances were calculated as shown in Fig. 7.

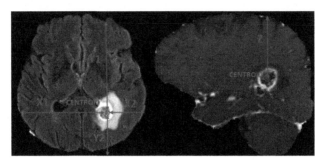

Fig. 7. Brain edges distance to tumor centroid

Textural. Matlab function *graycoprops* calculates contrast, energy, homogeneity and correlation. Mean values for every class region were added to data collection, including the complete tumor and the ring-boundaries structure. Moreover, some textural features based on biofilm analysis [25], such as porosity, diffusion distance, mean breadth, Euler connectivity number and fractal dimension were also calculated for every class.

Other. Organization provided also patient age and if the patient has gone through a gross total resection (GTR), a subtotal resection (STR) or no resection (NA). These three variables (GTR, STR and NA) were included as a categorical and binary features. All patients in test data were marked as GTR.

2.5   Feature Selection

After data collection, three different feature selection methods were used: *fscchi2* (ranking for classification using chi-square tests), *fscmrmr* (ranking for classification using minimum redundancy maximum relevance (MRMR) algorithm) and *fsrftest* (ranking for regression using F-tests).

Features were then ranked by its importance according to each algorithm. Training dataset was divided, 85% train and 15% validation. For every model, the number of features selected was chosen by tests performed on the validation dataset. These features were chosen in the order proposed by the feature selection methods.

2.6   Training Data and Models

In order to train a robust model, data was extracted from both, the ground truth segmentation and also the segmentation performed by the trained model. Therefore, the number of data entries was twice the number of patients. All patients with a known age were included in the training dataset. Basic Matlab machine learning models have been trained for the task, both classification and regression. Models and top-ranked features are shown in Table 3.

Classification models were trained up to three classes, short-term survivors (<300 days), mid-term (300–450) and long-term (>450). After classification, a survival number of days was assigned according to the model output. For each class, the average survival time for the training dataset was chosen, 150 days for those classified as short-time survivors, 376 days for those classified as mid-time and 796 days for long-time survivors.

Table 3. Model feature selection and principal features

Model	Feature selection	Type	Number of features	1st feature	2nd feature	3rd feature
1	Chi-Square	Classification	24	Principal Axis Length (1st)	Principal Axis Length (3rd)	Homogenity T1ce (WT)
2	MRMR	Classification	10	Subtotal Resection (STR)	Mode pxl value – Flair (Ring GTV)	Brain location (5)
3	F-Test	Regression	29	Principal Axis Length (1st)	Principal Axis Length (2nd)	Principal Axis Length (3rd)

Model-1: Classification. Decision tree algorithm, ensembled by RUSBoost method.
Model-2: Classification. Support vector machines (SVM) algorithm. Quadratic kernel function.
Model-3: Regression. Ensembled of regression trees.

Final prediction was calculated as the arithmetic mean of the three models.

2.7   Results

Final results for the survival task are shown in Table 4.

Table 4. Testing data results: survival task

Number of cases	Accuracy	MSE	Median SE	STD SE	Spearman R
107	0,617	382776,485	61504	1095051,41	0,396

2.8 Conclusions

An analysis on models and features reveals that some characteristics and volumes play a major role on survival prediction.

Geometrical features, like length of the principal axis (1st, 2nd and 3rd axe), for both, the whole tumor and especially the GTV (classes 1 and 4), GTV sphere equivalent diameter, volume and surface, or of brain volume as edema (class 2).

Location features as tumor location, specially positions 5, 10, 11 and 13 according to Fig. 6 or tumor centroid distance to brain edges: X1 and Y1 for GTV centroid and whole tumor centroid.

Other features like fractal dimension and porosity for GD-enhancing tumor volume (class 4), patient age and sum, standard deviation and mean pixel value for GD-enhancing tumor volume, calculated over segmentation transformed by Eq. 5.

A total of 52 features have been used for training. Models contain a total of 63 features, but nine of them are repeated in two models, and only one (patient age) in all three.

An analysis of the top-ten main features for each of the three models shows that tumor size and tumor location seem to be the most important characteristics in terms of survival. Tumor size has four features in model 1 and five in model 3. For tumor location three features appear on model 1, four in model 2, and two in model 3. Age is the only feature that appears in all three models.

GTV and GD-enhancing tumor (class 4) properties have a large representation in all models as more than half features have been calculated over these regions.

Some other features of importance are calculations over the segmentation mask, transformed by Eqs. 4 and 5, such as sum of pixel value, mean pixel value or standard deviation. Fifteen features were calculated over these matrixes.

Acknowledgments. This work was supported by the Fundación HM, under the grant "Beca intramural 2018".

References

1. Menze, B.H., et al.: The multimodal brain tumor image segmentation benchmark (BRATS). IEEE Trans. Med. Imaging **34**(10), 1993–2024 (2015). https://doi.org/10.1109/TMI.2014.2377694
2. Bakas, S., et al.: Advancing the cancer genome atlas glioma MRI collections with expert segmentation labels and radiomic features. Nat. Sci. Data **4**, 170117 (2017). https://doi.org/10.1038/sdata.2017.117

3. Bakas, S., et al.: Identifying the best machine learning algorithms for brain tumor segmentation, progression assessment, and overall survival prediction in the brats challenge. arXiv preprint arXiv:1811.02629 (2018)
4. Bakas, S., et al.: Segmentation labels and radiomic features for the pre-operative scans of the TCGA-GBM collection. The Cancer Imaging Archive (2017). https://doi.org/10.7937/K9/TCIA.2017.KLXWJJ1Q
5. Bakas, S., et al.: Segmentation labels and radiomic features for the pre-operative scans of the TCGA-LGG collection. The Cancer Imaging Archive (2017). https://doi.org/10.7937/K9/TCIA.2017.GJQ7R0EF
6. Ronneberger, O., Fischer, P., Brox, T.: U-net: convolutional networks for biomedical image segmentation. In: Navab, N., Hornegger, J., Wells, W., Frangi, A. (eds.) MICCAI 2015. LNCS, pp. 234–241. Springer, Heidelberg (2015). https://doi.org/10.1007/978-3-319-24574-4_28
7. Milletari, F., Navab, N., Ahmadi, S.-A.: V-net: fully convolutional neural networks for volumetric medical image segmentation. arXiv arXiv:1606.04797 (2016)
8. Pascanu, R., Mikolov, T., Bengio, Y.: On the difficulty of training recurrent neural networks. arXiv arXiv:1211.5063 (2013)
9. Iglovikov, V., Shvets, A.: TernausNet: U-Net with VGG11 Encoder pre-trained on imagenet for image segmentation. arXiv arXiv:1801.05746 (2018)
10. Deng, J., Dong, W., Socher, R., Li, L.-J., Li, K., Li, F.F.: ImageNet: a large-scale hierarchical image database. In: IEEE Conference on Computer Vision and Pattern Recognition, pp. 248–255 (2009). https://doi.org/10.1109/CVPR.2009.5206848.
11. Tan, M., Le, Q.: EfficientNet: rethinking model scaling for convolutional neural networks. In: Chaudhuri, K., Salakhutdinov, R. (eds.) Proceedings of the 36th International Conference on Machine Learning, volume 97 of Proceedings of Machine Learning Research, Long Beach, California, USA, 09–15 Jun 2019, pp. 6105–6114. PMLR (2019)
12. Buslaev, A., Parinov, A., Khvedchenya, E., Iglovikov, V.I., Kalinin, A.A.: Albumentations: fast and flexible image augmentations. arXiv:1809.06839 (2018)
13. Karimi, D., Salcudean, S.E.: Reducing the Hausdorff distance in medical image segmentation with convolutional neural networks. arXiv:1904.10030 (2019)
14. Kingma, D.P., Ba, J.L.: Adam: a method for stochastic optimization. arXiv:1412.6980 (2014)
15. A 3D multi-modal medical image segmentation library in PyTorch. https://github.com/black0017/MedicalZooPytorch
16. Hermans, J., Spanakis, G., Möckel, R.: Accumulated gradient normalization (2017)
17. Breiman, L.: Random forests. Mach. Learn. 45, 5–32 (2001). https://doi.org/10.1023/A:1010933404324
18. Swanson, K.R., Rostomily, R.C., Alvord, E.C.: A mathematical modelling tool for predicting survival of individual patients following resection of glioblastoma: a proof of principle. Br. J. Cancer 98, 113–119 (2008)
19. Swanson, K.R., Bridge, C., Murray, J.D., Ellsworth, C., Alvord Jr., E.C.: Virtual and real brain tumors: using mathematical modeling to quantify glioma growth and invasion. J. Neurol. Sci. 216, 1–10 (2003)
20. Lambin, P., et al.: Radiomics: the bridge between medical imaging and personalized medicine. Nat. Rev. Clin. Oncol. 14(12), 749–762 (2017). https://doi.org/10.1038/nrclinonc.2017.141. Epub 2017 Oct 4 PMID: 28975929
21. Zhou, J., et al.: Predicting the response to neoadjuvant chemotherapy for breast cancer: wavelet transforming radiomics in MRI. BMC Cancer 20(1), 100 (2020). https://doi.org/10.1186/s12885-020-6523-2.PMID:32024483;PMCID:PMC7003343
22. Wu, J., Tha, K.K., Xing, L., Li, R.: Radiomics and radiogenomics for precision radiotherapy. J. Radiat. Res. 59(suppl_1), i25–i31 (2018). https://doi.org/10.1093/jrr/rrx102. PMID: 29385618; PMCID: PMC5868194.

23. Jeong, J., Ali, A., Liu, T., Mao, H., Curran, W.J., Yang, X.: Radiomics in cancer radiotherapy: a review. arXiv:1910.02102v2 (2019)
24. Awad, A.-W., et al.: Impact of removed tumor volume and location on patient outcome in glioblastoma. J. Neurooncol. **135**(1), 161–171 (2017). https://doi.org/10.1007/s11060-017-2562-1
25. Lewandowski, Z., Beyenal, H.: Fundamentals of Biofilm Research. CRC Press (2007)

Multimodal Brain Tumor Segmentation and Survival Prediction Using a 3D Self-ensemble ResUNet

Linmin Pei[✉], A. K. Murat, and Rivka Colen

Department of Diagnostic Radiology, The University of Pittsburgh
Medical Center, Pittsburgh, PA 15232, USA
{peil,akm,colenrr}@upmc.edu

Abstract. In this paper, we propose a 3D self-ensemble ResUNet (srUNet) deep neural network architecture for brain tumor segmentation and machine learning-based method for overall survival prediction of patients with gliomas. UNet architecture has been using for semantic image segmentation. It also been used for medical imaging segmentation, including brain tumor segmentation. In this work, we utilize the srUNet to differentiate brain tumors, then the segmented tumors are used for survival prediction. We apply the proposed method to the Multimodal Brain Tumor Segmentation Challenge (BraTS) 2020 validation dataset for both tumor segmentation and survival prediction. The tumor segmentation result shows dice score coefficient (DSC) of 0.7634, 0.899, and 0.816 for enhancing tumor (ET), whole tumor (WT), and tumor core (TC), respectively. For the survival prediction method, we achieve 56.4% classification accuracy with mean square error (MSE) 101697, and 55.2% accuracy with MSE 56169 for training and validation, respectively. In the testing phase, the proposed method offers the DSC of 0.786, 0.881, and 0.823, for ET, WT, and TC, respectively. It also achieves an accuracy of 0.43 for overall survival prediction.

Keywords: Deep neural network · Tumor segmentation · Survival prediction · Feature fusion

1 Introduction

Brain tumors, originated from glioma cells in the central nervous system (CNS), are common cancers in adults. According to a report, there are 23 per 100,000 population diagnosed with CNS brain tumors annually in the US [1]. Among patients with brain tumors, the estimated five- and ten- year relative survival rates are 35.0% and 29.3% for patients with a malignant brain tumor, respectively [1]. The brain tumors can be categorized into four grades based on degrees of aggressiveness, variable prognosis and various heterogeneous histological sub-regions [2–4]. It is known that survival period of glioma patients highly depends on the tumor grade [5]. Patients with high-grade glioma (HGG) generally have longer survival periods than that of low-grade glioma (LGG) patients. Even though with modern treatment advancement, the median survival period

© Springer Nature Switzerland AG 2021
A. Crimi and S. Bakas (Eds.): BrainLes 2020, LNCS 12658, pp. 367–375, 2021.
https://doi.org/10.1007/978-3-030-72084-1_33

of patients with glioblastoma (GBM) still remains 12–16 months [4]. Brain tumor segmentation is crucial for brain tumor prognosis, treatment planning, and follow-up evaluation. An accurate tumor segmentation could lead to a better prognosis. Traditionally, brain tumor segmentation is done by radiologists, while the manual process suffers from low efficiency, long time, and error-prone to an observer [5]. Therefore, computer-aided automatic brain tumor segmentation is highly desired. Because of non-invasiveness and high resolution for soft tissues, structural Magnetic resonance imaging (MRI) is widely used for brain tumor study. However, using only one single structural MRI is insufficient to segment all type tumors due to the image artificial facts and complication of different tumors. Multimodal MRI (mMRI) offers complementary information for different tumors. The mMRI sequences include T1-weighted MRI (T1), T1-weighted MRI with contrast enhancement (T1ce), T2-weighted MRI (T2), and T2-weighted MRI with fluid-attenuated inversion recovery (T2-FLAIR). T1ce and T2-FLAIR are usually considered as good sources to identify enhancing tumor (ET)/necrosis (NC), and peritumoral edema (ED), respectively. There is still challenging for computer-aided automatic brain tumor segmentation because of the impact of signal noise, multi-image co-registration, and intensity inhomogeneity, etc.

There are many works on brain tumor segmentation in literature. Based on the methods, they can be grouped as threshold-based, region-based, to conventional machine learning-based method [6–11]. However, hand-crafted feature extraction is a prerequisite for those methods, which is a challenging process. Recent years, deep learning is becoming success because of the computer hardware improvement and availability of large dataset. It has been using in many domains, such as computer vision [12], medical imaging analysis [13], etc. One of the great advantages of using deep learning-based method is automatic feature extraction and selection, which common used in traditional machine learning-based methods [12, 14–17].

In this work, we use a 3D srUNet for brain tumor segmentation. The srUNet is composed two parts, a ResUNet, and a self-ensemble model. The self-ensemble model is attached to at the end of ResUNet to constrain the output. In addition, we believe survival period of patients with gliomas are highly related to the brain tumors. The segmented brain tumors are fed into another deep learning model for overall survival prediction.

2   Method

2.1   Brain Tumor Segmentation

There may have several brain tumors in brain of patients diagnosed with gliomas, enhancing tumor (ET), non-enhancing tumor (NET), necrosis (NC), and peritumoral edema (ED). They are reflected with different appearances on mMRI. The ET is showing brighter than other type of tumors, while ED appears brighter in T2 and T2-FLAIR. Even though mMRI offers comprehensive information, distinguishing all tumors is still a difficult task due to image artificial facts. For image semantic segmentation, deep learning-based methods usually outperform the traditional machine learning methods [2].

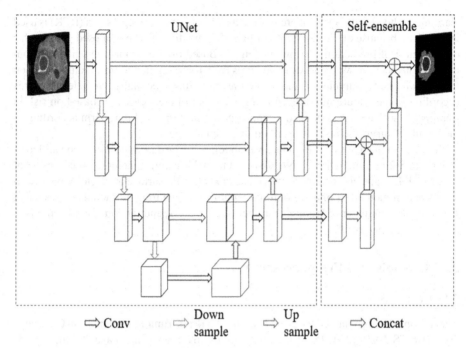

Fig. 1. The proposed srUNet for brain tumor segmentation.

To achieve accurate brain tumor segmentation, we propose a 3D reUNet deep learning-based method. The proposed architecture is showing in Fig. 1. The reUNet consists of a regular ResUNet and a self-ensemble model. The self-ensemble model prevents from gradient vanish and regularize the former ResUNet.

2.2 Survival Prediction

In the section, we describe a risk-guided method for overall patient survival prediction using Random Forest Regression (RFR). The tumor segmentation probabilities obtained from previous section are used for overall survival prediction. Even through deep learning-based methods offer good performances in semantic segmentation, they require large amount of training samples. It is inapplicable in the survival prediction task due to the limited number of cases in the challenge. Therefore, we use a regular machine

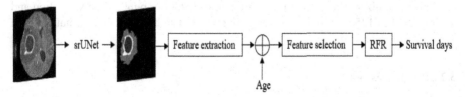

Fig. 2. The proposed pipeline for overall survival prediction.

learning method for the survival prediction task. Random forest regression (RFR) is used in this work because of the capability of handling overfitting. The proposed pipeline for overall survival prediction is shown in Fig. 2. Based on the tumor segmentation from the previous section, we then extract features from these segmentations, including shape features and non-radiomics features, such as age. Subsequentially, we select features according to the feature importance using a Random Forest classifier based on risk in training phase. The risk is defined as short-term, mid-term, and long-term according to the length of survival days, as described in the challenge.

To find out the feature importance, we random split the data as training and validation as ratio of 8:2 in training phase. We extract a total 34 features, including the number of NC, ED, ET, shape elongation, flatness, least axis length, surface area, etc. We perform a grid search method to find the optimal parameters. The training model suggests that the optimal for number of features, number of estimators, and random forest depth is 7, 31, and 50, respectively.

3 Materials and Pre-processing

3.1 Data

The training data is obtained from the Multimodal Brain Tumor Segmentation Challenge 2020 (BraTS 2020) [2–4, 18, 19], with a total of 369 cases which have 293 high-grade gliomas (HGG) and 76 low-grade gliomas (LGG). Each patient case contains multimodal MRI (mMRI), including T1, T1-ce, T2, and T2-FLAIR. Note that all images are co-registered, skull-stripped, and denoised [3]. The size of each image is 240 × 240 × 155 across cases. In training data, ground truth is available for public. There are several tumor sub-tissues: necrotic (NC), peritumoral edema (ED), and enhancing tumor (ET). However, the evaluation of the BraTS 2020 is based on three tumor subregions: enhancing tumor, tumor core (TC), and whole tumor (WT). TC is the union of ET and NC, while the WT is the combination of all abnormal tissues.

For survival prediction task, there are 235 cases available with overall survival (in days), age, and resection status. Based on the survival days, it is categorized as three risks: short-term, mid-term, and long-term. The short-term, mid-term, and long-term is defined as survival days within 10 months, 10–15 months, and more than 15 months, respectively. In the training data, there are 99 cases, 49 cases, and 87 cases for short-term, mid-term, and long-term, respectively. However, only cases with resection status Gross Total Resection (GTR) is evaluated in BraTS 2020 Challenge.

In addition, to estimate performance of the proposed method, we evaluate the method using the validation and testing data. Both of data are privately owned by the BraTS challenge organizer. The ground truths are not available for public. The validation and testing data has 125 and 166 cases, respectively. All data has the same format and size as in training data.

3.2 Pre-processing

In order to minimize the impact of intensity variance across cases and modalities, we perform z-score intensity normalization for brain region only in MRIs. The z-score

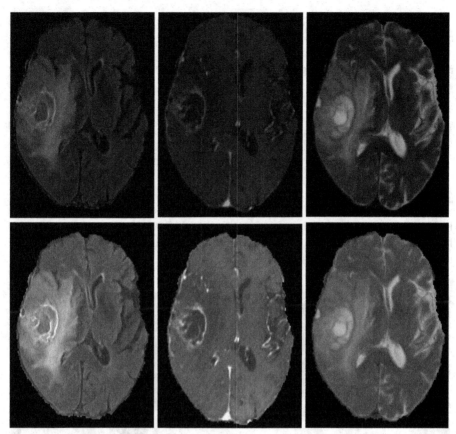

Fig. 3. An example of z-score normalization. Top from left to right: FLAIR, T1ce, and T2. Bottom from left to right: normalized FLAIR, T1ce, and T2.

normalization ensures intensity with zero mean and unit standard deviation (std) [20]. In the study, we apply the z-score normalization for brain region only. Figure 3 illustrates an example of image comparison before and after z-score normalization.

4 Experiments and Results

4.1 Hyper-parameter Setting

The original size of image volume is $240 \times 240 \times 155$. We empirically crop all mMRIs with the size as $160 \times 192 \times 128$, so that to reduce the computational burden of the graphics processing unit (GPU). With such size of the cropped image, it contains all abnormal tissues, without losing any tumor information. In addition, we set the batch size as 1 for the proposed 3D srUNet. The loss function is computed as follows:

$$L_{loss} = 1 - DSC, \tag{1}$$

where $DSC = \frac{2TP}{FP+2TP+FN}$ is dice similarity coefficient [21]. TP, FP, and FN are the numbers of true positive, false positive and false negative, respectively.

We set the training epoch as 300, and use Adam [22] optimizer with an initial learning rate of $lr_0 = 0.001$ in training phase, and the learning rate (lr_i) is gradually reduced by the following:

$$lr_i = lr_0 * \left(1 - \frac{i}{N}\right)^{0.9}, \tag{2}$$

where i is epoch counter, and N is a total number of epochs in training.

4.2 Training Stage

For brain tumor segmentation task, we randomly split 80% data for training, and 20% for validation based on HGG and LGG. Figure 4 shows a case with segmentation using the proposed method in multiple views.

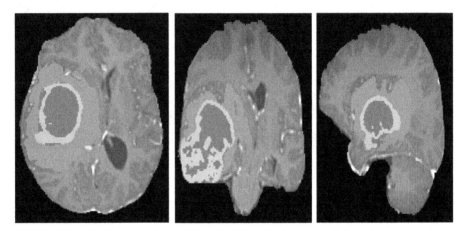

Fig. 4. Brain tumor segmentation using the proposed method. From left to right: segmentation over-laid over T1-ce in axial view, coronal view, and sagittal view, respectively.

For survival prediction, we first extract a total 34 number of features, such as, the amount of necrosis, edema, and enhancing tumor, original shape elongation, least axis length, etc. then we select top 7 number of features. we use 5-fold cross validation (CV), and the average accuracy of 56.4% ± 8. 3% and a mean square error of 101523.

4.3 Online Evaluation

We apply the proposed method to BraTS 2020 validation dataset and evaluate the performance through the online portal. There are 125 cases with unknown tumor grade, and the ground truths are privately owned by the challenge organizer, our result achieves average DSC as 0.7634, 0.899, and 0.816 for ET, WT, and TC, respectively. Hausdorff distance

(HD), a matric measuring the spacing distance between segmentation and ground truth, is also provided by the online evaluation. A smaller HD indicates a better segmentation. The average of HD at 95 percentiles is 33.26 mm, 5.28 mm, and 7.74 mm for ET, WT, and TC, respectively. The result is shown in Table 1. For survival prediction in the validation phase, we achieve 55.2% classification accuracy with MSE 95924, as shown in Table 2. In the testing phase, our proposed method achieves a DSC of 0.786, 0.823, and 0.881 for ET, TC, and WT, respectively. For the overall survival prediction, the online evaluation shows the accuracy of 0.43, 402528 of MSE.

Table 1. Brain tumor segmentation performance using the online evaluation of BraTS 2019 validation and testing dataset.

Phase	Dice_ET	Dice_WT	Dice_TC	Hausdorff95_ET	Hausdorff95_WT	Hausdorff95_TC
Validation	0.7634	0.899	0.816	33.26	5.28	7.74
Testing	0.786	0.881	0.823	18.00	7.183	4.987

Table 2. Performance of our fused-based survival prediction model using the online evaluation.

Phase	Accuracy	MSE	medianSE	stdSE	SpearmanR
Training	0.564	101523	23968	225744	0.51
Validation	0.552	101697	56169	116680	0.329
Testing	0.43	402528	66049	1189662	0.33

According to the comparison of tumor segmentation performance comparison in Table 1, it shows the proposed method has better performances in ET and TC in testing phase than that in validation phase. The Hausdorff distances are smaller in ET and TC in testing phase comparing to validation phase. However, it has a slightly poor performance in WT comparing the testing phase and validation phase.

As for overall survival prediction, the proposed method has worse performance in accuracy, MSE, medianSE, and stdSE in testing phase than that in validation phase. However, it has a slightly better in SpearmanR.

5   Conclusion

In the paper, we propose a deep learning-based method, namely srUNet for brain tumor segmentation. The srUNet is composed by a resUNet and self-ensemble model. The self-ensemble model prevents from gradient vanish issue and regularize the training model to achieve a better segmentation. Based on the segmentations, we use a Random Forest Regression for the overall survival prediction due to lack of data, instead of using a deep learning-based method. The online evaluation suggests a promising performance on both brain tumor segmentation and overall survival prediction.

References

1. Ostrom, Q.T., Gittleman, H., Truitt, G., Boscia, A., Kruchko, C., Barnholtz-Sloan, J.S.: CBTRUS statistical report: primary brain and other central nervous system tumors diagnosed in the United States in 2011–2015. Neuro-oncology **20**(suppl_4), iv1–iv86 (2018)
2. Bakas, S., et al.: Identifying the best machine learning algorithms for brain tumor segmentation, progression assessment, and overall survival prediction in the BRATS challenge. arXiv preprint arXiv:1811.02629 (2018)
3. Bakas, S., et al.: Advancing the cancer genome atlas glioma MRI collections with expert segmentation labels and radiomic features. Sci. Data **4**, 170117 (2017)
4. Menze, B.H., et al.: The multimodal brain tumor image segmentation benchmark (BRATS). IEEE Trans. Med. Imaging **34**(10), 1993–2024 (2014)
5. Shboul, Z.A., Alam, M., Vidyaratne, L., Pei, L., Elbakary, M.I., Iftekharuddin, K.M.: Feature-guided deep radiomics for glioblastoma patient survival prediction (in English). Front. Neurosci. Original Res. **13**, 966 (2019)
6. Mustaqeem, A., Javed, A., Fatima, T.: An efficient brain tumor detection algorithm using watershed & thresholding based segmentation. Int. J. Image Graph. Signal Process. **4**(10), 34 (2012)
7. Pei, L., Bakas, S., Vossough, A., Reza, S.M., Davatzikos, C., Iftekharuddin, K.M.: Longitudinal brain tumor segmentation prediction in MRI using feature and label fusion. Biomed. Signal Process. Control **55**, 101648 (2020)
8. Pei, L., Reza, S.M., Li, W., Davatzikos, C., Iftekharuddin, K.M.: Improved brain tumor segmentation by utilizing tumor growth model in longitudinal brain MRI. In: Medical Imaging 2017: Computer-Aided Diagnosis, vol. 10134, p. 101342L. International Society for Optics and Photonics (2017)
9. Pei, L., Reza, S.M., Iftekharuddin, K.M.: Improved brain tumor growth prediction and segmentation in longitudinal brain MRI. In: 2015 IEEE International Conference on Bioinformatics and Biomedicine (BIBM), pp. 421–424. IEEE (2015)
10. Prastawa, M., Bullitt, E., Ho, S., Gerig, G.: A brain tumor segmentation framework based on outlier detection. Med. Image Anal. **8**(3), 275–283 (2004)
11. Ho, S., Bullitt, E., Gerig, G.: Level-set evolution with region competition: automatic 3-D segmentation of brain tumors. In: Null, p. 10532. Citeseer (2002)
12. LeCun, Y., Bengio, Y., Hinton, G.: Deep learning. Nature **521**, 436 (2015)
13. Pereira, S., Meier, R., Alves, V., Reyes, M., Silva, C.A.: Automatic brain tumor grading from MRI data using convolutional neural networks and quality assessment. In: Stoyanov, D., et al. (eds.) MLCN/DLF/IMIMIC -2018. LNCS, vol. 11038, pp. 106–114. Springer, Cham (2018). https://doi.org/10.1007/978-3-030-02628-8_12
14. Goodfellow, I., Bengio, Y., Courville, A.: Deep Learning. MIT Press, Cambridge (2016)
15. Mohsen, H., El-Dahshan, E.-S.A., El-Horbaty, E.-S.M., Salem, A.-B.M.: Classification using deep learning neural networks for brain tumors. Future Comput. Inf. J. **3**(1), 68–71 (2018)
16. He, K., Zhang, X., Ren, S., Sun, J.: Deep residual learning for image recognition. In: Proceedings of the IEEE Conference on Computer Vision and Pattern Recognition, pp. 770–778 (2016)
17. Ronneberger, O., Fischer, P., Brox, T.: U-net: convolutional networks for biomedical image segmentation. In: Navab, N., Hornegger, J., Wells, W.M., Frangi, A.F. (eds.) MICCAI 2015. LNCS, vol. 9351, pp. 234–241. Springer, Cham (2015). https://doi.org/10.1007/978-3-319-24574-4_28
18. Bakas, S., et al.: Segmentation labels and radiomic features for the pre-operative scans of the TCGA-GBM collection. The Cancer Imaging Archive (2017) (2017)

19. Bakas, S., et al.: Segmentation labels and radiomic features for the pre-operative scans of the TCGA-LGG collection. The Cancer Imaging Archive, vol. 286 (2017)
20. Kreyszig, E.: Advanced Engineering Mathematics, 10th edn. Wiley, Hoboken (2009)
21. Dice, L.R.: Measures of the amount of ecologic association between species. Ecology **26**(3), 297–302 (1945)
22. Kingma, D.P., Ba, J.: Adam: a method for stochastic optimization. arXiv preprint arXiv:1412.6980 (2014)

MRI Brain Tumor Segmentation and Uncertainty Estimation Using 3D-UNet Architectures

Laura Mora Ballestar$^{(\boxtimes)}$ and Veronica Vilaplana$^{(\boxtimes)}$

Signal Theory and Communications Department, Universitat Politècnica
de Catalunya. BarcelonaTech, Barcelona, Spain
veronica.vilaplana@upc.edu

Abstract. Automation of brain tumor segmentation in 3D magnetic resonance images (MRIs) is key to assess the diagnostic and treatment of the disease. In recent years, convolutional neural networks (CNNs) have shown improved results in the task. However, high memory consumption is still a problem in 3D-CNNs. Moreover, most methods do not include uncertainty information, which is especially critical in medical diagnosis. This work studies 3D encoder-decoder architectures trained with patch-based techniques to reduce memory consumption and decrease the effect of unbalanced data. The different trained models are then used to create an ensemble that leverages the properties of each model, thus increasing the performance. We also introduce voxel-wise uncertainty information, both epistemic and aleatoric using test-time dropout (TTD) and data-augmentation (TTA) respectively. In addition, a hybrid approach is proposed that helps increase the accuracy of the segmentation. The model and uncertainty estimation measurements proposed in this work have been used in the BraTS'20 Challenge for task 1 and 3 regarding tumor segmentation and uncertainty estimation.

Keywords: Brain tumor segmentation · Deep learning · Uncertainty · 3d convolutional neural networks

1 Introduction

Brain tumors are categorized into primary, brain originated; and secondary, tumors that have spread from elsewhere and are known as brain metastasis tumors. Among malignant primary tumors, gliomas are the most common in adults, representing 81% of brain tumors [7]. The World Health Organization (WHO) categorizes gliomas into grades I-IV which can be simplified into two types (1) "low grade gliomas" (LGG), grades I-II, which are less common and are characterized by low blood concentration and slow growth and (2) "high grade gliomas" (HGG), grades III-IV, which have a faster growth rate and aggressiveness.

V. Vilaplana—This work has been partially supported by the project MALEGRA TEC2016-75976-R financed by the Spanish Ministerio de Economía y Competitividad.

The extend of the disease is composed of four heterogeneous histological sub-regions, i.e. the peritumoral edematous/invaded tissue, the necrotic core (fluid-filled), the enhancing and non-enhancing tumor (solid) core. Each region is described by varying intensity profiles across MRI modalities (T1-weighted, post-contrast T1-weighted, T2-weighted, and Fluid-Attenuated Inversion Recovery-FLAIR), which reflect the diverse tumor biological properties and are commonly used to assess the diagnosis, treatment and evaluation of the disease. These MRI modalities facilitate tumor analysis, but at the expense of performing manual delineation of the tumor regions which is a challenging and time-consuming process. For this reason, automatic mechanisms for region tumor segmentation have appeared in the last decade thanks to the adv ancement of deep learning models in computer vision tasks. Despite these recent advances, the segmentation of brain tumors in multimodal MRI scans is still a challenging task in medical image analysis due to the highly heterogeneous appearance and shape of the problem.

The Brain Tumor Segmentation (BraTS) [1–5] challenge started in 2012 with a focus on evaluating state-of-the-art methods for glioma segmentation in multi-modal MRI scans. BraTS 2020 training dataset includes 369 cases (293 HGG and 76 LGG), each with four 3D MRI modalities rigidly aligned, re-sampled to 1 mm^3 isotropic resolution and skull-stripped with size $240 \times 240 \times 155$. Each provides a manual segmentation approved by experienced neuro-radiologists. Training annotations comprise the enhancing tumor (ET, label 4), the peritumoral edema (ED, label 2), and the necrotic and non-enhancing tumor core (NCR/NET, label 1). The nested sub-regions considered for evaluation are: whole tumor WT (label 1, 2, 4), tumor core TC (label 1, 4) and enhancing tumor ET (label 4). The validation set includes 125 cases, with unknown grade nor ground truth annotation. The test set is composed of 166 cases.

The goal of this work is to develop a 3D convolutional neural network (CNN) for brain tumor segmentation from 3D MRIs and provide an uncertainty measure to assess the confidence on the model predictions. The proposed methods are used to participate in BraTS'20 Challenge for tasks 1 and 3, respectively. In task 1, we explore the use of two well-known 3D-CNN for medical imaging –V-Net [6] and 3D-UNet [27]– and apply some modifications to their baselines. With both networks, the usage of sampling techniques is necessary due memory limitations as well as data augmentation to prevent over-fitting. For task 3, the work provides voxel-wise uncertainty measures computed at test time, with global and per sub-region information. Uncertainty is estimated using both epistemic and aleatoric [8] uncertainties using test-time dropout (TTD) [9] and data augmentation, respectively.

2 Related Work

2.1 Semantic Segmentation

Brain tumor segmentation methods include generative and discriminative approaches. Generative methods try to incorporate prior knowledge and model probabilistic distributions whereas discriminative methods extract features from image representations. This latter approach has thrived in recent years thanks to the advancement in CNNs, as demonstrated in the winners of the previous BraTS. The biggest break through in this area was introduced by DeepMedic [10] a 3D CNN that exploits multi-scale features using parallel pathways and incorporates a fully connected conditional random field (CRF) to remove false positives. [11] compares the performances of three 3D CNN architectures showing the importance of the multi-resolution connections to obtain fine details in the segmentation of tumor sub-regions. More recently, EMMA [12] creates and ensemble at inference time which reduces overfitting but at high computational cost, and [13] proposes a cascade of two CNNs, where the first network produces raw tumor masks and the second network is trained on the vicinity of the tumor to predict tumor regions. BraTS 2018 winner [14] proposed an asymmetrically encoder-decoder architecture with a variational autoencoder to reconstruct the image during training, which is used as a regularizer. Isensee, F [15] uses a regular 3D-U-Net optimized on the evaluation metrics and co-trained with external data. BraTS 2019 winners [16] use a two-stage cascade U-Net trained end-to-end. Finally, [17] applies several tricks in three categories: data processing, model devising and optimization process to boost the model performance.

2.2 Uncertainty

Uncertainty information of segmentation results is important, specially in medical imaging, to guide the clinical decisions and help understand the reliability of the provided segmentation, hence being able to identify more challenging cases which may require expert review. Segmentation models for brain tumor MRIs tend to label voxels with less confidence in the surrounding tissues of the segmentation targets [19], thus indicating regions that may have been miss-segmented.

Last year's BraTS challenge already started introducing uncertainty measurements. [18] computes epistemic uncertainty using TTD. They obtain a posterior distribution generated after running several epochs for each image at test-time. Then, mean and variance are used to evaluate the model uncertainty. A different approach is proposed by Wang G [19], who uses TTD and data augmentation to estimate the voxel-wise uncertainty by computing the entropy instead of the variance. Finally, [20] proposes to incorporate uncertainty measures during training as they define a loss function that models label noise and uncertainty.

3   Method

3.1   Dataset Statistics

The biggest complexity for brain tumor segmentation is derived from the class imbalance. The tumor regions account for a 5–15% of the brain tissue and each tumor region is an even smaller portion. Figure 1 provides a graphical representation of the distribution per each tumor class: ET, NCR, ED; without healthy tissue. It can be seen, that ED is more probable than ET and NCR and that there is high variability between subjects in the NCR label. Another complexity is the difference between glioma grades as LGG patients are characterized by low blood concentration which is translated to low appearance of ET voxels and higher number of voxels for NCR and NET regions.

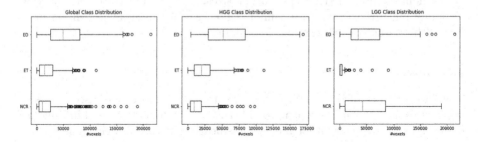

Fig. 1. Distribution of each class ED, ET, NCR. From left to right, (1) number of voxels in all cases, (2) number of voxels for the HGG and (3) number of voxels for the LGG

3.2 Data Pre-processing and Augmentation

MRI intensity values are not standardized as the data is obtained from different institutions, scanners and protocols. Therefore we normalize each modality of each patient independently to have zero mean and unit std based on non-zero voxels only, which represent the brain region.

We also apply data augmentation techniques to prevent over-fitting by trying to disrupt minimally the data. For this, we apply Random Flip (for all 3 axes) with a 50% probability, Random 90° Rotation on two axis with a 50% probability, Random Intensity Shift between ($-0.1..0.1$ of data std) and Random Intensity Scale on all input channels at range ($0.9..1.1$).

3.3 Sampling Strategy

3D-CNNs are computationally expensive and in many cases, the input data cannot be fed directly to the network. Patch-wise training helps to free memory resources so more images can be fed in one batch. However, there is a trade-off between patch size and batch size. Bigger batches will have a more accurate representation of the data but will require smaller patches (due to memory constraints) that provide local information but lack contextual knowledge.

Another key aspect to consider when selecting the patching strategy is to maintain the class distribution. Losing this distribution can generate a biased model, i.e. if the model only sees small patches with tumor it will likely miss-classify healthy tissue.

In this work, we have used two approaches depending on the patch size.

- Binary Distribution: Small patches, equal or lower than 64^3 are randomly selected with a 50% probability of being centred on healthy tissue and 50% probability on tumor [10].
- Random Tumor Distribution: Bigger sizes, 112^3 or 128^3, are selected randomly but always centred in tumor region, as the patches will contain more healthy tissue and background information.

3.4 Loss

The Dice score coefficient (DSC) is a measure of overlap widely used to assess segmentation performance when ground truth is available. Proposed in Milletari et al. [6] as a loss function for binary classification, it can be written as:

$$L_{dice} = 1 - \frac{2 * \sum_{i=1}^{N} p_i g_i}{\sum_{i}^{N} p_i^2 + \sum_{=1i}^{N} g_i^2 + \epsilon} \tag{1}$$

where N is the number of voxels, p_i and g_i correspond to the predicted and ground truth labels per voxel respectively, and ϵ is added to avoid zero division.

Many variations of the dice loss have been proposed in the literature. For instance, the Generalized Dice Loss (GDL) [26] which is based on the generalized dice score (GDS) [28] for multiple class evaluation. Its goal is to correct the correlation between region size and dice score, by weighting the contribution of each label with the inverse of its volume. It is described as:

$$L_{diceGDL} = 1 - 2 \frac{\sum_{l=1}^{L} w_l \sum_{i=1}^{N} p_{li} g_{li}}{\sum_{l=1}^{L} w_l \sum_{i=1}^{N} p_{li} + g_{li} + \epsilon} \tag{2}$$

where L represents the number of classes and w_l the weight given to each class. We use the GDL variant as it is more suited for unbalanced segmentation problems.

3.5 Network Architecture

This work proposes three networks, variations of V-Net [6] and 3U-Net [27] architectures, for brain tumor segmentation and creates an ensemble to mitigate the bias in each independent model.

The different models are trained using the ADAM optimizer, with start learning rate of $1e-4$, decreased by a factor of 5 whenever the validation loss has not improved in the past 30 epochs and regularized with a l2 weight decay of $1e-5$. They all use the GDL loss.

V-Net. The V-Net implementation has been adapted to use four output channels (Non-Tumor, ED, NCR/NET, ET) and uses Instance Normalization [21] in contrast to Batch Normalization, which normalizes across each channel for each training example instead of the whole batch. Also, as proposed in [15], we have increased the number of feature maps to 32 at the highest resolution, instead of 16 as proposed by the original implementation. Figure 2 shows the network architecture with an input patch size of $64 \times 64 \times 64$.

The network has been trained using a patch size of 96^3 and the random tumor distribution strategy (see Sect. 3.3). The maximum batch size due to memory constraints is 2.

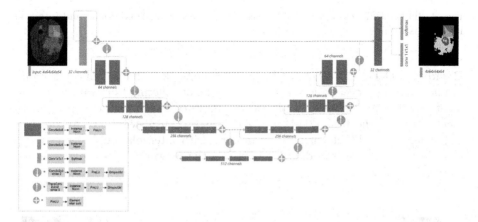

Fig. 2. V-Net [6] architecture with instance normalization, PreLU non-linearities, 32 feature channels at the highest resolution. Feature dimensionality is denoted at each block. The network outputs the segmentation and the softmax prediction.

3D-UNet. We use the original implementation with some minor modifications. Batch Normalization is changed for Group Normalization and, as in V-Net, we use 32 feature maps at the highest resolution.

The network architecture is divided into symmetric Encoder and Decoder parts. The Encoder is composed of two convolutional blocks - with 3DConv + ReLu + GroupNorm structure. The downsampling is performed with 2^3 Max-Pooling and the corresponding upsampling is performed with interpolation. All convolutional layers have kernel size 3^3, except for the last one that has $1 \times 1 \times 1$ kernel and 4 feature maps as output. In this case, we use ReLu non-linearity and the skip-connections are joined with a concatenation step. The network outputs a four-channel segmentation map with the training labels as well as a softmax. The detailed architecture can be seen in Fig. 3.

The *Basic 3D-UNet* is trained with a patch size of 112^3 and a batch size of 2.

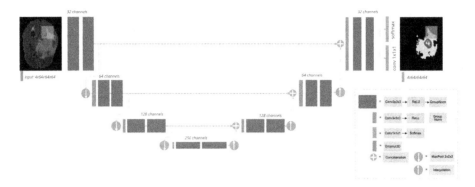

Fig. 3. 3D-Unet[27] architecture with Group Normalization, MaxPooling and Interpolation Upsampling and ReLU non-linearity

Residual 3D-UNet. Expands the previous network with residual connections to allow having a deeper network with less risk of suffering from vanishing gradient. Adding to the residual blocks, the network also introduces some modifications w.r.t the basic 3D-UNet: (1) it uses element-wise sum to join the skip-connections, (2) it changes upsampling with interpolation for transposed convolutions and (3) it adds more depth to the network thanks to the resnet connections (Fig. 4).

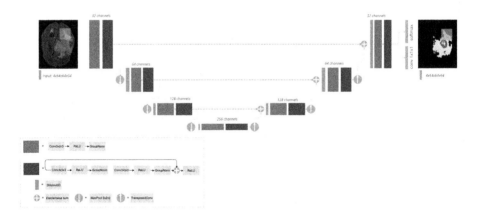

Fig. 4. 3D-Unet [27] architecture with RestNet blocks at each level, MaxPooling, TransposedConvolutions and ReLU non-linearity

This network is trained following two different strategies. The first one, *3D-UNet-residual* uses a patch size of 112^3 and a batch size of 2 for the whole training, whereas *3D-UNet-residual-multiscale* varies the sampling strategy so the network sees local and global information. For that, the first half of the training uses a patch size of 128^3 with a batch size of 1. Then, the patch size is reduced to 112^3 and the batch increased to 2.

3.6   Post-processing

In order to correct the appearance of false positives in the form of small and separated connected components, this work uses a post-processing step that keeps the two biggest connected components if their proportion is bigger than some threshold -obtained by analysing the training set. With this process, small connected components that may be false positives are removed but big enough components are kept as some of the subjects may have several tumors.

Moreover, one of the biggest difficulties of this challenge is to provide an accurate segmentation of the smallest sub-region, ET, which is particularly difficult to segment in LGG patients, as almost 40% have no enhancing tumor in the training set. In the evaluation step, BraTS awards a Dice score of 1 if a label is absent in both the ground truth and the prediction. Conversely, only a single false positive voxel in a patient where no enhancing tumor is present in the ground truth will result in a Dice score of 0. Therefore, some previous works [15,16] propose to replace enhancing tumor voxels for necrosis if the total number of enhancing voxels is smaller than some threshold, which is found for each experiment independently. However, we were not able to find a threshold that improved the performance as it helped for some subjects but made some other results worse.

3.7   Uncertainty

This year's BraTS includes a third task to evaluate the model uncertainty and reward methods with predictions that are: (a) confident when correct and (b) uncertain when incorrect. In this work, we model the voxel-wise uncertainty of our method at test time, using test time dropout (TTD) and test-time data augmentation (TTA) for epistemic and aleatoric uncertainty respectively.

We compute epistemic uncertainty as proposed in Gal et al. [23], who uses dropout as a Bayesian Approximation in order to simplify the task. Therefore, the idea is to use dropout both at training and testing time. The paper suggests to repeat the prediction a few hundred times with random dropout. Then, the final prediction is the average of all estimations and the uncertainty is modelled by computing the variance of the predictions. In this work, we perform $B = 20$ iterations and use dropout with a 50% probability to zero out a channel. The uncertainty map is estimated with the variance for each sub-region independently. Let $Y^i = \{y_1^i, y_2^i...y_B^i\}$ be the vector that represents the i-th voxel's predicted labels, the voxel-wise uncertainty map, for each evaluation region, is obtained as the variance:

$$var = \frac{1}{B} \sum_{b=1}^{B} (y_b^i - y_{mean}^i)^2 \qquad (3)$$

Uncertainty can also be estimated with the entropy, as [19] showed. However, the entropy will provide a global measure instead of map for each sub-region. In this case, the voxel-wise uncertainty is calculated as:

$$H(Y^i|X) \approx - \sum_{m=1}^{M} \hat{p}_m^i \ln(\hat{p}_m^i) \tag{4}$$

where $\hat{p}_m^i$ is the frequency of the m-th unique value in Y^i and X represents the input image.

To model aleatoric uncertainty we apply the same augmentation techniques from the training step plus random Gaussian noise, in order to add modifications not previously seen by the network. The final prediction and uncertainty maps are computed following the same strategies as in the epistemic uncertainty.

All that begin said, we hope to evaluate the model's behaviour w.r.t to input and model variability by defining the several experiments:

- Aleatoric Uncertainty: model aleatoric uncertainty with (1) TTA-variance, providing three uncertainty maps (ET, TC, WT) and (2) TTA-entropy, with one global map.
- Epistemic Uncertainty: model epistemic uncertainty with (1) TTD-variance, providing three uncertainty maps (ET, TC, WT) and (2) TTD-entropy, with one global map.
- Hybrid (Aleatoric + Epistemic) Uncertainty: model both aleatoric and epistemic uncertainty together with (1) TTD+TTA-variance, providing three uncertainty maps (ET, TC, WT) and (2) TTD+TTA-entropy, with one global map.

4   Results

The code[1] has been implemented in Pytorch [24] and trained on the GPI[2] servers, based on 2 Intel(R) Xeon(R) @ 2.40 GHz CPUs using 16 GB RAM and a 12 GB NVIDIA GPU, using BraTS 2020 training dataset. We report results on training, validation and test datasets. All results, prediction and uncertainty maps, are uploaded to the CBICA's Image Processing Portal (IPP) for evaluation of Dice score, Hausdorff distance (95th percentile), sensitivity and specificity per each class. Specific uncertainty evaluation metrics are the ratio of filtered TN (FTN) and the ratio of filtered TP (FTP).

4.1   Segmentation

The principal metrics to evaluate the segmentation performance are the Dice Score, which is an overlap measure for pairwise comparison of segmentation mask X and ground truth Y:

---

[1] Github repository: https://github.com/imatge-upc/mri-braintumor-segmentation.

[2] The Image and Video Processing Group (GPI) is a research group of the Signal Theory and Communications Department, Universitat Politècnica de Catalunya.

$$DSC = 2 * \frac{|X \cap Y|}{|X| + |Y|} \tag{5}$$

and the Hausdorff distance, which is the maximum distance of a set to the nearest point in the other set, defined as:

$$D_H(X,Y) = max \left\{ sup_{x \in X} \inf_{y \in Y} d(x,y)), sup_{y \in Y} \inf_{x \in X} d(x,y)) \right\} \tag{6}$$

where sup represents the supremum and inf the infimum. In order to have more robust results and to avoid issues with noisy segmentation, the evaluation scheme uses the 95th percentile.

Tables 1 and 2 show Dice and Hausdorff Distance (95th percentile) scores for training and validation sets respectively.

Table 1. Segmentation results on training dataset (369 cases).

Method	Dice			Hausdorff (mm)		
	WT	TC	ET	WT	TC	ET
V-Net	**0.87**	0.83	0.74	10.19	12.89	35.96
Basic 3D-UNet	0.85	0.84	0.76	**6.97**	10.13	28.23
Residual 3D-UNet	0.82	0.82	0.76	8.56	12.11	28.93
Residual 3D-UNet-multiscale	0.84	0.84	0.76	7.43	12.37	**27.09**
Ensemble - mean	0.85	**0.85**	**0.77**	10.46	**6.90**	29.03

Table 2. Segmentation results on validation dataset (125 cases)

Method	Dice			Hausdorff (mm)		
	WT	TC	ET	WT	TC	ET
V-Net + post	**0.86**	0.78	0.69	14.50	16.15	43.52
Basic 3D-UNet +post	0.81	0.78	0.67	13.10	14.01	43.89
Residual 3D-UNet + post	0.81	0.78	0.71	11.85	18.82	**34.97**
Residual 3D-UNet-multiscale + post	0.83	0.77	0.72	12.34	13.11	37.42
Ensemble mean + post	0.84	**0.79**	**0.72**	10.93	**12.24**	37.97

The model used with the test set is the Residual 3D-UNet-multiscale with post-processing. Table 3 shows the results in the training, validation and test sets for comparison.

Table 3. Segmentation Results for model Residual 3D-UNet-multiscale + post on the three datasets

Dataset	Dice			Hausdorff (mm)		
	WT	TC	ET	WT	TC	ET
Train	0.84	0.84	0.76	7.43	12.37	27.09
Valid	0.83	0.77	0.72	12.34	13.11	37.42
Test	0.81	0.82	0.77	12.59	19.73	21.96

All the proposed models are greatly penalized when no ET is present on the ground truth. In addition, the V-Net suffers more from false positives and 3D-UNet based models from false negatives. The excess of false positives may be caused due to the usage of small patches instead of using the whole volume, which provokes a variation in the proportion of healthy tissue against tumor regions. On the other hand, 3D-UNet models use bigger patch sizes and pooling layers instead of strided convolutions which may be the cause of having a larger number of false negatives. Increasing the patch size helps reduce false positives but it misses local information, which is reflected in label miss-classification on the region's boundaries. Figure 5 shows a visual comparison of the models with a representation on the explained behaviours.

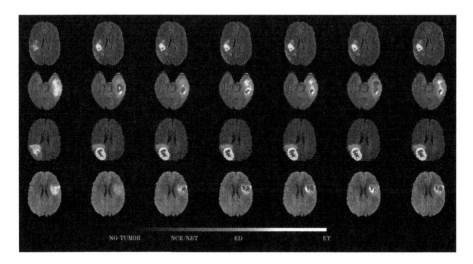

Fig. 5. Training results on patients: 280, 010, 331 and 178 (top-bottom). Image order: (1) Flair (2) GT (3) Residual 3D-UNet-multiscale (4) Residual 3D-UNet (5) Basic 3D-UNet (6) V-Net (7) Ensemble mean

4.2 Uncertainty

BraTS requires to upload three uncertainty maps, one for each subregion (WT, TC, ET) together with the prediction map. Values must be normalized between 0–100 such that "0" represents the most certain prediction and "100" represents the most uncertain. The metrics used are the FTP ratio defined as $FTP = (TP_{100} - TP_T)/TP_{100}$, where T represents the threshold used to filter the more uncertain values. The ratio of filtered true negatives (FTN) is calculated in a similar manner. The integrated score will be calculated as follows:

$$score = AUC_1 + (1 - AUC_2) + (1 - AUC_3). \tag{7}$$

From this point forward all experiments are performed on the model *Residual 3D-UNet-multiscale*, as it is the one with more balanced results across the different regions. Table 4 shows the results for the epistemic, aleatoric and hybrid uncertainties when computed with entropy or variance. As a general overview, we can see that the AUC-Dice, which is computed by averaging the segmentation results for several thresholds that filter uncertain predictions, improves 2 to 3 points w.r.t the results obtained in the segmentation task ($WT : 0.8172$, $TC : 0.7664$, $ET : 0.7071$). Although the metrics are not the same, it indicates that the model is more certain on the TP and less certain on FP and FN. Moreover, the AUC-Dice is higher when using entropy as the uncertainty measure.

Our results show that the model is more uncertain in LGG patients, particularly on epistemic uncertainty; meaning the model requires more data to achieve a more confident prediction. If we compare the behaviour between the uncertainty types, we see that (1) aleatoric focuses on the region boundaries, with small variations (2) epistemic improves results on the ET region but filters more TP and TN and(2) the hybrid approach achieves the best Dice-AUC results when using entropy as the uncertainty measurement.

Table 4. Validation results on the *Residual 3D-UNet-multiscale* for the followed approaches to estimate uncertainty.

Measure	Method	Dice Score			Ratio FTP			Ratio FTN		
		WT	TC	ET	WT	TC	ET	WT	TC	ET
Variance	TTA	0.83	**0,77**	0,71	**0,05**	**0,05**	**0,04**	9.0e−4	2.0e−4	**1.0e−4**
	TTD	0.83	0,76	**0,73**	0,17	0,16	0,09	2.4e−3	1.5e−3	4.0e−4
	Hybrid	**0,83**	0,76	0,73	0,18	0,16	0,10	3.6e−3	2.0e−3	5.0e−4
Entropy	TTA	0,83	0,78	0,71	0,06	**0,05**	**0,04**	1.1e−3	4.7e−3	6.3e−3
	TTD	0,82	0,78	0,74	0,15	0,13	0,07	2.1e−3	8.2e−3	1.22e−2
	Hybrid	**0,83**	**0,79**	**0,77**	0,15	0,12	0,07	3.0e−3	1.01e−3	1.39e−2

We participate in the challenge using test time augmentation (TTA) when uncertain values are computed using variance as it achieves the highest integrated scores per sub-region on the validation set. Table 5 shows the obtained results in both validation and test sets. The achieved integrated scores for validation are

0.93, 0.91 and 0.89 and for test 0.93, 0.93, 0.91 for WT, TC and ET respectively. We see a two point improvement on the ET and TC sub-regions for the test set.

Table 5. Uncertainty Results for the *Residual 3D-UNet-multiscale* model computed using TTA and variance for each sub-region independently. We show results for validation and test set for comparison

Dataset	DICE AUC			FTP RATIO AUC			FTN RATIO AUC		
	WT	TC	ET	WT	TC	ET	WT	TC	ET
Valid	**0.8316**	0.7715	0.7088	0.0449	**0.0538**	**0.0380**	**0.0009**	**0.0002**	**0.0001**
Test	0.8299	**0.8124**	**0.7654**	**0.0332**	0.0537	0.0395	0.0020	0.0005	0.0003

5   Discussion and Conclusions

This work proposes a set of models based on two 3D-CNNs specialized in medical imaging, V-Net and 3D-UNet. As each of the trained models performs better in a particular tumor region, we define an ensemble of those models in order to increase the performance. Moreover, we analyze the implication of uncertainty estimation on the predicted segmentation in order to understand the reliability of the provided segmentation and identify challenging cases, but also as a means of improving the model accuracy by filtering uncertain voxels that should refer to wrong predictions. We use the Residual 3D-UNet-multiscale as our model to participate at the BraTS'20 challenge.

The best results in the validation set are obtained when creating an ensemble of the proposed models, as we can leverage the biases of each model, but are still far from the current state the art. These results may be caused by a bad training strategy where the sampling technique does not reflect the correct label distribution, thus providing more false detections. This is reflected more in the ET region as all models predict more tumor voxels of this label, which is greatly penalized when the ground truth does not contain it. In order to improve results, future work should try to provide a better representation of the labels, not just increase the patch size, but maybe let the network see both local and more global information.

Another potential problem is the model's simplicity. Although previous works achieve good results using a 3D-UNet, i.e. [15], adding more complexity to the network may help boost the performance. Therefore a possible line of work would be to extend the proposed models into a cascaded network, where each nested evaluation region –WT, TC and ET– is learnt as a binary problem. Also, LGG subjects usually achieve lower accuracy on the prediction. In order to improve the results, we could research other post processing techniques and design them specifically to target each one of the glioma grades, as they may be differentiated by the sub-region distribution.

For uncertainty estimation, the work evaluates the usage of aleatoric, epistemic and a hybrid approach using the entropy as a global measure and variance to evaluate uncertainty on each evaluation region. In the provided results, it has been seen that using uncertainty information actually helps improve the accuracy of the network, achieving the best Dice Score (AUC, estimated from filtering uncertain voxels) when using the hybrid approach and entropy as the uncertainty measure. Our method achieves a score of 0.93, 0.93, 0.91 for WT, TC and ET respectively on the test set.

References

1. Menze, B.H., Jakab, A., Bauer, S., Kalpathy-Cramer, J., Farahani, K., Kirby, J., et al.: The multimodal brain tumor image segmentation benchmark (BRATS). IEEE Trans. Med. Imaging **34**(10), 1993–2024 (2015). https://doi.org/10.1109/TMI.2014.2377694
2. Bakas, S., Akbari, H., Sotiras, A., Bilello, M., Rozycki, M., Kirby, J.S., et al.: Advancing The Cancer Genome Atlas glioma MRI collections with expert segmentation labels and radiomic features. Nat. Sci. Data **4**, 170117 (2017). https://doi.org/10.1038/sdata.2017.117
3. Bakas, S., Reyes, M., Jakab, A., Bauer, S., Rempfler, M., Crimi, A., et al.: Identifying the best machine learning algorithms for brain tumor segmentation, progression assessment, and overall survival prediction in the BRATS challenge, arXiv preprint arXiv:1811.02629 (2018)
4. Bakas, S., Akbari, H., Sotiras, A., Bilello, M., Rozycki, M., Kirby, J., et al.: Segmentation labels and radiomic features for the pre-operative scans of the TCGA-GBM collection. Cancer Imaging Arch. (2017). https://doi.org/10.7937/K9/TCIA.2017.KLXWJJ1Q
5. Bakas, S., Akbari, H., Sotiras, A., Bilello, M., Rozycki, M., Kirby, J., et al.: Segmentation labels and radiomic features for the pre-operative scans of the TCGA-LGG collection. Cancer Imaging Arch. (2017). https://doi.org/10.7937/K9/TCIA.2017.GJQ7R0EF
6. Milletari, F., Navab, N., Ahmadi, S.-A.: V-Net: fully convolutional neural networks for volumetric medical image segmentation. In: 2016 Fourth International Conference on 3D Vision (3DV). IEEE (2016)
7. Morgan, L.L.: The epidemiology of glioma in adults: a "state of the science" review. Neuro-oncology **17** (2015). https://doi.org/10.1093/neuonc/nou358
8. Der Kiureghian, A., Ditlevsen, O.: Aleatory or epistemic? Does it matter? Struct. Saf. **31**(2), 105–112 (2009)
9. Gal, Y., Ghahramani, Z.: Dropout as a Bayesian approximation: representing model uncertainty in deep learning. arXiv preprint arXiv:1506.02142 (2015)
10. Kamnitsas, K., et al.: Efficient multi-scale 3D CNN with fully connected CRF for accurate brain lesion segmentation. Med. Image Anal. **36**, 61–78 (2017). https://doi.org/10.1016/j.media.2016.10.004
11. Casamitjana, A., Puch, S., Aduriz, A., Vilaplana, V.: 3D Convolutional Neural Networks for Brain Tumor Segmentation: A Comparison of Multi-resolution Architectures. In: Crimi, A., Menze, B., Maier, O., Reyes, M., Handels, H. (eds.) BrainLes 2016. LNCS, vol. 10154, pp. 150–161. Springer, Cham (2016). https://doi.org/10.1007/978-3-319-55524-9_15
12. Kamnitsas, K., Bai, W., Ferrante, E., McDonagh, S., Sinclair, M., Pawlowski, N: Ensembles of multiple models and architectures for robust brain tumour segmentation. In: International MICCAI Brainlesion Workshop, Quebec, QC, pp. 450–462 arXiv preprint arXiv:1711.01468 (2017)

13. Casamitjana, A., Catà, M., Sánchez, I., Combalia, M., Vilaplana, V.: Cascaded V-Net using ROI masks for brain tumor segmentation. In: Crimi, A., Bakas, S., Kuijf, H., Menze, B., Reyes, M. (eds.) BrainLes 2017. LNCS, vol. 10670, pp. 381–391. Springer, Cham (2018). https://doi.org/10.1007/978-3-319-75238-9_33

14. Myronenko, A.: 3D MRI brain tumor segmentation using autoencoder regularization. arXiv preprint arXiv:1810.11654 (2016)

15. Isensee, F., Kickingereder, P., Wick, W., Bendszus, M., Maier-Hein, K.H.: No new-net. In: Crimi, A., Bakas, S., Kuijf, H., Keyvan, F., Reyes, M., van Walsum, T. (eds.) BrainLes 2018. LNCS, vol. 11384, pp. 234–244. Springer, Cham (2019). https://doi.org/10.1007/978-3-030-11726-9_21

16. Jiang, Z., Ding, C., Liu, M., Tao, D.: Two-stage cascaded U-Net: 1st place solution to BraTS challenge 2019 segmentation task. In: Crimi, A., Bakas, S. (eds.) BrainLes 2019. LNCS, vol. 11992, pp. 231–241. Springer, Cham (2020). https://doi.org/10.1007/978-3-030-46640-4_22

17. Zhao, Y.-X., Zhang, Y.-M., Liu, C.-L.: Bag of tricks for 3D MRI brain tumor segmentation. In: Crimi, A., Bakas, S. (eds.) BrainLes 2019. LNCS, vol. 11992, pp. 210–220. Springer, Cham (2020). https://doi.org/10.1007/978-3-030-46640-4_20

18. Parth, N., Avinash, K., Ganapathy, K.: Demystifying brain tumor segmentation networks: interpretability and uncertainty analysis. Front. Comput. Neurosci. **14**, 6 (2020). https://doi.org/10.3389/fncom.2020.00006

19. Wang, G., Li, W., Ourselin, S., Vercauteren, T.: Automatic Brain tumor segmentation based on cascaded convolutional neural networks with uncertainty estimation. Front. Comput. Neurosci. **13**, 56 (2019). https://doi.org/10.3389/fncom.2019.00056

20. McKinley, R., Meier, R., Wiest, R.: Ensembles of densely-connected CNNs with label-uncertainty for brain tumor segmentation. In: Crimi, A., Bakas, S., Kuijf, H., Keyvan, F., Reyes, M., van Walsum, T. (eds.) BrainLes 2018. LNCS, vol. 11384, pp. 456–465. Springer, Cham (2019). https://doi.org/10.1007/978-3-030-11726-9_40

21. Ulyanov, D., Vedaldi, A., Lempitsky, V.: Instance normalization: the missing ingredient for fast stylization. arXiv preprint arXiv:1607.08022 (2016)

22. Kamnitsas, K., et al.: Efficient multi-scale 3D CNN with fully connected CRF for accurate brain lesion segmentation. Med. Image Anal. **36**, 61–78 (2017)

23. Gal, Y., Ghahraman, Z.: Dropout as a Bayesian approximation: representing model uncertainty in deep learning. arXiv preprint arXiv:1506.02142 (2015)

24. Paszke, A., et al.: Automatic differentiation in PyTorch. In: NIPS-W (2017)

25. Ronneberger, O., Fischer, P., Brox, T.: U-Net: convolutional networks for biomedical image segmentation. In: Navab, N., Hornegger, J., Wells, W.M., Frangi, A.F. (eds.) MICCAI 2015. LNCS, vol. 9351, pp. 234–241. Springer, Cham (2015). https://doi.org/10.1007/978-3-319-24574-4_28

26. Sudre, C.H., Li, W., Vercauteren, T., Ourselin, S., Jorge Cardoso, M.: Generalised dice overlap as a deep learning loss function for highly unbalanced segmentations. In: Cardoso, M.J., et al. (eds.) DLMIA/ML-CDS -2017. LNCS, vol. 10553, pp. 240–248. Springer, Cham (2017). https://doi.org/10.1007/978-3-319-67558-9_28

27. Çiçek, Ö., Abdulkadir, A., Lienkamp, S.S., Brox, T., Ronneberger, O.: 3D U-Net: learning dense volumetric segmentation from sparse annotation, arXiv preprint arXiv:1606.06650 (2016)

28. Crum, W.R., Camara, O., Hill, D.L.G.: Generalized overlap measures for evaluation and validation in medical image analysis. IEEE Trans. Med. Imaging **25**(11), 1451–1461 (2006)

Utility of Brain Parcellation in Enhancing Brain Tumor Segmentation and Survival Prediction

Yue Zhang[1,2], Jiewei Wu[3], Weikai Huang[1], Yifan Chen[4], Ed. X. Wu[2], and Xiaoying Tang[1(✉)]

[1] Department of Electrical and Electronic Engineering,
Southern University of Science and Technology, Shenzhen, China
tangxy@sustech.edu.cn
[2] Department of Electrical and Electronic Engineering,
The University of Hong Kong, Hong Kong, China
[3] School of Electronics and Information Technology, Sun Yat-Sen University,
Guangzhou, China
[4] School of Life Science and Technology,
University of Electronic Science and Technology, Chengdu, China

Abstract. In this paper, we proposed a UNet-based brain tumor segmentation method and a linear model-based survival prediction method. The effectiveness of UNet has been validated in automatically segmenting brain tumors from multimodal magnetic resonance (MR) images. Rather than network architecture, we focused more on making use of additional information (brain parcellation), training and testing strategy (coarse-to-fine), and ensemble technique to improve the segmentation performance. We then developed a linear classification model for survival prediction. Different from previous studies that mainly employ features from brain tumor segmentation, we also extracted features from brain parcellation, which further improved the prediction accuracy. On the challenge testing dataset, the proposed approach yielded average Dice scores of 88.43%, 84.51%, and 78.93% for the whole tumor, tumor core, and enhancing tumor in the segmentation task and an overall accuracy of 0.533 in the survival prediction task.

Keywords: UNet · Brain parcellation · Brain tumor segmentation · Survival prediction

1 Introduction

Glioma tumors are the most common brain malignancies that seriously endanger the health of patients [1]. The incidence rate of glioma tumors is high, and early diagnosis is crucial for treatments and interventions. Magnetic resonance

---

Y. Zhang and J. Wu—Equal contribution.

A. Crimi and S. Bakas (Eds.): BrainLes 2020, LNCS 12658, pp. 391–400, 2021.
https://doi.org/10.1007/978-3-030-72084-1_35

imaging (MRI) is one of the most widely-employed noninvasive ways in detecting brain tumors [2]. Quantitative analyses of brain tumors provide important information for disease diagnosis, surgical planning and prognosis [3], wherein accurate segmentations of brain tumors and their sub-regions from multimodal MR images are essential. Labels manually traced by professional radiologists are regarded as the gold standard. However, fully-automated brain tumor segmentation approaches are urged given that manual tracing is excessively labor-intensive.

Deep learning has proved its effectiveness in various medical image segmentation tasks when a large amount of training data is available. The multimodal Brain Tumor Segmentation (BraTS) challenge has released a large amount of pre-operative MR images and the corresponding manual annotations [3–7]. Benefited from this dataset, deep learning has quickly become the mainstream for brain tumor segmentation [8–11]. Isensee et al. suggested that a well trained UNet is hard to beat and their released code on UNet performs much better than existing state-of-the-art methods on many segmentation tasks [12]. Kao et al. proposed a pipeline using an existing atlas to obtain brain parcellations and demonstrated that brain parcellation is beneficial for improving the brain tumor segmentation accuracy [11]. Inspired by these previous works, we propose a UNet-based brain tumor segmentation pipeline making use of brain parcellation in the brain tumors segmentation task. In addition, we employ coarse-to-fine and ensemble strategies to further improve the segmentation accuracy.

Survival prediction is very important in tumor prognosis and patient management [13–15]. In addition to brain tumor segmentation, the BraTS challenge also included a survival prediction task, aiming to predict survival days based on multimodal MR images, segmentation results and ages. Xue et al. proposed a multivariate linear regression model and this simple strategy outperformed many radiomics and/or machine learning-based methods in the final testing stage of BraTS 2018 [13]. Adopting a similar linear regression model, we demonstrate that brain parcellation is also useful for the survival prediction task. We also use a classification model and feature selection strategy to further improve our survival prediction accuracy.

2   Materials and Methods

2.1   Dataset

The dataset used in this study is provided by the BraTS 2020 challenge organizers [3–7]. For the segmentation task, there are 369 training cases, 125 validation cases, and 166 testing cases. Each case includes the following MR modals: T1-weighted, contrast enhanced T1-weighted (T1c), T2-weighted, and Fluid Attenuation Inversion Recovery (FLAIR) images. There are three tumor tissue labels: necrotic core and non enhancing tumor (NCR/NET—label 1), edema (ED—label 2), and enhancing tumor (ET—label 4). Label 3 is not provided. Whole tumor (WT) is the union of all foregrounds, including label 1, label 2 and label 4. Tumor core (TC) is defined to cover NCR/NET and ET, i.e., label 1 and label

4. The Dice scores and 95% Hausdorff distances (H95) of WT, TC, and ET are calculated to evaluate the segmentation performance [3]. For the segmentation task, we conduct five fold cross-validation on the training data and then test the proposed pipeline on the validation data ground truth labels of which are nevertheless unavailable. For the survival prediction task, BraTS 2020 consists of 236 training cases, 29 validation cases and 164 testing cases. Based on the number of survival days, the subjects are grouped into three classes, e.g. long-survivors (> 15 months), short-survivors (< 10 months), and mid-survivors (between 10 to 15 months). The classification accuracy (i.e. the number of correctly classified patients) is used to evaluate the prediction performance. For the survival prediction task, we perform leave-one-out cross-validation (LOOCV) analysis on the training dataset and also test the proposed method on the validation data.

2.2 Tumor Segmentation

The proposed method utilizes UNet as the baseline architecture [16]. The following four UNets are separately trained for the proposed method.

UNet^1. Given the fact that different tumor regions are better visible in different MR modals [17], four different-modal MR images are jointly inputted to UNet as multiple channels and the corresponding human annotation is regarded as the expected output.

UNet^2. A brain parcellation model is trained using a dataset published in the previous work [18], which can output a brain parcellation given a T1 MR image as the input. We use this model to obtain the brain parcellations for all training data and validation data of BraTS 2020.

UNet^3. Brain parcellations and multi-modal MR images are jointly used to train a model.

UNet^4. The fact that a smaller input region may lead to a more accurate segmentation motivates coarse-to-fine methods [19]. Therefore, we train a model using cropped multi-modal MR images and brain parcellations according to the WT masks.

In the testing stage, the segmentations are obtained using the flowchart shown in Fig. 1. Firstly, **UNet^1** is used to obtain a coarse segmentation S_1. **UNet^2** is used the obtain the brain parcellation which is then inputted to **UNet^3** jointly with the corresponding multi-modal MR images to obtain another coarse segmentation S_2. The two coarse segmentations S_1 and S_2 are then fused and used to localize the WT. Afterwards, all MR images and the brain parcellation are cropped and fed into **UNet^4** to predict a fine segmentation S_3. The final segmentation result is obtained by majority voting of S_1, S_2, S_3.

2.3 Survival Prediction

The proposed survival prediction pipeline is based on a previously-published multivariate linear regression model [13]. We use 1, 2, 3 to respectively represent

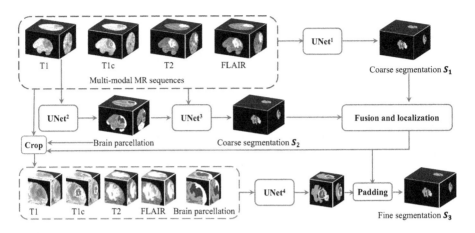

Fig. 1. Illustration of the testing workflow in the proposed method.

three resection status, i.e., Not Available (NA), Gross Total Resection (GTR) and Subtotal Resection (STR). In addition, there are three novel strategies used in our proposed model: 1) extracting additional features from brain parcellation, 2) combining classification and regression models, 3) feature selection. The entire prediction procedure is illustrated in Fig. 2.

Additional Features Obtained from 3D Brain Parcellation. In addition to the features extracted by Xue et al. [13], we extract five more features from brain parcellation. Specifically, we first use the aforementioned brain parcellation model to segment out tissue labels, namely corticospinal tract (CSF), white matter (WM), and gray matter (GM). Then we calculate the surface area of every overlapped region between two brain structures (i.e. overlapped regions between ET and CSF, ET and WM, ET and GM, ED and CSF, ED and WM, as well as NCR/NET and GM). The surface area calculation method is the same as that in Xue's work [13]. Please note other combinations are excluded because they are not overlapped.

Combination of Classification and Regression. Due to the limited sample size, it's relatively challenging to build a regression model to predict accurate survival days. Rather than a direct regression, we combine classification and regression to jointly predict survival days. We first apply a linear classification model to assign subjects to one of the three classes and calculate the probability of each class. After that, a linear regression model is used to identify the relationship between the aforementioned probabilities and survival day. The relationship between the three classes and the survival days should be: short-survivors – [0, 300), mid-survivors – [300, 450], and long-survivors – (450, +∞). However, the class derived from the predicted survival day D may disagree with the predicted class C. To solve this problem, we design a discriminator to check the consistency between C and D, i.e., we use our classification model to guide our regression model. If D lies in the range of C, the finally predicted survival days $D^* = D$. Otherwise, D^* will be set to be its closest integer that lies in the range of the predicted class. For example, the predicted survival day 401.1 whose predicted class is short-survivors will be updated to be 299.

Feature Selection. If used appropriately, feature selection can automatically remove irrelevant features and improve the overall performance. In this study, univariate linear regression is performed to obtain the correlation between features and survival day (i.e. the weight of each feature), and then K features with the highest correlations are selected as the input features. $K = 5$ is chosen based on LOOCV experiments.

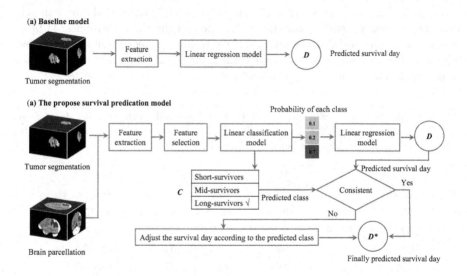

Fig. 2. The entire procedure of the proposed survival prediction pipeline.

2.4   Implementation Details

All procedures are implemented in python. The segmentation model is based on nnUNet[1] [9]. We use the original hyper-parameters of nnUNet and the epochs number is set to be 200. The survival prediction model builds its basis on Xue's work[2] [13].

---

[1] https://github.com/MIC-DKFZ/nnUNet.
[2] https://github.com/xf4j/brats18.

3   Results

3.1   Segmentation Results

Table 1 shows the mean Dice scores and H95 of WT, TC and ET segmentations from $UNet^1$ (3D UNet), $UNet^3$ (3D UNet + BP), $UNet^4$ (3D UNet + BP + C2F) and the ensemble of them. The model with the best performance of each metric is highlighted. For WT, $UNet^1$ and $UNet^3$ perform similarly, but the coarse-to-fine strategy can significantly improve the Dice score and reduce the H95. For TC, $UNet^3$ can improve the segmentation performance over $UNet^1$, and $UNet^4$ can further improve $UNet^3$ segmentation in terms of both Dice and H95. For ET, it is challenging for all three models, but the ensemble of them has the best Dice score compared with each individual model. The last column of Table 1 averages the numbers listed in the second, the third, and the fourth columns. It can be observed that the ensemble model performs the best in general. These results clearly show the effectiveness of the three strategies: 1) brain parcellation as an additional input; 2) coarse-to-fine strategy; 3) model ensemble. Figure 3 shows one representative MR image slice and its corresponding human annotation and automatic segmentations.

Table 1. Ablation analysis results of the proposed method on the training data. Keys: BP - brain parcellation, C2F- coarse-to-fine, H95–95% Hausdorff distances. ↓ indicates that a smaller value represents better performance.

Dice score	WT	TC	ET	Overall
3D UNet	91.42%	86.75%	78.76%	85.64%
3D UNet + BP	91.37%	86.98%	78.97%	85.77%
3D UNet + BP + C2F	**92.32%**	87.23%	78.66%	86.07%
Ensemble	92.29%	**87.70%**	**79.41%**	**86.47%**
H95↓	WT	TC	ET	Overall
3D UNet	4.46 mm	6.49 mm	25.32 mm	12.09 mm
3D UNet + BP	4.62 mm	5.44 mm	**23.33** mm	11.13 mm
3D UNet + BP + C2F	**3.56** mm	**4.29** mm	25.51 mm	11.12 mm
Ensemble	4.06 mm	4.44 mm	24.26 mm	**10.92** mm

The proposed method achieves an average Dice score of 91.10%, 86.02%, and 75.33% for WT, TC, and ET on the validation dataset, which has been evaluated by the organizer of BraTS 2020[3]. Figure 4 shows a representative slice of the final segmentation results of the validation dataset. It is challenging to identify the boundaries of WT, TC, ET from a single-modal MR image. The automated segmentation results generally match the prior knowledge that WT, TC, and ET are most visible respectively on FLAIR, T2, and T1c [3]. The proposed method

---

[3] https://www.cbica.upenn.edu/BraTS20/lboardValidation.html.

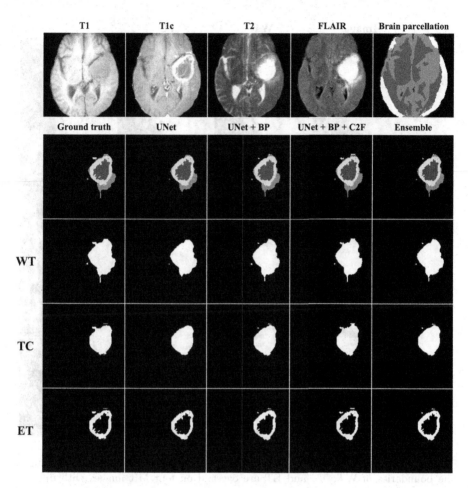

Fig. 3. A representative example of segmentation results of the training dataset.

achieves an average Dice score of 88.43%, 84.51%, and 78.93% for WT, TC, and ET on the testing dataset.

3.2   Survival Prediction Results

Table 2 shows the results of all prediction experiments. The best performance in terms of each metric is highlighted. Compared with the baseline method, adding additional features from brain parcellation can increase the accuracy from 0.436 to 0.445, which indicates brain parcellation can provide critical information for survival predication. Moreover, when incorporating classification, the accuracy is further promoted from 0.445 to 0.525, indicating a combination of classification and regression is effective for the prediction task. In addition to the above strategies, feature selection also increases the accuracy from 0.525 to 0.534. Those observations may well explain the superiority of the proposed pipeline.

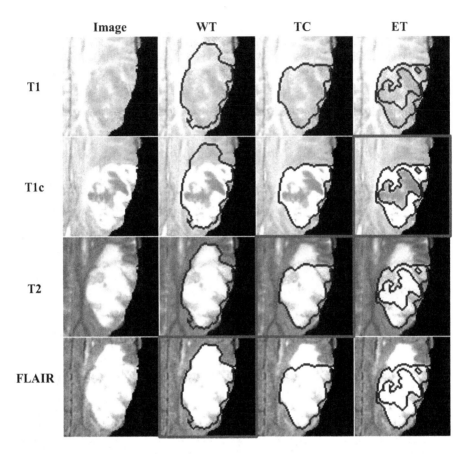

Fig. 4. A representative example of segmentation results on the validation dataset. The boundaries of WT, TC and ET are overlaid on four MR images with different modalities.

Table 2. Ablation analysis of the proposed survival prediction method on the training data. Keys: BP - brain parcellation, FS - feature selection. ↓ indicates that a smaller value represents better performance.

Baseline [13]	BP	Classification	FS	Accuracy	Mean error ↓
√				0.436	323.75 days
√	√			0.445	**322.21** days
√	√	√		0.525	325.89 days
√	√	√	√	**0.534**	328.60 days

For the validation dataset, the proposed method achieves an accuracy of 0.379, a mean error of 322.75 days, and a median error of 236.50 days. For the testing dataset, the proposed method achieves an accuracy of 0.533, a mean error of 640.58 days, and a median error of 228 days.

4 Discussion and Conclusion

In this work, to demonstrate that brain parcellation is useful for brain tumor segmentation and survival prediction, we proposed and validated a UNet-based segmentation method and a linear model-based survival prediction method. For the tumor segmentation task, a coarse-to-fine strategy incorporating suitable ensemble techniques has been proposed to further improve the segmentation performance. For the survival prediction task, a combination of classification and regression, together with feature selection, were employed to further enhance the prediction accuracy.

A potential limitation of this work is that the brain parcellation may be inaccurate around the tumor regions, because there are intensity changes surrounding the tumor regions. In the future, we aim to combine atlas-based methods with deep learning to get a more accurate brain parcellation [20], which may be helpful in further improving the accuracy of our brain tumor segmentation and survival prediction.

References

1. Louis, D.N., Perry, A., Reifenberger, G., Von Deimling, A., et al.: The 2016 World Health Organization classification of tumors of the central nervous system: a summary. Acta Neuropathol. **131**(6), 803–820 (2016). https://doi.org/10.1007/s00401-016-1545-1
2. Kumar, V., Gu, Y., Basu, S., Berglund, A., et al.: Radiomics: the process and the challenges. Magn. Reson. Imaging **30**(9), 1234–1248 (2012)
3. Bakas, S., et al.: Advancing the cancer genome atlas glioma MRI collections with expert segmentation labels and radiomic features. Sci. Data **4**, 170117 (2017)
4. Menze, B.H., et al.: The multimodal brain tumor image segmentation benchmark (BraTS). IEEE Trans. Med. Imaging **34**(10), 1993–2024 (2015)
5. Bakas, S., Reyes, M., Jakab, A., et al.: Identifying the best machine learning algorithms for brain tumor segmentation, progression assessment, and overall survival prediction in the BRATS challenge. arXiv preprint arXiv:1811.02629 (2018)
6. Bakas, S., Akbari, H., Sotiras, A., et al.: Segmentation labels and radiomic features for the pre-operative scans of the TCGA-GBM collection. Cancer Imaging Arch. (2017). https://doi.org/10.7937/K9/TCIA.2017.KLXWJJ1Q
7. Bakas, S., Akbari, H., Sotiras, A., et al.: Segmentation labels and radiomic features for the pre-operative scans of the TCGA-LGG collection. Cancer Imaging Arch. (2017). https://doi.org/10.7937/K9/TCIA.2017.GJQ7R0EF
8. Myronenko, A.: 3D MRI brain tumor segmentation using autoencoder regularization. In: Crimi, A., Bakas, S., Kuijf, H., Keyvan, F., Reyes, M., van Walsum, T. (eds.) BrainLes 2018. LNCS, vol. 11384, pp. 311–320. Springer, Cham (2019). https://doi.org/10.1007/978-3-030-11726-9_28

9. Isensee, F., Kickingereder, P., Wick, W., Bendszus, M., Maier-Hein, K.H.: No new-net. In: Crimi, A., Bakas, S., Kuijf, H., Keyvan, F., Reyes, M., van Walsum, T. (eds.) BrainLes 2018. LNCS, vol. 11384, pp. 234–244. Springer, Cham (2019). https://doi.org/10.1007/978-3-030-11726-9_21

10. McKinley, R., Meier, R., Wiest, R.: Ensembles of densely-connected CNNs with label-uncertainty for brain tumor segmentation. In: Crimi, A., Bakas, S., Kuijf, H., Keyvan, F., Reyes, M., van Walsum, T. (eds.) BrainLes 2018. LNCS, vol. 11384, pp. 456–465. Springer, Cham (2019). https://doi.org/10.1007/978-3-030-11726-9_40

11. Kao, P.-Y., Ngo, T., Zhang, A., Chen, J.W., Manjunath, B.S.: Brain tumor segmentation and tractographic feature extraction from structural MR images for overall survival prediction. In: Crimi, A., Bakas, S., Kuijf, H., Keyvan, F., Reyes, M., van Walsum, T. (eds.) BrainLes 2018. LNCS, vol. 11384, pp. 128–141. Springer, Cham (2019). https://doi.org/10.1007/978-3-030-11726-9_12

12. Isensee, F., Jäger, P.F., Kohl, S.A., et al.: Automated design of deep learning methods for biomedical image segmentation. arXiv preprint arXiv:1904.08128 (2020)

13. Feng, X., Tustison, J., Patel, H., et al.: Brain tumor segmentation using an ensemble of 3D U-nets and overall survival prediction using radiomic features. Front. Comput. Neurosci. **14**(25) (2020)

14. Puybareau, E., Tochon, G., Chazalon, J., Fabrizio, J.: Segmentation of gliomas and prediction of patient overall survival: a simple and fast procedure. In: Crimi, A., Bakas, S., Kuijf, H., Keyvan, F., Reyes, M., van Walsum, T. (eds.) BrainLes 2018. LNCS, vol. 11384, pp. 199–209. Springer, Cham (2019). https://doi.org/10.1007/978-3-030-11726-9_18

15. Sun, L., Zhang, S., Luo, L.: Tumor segmentation and survival prediction in glioma with deep learning. In: Crimi, A., Bakas, S., Kuijf, H., Keyvan, F., Reyes, M., van Walsum, T. (eds.) BrainLes 2018. LNCS, vol. 11384, pp. 83–93. Springer, Cham (2019). https://doi.org/10.1007/978-3-030-11726-9_8

16. Ronneberger, O., Fischer, P., Brox, T.: U-Net: convolutional networks for biomedical image segmentation. In: Navab, N., Hornegger, J., Wells, W.M., Frangi, A.F. (eds.) MICCAI 2015. LNCS, vol. 9351, pp. 234–241. Springer, Cham (2015). https://doi.org/10.1007/978-3-319-24574-4_28

17. Banerjee, S., Mitra, S., Uma Shankar, B.: Single seed delineation of brain tumor sing multi-thresholding. Inf. Sci. **330**, 88–103 (2016)

18. Wu, J., Zhang, Y., Tang, X.: Simultaneous tissue classification and lateral ventricle segmentation via a 2D U-net driven by a 3D fully convolutional neural network. In: EMBC 2019, pp. 5928–5931 (2019)

19. Zhou, Y., Xie, L., Shen, W., Wang, Y., Fishman, E.K., Yuille, A.L.: A fixed-point model for pancreas segmentation in abdominal CT scans. In: Descoteaux, M., Maier-Hein, L., Franz, A., Jannin, P., Collins, D.L., Duchesne, S. (eds.) MICCAI 2017. LNCS, vol. 10433, pp. 693–701. Springer, Cham (2017). https://doi.org/10.1007/978-3-319-66182-7_79

20. Zhang, Y., Wu, J., Liu, Y., Chen, Y., Wu, X., Tang, X.: MI-UNet: multi-inputs UNet incorporating brain parcellation for stroke lesion segmentation from T1-weighted magnetic resonance images. IEEE J. Biomed. Health Inform. **25**, 526–535 (2020)

Uncertainty-Driven Refinement of Tumor-Core Segmentation Using 3D-to-2D Networks with Label Uncertainty

Richard McKinley[✉], Micheal Rebsamen, Katrin Dätwyler, Raphael Meier, Piotr Radojewski, and Roland Wiest

Support Centre for Advanced Neuroimaging, University Institute of Diagnostic and Interventional Neuroradiology, Inselspital, Bern University Hospital, Bern, Switzerland
richard.mckinley@insel.ch

Abstract. The BraTS dataset contains a mixture of high-grade and low-grade gliomas, which have a rather different appearance: previous studies have shown that performance can be improved by separated training on low-grade gliomas (LGGs) and high-grade gliomas (HGGs), but in practice this information is not available at test time to decide which model to use. By contrast with HGGs, LGGs often present no sharp boundary between the tumor core and the surrounding edema, but rather a gradual reduction of tumor-cell density.

Utilizing our 3D-to-2D fully convolutional architecture, DeepSCAN, which ranked highly in the 2019 BraTS challenge and was trained using an uncertainty-aware loss, we separate cases into those with a confidently segmented core, and those with a vaguely segmented or missing core. Since by assumption every tumor has a core, we reduce the threshold for classification of core tissue in those cases where the core, as segmented by the classifier, is vaguely defined or missing.

We then predict survival of high-grade glioma patients using a fusion of linear regression and random forest classification, based on age, number of distinct tumor components, and number of distinct tumor cores.

We present results on the validation dataset of the Multimodal Brain Tumor Segmentation Challenge 2020 (segmentation and uncertainty challenge), and on the testing set, where the method achieved 4th place in Segmentation, 1st place in uncertainty estimation, and 1st place in Survival prediction.

1 Introduction

The BRATS challenge [4,15], and its accompanying dataset [1–3] of annotated glioma images have driven substantial amounts of research in brain tumor segmentation [5,6,18]. The data is taken from various sources/centers/scanners and incorporates both low-grade gliomas (LGGs) and high-grade gliomas (HGGs).

© Springer Nature Switzerland AG 2021
A. Crimi and S. Bakas (Eds.): BrainLes 2020, LNCS 12658, pp. 401–411, 2021.
https://doi.org/10.1007/978-3-030-72084-1_36

While this may enhance the applicability of models trained on the dataset, previous work has shown that the inclusion of both low- and high-grade tumors (which have rather different visual appearance) leads to lower performance than training on only low- or high-grade tumors [19].

Based on our 3rd-place entry to the BRaTS 2018/2019 segmentation challenge [11,13], which was trained using an uncertainty-aware loss function which outputs confidence maps for each voxel/tissue type, we attempt to solve the problem of uncertain core segmentation in LGGs. We observe that in the subclass of LGGs which have poorly a delineated core, the whole core is relatively uncertain, whereas in cases with a well-segmented core, the center of the core is predicted with very high certainty. We hypothesize that this arises in diffuse LGGs that have practically no visible boundary between the solid tumor core and the surrounding edema. This also reflects tumor biology: the cell density of tumor cells does not abruptly fall to zero, but rather gradually reduces. We attempt to compensate for this effect by reducing the threshold at which voxels are classified as tumor core, in cases where the core is poorly delineated (as defined by the confidence of the core segmentation). This is essentially the opposite approach to Nair et al. [16,17], where lesions with high uncertainty are deleted from the lesion mask.

2   Heteroscedastic Label-Flip Loss and Focal KL-divergence

In our previous BraTS submission [13], for the training of our model we used a *label-flip* loss: for each voxel and each tissue type, the classifier produces an output $p \in (0, 1)$ and an output $q \in (0, 0.5)$ which represents the probability that the classifier output differs from the ground truth.[1] The label flip loss function used in our previous Brats submission had the form:

$$\text{Focal}(p, (1 - x) * q + x * (1 - q)) + \text{BCE}(q, z) \tag{1}$$

where z is the indicator function for disagreement between the classifier (thresholded at the $p = 0.5$ level) and the ground truth, BCE is binary cross-entropy, and Focal is *focal loss* [9].

This can be understood as a form of heteroscedastic classification networks (a network which predicts the variance of their preactivation outputs) as introduced in [7]. The loss function has the advantage that (unlike predicting the variance of logit outputs as in Kendall et al.) it is differentiable in p and q and so can be backpropagated through. The focal loss term aims to separate tissue from non-tissue voxels, with attention paid to those close to the decision boundary (focal loss attenuates gradient contributions from well-classified examples far

---

[1] The range of values allowed for q tends to cause confusion, with many readers expecting to find values in a range $(0, 1)$, since q is framed as a probability. However, for a binary classification problem, an uncertainty of 0.5 indicates that the classifier is, in effect, randomly guessing. A q greater than 0.5 would correspond to a classifier which predicts e.g. label 1 but believes that the label in the ground truth is 0.

from the decision boundary). This allows learning to focus on difficult-to-classify cases, meaning that the effect of class imbalance between foreground and background is lessened. However, voxels close to the tumor border are inherently uncertain and are typically not consistently labeled by human raters. We want learning to concentrate on examples which are incorrectly classified, and which are not inherently uncertain (voxels with a high probability of deviation from the reference 'ground truth'). Focal loss is therefore attenuated according to the probability of a 'label-flip' between the output of the classifier and the ground truth.

In practice, this loss function rarely produces values of q close to 0.5. We believe this is because the loss function cannot actively encourage examples to remain (correctly) classified as fundamentally uncertain: the loss function always produces a gradient, however weak, towards the human-provided ground truth. For targets w in $[0, 1]$, the minimum value of BCE is the entropy $-w \ log(w)$ of w. Therefore, if w is close to 0.5, the loss can only be lowered by lowering the uncertainty. However, many examples will indeed have uncertainty close to 0.5, in particular along the boundaries between tissue types. Subtracting the entropy yields the Kullback-Leibler divergence, which does not provide any incentive to reduce the uncertainty measurement.

$$\mathrm{KL}(w \parallel p) = w \ log(w) - w \ log(p)$$

KL divergence is typically not used as a loss in classification problems, because the entropy term $w \ log(w)$ is fixed: however, in our setting w can vary. In this year's submission we amended the loss function to incorporate a form of *focal Kullback Leibler divergence*, which encourages highly uncertain examples to remain close to the decision boundary:

$$\mathrm{Focal_{KL}}(w \parallel p) = (p - w)^2(w \ log(w) - w \ log(p))$$

The new loss function combines focal KL-divergence with binary cross-entropy for estimation of label uncertainty:

$$\mathrm{Focal_{KL}}(w \parallel p) + \ \mathrm{BCE}(q, z)$$

where z is the indicator function for disagreement between the classifier (thresholded at the $p = 0.5$ level) and w is $(1 - x) * q + x * (1 - q)$. In this loss function (unlike the previous) the first term goes to zero as p tend to $w = (1 - x) * q + x * (1 - q)$. With this loss function we observe uncertainty values close to 0.5, representing total uncertainty (essentially uncertain examples which should not be further learned from).

In practical experiments we found that a combination of focal loss and the label-flip loss performed best: our final loss function was of the form:

$$\lambda \ \mathrm{Focal}(p, x) + (1 - \lambda) \ \mathrm{Focal_{KL}}(w \parallel p) + (1 - \lambda) \ \mathrm{BCE}(q, z)$$

with $\lambda = 0.1$ giving good results for retraining models already trained with the previous loss function.

2.1 Ensembling and Label Uncertainty

To ensemble the output of several models, we need to combine the output of the model into a single score. Simply taking the mean of the p_i and the mean of the q_i does not adequately reflect the joint opinion of the models. To see this, observe that for ensembling two predictions, if $p_1 = 0, q_1 = 0$ and $p_2 = 1, q_1 = 0$, the plain ensemble (by averaging) will yield a prediction of 0.5 with probability 0 of disagreeing with the human ground truth. For a single model output, the function $f(p, q) = (1-x) * I_{p <= 0.5} + x * (1-q) I_{p >= 0.5}$ gives a value between 0 and 1 which combines the prediction and the uncertainty of the model and can be ensembled with other predictions in an ordinary fashion: given n predictions $p_1 \ldots p_n$ and $q_1 \ldots q_n$, the mean of $f(p_i, q_i)$ provides a single ensembled prediction.

3 Application to Brain Tumor Segmentation

3.1 Data Preparation and Homogenization

The raw values of MRI sequences cannot be compared across scanners and sequences, and therefore a homogenization is necessary across the training examples. In addition, learning in CNNs proceeds best when the inputs are standardized (i.e. mean zero, and unit variance). To this end, the nonzero intensities in the training, validation and testing sets were standardized, this being done across individual volumes rather than across the training set. This achieves both standardization and homogenization.

3.2 The DeepSCAN Architecture with Attention

Our model architecture (shown in Fig. 1) is identical to our entry to the 2019 challenge: this model has also been applied to the segmentation of MS lesions [14], segmentation of brain anatomy [12], and measurement of cortical thickness [20]. The network was implemented in Pytorch: it consists of an initial phase of 3D convolutions to reduce a non-isotropic 3D patch to 2D, followed by a shallow encoder/decoder network using densely connected dilated convolutions in the bottleneck. This architecture is very similar to that used in our BraTS 2018 submission: principal differences are that we use Instance normalization rather than Batch normalization, and that we add a simple local attention mechanism of our own design (which element-wise multiplies the dilated feature maps with a mask calculated using non-dilated convolutions followed by a sigmoid) between dilated dense blocks. In testing in 2019 we found that adding this attention mechanism was more effective than adding additional dense block, or making dense blocks wider.

We use multi-task rather than multi-class classification: each tumor region (Whole tumor, tumor core, enhancing tumor) is treated as a separate binary classification problem. Inputs to the network (5*196*196 patches) were sampled randomly from either axial, sagittal or coronal direction. We perform simple data augmentation: reflection about the (approximate) midline, rotation around

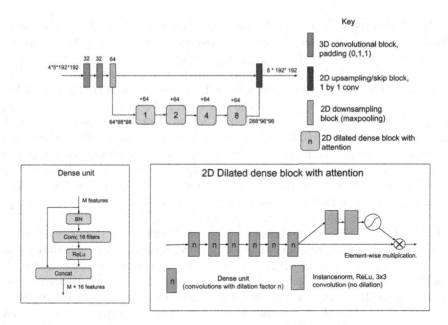

Fig. 1. The DeepSCAN classifier, as applied in this paper to Brain Tumor Segmentation

a random principal axis through a random angle, and global shifting/rescaling of voxel intensities. The network was trained with ADAM, using a batch size of 2 and weight decay 10^{-5}. Models were trained using five-fold cross-validation: to reduce the training time, we used the pre-trained models from our 2019 challenge entry as initial weights, ensuing that the same cases to train each model as in previous training rounds (new training examples were distributed evenly across the folds). We employed cosine annealing learning rate schedule with restarts [10], restarting every 20000 gradient steps (we refer to this rather loosely as an epoch). We trained our model for 10 epochs with learning rate annealing from 10^{-4} to 10^{-7}, and then for 10 further epochs with learning rate annealing from 10^{-5} to 10^{-8}.

Final segmentations were derived by ensembling (as described above) the axial, sagittal and coronal views from the five models obtained from 5-fold cross-validation, with test-time augmentation provided by flipping in the saggital direction and rotating input images through $45°$ in each plane.

4   Uncertainty-Based Filtering

In common with many previous entries, we perform minimal filtering of the final segmentation (Small components (<10voxels) of any tissue class were deleted). In addition, we observed (in common with our 2019 contribution) that for some low-grade gliomas the tumor core was completely missed, or that small isolated sections of the solid tumor were correctly segmented, but the majority was missed. Lowering the threshold for segmentation to 0.05 (from 0.5) meant that these solid tumors were correctly segmented, but led to a lot of false positive tumor core identification in high grade gliomas. Since the challenge does not provide the grade of gliomas, we require a proxy for the grade: we observed that in tumors which were well segmented (mostly HGGs) the mean of the ensembled model output inside the segmented core tumor was above 0.9, whereas in tumors with a poorly segmented core (mostly LGGs) the mean of the ensembled model output inside the tumor (if any tissue was identified as core at all) was typically below 0.75. Thinking of the mean of the ensembled model output as a confidence in the global performance of the model, we therefore set 0.75 as a mean tumor core confidence threshold, and in cases with mean core confidence below that threshold, we reset the threshold for core segmentation to 0.05, thus capturing more tissue as tumor core in those cases. Similar filtering was also applied to whole tumor (confidence threshold of .90) and enhancing tumor (confidence threshold of 0.8). If no tumor core was found after filtering, the whole tumor was set to be tumor core (as in our 2019 submission), on the biological basis that all gliomas should contain a solid core of tumor tissue. Finally, we have seen in our clinical cases that certain low-grade gliomas may be detected very poorly, with no voxels exceeding the 0.5 threshold for detection. If no tumor was detected at all, the threshold for whole tumor detection was lowered until at least 1000 voxels of tumor were detected, on the assumption that the model is only applied to glioma cases. This final 'failsafe' (which we did not observe being triggered in training or testing our model) means that the algorithm is only suitable for outlining tumors in patients known to have a glioma (or similar tumor), not for automatically *detecting* tumors.

4.1   Results

Results of our classifier, as applied to the official BraTS validation data from the 2020 challenge, as generated by the official BraTS validation tool, before and after filtering for uncertainty are shown in Table 1.

Table 1. Results on the BRATS 2020 validation set using the online validation tool. Raw output denotes the ensembled results of the five classifiers derived from cross-validation.

	Dice			Hausdorff 95		
	ET	WT	TC	ET	WT	TC
Raw output	0.76	0.90	0.80	26.8	5.25	12.4
Filtered output	0.76	0.91	0.85	26.8	3.91	5.61

5   Uncertainty Challenge

The BraTS uncertainty challenge requires the submission of an uncertainty score (from 0 to 100) per voxel and tissue type, denoting the certainty of classification at that voxel from most certain (100) to least certain (0). The uncertainty output of our network (denoted q above) can be directly converted to such an uncertainty score by taking

$$100 * (1 - 2q).$$

The uncertainty in our ensemble can likewise be extracted as for any ordinary model with a sigmoid output x as

$$100 * (1 - 2|0.5 - x|))$$

While this uncertainty measure gives a measure of uncertainty both inside and outside the provided segmentation, in practice the following uncertainty, which treats all positive predictions as certain, and only assigns uncertain values to negative predictions, performs better according to the challenge metrics (see Table 2):

$$200 * (max(0.5 - x, 0))$$

Table 2. Results on the BRATS 2020 validation set using the online validation tool: uncertainty challenge.

	Dice AUC			TP AUC			TN AUC		
	WT	TC	ET	WT	TC	ET	WT	TC	ET
Baseline	0.91	0.85	0.76	0	0	0	0	0	0
All uncertainties	0.93	0.86	0.8	0.07	0.13	0.1	0.004	0.003	0.001
Negative uncertainties	0.93	0.87	0.78	0	0	0	0.004	0.003	0.001

6   Survival Prediction

The survival prediction challenge requires a prediction of overall survival (in days) for subjects with "Gross Tumor Resection": while the output should be a number of days, the validation is based on accuracy of classification into three classes: long-survivors (>15 months), short-survivors (<10 months), and mid-survivors (e.g., between 10 and 15 months). The age of the cases is provided: all other predictors must be derived from the imaging. Previous work has shown that patient age alone can predict patient outcome relatively well, and outperform approaches integrating more complicated radiomic features [8, 21].

We worked on the basis that any survival prediction will be approximate, and that the number of training cases is relatively small, from heterogeneous sources, and rather degraded owing to registration artifacts. We therefore picked two segmentation-derived features of the image which, based on the performance of our algorithm in segmentation, should be robust between centers: number of disconnected tumor core regions (abbreviated below to 'number of cores') and number of disconnected whole tumor regions (abbreviated below to 'number of tumors'), as segmented by our model. As can be seen in Fig. 2, these measures are each predictive of survival on the training dataset, with increasing age, number of tumor components and number of tumor cores leading to reduced survival.

Initially, we built a least squares regression model predicting the survival time from age, number of disconnected cores and number of tumors. To avoid the survival prediction being biased by the longer survival times of especially long survivors, we replaced all survival times greater than 1000 d with 1000 (since these survival times are all far in excess of the 15 month cutoff relevant for the challenge). Here we found that the image-derived features were largely unhelpful: ordinary least squares models built on Age, number of tumor cores and number of tumors had coefficients for number of tumor cores and number of tumor compartments which were not significantly different from zero.

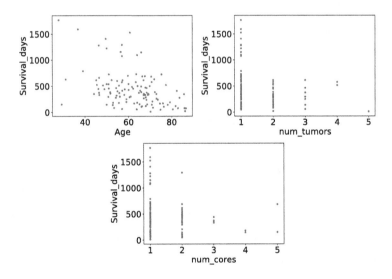

Fig. 2. Scatter plots of overall survival (in days) versus age, number of tumor components, and number of tumor cores, on the BraTS 2020 training dataset

The actual validation for this challenge is based on classification accuracy (into the three survival classes) rather than regression performance. The OLS regression models we built had an accuracy of 44 %. We subsequently trained a Random Forest model with a rather large number of trees (1000) with maximum depth 3 to predict the survival class, using Age alone, and also Age plus our image-derived features. The random forest classifier on age alone also achieved

44 % accuracy in cross-validation, while the model incorporating number of cores and number of tumors achieved 53 % accuracy. This increased can be attributed to the tendency of the random forest model to predict more short-term survivors than the linear regression model.

Our final model was a form of ensemble between the linear regression model and the classification model, in which we allow the random forest classifier to override the prediction of the linear model. To increase robustness, we only allow this when the RF is confident in its prediction: the predicted survival time was the output of the linear regression model, unless the classification model predicted (with probability 50 % or more) a different class to the regression model. In that case the predicted overall survival was moved to a fixed survival time corresponding to 10 months (in the case of low predicted survival) or 15 months (in the case of long predicted survival).

Table 3. Results on the BRATS 2020 test set (segmentation task), as delivered by the challenge organizers.

	Dice			Hausdorff95		
	ET	WT	TC	ET	WT	TC
Mean	0.82	0.90	0.83	13.6	4.7	22.0
StdDev	0.18	0.10	0.26	63.7	6.6	79.6

Table 4. Results on the BRATS 2020 test set (survival task), as delivered by the challenge organizers.

	Accuracy	MSE	medianSE	stdSE	SpearmanR
Value	0.617	391589	57600	1194915	0.378

Table 5. Results on the BRATS 2020 test set (uncertainty task), as delivered by the challenge organizers.

	Dice AUC			Score		
	WT	TC	ET	WT	TC	ET
Mean	0.91	0.85	0.84	0.97	0.95	0.95
StdDev	0.10	0.26	0.18	0.033	0.086	0.059

7   Results and Conclusions

Our results on the BraTS testing set are presented in Tables 3, 4 and 5. The mean Dice coefficients for our model were slightly improved over our 2019 method (the methodology for calculating Hausdorff distance changed from 2019 to 2020), which was ranked third in the 2019 challenge. In terms of raw numeric challenge score, our method ranked joint 4th together with three other methods: however, no statistical significance was found between the score of the 4th ranked teams and the team ranked 3rd.

Our accuracy of 0.617 in predicting patient survival ranked joint first in the survival challenge. This is striking, as our method made use of only simple features, rather than a complex radiomic pipeline. In particular, it is surprising that the number of tumor cores and number of tumor compartments has not been used by previous challenge participants though it may have been reflected in surrogate features defined by radiomic pipelines (for example, ratio of volume to surface area). We believe our fusion of a prediction and classification model is also novel.

Our method for uncertainty estimation ranked first in the challenge, and was significantly better than the other methods entered in the challenge ($p < 10^{-50}$). Since this result arises from a challenge, it is difficult to analyse whether this dramatic difference arises from the superior uncertainty estimation of our model, or the observation that the challenge metric asymmetrically rewarded uncertainties inside and outside of the predicted volume. Given the diminishing returns of comparing hard segmentations to manually derived ground truths, and the additional information that soft outputs characterizing uncertainty provide, this topic certainly merits further investigation.

Acknowledgements. This work was supported by the Swiss Personalized Health Network (SPHN, project number 2018DRI10). Calculations were performed on UBELIX (http://www.id.unibe.ch/hpc), the HPC cluster at the University of Bern.

References

1. Bakas, S., et al.: Advancing the cancer genome atlas glioma MRI collections with expert segmentation labels and radiomic features. Nat. Sci. Data **4**(1), 1–13 (2017)
2. Bakas, S., et al.: Segmentation labels and radiomic features for the pre-operative scans of the TCGA-GBM collection. The Cancer Imaging Archive (2017)
3. Bakas, S., et al.: Segmentation labels and radiomic features for the pre-operative scans of the TCGA-LGG collection. The Cancer Imaging Archive (2017)
4. Bakas, S., et al.: Identifying the best machine learning algorithms for brain tumor segmentation, progression assessment, and overall survival prediction in the brats challenge. ArXiv abs/1811.02629 (2018)
5. Havaei, M., et al.: Brain tumor segmentation with deep neural networks. Med. Image Anal. **35**, 18–31 (2017)
6. Kamnitsas, K.: Efficient multi-scale 3D CNN with fully connected CRF for accurate brain lesion segmentation. Med. Image Anal. **36**, 61–78 (2017)

7. Kendall, A., Gal, Y.: What uncertainties do we need in Bayesian deep learning for computer vision? In: NIPS (2017)
8. Kofler, F., et al.: A baseline for predicting glioblastoma patient survival time with *classical statistical* models and *primitive* features ignoring image information. In: Crimi, A., Bakas, S. (eds.) BrainLes 2019. LNCS, vol. 11992, pp. 254–261. Springer, Cham (2020). https://doi.org/10.1007/978-3-030-46640-4_24
9. Lin, T.Y., Goyal, P., Girshick, R.B., He, K., Dollár, P.: Focal loss for dense object detection. In: 2017 IEEE International Conference on Computer Vision (ICCV), pp. 2999–3007 (2017)
10. Loshchilov, I., Hutter, F.: SGDR: stochastic gradient descent with warm restarts. In: ICLR (2017)
11. McKinley, R., Meier, R., Wiest, R.: Ensembles of densely connected CNNs with label-uncertainty for brain tumor segmentation. In: Crimi, A., Bakas, S., Menze, B. (eds.) Brainlesion: Glioma, Multiple Sclerosis. Stroke and Traumatic Brain Injuries. Springer International Publishing, Cham (2019)
12. McKinley, R., Rebsamen, M., Meier, R., Reyes, M., Rummel, C., Wiest, R.: Few-shot brain segmentation from weakly labeled data with deep heteroscedastic multi-task networks. arXiv e-print. https://arxiv.org/abs/1904.02436, April 2019
13. McKinley, R., Rebsamen, M., Meier, R., Wiest, R.: Triplanar Ensemble of 3D-to-2D CNNs with Label-Uncertainty for Brain Tumor Segmentation, pp. 379–387, May 2020
14. McKinley, R., et al.: Simultaneous lesion and neuroanatomy segmentation in multiple sclerosis using deep neural networks. ArXiv abs/1901.07419 (2019)
15. Menze, B.H., et al.: The multimodal brain tumor image segmentation benchmark (BRATS). IEEE Trans. Med. Imaging **34**(10), 1993–2024 (2015)
16. Nair, T., Precup, D., Arnold, D.L., Arbel, T.: Exploring uncertainty measures in deep networks for multiple sclerosis lesion detection and segmentation. Med. Image Anal. **59**, 101557 (2020)
17. Nair, T., et al.: Exploring uncertainty measures in deep networks for multiple sclerosis lesion detection and segmentation. In: Proceedings of MICCAI (2018)
18. Pereira, S., Pinto, A., Alves, V., Silva, C.A.: Brain tumor segmentation using convolutional neural networks in MRI images. IEEE Trans. Med. Imaging **35**(5), 1240–1251 (2016)
19. Rebsamen, M., Knecht, U., Reyes, M., Wiest, R., Meier, R., McKinley, R.: Divide and conquer: stratifying training data by tumor grade improves deep learning-based brain tumor segmentation. Front. Neurosci. **13**, 1182 (2019)
20. Rebsamen, M., Rummel, C., Reyes, M., Wiest, R., McKinley, R.: Direct cortical thickness estimation using deep learning-based anatomy segmentation and cortex parcellation. Hum. Brain Mapping **41**(17), 4804–4814 (2020)
21. Weninger, L., Rippel, O., Koppers, S., Merhof, D.: Segmentation of brain tumors and patient survival prediction: methods for the BraTS 2018 challenge. In: Crimi, A., Bakas, S., Kuijf, H., Keyvan, F., Reyes, M., van Walsum, T. (eds.) BrainLes 2018. LNCS, vol. 11384, pp. 3–12. Springer, Cham (2019). https://doi.org/10.1007/978-3-030-11726-9_1

Multi-decoder Networks with Multi-denoising Inputs for Tumor Segmentation

Minh H. Vu[1]([✉]), Tufve Nyholm[1], and Tommy Löfstedt[2]

[1] Department of Radiation Sciences, Umeå University, Umeå, Sweden
minh.vu@umu.se
[2] Department of Computing Science, Umeå University, Umeå, Sweden
tommy@cs.umu.se

Abstract. Automatic segmentation of brain glioma from multimodal MRI scans plays a key role in clinical trials and practice. Unfortunately, manual segmentation is very challenging, time-consuming, costly, and often inaccurate despite human expertise due to the high variance and high uncertainty in the human annotations. In the present work, we develop an end-to-end deep-learning-based segmentation method using a multi-decoder architecture by jointly learning three separate sub-problems using a partly shared encoder. We also propose to apply smoothing methods to the input images to generate denoised versions as additional inputs to the network. The validation performance indicates an improvement when using the proposed method. The proposed method was ranked 2nd in the task of Quantification of Uncertainty in Segmentation in the Brain Tumors in Multimodal Magnetic Resonance Imaging Challenge 2020.

Keywords: Brain tumor segmentation · Uncertainty estimation · Medical imaging · MRI · Ensemble · Deep learning

1 Introduction

Glioma is a particular kind of brain tumor that develops from glial cells. It is the most frequently occurring type of brain tumor and the one with the highest mortality rate. Glioma is categorized by the World Health Organization (WHO) into four grades: low-grade glioma (LGG) (class I and II), and high-grade glioma (HGG) (class III and IV), where HGG is being considered a dangerous and life-threatening tumor. Specifically, about 190,000 cases occur annually worldwide [6], and around 90 % [18] of patients die within 24 months of surgical resection. Segmentation of the tumor plays a role both for radiotherapy treatment planning and for diagnostic follow-up of the disease. Manual segmentation is time-consuming, subjective, and associated with uncertainties due to the variation of shape, location, and appearance of the tumors. Hence, decision

© Springer Nature Switzerland AG 2021
A. Crimi and S. Bakas (Eds.): BrainLes 2020, LNCS 12658, pp. 412–423, 2021.
https://doi.org/10.1007/978-3-030-72084-1_37

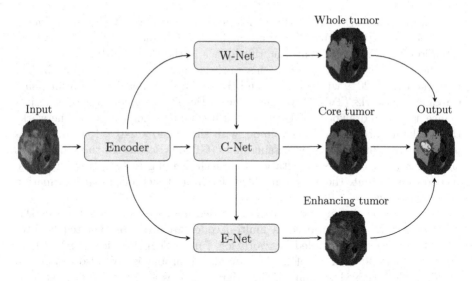

Fig. 1. Schematic visualization of the MDNet architecture.

support or automating the segmentation may improve the treatment quality as well as enhancing the efficiency when handling this patient group.

Inspired by a need of automatic segmentation of brain tumors in multimodal magnetic resonance imaging (MRI) scans, the Brain Tumors in Multimodal Magnetic Resonance Imaging Challenge 2020 (BraTS 2020) [2–5,14] is a yearly challenge (associated with the International Conference on Medical Image Computing and Computer Assisted Intervention (MICCAI)) that aims to evaluate state-of-the-art methods for brain tumor segmentation. BraTS 2020 provides the participants with images from four structural MRI modalities: post-contrast T1-weighted (T1c), T2-weighted (T2w), T1-weighted (T1w), and T2 Fluid Attenuated Inversion Recovery (FLAIR) for brain tumor analysis and segmentation. Masks were annotated manually by one to four raters followed by improvements by expert raters. The segmentation performances of the participants were evaluated using the Sørensen-Dice coefficient (DSC), sensitivity, specificity, and the 95th percentile of the Hausdorff distance (HD95).

Since the introduction of the U-Net by Ronneberger *et al.* [16], Convolutional Neural Networks (CNNs) incorporating skip connections have become the baseline architecture for medical image segmentation. Various architectures, often building on or extending this baseline, have been proposed to address the brain tumor segmentation problem. In BraTS 2019, Jiang *et al.* [11], who was the first-place winner of the challenge, proposed an end-to-end two-stage cascaded U-Net to segment the substructures of brain tumors from coarse (in the first stage) to fine (in the second stage) prediction. In the same challenge, Zhao *et al.* [21], who won the second place, introduced numerous tricks for 3D MRI brain tumor segmentation including processing methods, model designing methods, and optimizing methods. McKinley *et al.* [13] proposed DeepSCAN, which is a modification of their previous 3D-to-2D Fully Convolutional Network (FCN), by replacing batch normalization with instance normalization and adding a lightweight local attention mechanism to secure the third place in the BraTS 2019.

The architecture proposed in this work is an extension of the one in [20] from the BraTS 2019 where End-to-end Hierarchical Tumor Segmentation using Cascaded Networks (TuNet) was introduced. Despite achieving a decent performance, the main drawback of TuNet is that it comprises three cascaded networks that make it hard to fit a full volume, with shape $240 \times 240 \times 155$, into memory on any recent graphics processing units (GPUs). Because of this, the TuNet adapted a patch-based segmentation approach, leading to long training times. In addition to that, the TuNet might suffer from a lack of global information about the image.

Motivated by the successes of the cascaded networks, presented in *e.g.* [11, 20], the present work proposes a multi-decoder architecture, denoted End-to-end Multi-Decoder Cascaded Network for Tumor Segmentation (MDNet), to separate a complicated problem into simple sub-problems. We also propose to use multiple denoised versions of the original images as inputs to the network. The hypothesis was that this would counteract the salt and pepper noise often seen in MRI scans [1]. To the best of our knowledge, this is the first use of this technique.

The authors hypothesize that the MDNet will reduce overfitting problems by employing a shared encoder between three different decoders, while denoised MRI images will help the network to gain more insight into the multimodal input images with the presence of another two versions of the images: (i) a salt and pepper-free one from the use of a median filter, and (ii) one with reduced high-frequency components by employing a low-pass Gaussian filter.

2   Methods

Inspired by the drawbacks of the method proposed in [20], the authors here also propose an end-to-end framework that separates the complicated multi-class tumor segmentation problem into three simpler binary segmentation problems, but with a major change in the design. The MDNet consumes much less memory compared to the TuNet, which means that whole input volumes can be fit into the GPU memory. Hence, the proposed MDNet can take advantage of global details. In addition to that, the design of MDNet results in shorter training times since it uses whole volumes instead of patches, as was the case with the TuNet.

2.1   Encoder Network

The encoder network consists of conventional convolution blocks [16], where each block includes a convolution layer with batch normalization and a leaky rectified linear unit (LeakyReLU) activation function. Each convolutional block is then followed by a Squeeze-and-Excitation block (SEB) (see Sect. 2.3). Max-pooling layers were used for downsampling. All convolutional filters had the size of 3 × 3 × 3, and the initial numbers of filters were set to twelve, which in the proposed architecture is equivalent to three denoising methods applied to the four given modalities (see Sect. 2.4). The encoder output has shape 96 × 20 × 24 × 16. The complete architecture of the proposed encoder network is detailed in Table 1.

Table 1. The encoder architecture. "Conv3" denotes a 3 × 3 × 3 convolution, "BN" stands for batch normalization, "LeakyReLU" is the leaky rectified linear unit, and "SEB" denotes the Squeeze-and-Excitation block (see Sect. 2.3).

Name	Layers	Repeat	Output size
Input			12 × 160 × 192 × 128
EncBlk–0	Conv3, BN, LeakyReLU, SEB	2	12 × 160 × 192 × 128
EncDwn–1	MaxPooling	1	12 × 80 × 96 × 64
EncBlk–1	Conv3, BN, LeakyReLU, SEB	2	24 × 80 × 96 × 64
EncDwn–2	MaxPooling	1	24 × 40 × 48 × 32
EncBlk–2	Conv3, BN, LeakyReLU, SEB	2	48 × 40 × 48 × 32
EncDwn–3	MaxPooling	1	48 × 20 × 24 × 16
EncBlk–3	Conv3, BN, LeakyReLU, SEB	2	96 × 20 × 24 × 16

2.2 Multi-decoder Networks

Table 2 illustrates the proposed multi-decoder networks. The decoder networks include three separate paths, where each path is employed to cope with a specific aforementioned tumor region including whole, core, and enhancing, that are denoted by W-Net, C-Net, and E-Net, respectively. Each decoder path comprises skip connections as in U-Net. There was also a SEB after each convolution block and a concatenation operation of the output of the spatial upsampling layers with the feature maps from the encoder at the same level. To enrich the feature maps at the beginning of each level in the C-Net, the feature map at the end of the W-Net on the same level is used. A similar approach is employed in the decoder network of the E-Net and C-Net. By utilizing these, we hypothesize that the W-Net will constrain the C-Net, while the C-Net will constrain the E-Net. Figure 1 illustrates the proposed architecture.

2.3 Squeeze-and-Excitation Block

We added a channel-based SEB as proposed by Hu *et al.* [8] after each convolution block or concatenation operation. The idea of SEB is to adapt the weight of each channel in a feature map by adding a content-aware mechanism at almost no computational cost. In recent days, SEB has been widely employed to achieve a huge boost in performance. A conventional SEB includes the following layers in

Table 2. Decoder architectures. Here, "Conv3" means a $3 \times 3 \times 3$ convolution, "Conv1" a $1 \times 1 \times 1$ convolution, "BN" denotes for batch normalization, "LeakyReLU" means the leaky rectified linear unit, "SEB" denotes the Squeeze-and-Excitation block (see Sect. 2.3), "Up–{X}" represents the 3D linear spatial upsampling of block X, (+) denotes the concatenation operation. In the name column, $W-$, $C-$ and $E-$ correspond to the whole, core, and enhancing tumor regions, respectively.

Name	Layers	Repeat	Output size
W–DecCat–2	Up–EncBlk–3 + EncBlk–2	1	$144 \times 40 \times 48 \times 32$
W–DecSae–2	SEB	1	$144 \times 40 \times 48 \times 32$
W–DecBlk–2	Conv3, BN, LeakyReLU, SEB	2	$48 \times 40 \times 48 \times 32$
W–DecCat–1	Up–DecBlk–2 + EncBlk–1	1	$72 \times 80 \times 96 \times 64$
W–DecSae–1	SEB	1	$72 \times 80 \times 96 \times 64$
W–DecBlk–1	Conv3, BN, LeakyReLU, SEB	2	$24 \times 80 \times 96 \times 64$
W–DecCat–0	Up–DecBlk–1 + EncBlk–0	1	$36 \times 160 \times 192 \times 128$
W–DecSae–0	SEB	1	$36 \times 160 \times 192 \times 128$
W–DecBlk–0	Conv3, BN, LeakyReLU, SEB	2	$12 \times 160 \times 192 \times 128$
W–Output	Conv1, Sigmoid	1	$1 \times 160 \times 192 \times 128$
C–DecCat–2	W–DecBlk–2 + W–DecCat–2	1	$192 \times 40 \times 48 \times 32$
C–DecSae–2	SEB	1	$192 \times 40 \times 48 \times 32$
C–DecBlk–2	Conv3, BN, LeakyReLU, SEB	2	$48 \times 40 \times 48 \times 32$
C–DecCat–1	W–DecBlk–1 + W–DecCat–1	1	$96 \times 80 \times 96 \times 64$
C–DecSae–1	SEB	1	$96 \times 80 \times 96 \times 64$
C–DecBlk–1	Conv3, BN, LeakyReLU, SEB	2	$24 \times 80 \times 96 \times 64$
C–DecCat–0	W–DecBlk–0 + W–DecCat–0	1	$48 \times 160 \times 192 \times 128$
C–DecSae–0	SEB	1	$48 \times 160 \times 192 \times 128$
C–DecBlk–0	Conv3, BN, LeakyReLU, SEB	2	$12 \times 160 \times 192 \times 128$
C–Output	Conv1, Sigmoid	1	$1 \times 160 \times 192 \times 128$
E–DecCat–2	C–DecBlk–2 + W–DecCat–2	1	$240 \times 40 \times 48 \times 32$
E–DecSae–2	SEB	1	$240 \times 40 \times 48 \times 32$
E–DecBlk–2	Conv3, BN, LeakyReLU, SEB	2	$48 \times 40 \times 48 \times 32$
E–DecCat–1	C–DecBlk–1 + W–DecCat–1	1	$96 \times 80 \times 96 \times 64$
E–DecSae–1	SEB	1	$96 \times 80 \times 96 \times 64$
E–DecBlk–1	Conv3, BN, LeakyReLU, SEB	2	$24 \times 80 \times 96 \times 64$
E–DecCat–0	C–DecBlk–0 + W–DecCat–0	1	$48 \times 160 \times 192 \times 128$
E–DecSae–0	SEB	1	$48 \times 160 \times 192 \times 128$
E–DecBlk–0	Conv3, BN, LeakyReLU, SEB	2	$12 \times 160 \times 192 \times 128$
E–Output	Conv1, Sigmoid	1	$1 \times 160 \times 192 \times 128$

sequence: global pooling, fully connected, rectified linear unit (ReLU) activation function, fully connected, and a sigmoid activation function [8].

2.4   Denoising the Inputs

The inputs to the network were the MRI modalities, and also each modality after denoising using two different methods: median denoising and Gaussian smoothing. The authors then concatenated the three versions of the images for each modality (the raw image, and the two denoised versions) to obtain a total of twelve images, that were input as different channels. For the median denoising, we used a $3 \times 3 \times 3$ median filter; the Gaussian smoothing used a $3 \times 3 \times 3$ Gaussian filter with a standard deviation of 0.5. In this sense, adding a Gaussian smoothed version of the input is similar to adding a down-scaled version of the input image as was proposed for the TuNet [20].

2.5   Preprocessing and Augmentation

All input images were normalized to have a mean zero and unit variance. In order to reduce overfitting and increase the diversity of data available for training models, we used on-the-fly data augmentation [9] comprising: (1) randomly rotating the images in the range $[-1, 1]$ degrees on all three axes, (2) random mirror flipping with a probability of 0.5 on all three axes, (3) elastic transformation with a probability of 0.3, (4) random scaling in the range $[0.9, 1.1]$ with a probability of 0.3, and (5) random cropping with subsequent resizing with a probability of 0.3.

As in [17], the elastic transformations used a random displacement field, Δ, such that

$$R_w = R_o + \alpha \Delta, \tag{1}$$

where α is the strength of the displacement, while R_w and R_o denote the location of a voxel in the warped and original image, respectively. For each axis, a random number was drawn uniformly in $[-1, 1]$ such that $\Delta_x \sim \mathcal{U}(-1, 1)$, $\Delta_y \sim \mathcal{U}(-1, 1)$, and $\Delta_z \sim \mathcal{U}(-1, 1)$. The displacement field was finally convolved with a Gaussian kernel having standard deviation σ. In the present case, $\alpha = 1$ and $\sigma = 0.25$.

2.6   Post-processing

The most challenging task of BraTS 2020 specifically, and BraTS challenges in general, is to distinguish between LGG and HGG patients by labeling small vessels lying in the tumor core as edema or necrosis. In order to tackle this problem, we used the same strategy as proposed in our previous work [20]. In specific, we labeled all small enhancing tumor region with less than 500 connected voxels as necrosis. The proposed post-processing step aims to handle a few cases where the proposed networks fail to differentiate between the whole and core tumor regions.

2.7   Task 3: Quantification of Uncertainty in Segmentation

The organizers of the BraTS challenge introduced the task of "Quantification of Uncertainty in Segmentation" in BraTS 2019 and was held again in BraTS

2020. This task is aimed to measure the uncertainty in the context of glioma region segmentation by rewarding predictions that are (a) confident when correct and (b) uncertain when incorrect. Participants were expected to generate uncertainty maps in the range of $[0, 100]$, where 0 represents the most certain and 100 represents the most uncertain. The performance was evaluated based on three metrics: Dice Area Under Curve (DAUC), Ratio of Filtered True Positives (RFTPs), and Ratio of Filtered True Negatives (RFTNs).

Similar to [20], the proposed network, MDNet, predicts the probability of three tumor regions, it thus benefits from this task. Following [20], an uncertainty score, $u^r_{i,j,k}$, at voxel (i, j, k) is defined by

$$u^r_{i,j,k} = \begin{cases} 200(1 - p^r_{i,j,k}), & \text{if } p^r_{i,j,k} \geq 0.5, \\ 200 p^r_{i,j,k}, & \text{if } p^r_{i,j,k} < 0.5, \end{cases} \tag{2}$$

where $u^r_{i,j,k} \in [0, 100]^{|\mathcal{R}|}$ and $p^r_{i,j,k} \in [0, 1]^{|\mathcal{R}|}$ are the uncertainty score map and probability map, respectively. Here, $r \in \mathcal{R}$, where $\mathcal{R}$ is the set of tumor regions, *i.e.* whole, core, and enhancing region.

3   Experiments

3.1   Implementation Details and Training

The proposed method was implemented in Keras 2.2.4[1] with TensorFlow 1.12.0[2] as the backend. The experiments were trained on NVIDIA Tesla V100 GPUs from the High Performance Computer Center North (HPC2N) at Umeå University, Sweden. Seven models were trained from scratch for $N_e = 200$ epochs, with a mini-batch size of one. The training time for a single model was about six days.

3.2   Loss

For evaluation of the segmentation performance, we used a combination of the DSC loss and categorical cross–entropy (CE) as the loss function. The DSC is defined as [19,20]

$$D(u, v) = \frac{2 \cdot |u \cap v|}{|u| + |v|}, \tag{3}$$

where u and v are the output segmentation and its corresponding ground truth, respectively. To include the the DSC in the loss function, we employed the soft DSC loss, which is defined as [10,19,20]

$$\mathcal{L}_{DSC}(u, v) = \frac{-2 \sum_i u_i v_i}{\sum_i u_i + \sum_i v_i + \epsilon}, \tag{4}$$

---

[1] https://keras.io.
[2] https://tensorflow.org.

where for each label i, the u_i is the softmax output of the proposed network for label i, v is a one-hot encoding of the ground truth labels (segmentation maps in this case), and $\epsilon = 1 \cdot 10^{-5}$ is a small constant added to avoid division by zero.

Following [10,19], for unbalanced data sets with small structures like in the BraTS 2020 data, we added the CE term to our loss function to make the loss surface smoother. The CE is defined as

$$\mathcal{L}_{CE}(u, v) = -\sum_i u_i \cdot \log(v_i). \tag{5}$$

The combination of the DSC loss and CE (denoted a *hybrid loss*) is simply defined as the sum of the two losses, as

$$\mathcal{L}_{\text{hybrid}}(u, v) = \mathcal{L}_{DSC}(u, v) + \mathcal{L}_{CE}(u, v). \tag{6}$$

The final loss function that was used for training contained one hybrid loss for each tumor region, and was thus

$$\mathcal{L}(u, v) = \sum_{r \in \mathcal{R}} \mathcal{L}_{\text{hybrid}}(u_r, v_r), \tag{7}$$

where $\mathcal{R}$ again is the set of tumor regions (the whole, core, and enhancing regions) and $\mathcal{L}_{\text{hybrid}}(u_r, v_r)$ is the hybrid loss for a particular tumor region.

The segmentation performance was also evaluated using the HD95, a common metric for evaluating segmentation performances. The Hausdorff distance (HD) is defined as [7]

$$H(u, v) = \max\{d(u, v), d(v, u)\}, \tag{8}$$

where

$$d(u, v) = \max_{u_i \in u} \min_{v_i \in v} \|u_i - v_i\|_2, \tag{9}$$

in which $\|u_i - v_i\|_2$ is the spatial Euclidean distance between points u_i and v_i on the boundaries of output segmentation u and ground truth v.

3.3   Optimization

The authors used the Adam optimizer [12] with an initial learning rate of $\alpha_0 = 1 \cdot 10^{-4}$ and momentum parameters of $\beta_1 = 0.9$ and $\beta_2 = 0.999$. Following Myronenko *et al.* in [15], the learning rate was decayed as

$$\alpha_e = \alpha_0 \cdot \left(1 - \frac{e}{N_e}\right)^3, \tag{10}$$

where e and $N_e = 200$ are epoch counter and total number of epochs, respectively.

The authors also used L_2 regularization with a penalty parameter of $1 \cdot 10^{-5}$, which was applied to the kernel weight matrices, for all convolutional layers to counter overfitting. The activation function of the final layer was the logistic sigmoid function.

4   Results and Discussion

Table 3 shows the mean DSC and HD95 scores and standard deviations (SDs) computed from the five-folds of cross-validation on 369 cases of the training set. From Table 3 we see that: (i) the U-Net with denoised input improved the DSC and HD95 on all tumor regions, and (ii) the proposed model with denoising boosted the performance in both metrics (DSC and HD95) by a large margin.

Table 3. Mean DSC (higher is better) and HD95 (lower is better) and their SEs (in parentheses) computed from the five-folds of cross-validation on the training set (369 cases) for the different models.

Model	DSC			HD95		
	Whole	Core	Enh.	Whole	Core	Enh.
U-Net without denoising	90.66 (0.38)	86.93 (0.71)	76.16 (1.37)	4.91 (0.41)	4.78 (0.42)	3.46 (0.31)
U-Net with denoising	90.98 (0.31)	87.53 (0.68)	76.55 (1.36)	4.49 (0.26)	4.32 (0.29)	3.41 (0.29)
Proposed with denoising	92.75 (0.25)	88.34 (0.70)	78.13 (1.32)	4.32 (0.29)	4.30 (0.31)	3.29 (0.24)

Table 4 shows the mean DSC and HD95 scores on the validation set, computed on the predicted masks by the evaluation server[3] (team name UmU). The BraTS 2020 final validation dataset results were 90.55, 82.67 and 77.17 for the average DSC, and 4.99, 8.63 and 27.04 for the average HD95, for whole tumor, tumor core and enhanced tumor core, respectively. These results were slightly lower than the top-ranking teams.

Table 5 provides the mean DAUC, RFTPs, and RFTNs scores on the validation set obtained after uploading the predicted masks and corresponding uncertainty maps to the evaluation server[4]. As can be seen from Table 5, the RFTNs scores were the best amongst the best-ranking participants.

Table 6 and Table 7 show the mean DSC and HD95, and the mean DAUC, RFTPs, and RFTNs scores on the test set, respectively. In the task of Quantification of Uncertainty in Segmentation, our proposed method was ranked 2nd.

---

[3] https://www.cbica.upenn.edu/BraTS20/lboardValidation.html.
[4] https://www.cbica.upenn.edu/BraTS20/lboardValidationUncertainty.html.

Table 4. Results of Segmentation Task on BraTS 2020 validation data (125 cases). The results were obtained by computing the mean of predictions of seven models trained from the scratch. "UmU" denotes the name of our team. The metrics were computed by the online evaluation platform. All the predictions were post-processed before submitting to the server. The top rows correspond to the top-ranking teams from the online system retrieved at 11:38:02 EDT on August 3, 2020.

Team	DSC			HD95		
	Whole	Core	Enh.	Whole	Core	Enh.
deepX	91.02	85.00	78.53	4.44	5.90	24.06
Radicals	90.82	84.96	78.69	4.71	8.56	35.01
WassersteinDice	90.58	83.79	78.01	4.74	8.96	27.02
CKM	90.83	83.82	78.59	4.87	5.97	26.57
UmU	90.55	82.67	77.17	4.99	8.63	27.04

Table 5. Results of Quantification of Uncertainty Task on BraTS 2020 validation data (125 cases) including mean DAUC (higher is better), RFTPs (lower is better) and RFTNs (lower is better). The results were obtained by computing the mean of predictions of seven models trained from scratch. "UmU" denotes the name of our team and the ensemble of seven models that were trained from the scratch. The metrics were computed by the online evaluation platform. The top rows correspond to the top-ranking teams from the online system retrieved at 11:38:02 EDT on August 3, 2020.

Team	DAUC			RFTPs			RFTNs		
	Whole	Core	Enh.	Whole	Core	Enh.	Whole	Core	Enh.
med_vision	95.24	92.23	83.24	0.28	0.62	0.93	87.74	98.74	98.74
nsu_btr	93.58	90.04	85.14	35.72	48.18	9.59	98.44	98.60	98.64
SCAN	93.46	82.98	80.64	12.40	19.95	21.53	0.87	0.42	0.24
UmU	92.59	83.61	78.83	4.48	10.13	7.95	0.27	0.17	0.08

Table 6. Results of Segmentation Task on BraTS 2020 test data (166 cases). The results were obtained by computing the mean of predictions of seven models trained from the scratch. The metrics were computed by the online evaluation platform. All the predictions were post-processed before submitting to the server.

Team	DSC			HD95		
	Whole	Core	Enh.	Whole	Core	Enh.
UmU	88.26	82.49	80.84	6.30	22.27	20.06

Table 7. Results of Quantification of Uncertainty Task on BraTS 2020 test data (166 cases) including mean DAUC (higher is better), RFTPs (lower is better) and RFTNs (lower is better). The results were obtained by computing the mean of predictions of seven models trained from scratch. The metrics were computed by the online evaluation platform.

Team	DAUC			RFTPs			RFTNs		
	Whole	Core	Enh.	Whole	Core	Enh.	Whole	Core	Enh.
UmU	90.61	85.83	83.03	4.18	5.49	4.45	0.31	1.68	0.07

5 Conclusion

In this work, we proposed a multi-decoder network for segmenting tumor substructures from multimodal brain MRI images by separating a complex problem into simpler sub-tasks. The proposed network adopted a U-Net-like structure with Squeeze-and-Excitation blocks after each convolution and concatenation operation. We also proposed to stack original images with their denoised versions to enrich the input and demonstrated that the performance was boosted in both DSC and HD95 metrics by a large margin. The results on the test set indicated that: (i) the proposed method performed competitively in the task of Segmentation, with DSC scores of 88.26/82.49/80.84 and HD95 scores of 6.30/22.27/20.06 for the whole tumor, tumor core, and enhancing tumor core, respectively, (ii) the proposed method was top 2 performing ones in the task of Quantification of Uncertainty in Segmentation.

Acknowledgement. The computations were performed on resources provided by the Swedish National Infrastructure for Computing (SNIC) at the HPC2N in Umeå, Sweden. We are grateful for the financial support obtained from the Cancer Research Fund in Northern Sweden, Karin and Krister Olsson, Umeå University, The Västerbotten regional county, and Vinnova, the Swedish innovation agency.

References

1. Ali, H.M.: A new method to remove salt pepper noise in magnetic resonance images. In: 2016 11th International Conference on Computer Engineering Systems (ICCES), pp. 155–160 (2016)
2. Bakas, S., et al.: Segmentation labels and radiomic features for the pre-operative scans of the TCGA-GBM collection. The cancer imaging archive (2017) (2017)
3. Bakas, S., et al.: Segmentation labels and radiomic features for the pre-operative scans of the TCGA-LGG collection. Cancer Imaging Archive **286** (2017)
4. Bakas, S., et al.: Advancing the cancer genome atlas glioma MRI collections with expert segmentation labels and radiomic features. Sci. Data **4**, 170117 (2017)
5. Bakas, S., et al.: Identifying the best machine learning algorithms for brain tumor segmentation, progression assessment, and overall survival prediction in the BRATS challenge. arXiv preprint arXiv:1811.02629 (2018)

6. Castells, X., et al.: Automated brain tumor biopsy prediction using single-labeling CDNA microarrays-based gene expression profiling. Diagn. Mol. Pathol. **18**, 206–218 (2009)

7. Hausdorff, F.: Erweiterung einer stetigen Abbildung, pp. 555–568. Springer, Heidelberg (2008). https://doi.org/10.1007/978-3-540-76807-4_16

8. Hu, J., Shen, L., Sun, G.: Squeeze-and-excitation networks. In: Proceedings of the IEEE Conference on Computer Vision and Pattern Recognition, pp. 7132–7141 (2018)

9. Isensee, F., et al.: batchgenerators–a python framework for data augmentation, January 2020

10. Isensee, F., Kickingereder, P., Wick, W., Bendszus, M., Maier-Hein, K.H.: No new-net. In: Crimi, A., Bakas, S., Kuijf, H., Keyvan, F., Reyes, M., van Walsum, T. (eds.) BrainLes 2018. LNCS, vol. 11384, pp. 234–244. Springer, Cham (2019). https://doi.org/10.1007/978-3-030-11726-9_21

11. Jiang, Z., Ding, C., Liu, M., Tao, D.: Two-stage cascaded U-Net: 1st place solution to BraTS challenge 2019 segmentation task. In: Crimi, A., Bakas, S. (eds.) BrainLes 2019. LNCS, vol. 11992, pp. 231–241. Springer, Cham (2020). https://doi.org/10.1007/978-3-030-46640-4_22

12. Kingma, D.P., Ba, J.: Adam: a method for stochastic optimization. arXiv preprint arXiv:1412.6980 (2014)

13. McKinley, R., Rebsamen, M., Meier, R., Wiest, R.: Triplanar ensemble of 3D-to-2D CNNs with label-uncertainty for brain tumor segmentation. In: Crimi, A., Bakas, S. (eds.) BrainLes 2019. LNCS, vol. 11992, pp. 379–387. Springer, Cham (2020). https://doi.org/10.1007/978-3-030-46640-4_36

14. Menze, B.H., et al.: The multimodal brain tumor image segmentation benchmark (BRATS). IEEE TRans. Med. Imaging **34**(10), 1993–2024 (2014)

15. Myronenko, A.: 3D MRI brain tumor segmentation using autoencoder regularization (2018). http://arxiv.org/abs/1810.11654

16. Ronneberger, O., Fischer, P., Brox, T.: U-Net: convolutional networks for biomedical image segmentation. In: Navab, N., Hornegger, J., Wells, W.M., Frangi, A.F. (eds.) MICCAI 2015. LNCS, vol. 9351, pp. 234–241. Springer, Cham (2015). https://doi.org/10.1007/978-3-319-24574-4_28

17. Simard, P.Y., Steinkraus, D., Platt, J.C.: Best practices for convolutional neural networks applied to visual document analysis. In: Seventh International Conference on Document Analysis and Recognition, 2003. Proceedings, pp. 958–963 (2003)

18. Thurnher, M.: The 2007 WHO classification of tumors of the central nervous system–what has changed? Am. J. Neuroradiol. (2012)

19. Vu, M.H., Grimbergen, G., Nyholm, T., Löfstedt, T.: Evaluation of multi-slice inputs to convolutional neural networks for medical image segmentation. arXiv preprint arXiv:1912.09287 (2019)

20. Vu, M.H., Nyholm, T., Löfstedt, T.: TuNet: end-to-end hierarchical brain tumor segmentation using cascaded networks. In: Crimi, A., Bakas, S. (eds.) BrainLes 2019. LNCS, vol. 11992, pp. 174–186. Springer, Cham (2020). https://doi.org/10.1007/978-3-030-46640-4_17

21. Zhao, Y.-X., Zhang, Y.-M., Liu, C.-L.: Bag of tricks for 3D MRI brain tumor segmentation. In: Crimi, A., Bakas, S. (eds.) BrainLes 2019. LNCS, vol. 11992, pp. 210–220. Springer, Cham (2020). https://doi.org/10.1007/978-3-030-46640-4_20

MultiATTUNet: Brain Tumor Segmentation and Survival Multitasking

Diedre Carmo$^{(\boxtimes)}$, Leticia Rittner , and Roberto Lotufo

MICLab, School of Electrical and Computing Engineering,
University of Campinas, São Paulo, Brazil
{diedre,lrittner,lotufo}@dca.fee.unicamp.br
http://miclab.fee.unicamp.br/

Abstract. Segmentation of Glioma from three dimensional magnetic resonance imaging (MRI) is useful for diagnosis and surgical treatment of patients with brain tumor. Manual segmentation is expensive, requiring medical specialists. In the recent years, the Brain Tumor Segmentation Challenge (BraTS) has been calling researchers to submit automated glioma segmentation and survival prediction methods for evaluation and discussion over their public, multimodality MRI dataset, with manual annotations. This work presents an exploration of different solutions to the problem, using 3D UNets and self attention for multitasking both predictions and also training (2D) EfficientDet derived segmentations, with the best results submitted for the official challenge leaderboard. We show that end-to-end multitasking survival and segmentation, in this case, led to better results.

Keywords: Deep learning · Attention · Multitask · Glioma · Segmentation · Survival prediction

1 Introduction

Assessment of brain tumors is important in the diagnostic of Cancer [3]. Automatic segmentation can aid in this assessment, allowing for description of relevant tumor features such as its volume. However, tumors are very heterogeneous in shape, having different associated grades and classifications. Due to this variability, automatic segmentation of brain tumors is still a challenge [4].

A public data source of glioma type brain tumors is the base of the BraTS challenge [12]. This challenge provides manual segmentations of glioma regions, annotated over the provided four modalities of MRI, T1, contrast enhanced T1, T2 and FLAIR. Submitted methods have to provide three segmentation maps: Whole Tumor (WT), Tumor Core (TC) and Enhancing Tumor (ET) (Fig. 1). An additional task consists of predicting the survival (in days) of patients, using the tumor segmentations generated by your model. Most top-ranking methods in past challenges used deep learning based methods, with one specific type of architecture having great presence: 3D UNets. Isensee et al. adapted the UNet for 3D convolutions. Interestingly, this is one of the leading methods of the 2017 [8]

© Springer Nature Switzerland AG 2021
A. Crimi and S. Bakas (Eds.): BrainLes 2020, LNCS 12658, pp. 424–434, 2021.
https://doi.org/10.1007/978-3-030-72084-1_38

and 2018 [9] challenges using mostly a single UNet architecture, showing that a well trained UNet can be superior to complex ensemble approaches. Isensee's work seems to have inspired a large part of submissions in later years, which used similar hyperparameters and also attempted to use a modified 3D UNet or ensembles of 3D UNets [10,13,14].

Our proposal also explores the use of multiple 3D UNets, for multitasking segmentation and survival prediction with self-attention maps. Additionally, we investigate deviating from the UNet approach for segmentation with the EfficientDet [17], a recently proposed state-of-the-art semantic segmentation architecture.

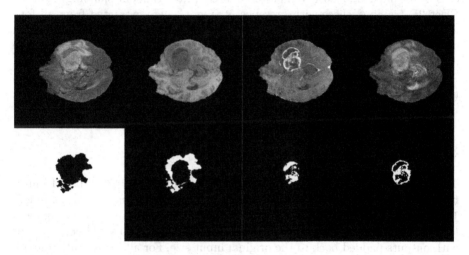

Fig. 1. The four data modalities are showcased, in order: FLAIR, T1, T1 with Contrast and T2. Also displayed in the bottom row are manual annotations, in order: background, edema (ED), non-enhancing tumor (NET) and enhancing tumor (ET).

2   Data

The BraTS 2020 training dataset contains 369 MRI scans of various modalities: T1, post-contrast T1, T2, and FLAIR volumes (Fig. 1). All scans are of Low or High grade gliomas (LGG/HGG) [1,2], acquired with different clinical protocols and various scanners from multiple institutions. Manual annotations comprise the GD-enhancing tumor (ET), the peritumoral edema (ED), and the necrotic and non-enhancing tumor core (NET). More details in [3,4]. Note that the challenge's evaluation is performed over three targets: the enhancing tumor (ET), the tumor core (TC) composed of ET + NET, and the whole tumor (WT) composed of ET + NET + ED. The provided data are distributed after their pre-processing: co-registration to the same anatomical template, interpolation to the same resolution and skull-stripping.

We apply additional processing inspired by Isensee et al. [8]. The images are subtracted by the mean and divided by the standard deviation of the brain region, and clipped inside the interval -5 to 5. Finally, they are min-max normalized to the interval 0 to 1.

Segmentation targets were organized into nested sigmoid targets, resulting in optimization being performed directly over the three evaluation targets (WT, TC and ET). Another target also present to some of the subjects is survival prediction in days, which will be used for our survival prediction path while multitasking.

For the reported validation comparisons between architectures, a holdout of 80% training and 20% validation is used (after random shuffling). For T-EDet 2D training, slices with presence of tumor where extracted from all four modalities, generating approximately 20000 training slices and 5000 validation slices. The official validation with 125 subjects is blind and performed by the challenge runners, treated here as our internal test set for method development. Additionally, on the workshop itself, results were revealed for the hidden official test set of 166 subjects, that resulted in the final decision of top performing methods.

3 Methodology

This section explains the methodology of each part of our UNet based multi-task approach and segmentation with T-EDet. All experiments used random $128 \times 128 \times 128$ patches (128×128 for T-EDet) and random intensity augmentation of 0.1. Validation/testing input consists of $160 \times 160 \times 128$ center crops with outputs padded back to the original input size. For all segmentation architectures, the loss function consists of $1 - (WT_{dice} + TC_{dice} + ET_{dice})/3$, in other words, 1 subtracted by the mean of Dices for each target.

3.1   T-EDet

T-EDet is a modification of EfficientDet [17], a recent state-of-the-art semantic segmentation architecture (Fig. 2). We refer to this architecture as Tumor-EfficientDet (T-EDet).

The starting point of EfficientDet is the computation of backbone features using a pre-trained EfficientNet [16], of which we use the D4 variation. A important problem arises in the form of incompatible number of input channels, as the original network was trained with a 3-channel input. To circumvent this, an adaptation 1×1 convolution was introduced, converting the 4-channel (4 modalities) input into a 3-channel input. Additionally, we changed which features from the pre-trained EfficientNet are used by the bi-directional feature pyramid network (BiFPN), to access initial features and a BiFPN output of half the size of the original image. To bring this representation to the size of the input image, a transposed convolution followed by batch normalization and the same swish

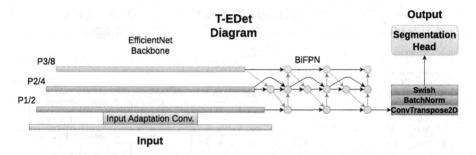

Fig. 2. Illustration of the modified EfficientDet architecture (T-EDet), where the main differences are the feature maps used as BiFPN input and the transposed upsample before the segmentation head. Consult the original paper [17] for a Figure for comparison.

activation [15] used by the original network were added. This new representation goes to the segmentation head. The segmentation head is composed by three blocks of depthwise convolutions [6], batch norm and swish, followed by a final convolution for channel reduction to the number of classes. In total this architecture has 15M parameters.

3.2   UNet 3D

UNet 3D is used both in isolation and as the backbone of our multitask experiment. Its architecture is based on previous experiments with hippocampus segmentation and 2D UNets [5]. We basically extended the same architecture to 3D convolutions, compensating the increased memory overhead with less channels and replacing batch normalization with group normalization [18]. The architecture after adaptation to 3D is depicted in Fig. 3.

In regards to training methodology, we employ some similar hyperparameters to Isensee's work [8]: An initial learning rate of 5e−4, weight decay of 1e−5, max epochs of 300 and exponential LR decay by 0.985. The optimizer is changed from Adam to RAdam [11], due to slightly better results in early experiments. Experiments were performed with using mixed-precision, and no difference in Dice was noted.

3.3   Multitask with MultiATTUNet

Our main experiment consists of end-to-end refinement of the segmentations produced by a frozen pre-trained backbone UNet 3D, and the addition of an attention based survival prediction branch, with the goal of multitasking segmentation and survival prediction. Outputs of the backbone and survival attention are fed to individual additional UNets specialized in each one of the three labels (WT, TC, ET), called output UNets. This whole architecture is named MultiATTUNet (Fig. 4).

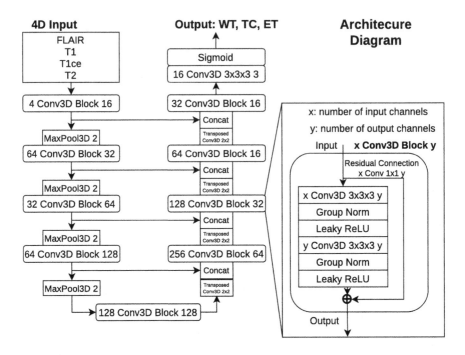

Fig. 3. Representation of the 3D UNet architecture used in this work.

Survival prediction for multitasking is performed with the addition of a branch we named CNN3DAtt (Fig. 5). Four MRI modalities from the original input plus three channels of the backbone segmentation result in seven input channels. CNN3DAtt, is inspired by an Attention based CNN by Gorriz et al. [7], originally a 2D network. Here, it is adapted to a 3D network with the introduction of Group Normalization and 3D convolutions. The number of channels is reduced in comparison to the original work when transforming from 2D to 3D. The network produces unsupervised positional mappings called attention maps (Att), using the Sigmoid activation function. These function as a heat map and a "gate" that allows only the convolutional features desired for optimization to pass through, in that position. Attention blocks derive attention maps from convolutional non-linear activations in different stages, converging into fully connected layers. The conversion of 3D features from the attention map to 1D features for the fully connected layers is performed with Global Average Pooling, resulting in an average value per channel. The age value is added at this stage as an extra neuron. The final output consists of a single neuron, used as an direct activation for days of survival. The activation of this neuron is limited between 1 and 2000, using a sigmoid activation. The interval was chosen based on the interval observed in the training set.

The loss function for survival is Smooth L1 Loss. The survival loss is applied before the segmentation loss, only when age and survival data is present for the

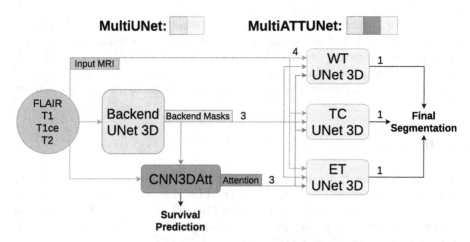

Fig. 4. Illustration of MultiUNet and the full MultiATTUnet architecture. Numbers are number of channels. Note that the MultiUNet architecture doesn't include the red parts.

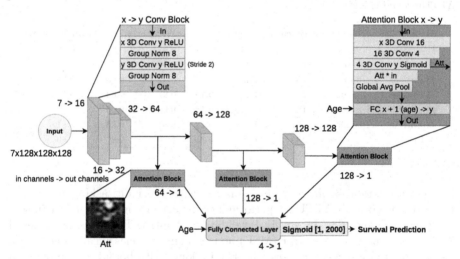

Fig. 5. Illustration of CNN3DAtt, for survival prediction based on Attention maps. A sample attention map slice of the first attention block is displayed.

subject, with retention of the computational graph. This way survival gradients will have an influence in the segmentation loss applied later. This called for a batch size of 1, and batch accumulation is used to produce optimizations over more than 1 example. Learning rate is also slowed by a factor of 10, and other hyperparameters are equal to backbone training.

This multitask approach is labeled MultiATTUNet, and the same architecture without the survival branch, is labeled MultiUNet. Since this approach multiply the total number of weights to be trained by 3 in relation to only train-

ing a UNet, we experimented with varying the number of channels of the output
UNets. Consider "Base" as the channel configuration of the backbone UNet 3D.
We define an additional size "Small", with half the number of channels for each
convolution. Using the "Base" size resulted in a total of 12M trainable parameters and 6M trainable parameters are found with Small size for the output
unets.

4   Results

We compare to check if using multiple 3D UNets is really beneficial in relation
to a single architecture UNet 3D or T-EDet (Table 1). Additionally we add the
task of survival prediction to the mix with MultiATTUnet. Note that results in
validation do not consider any post-processing to the network's output. Experiments were run in a Titan X GPU and a E3-1220 v3 CPU.

Table 1. Comparison of validaton results with different sizes MultiUNet and Multi-
ATTUnet and single UNet 3D.

Method	Output size	Batch size	Precision	Val loss	WT (Dice)	TC (Dice)	ET (Dice)
UNet 3D	Base	4	Full	0.15	0.91	0.86	0.80
UNet 3D	Base	4	Mixed	0.15	0.91	0.85	0.80
MultiUNet	Base	1*4	Full	0.16	0.91	0.83	0.79
MultiUNet	Small	2*2	Full	0.16	0.91	0.83	0.78
MultiATTUNet	Base	1*4	Full	0.14	0.91	0.86	0.81
MultiATTUNet	Small	1*4	Full	0.14	0.91	0.86	0.81
T-EDet	-	32	full	0.15	0.89	0.87	0.80

Best validation results were encountered with the multitasking model. T-
EDet surpassed MultiATTUnet performance in segmenting TC but lost in segmenting WT and ET. In survival validation, MultiATTUnet Base achieved
73755 $days^2$ mean square error (MSE) and 0.74 survival classification accuracy,
considering three survival classes (short, mid, long) with boundaries in 300 and
450 days.

4.1   Test: Official Validation

We considered the official validation, as computed by the BraTS 2020 challenge
runners (Table 2) as a test set. Only for these experiments, we employed as a
post-processing step an specific threshold in Enhancing Tumor segmentations,
where outputs with volume smaller than 300 were completely zeroed out. This
was done to avoid false positive ET predictions that could bring Dice to 0 in
subjects that do not have presence of Enhancing Tumor.

We trained an off-the-shelf implementation of Isensee's 3D UNet in this year's
training set, using the same parameters reported in the original paper [8]. Our

Table 2. Official validation results for the best 3D UNet and EfficientDet models, computed by the BraTS2020 challenge organizers.

Official validation	WT (Dice)	TC (Dice)	ET (Dice)
Isensee [8]	0.88	0.76	0.71
UNet 3D Backbone	0.89	0.81	0.75
MultiUNet	0.89	0.79	0.76
MultiATTUNet	0.89	0.82	0.77
T-EDet	0.87	0.80	0.76
		Survival accuracy	**Survival MSE**
MultiATTUNet		0.552	106659.897

method returned better Dices, with MultiATTUNet outperforming all of our other approaches. In survival assessment with the multitask approach our results are close to the top performing methods.

4.2 Qualitative Results

Qualitative comparisons between manual annotations and the segmentations of MultiATTUNet show similar shape, with the prediction performed by the network assuming a softer appearance (Fig. 6). All tested models returned a visually softer segmentation appearance than what is marked in the groundtruth.

4.3 Test 2: Official Test

The final hidden test performed by the BraTS challenge runners, which decided the winners of the challenge, showcased consistent performance in relation to our internal evaluations and official validation. However, results from ou best model were not enough to reach the top 5 performance for this year. Results are in Table 3.

Table 3. Official test results for our best methodology, MultiATTUnet.

WT (Dice)	TC (Dice)	ET (Dice)
0.87	0.82	0.80
	Survival accuracy	**Survival MSE**
	0.467	461672

5   Discussion and Future Work

Our UNet 3D architecture surpassed the performance of Isensee's architecture, in this year's validation, also using more memory. From the results, one can

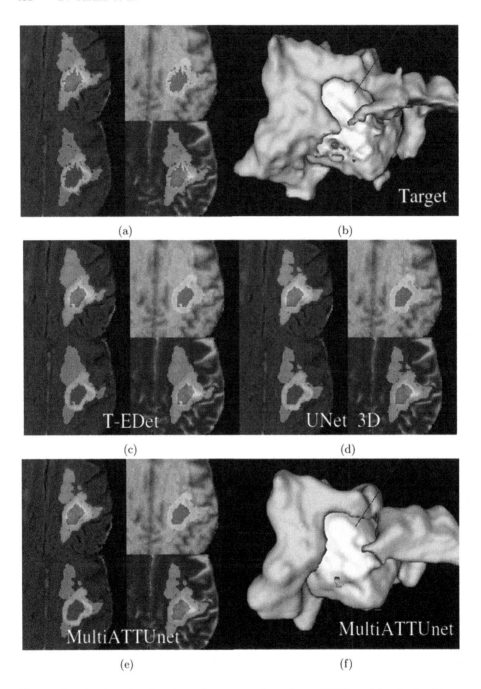

Fig. 6. Qualitative visualization of segmentations from: T-EDet (c); the backbone UNet 3D (d) and MultiATTUnet (e, f), in the same internal validation subject (ID: BraTS20_Training_023). Manual annotations are displayed on (a, b). Visualization produced with ITK-Snap [19].

conclude that the multitask approach with survival attention slightly improved segmentation performance, being our best result. Visualization of attention maps show they are, in fact, focusing on the tumor area, and the inclusion of its gradients from survival loss application seems to have improved the optimization of segmentation from the output UNets. Additionally, using output UNets half the size of the backbone UNet didn't impact validation Dice.

T-EDet achieved superior performance than the UNet 3D backbone alone. Further exploring T-EDet, or even mixing both 3D and 2D methods might lead to even better performance in future work. We believe aligning volumetric information from 3D networks with 2D slice-wise predictions with transfer learning might be the best way forward. Survival prediction results did fall down between official validation and testing, with special attention to the large increase in Survival MSE, although accuracy didn't fall down as much. This suggests more exploration of the survival multitasking is needed, to achieve a conclusion of why this large deviation happened. From other methods results, its notable that no method is achieving much higher accuracy using only image features.

6   Conclusion

This paper describes our proposed models and their most relevant results on the BraTS 2020 Challenge, specifically on the survival and segmentation tasks. Our models included: a modification of a state-of-art 2D semantic segmentation architecture, EfficientDet; the UNet architecture in its 3D variation, and; a multitask usage of multiple UNets and a custom attention based CNN for survival prediction. We showed that our multitask model returned the best results compared to the other experimented methods, suggesting that multitask between survival and segmentation has potential to improve performance. Our final best results consisted of Dices of 0.89, 0.82, 0.77 for Whole Tumor, Tumor Core and Enhancing Tumor respectively, and 0.55 validation accuracy (using short, mid and long survival classes) on the official leaderboard, close to the top performing methods. For the official final testing, which decided the winners, we achieved 0.87, 0.82, 0.80 WT, TC and ET mean Dice; and 0.467 survival accuracy. Note that we achieved those results with a multitask model both for segmentation and survival, different from other approaches that have specialized models for each challenge track.

Acknowledgement. We thank Israel Campiotti from NeuralMind for his help in the implementation of the modified EfficientDet architecture. D Carmo thanks the support from São Paulo Research Foundation (FAPESP) grant 2019/21964-4. R Lotufo thanks CNPq for the grant 310828/2018-0. This work is also supported by grant 2013/07559-3 (FAPESP-BRAINN).

References

1. Bakas, S., et al.: Segmentation labels and radiomic features for the pre-operative scans of the TCGA-LGG collection. The cancer imaging archive **286** (2017)

2. Bakas, S., et al.: Segmentation labels and radiomic features for the pre-operative scans of the TCGA-GBM collection. The cancer imaging archive. Nat. Sci .Data **4**, 170117 (2017)
3. Bakas, S., et al.: Advancing the cancer genome atlas glioma MRI collections with expert segmentation labels and radiomic features. Sci. Data **4**, 170117 (2017)
4. Bakas, S., et al.: Identifying the best machine learning algorithms for brain tumor segmentation, progression assessment, and overall survival prediction in the brats challenge. arXiv preprint arXiv:1811.02629 (2018)
5. Carmo, D., Silva, B., Yasuda, C., Rittner, L., Lotufo, R.: Hippocampus Segmentation on Epilepsy and Alzheimer's Disease Studies with Multiple Convolutional Neural Networks. arXiv:2001.05058 [cs, eess], January 2020
6. Chollet, F.: Xception: deep learning with depthwise separable convolutions. In: Proceedings of the IEEE Conference on Computer Vision and Pattern Recognition, pp. 1251–1258 (2017)
7. Górriz, M., Antony, J., McGuinness, K., Giró-i Nieto, X., O'Connor, N.E.: Assessing Knee OA Severity with CNN attention-based end-to-end architectures. arXiv:1908.08856 [cs, eess], August 2019
8. Isensee, F., Kickingereder, P., Wick, W., Bendszus, M., Maier-Hein, K.H.: Brain Tumor Segmentation and Radiomics Survival Prediction: Contribution to the BRATS 2017 Challenge. arXiv:1802.10508 [cs], February 2018
9. Isensee, F., et al. (eds.): Brainlesion: Glioma, Multiple Sclerosis, Stroke and Traumatic Brain Injuries, pp. 234–244. Lecture Notes in Computer Science, Springer International Publishing, Cham (2019). https://doi.org/10.1007/978-3-030-11726-9_21
10. Jiang, Z., Ding, C., Liu, M., Tao, D.: Two-stage cascaded U-Net: 1st place solution to BraTS challenge 2019 segmentation task. In: Crimi, A., Bakas, S. (eds.) BrainLes 2019. LNCS, vol. 11992, pp. 231–241. Springer, Cham (2020). https://doi.org/10.1007/978-3-030-46640-4_22
11. Liu, L., et al.: On the Variance of the Adaptive Learning Rate and Beyond. arXiv:1908.03265 [cs, stat], April 2020
12. Menze, B.H., et al.: The multimodal brain tumor image segmentation benchmark (brats). IEEE Trans. Med. Imaging **34**(10), 1993–2024 (2014)
13. Myronenko, A.: 3D MRI brain tumor segmentation using autoencoder regularization. In: Crimi, A., Bakas, S., Kuijf, H., Keyvan, F., Reyes, M., van Walsum, T. (eds.) BrainLes 2018. LNCS, vol. 11384, pp. 311–320. Springer, Cham (2019). https://doi.org/10.1007/978-3-030-11726-9_28
14. Myronenko, A., Hatamizadeh, A.: Robust semantic segmentation of brain tumor regions from 3D MRIs. In: Crimi, A., Bakas, S. (eds.) BrainLes 2019. LNCS, vol. 11993, pp. 82–89. Springer, Cham (2020). https://doi.org/10.1007/978-3-030-46643-5_8
15. Ramachandran, P., Zoph, B., Le, Q.V.: Searching for activation functions. arXiv preprint arXiv:1710.05941 (2017)
16. Tan, M., Le, Q.V.: Efficientnet: rethinking model scaling for convolutional neural networks. arXiv preprint arXiv:1905.11946 (2019)
17. Tan, M., Pang, R., Le, Q.V.: Efficientdet: scalable and efficient object detection. In: Proceedings of the IEEE/CVF Conference on Computer Vision and Pattern Recognition, pp. 10781–10790 (2020)
18. Wu, Y., He, K.: Group normalization. In: Proceedings of the European conference on computer vision (ECCV), pp. 3–19 (2018)
19. Yushkevich, P.A., et al.: User-guided 3D active contour segmentation of anatomical structures: significantly improved efficiency and reliability. Neuroimage **31**(3), 1116–1128 (2006)

A Two-Stage Cascade Model with Variational Autoencoders and Attention Gates for MRI Brain Tumor Segmentation

Chenggang Lyu and Hai Shu[✉]

Department of Biostatistics, School of Global Public Health,
New York University, New York, NY 10003, USA
hs120@nyu.edu

Abstract. Automatic MRI brain tumor segmentation is of vital importance for the disease diagnosis, monitoring, and treatment planning. In this paper, we propose a two-stage encoder-decoder based model for brain tumor subregional segmentation. Variational autoencoder regularization is utilized in both stages to prevent the overfitting issue. The second-stage network adopts attention gates and is trained additionally using an expanded dataset formed by the first-stage outputs. On the BraTS 2020 validation dataset, the proposed method achieves the mean Dice score of 0.9041, 0.8350, and 0.7958, and Hausdorff distance (95%) of 4.953, 6.299, 23.608 for the whole tumor, tumor core, and enhancing tumor, respectively. The corresponding results on the BraTS 2020 testing dataset are 0.8729, 0.8357, and 0.8205 for Dice score, and 11.4288, 19.9690, and 15.6711 for Hausdorff distance. The code is publicly available at https://github.com/shu-hai/two-stage-VAE-Attention-gate-BraTS2020.

Keywords: Attention gate · Brain tumor segmentation ·
Encoder-decoder network · Variational autoencoder

1 Introduction

Brain tumors can be categorized into primary tumors and secondary tumors depending on where they originate. Glioma, the most common type of primary brain tumor, can be further categorized into low-grade gliomas (LGG) and high-grade gliomas (HGG). HGG is a malignant brain tumor type with a high degree of aggressiveness that often requires surgery. Usually, several complimentary 3D Magnetic Resonance Imaging (MRI) modalities are acquired to highlight different tissue properties and areas of tumor spread. Compared to traditional methods that rely on physicians' professional knowledge and experience, automatic 3D brain tumor segmentation is time-efficient and can provide objective and reproducible results for further tumor analysis and monitoring. In recent years, deep-learning based segmentation approaches have exhibited superior performance than traditional methods.

© Springer Nature Switzerland AG 2021
A. Crimi and S. Bakas (Eds.): BrainLes 2020, LNCS 12658, pp. 435–447, 2021.
https://doi.org/10.1007/978-3-030-72084-1_39

The Multimodal Brain Tumor Segmentation Challenge (BraTS) is an annual international competition that aims to evaluate state-of-the-art methods of brain tumor segmentation [1–3,13]. The organizer provides a 3D multimodal MRI dataset with "ground-truth" tumor segmentation labels annotated by physicians and radiologists. For each patient, four 3D MRI modalities are provided including native T1-weighted (T1), post-contrast T1-weighted (T1c), T2-weighted (T2), and T2 Fluid Attenuated Inversion Recovery (T2-FLAIR) volumes. The brain tumor segmentation task concentrates on three tumor sub-regions: the necrotic and non-enhancing tumor (NCR/NET, labeled 1), the peritumoral edema (ED, labeled 2) and the GD-enhancing tumor (ET, labeled 4). Figure 1 shows an image set of a patient. The rankings of competing methods for this segmentation task are determined by metrics, including Dice score, Hausdorff distance (95%), Sensitivity, and Specificity, evaluated on the testing dataset for ET, tumor core (TC = ET + NCR/NET), and whole tumor (WT = TC + ED) [4].

In BraTS 2018, Myronenko [14] proposed an asymmetrical U-Net with a larger encoder for feature extraction and a smaller decoder for label reconstruction, and won the first place of the challenge. An encouraging innovation of the method is utilizing a variational autoencoder (VAE) branch to regularize the encoder and boost generalization performance. The champion team of BraTS 2019, Jiang et al. [12], proposed a two-stage network, which used an asymmetrical U-Net, similar to Myronenko [14], in the first stage to obtain a coarse prediction, and then employed a similar but wider network in the second stage to refine the prediction. An additional branch was adopted in the decoder of the second-stage network to regularize the associated encoder. The success of the above two models indicates the feasibility and the importance of adding a branch to the decoder to reduce overfitting and boost the model performance.

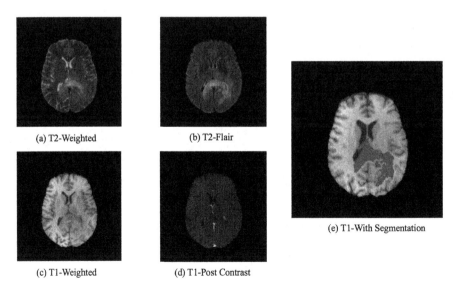

(a) T2-Weighted

(b) T2-Flair

(c) T1-Weighted

(d) T1-Post Contrast

(e) T1-With Segmentation

Fig. 1. An example image set. Subfigure (e) highlights three tumor subregions: ED (orange), NCR/NET (yellow), and ET (red). (Color figure online)

Compared with general computer vision problems, 3D MRI image segmentation tasks generally face two special challenges: the scarcity of training data and the class imbalance [20]. To alleviate the shortage of training data, Isensee et al. [11] took the advantage of additional labeled data by using a co-training strategy. Zhou et al. [23] combined several performance-boosting tricks, such as introducing a focal loss to alleviate the class imbalance, to achieve further improvements.

For brain tumor segmentation tasks specifically, another challenging difficulty is the variability of tumor morphology and location across different tumor development stages and different cases. To improve the prediction accuracy, many segmentation methods [16,18,22,23] decompose the task into separate localization and subsequent segmentation steps, with additional preceding models for object localization. For instance, Wang et al. [18] sequentially trained three networks according to the tumor subregion hierarchy. Oktay et al. [15] demonstrated that the same objective can be achieved by introducing attention gates (AGs) into the standard convolutional-neural-network framework in pancreas tumor segmentation tasks.

Inspired by aforementioned works, in this paper we propose a two-stage cascade network for brain tumor segmentation. We borrow the network structure of Myronenko [14] as the first-stage network to obtain relatively rough segmentation results. The second stage network uses the concatenation of the preliminary segmentation maps from the first-stage network and the MRI images as the input, with the aim to refine the prediction of the NCR/NET and ET subregions. We apply AGs [15] to further suppress the feature responses in irrelevant background regions. Our second-stage network exhibits the capabilities to (i) provide more model candidates with competitive performance for model ensembling, (ii) stabilize the predictions across models of different epochs, and (iii) improve the performance of each single model, particularly for NCR/NET and ET. The implementation details and segmentation results are provided in Sects. 3 and 4.

2   Method

The proposed two-stage network structure consists of two cascaded networks. The first-stage network takes the multimodal MRI images as input and predicts coarse segmentation maps. The concatenation of the preliminary segmentation maps and the MRI images is passed into the second-stage network to generate improved segmentation results.

2.1   The First-Stage Network: Asymmetrical U-Net with a VAE Branch

The network architecture (Fig. 2) consists of a larger encoding path for semantic feature extraction, a smaller decoding path for segmentation map prediction, and a VAE branch for input images reconstruction. This part is identical to the network proposed in [14].

Encoder. The encoder consists of ResNet [7,8] blocks for four spatial levels, with the number of blocks 1, 2, 2, and 4, respectively. Each ResNet block has two convolutions with Group Normalization and ReLU, followed by an additive identity skip connection. The input of the encoder is an MRI crop of size $4 \times 160 \times 192 \times 128$, with the first channel referring to the four MRI modalities. The input is processed by a $3 \times 3 \times 3$ convolution layer with 32 filters and a dropout layer with a rate of 0.2, and then passed through a series of ResNet blocks. Between every two blocks with different spatial levels, a $3 \times 3 \times 3$ convolution with a stride of 2 is used to reduce the resolution of the feature maps by 2 and double the number of feature channels simultaneously. The endpoint of the encoder has size $256 \times 20 \times 24 \times 16$, which is 1/8 of the spatial size of the input data.

Decoder. The decoder has an almost symmetrical architecture with the encoder, except for the number of ResNet blocks within each spatial level is 1. After each block, we use a trilinear up-sampler to recover the spatial size by 2 and a $1 \times 1 \times 1$ convolution to reduce the number of feature channels by 2, followed by an additive skip connection from the encoder output of the corresponding spatial level. The operations within each block are the same as those in the encoder. At the end of the decoder, a $1 \times 1 \times 1$ convolution is used to reduce the number of feature channels from 32 to 3, followed by a sigmoid function to convert feature maps into probability maps.

VAE Branch. This decoder branch receives the output of the encoder and produces a reconstructed image of the original input. In the beginning, the decoder endpoint output is reduced to a lower-dimensional space of 256 using a fully connected layer, where 256 represents 128 means and 128 standard deviations of Gaussian distributions, from which a sample of size 128 is drawn. Then the drawn vector is mapped back to the high-dimensional space with the same spatial property and reconstructed into the input image dimensions gradually following the same strategy as the decoder. Notice that there is no additive skip connection between encoder and the VAE branch.

2.2 The Second-Stage Network: Attention-Gated Asymmetrical U-Net with the VAE Branch

The input of the second-stage network (Fig. 2) is constructed based on the segmentation maps produced by the first-stage network. To alleviate the label imbalance problem, we crop the output of the first-stage network into a spatial size of $128 \times 128 \times 128$ voxels concentrating on the tumor area. The cropped segmentation maps are then concatenated to the original MRI images (cropped to the same area).

Encoder. The encoder part of the second-stage network has the same structure as in the first-stage network, whereas the input has 7 channels (3 for segmentation maps and 4 for multimodal MRI images), and has a spatial size of
$128 \times 128 \times 128$ voxels.

Decoder. Different from the first-stage network, we add the AGs of [15] in
the decoder part. The architecture of the AGs is demonstrated in the next sub-
section. At each spatial level, the gating signal from the coarser scale is passed
into the attention gate to determine the attention coefficients. The output of
an AG is the Hadamard product of input features from encoder through skip
connection and attention coefficients. The output of AG at each spatial level is
then integrated with the 2-times up-sampled features from the coarser scale by
an element-wise summation. The rest of the network architecture remains the
same as the decoder in the first-stage network.

Attention Gate. Instead of using a single identical scalar value to represent
attention level for each voxel vector, a gating vector g_i is computed to determine

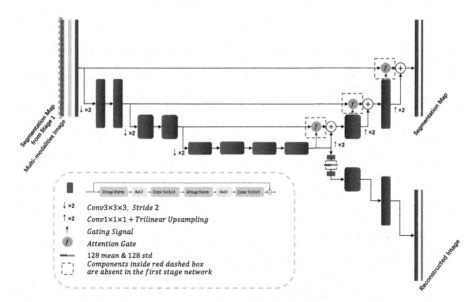

Fig. 2. The network architecture of both stages. In the first stage, input (orange strip)
is the cropped MRI images ($4 \times 160 \times 192 \times 128$), followed by a $3 \times 3 \times 3$ convolution
with 32 filters and a dropout layer (yellow strip). The output of the decoder is a
segmentation map of size $3 \times 160 \times 192 \times 128$ with three channels indicating three
tumor subregions (WT, TC, and ET). The VAE branch is in charge of input image
reconstruction and is disabled while doing inference. In the second stage, the input is
the cropped concatenation of the first-stage segmentation map (blue strip) and MRI
images (orange strip) (total $7 \times 128 \times 128 \times 128$), and the output is a segmentation
map ($3 \times 128 \times 128 \times 128$). Note that there is no input concatenation and attention
gates in the first-stage network. (Color figure online)

focus regions for each voxel i. Within the l-th spatial level, the AG is formulated as follows:

$$q_{att}^l = W_{int}^T \sigma_1(W_X^T x_i^l + W_g^T g_i^{l+1} + b_g) + b_{W_{int}} \tag{1}$$

$$\alpha_i^l = \sigma_2(q_{att}^l(x_i^l, g_i^{l+1}; \theta_{att})) \tag{2}$$

$$\hat{x}_i^l = \alpha_i^l \times x_i^l \tag{3}$$

In each AG (Fig. 3), complementary information is extracted from the gating signal g_i^{l+1} from the coarser scale. To reduce the computational cost, linear transformations W_x^T and W_g^T ($1 \times 1 \times 1$ convolutions) are performed on the input features x_i^l and gating signals g_i^{l+1}, to downsize the feature size by 2, and to reduce the number of channels by 2, respectively. The transformed input features and gating signals therefore have the same spatial shape. The sum of them through element-wise summation is activated by the ReLU function σ_1 and mapped by W_{int}^T into a lower dimensional space for gating operation, followed by the sigmoid function σ_2 and a trilinear up-sampler to restore the size of attention coefficients matrix α_i^l to match the resolution of the input features. The output $\hat{x}_i^l$ of the AG is obtained by element-wise multiplication of the input features x_i^l and the attention coefficient matrix α_i^l.

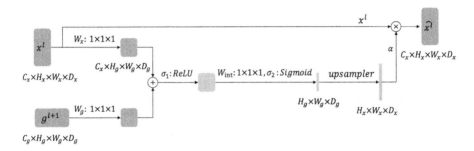

Fig. 3. Attention gate.

2.3   Loss Function

For both stages, the loss function has 3 parts:

$$L = L_{dice} + 0.1 \times L_{L2} + 0.1 \times L_{KL}. \tag{4}$$

L_{dice} is the soft dice loss that encourages the decoder output p_{pred} to match the ground-truth segmentation mask p_{true}:

$$L_{dice} = 1 - \frac{2 \times \sum p_{pred} \times p_{true}}{\sum p_{pred}^2 + \sum p_{true}^2}. \tag{5}$$

L_{L2} is the L2 loss that is applied to the VAE branch output I_{pred} to match the input image I_{input}:

$$L_{L2} = \sum (I_{pred} - I_{input})^2. \tag{6}$$

L_{KL} is the KL divergence that is used as a VAE penalty term to induce the estimated Gaussian distribution to approach the standard Gaussian distribution:

$$L_{KL} = \frac{1}{N} \sum \mu^2 + \sigma^2 - \log \sigma^2 - 1, \tag{7}$$

where N is the number of the voxels. As suggested in [14], we set the hyper-parameter weight to be 0.1 to reach a good balance between the dice and VAE loss terms.

3   Expriment

3.1   Data Description

The BraTS 2020 training dataset includes 259 cases of HGG and 110 cases of LGG. All image modalities (T1, T1c, T2, and T2-FLAIR) are co-registered with image size of $240 \times 240 \times 155$ voxels and 1 mm isotropic resolution. The training data are provided with annotations, while the validation dataset (125 cases) and testing dataset (166 cases) are provided without annotations. Participants can evaluate their methods by uploading predicted segmentation volumes to the organizer's server. Multiple times of submission for the validation evaluation are permitted, whereas only one submission is allowed for the final testing evaluation.

3.2   Implementation Details

Our network is implemented in Pytorch and trained on four NVIDIA P40 GPUs.

Optimization. We use Adam optimizer with initial learning rate of $lr_0 = 10^{-4}$ for weights updating. We progressively decay the learning rate according to the following formula:

$$lr = lr_0 \times (1 - \frac{e}{N_e})^{0.9}, \tag{8}$$

where e is an epoch counter, and N_e is the total number of the epochs during training. In our case, N_e is set to 300.

Data Preprocessing. Before feeding input data into the first-stage network, we preprocessed the input data by applying intensity normalization to each MRI modality for each patient. The data is subtracted by the mean and divided by the standard deviation of the non-zero region. In the second stage, we crop the segmentation maps from the first-stage network into $128 \times 128 \times 128$-sized patches for each patient while ensuring that the patch includes most tumor voxels. The patches are concatenated with the normalized MRI images (after data augmentation, cropped at the same position) and fed to the second-stage network for training.

Data Augmentation. To reduce the risk of overfitting, three data augmentation strategies are used. First, the training data is randomly cropped into size of $160 \times 192 \times 128$ before fed into the first-stage network. In addition, in both stages, we randomly shift the intensity of the input data by a value in $[-0.1, 0.1]$ of the standard deviation of each channel, and randomly scale intensity of the input data by a factor in $[0.9, 1.1]$. Finally, we apply random flipping along each 3D axis with a probability of 50%, in both stages.

Expanded Training Data. Since the training processes of the two stages are independent, we can select several first-stage trained models of competitive performance and use their segmentation results as the training data for training the second-stage network. Such a strategy trades a longer training process for better model performance and stability of results. Specifically, we select 6 individual first-stage models (of different epochs with different train-validation divisions) and combined their segmentation results into an extensive dataset to train the second-stage network (Fig. 4). Note that the train-validation division is based on patient IDs. The 6 segmentation results belonging to the same patient

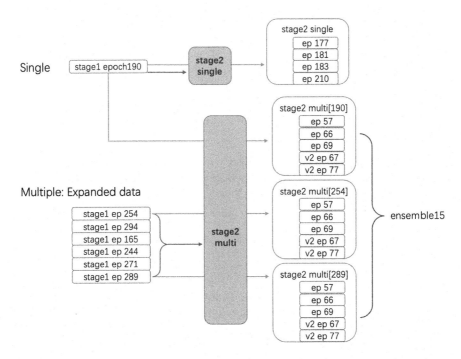

Fig. 4. Stage-2 network training schema. In the first case, the training (blue) and inference (yellow) of the stage-2 network are based on the segmentation results of the same stage-1 model; in the second case, the segmentation results from six stage-1 models are combined into an extensive training dataset for training the stage-2 network (blue). The segmentation results of three stage-1 models are respectively input into five stage-2 models of different epochs for inference (yellow), and finally 15 segmentation results are obtained for ensemble (red). (Color figure online)

are consequentially grouped into the same set. We also have tried training the second-stage network using one single model's segmentation result, but obtained only slight improvement compared to the first-stage network.

Postprocess. It is observed that when the predicted volume of ET is particularly small, the algorithm tends to predict TC voxels as ET falsely. In postprocessing, based on our experience we replace ET with TC when the volume of predicted ET is less than 500 voxels.

Ensemble. We use majority voting to conduct model ensembling. In particular, if a voxel has equal votes in multiple categories, the final predicted category of the voxel is determined based on the average probability of each category.

4   Results

4.1   Quantitative Results

The validation dataset for BraTS 2020 includes 125 cases without providing tumor subtypes (HGG/LGG) or tumor subregion annotations. Table 1 reports the segmentation result of per-class Dice score and Hausdorff distance for the validation dataset evaluated by the official platform (https://ipp.cbica.upenn.edu/).

By comparing the segmentation performance of the 190th-epoch models of the two stages, we see that the improvement on accuracy brought by the presence of the second-stage network is more evident for TC than that for WT, and training the second-stage network with expanded training data further improves the Dice score for TC.

Table 1. Segmentation results on validation data.

Stage	Method	Dice			Hausdorff (mm)		
		ET	WT	TC	ET	WT	TC
1	Model ep190	0.7881	0.8992	0.8206	23.716	5.657	6.664
	Model ep254	0.7930	0.8980	0.8258	26.516	5.958	6.565
	Model ep289	0.7902	0.8973	0.8286	24.146	6.174	7.042
	Ensemble 9	0.7946	0.9022	0.8282	23.651	5.176	6.307
2	Model ep190: Single	0.7925	0.9012	0.8241	23.752	5.159	6.692
	Model ep190: Multiple	0.7897	0.9016	0.8291	29.327	5.288	6.632
	Model ep254: Multiple	0.7769	0.9010	0.8316	32.505	5.558	6.557
	Model ep289: Multiple	0.7896	0.9002	0.8361	21.383	5.521	6.459
	Ensemble 15: Multiple	0.7960	0.9039	0.8345	23.630	4.959	6.331
	Ensemble 21: Multiple	0.7958	0.9041	0.8350	23.608	4.953	6.299
	Ensemble 27: Multiple	0.7952	0.9039	0.8350	23.590	4.962	6.303

As a performance-boosting component, the second-stage network trained with expanded data can be added to any first-stage model to enhance the segmentation performance. The second-stage network with expanded data also reduces the performance variation across models of different epochs. Table 2 shows that the standard deviation (SD) of the TC's Dice score and Hausdorff distance are reduced by 68% and 93% in the second-stage, respectively. The SDs are calculated based on the performance of all trained non-ensembled models. We also observe that the second-stage network remarkably reduces the variation of ET's Dice score and Hausdorff distance, but this improvement no longer exists after post-processing.

The BraTS 2020 testing dataset contains 166 cases without providing tumor annotations. Our segmentation results on this dataset are presented in Table 3.

Table 2. The performance variation across non-ensembled models on validation data.

Stage	Metric	Dice			Hausdorff (mm)		
		ET	WT	TC	ET	WT	TC
1	SD	0.0128	0.0013	0.0110	4.097	0.236	1.427
	Range	[0.714, 0.750]	[0.894, 0.899]	[0.798, 0.835]	[30.535, 42.452]	[5.657, 6.433]	[6.566, 10.502]
2	SD	0.0070	0.0012	0.0035	2.337	0.184	0.102
	Range	[0.715, 0.742]	[0.898, 0.902]	[0.822, 0.838]	[33.457, 42.546]	[5.191, 6.024]	[6.168, 6.741]

Note: The variation metrics are calculated based on the results from 9 stage-1 models and 37 stage-2 models without post-processing.

Table 3. Segmentation results on testing data.

Stage	Method	Dice			Hausdorff (mm)		
		ET	WT	TC	ET	WT	TC
2	Ensemble 21: Multiple	0.8205	0.8729	0.8357	15.6711	11.4288	19.9690

4.2 Attention Map

The attention matrices in the finest scale are visualised in the form of heatmap with red indicating higher weights and blue indicating lower weights (Fig. 5). In the first few training epochs, we observe that AGs grasp the tumor's location and meanwhile assign a high weight to gray matter. As the training progresses, the weights assigned to non-tumor regions gradually decrease. AGs also suggest the model avoid misclassification of voxels around the tumor boundary by gradually decreasing weights assigned to those voxels.

Epoch 3	Epoch 20	Epoch 115	Ground-truth Annotation

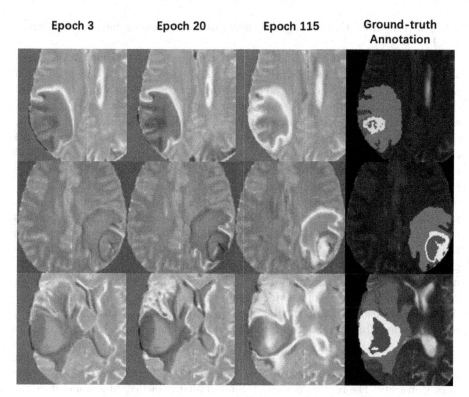

Fig. 5. The first three columns show the attention maps at training epochs 3, 20, and 115, respectively. The fourth column shows the example images of T2-modality with ground-truth annotations extracted from the BraTS 2020 training dataset. The model gradually learns to assign lower weights to non-tumor areas and the tumor boundary.

5 Concluding Remarks

This paper proposes a two-stage cascade network with VAEs and AGs for 3D MRI brain tumor segmentation. The results indicate the second-stage network improves and stabilizes the prediction for all three tumor subregions, particularly for TC and ET (before post-processing). The second-stage network can also produce more qualified model candidates for further model ensembling. In this study, we use the segmentation results of multiple first-stage models to train the second-stage network. Though this helps improve the model's prediction performance, it noticeably increases the training time as a trade-off. Consequentially, this technique may not be suitable for occasions with limited computing resources and research time. In addition, we can see from Table 1 that even if the expanded training data does not include the output of the first-stage 190th-epoch model, we can still use the trained second-stage models to obtain a better result based on the first-stage prediction. This indicates that the second-stage network trained by this strategy has generalizability among models of different epochs.

Since first proposed in natural language processing [17], the attention mechanism has been extensively studied and widely used in image segmentation problems. Technically speaking, the attention mechanism in image segmentation tasks can be divided into the spatial attention, such as the AGs used in our method, and the channel attention, e.g., the "squeeze and excitation" block in [9,22]. It was proposed in [6] to combine the two kinds of attention in 2D problems, but multiplications between huge matrices involved in the method will likely exceed the computational limits in 3D scenarios. Further research is expected to include the appropriate combination of the two attention mechanisms into the brain tumor segmentation to enhance the segmentation accuracy. Besides, Dai et al. [5] utilized the extreme gradient boosting (XGboost) in model ensemble and gained extra improvement on accuracy as compared with the majority voting and probability averaging approaches. It may be worth integrating XGboost into our method, as the existence of the second-stage provides more models to be chosen from for the XGboost training. Moreover, Zhong et al. [21] has recently developed a segmentation network model that incorporates the dilated convolution [19] and the dense block [10]. The two popular deep-learning techniques may be valuable to be combined into our network structure.

Acknowledgements. This research was partially supported by the grant R21AG070303 from the National Institutes of Health and a startup fund from New York University. The content is solely the responsibility of the authors and does not necessarily represent the official views of the National Institutes of Health or New York University.

References

1. Bakas, S., et al.: Segmentation labels and radiomic features for the pre-operative scans of the TCGA-GBM collection. The Cancer Imaging Archive (2017)
2. Bakas, S., et al.: Segmentation labels and radiomic features for the pre-operative scans of the TCGA-LGG collection. The Cancer Imaging Archive (2017)
3. Bakas, S., et al.: Advancing the cancer genome atlas glioma MRI collections with expert segmentation labels and radiomic features. Nat. Sci. Data **4**, 170117 (2017)
4. Bakas, S., et al.: Identifying the best machine learning algorithms for brain tumor segmentation, progression assessment, and overall survival prediction in the BRATS challenge. arXiv preprint arXiv:1811.02629 (2018)
5. Dai, L., Li, T., Shu, H., Zhong, L., Shen, H., Zhu, H.: Automatic brain tumor segmentation with domain adaptation. In: Crimi, A., Bakas, S., Kuijf, H., Keyvan, F., Reyes, M., van Walsum, T. (eds.) BrainLes 2018. LNCS, vol. 11384, pp. 380–392. Springer, Cham (2019). https://doi.org/10.1007/978-3-030-11726-9_34
6. Fu, J., et al.: Dual attention network for scene segmentation. In: Proceedings of the IEEE Conference on Computer Vision and Pattern Recognition, pp. 3146–3154 (2019)
7. He, K., Zhang, X., Ren, S., Sun, J.: Deep residual learning for image recognition. In: Proceedings of the IEEE Conference on Computer Vision and Pattern Recognition, pp. 770–778 (2016)

8. He, K., Zhang, X., Ren, S., Sun, J.: Identity mappings in deep residual networks. In: Leibe, B., Matas, J., Sebe, N., Welling, M. (eds.) ECCV 2016. LNCS, vol. 9908, pp. 630–645. Springer, Cham (2016). https://doi.org/10.1007/978-3-319-46493-0_38

9. Hu, J., Shen, L., Sun, G.: Squeeze-and-excitation networks. In: Proceedings of the IEEE Conference on Computer Vision and Pattern Recognition, pp. 7132–7141 (2018)

10. Huang, G., Liu, Z., Van Der Maaten, L., Weinberger, K.Q.: Densely connected convolutional networks. In: Proceedings of the IEEE Conference on Computer Vision and Pattern Recognition, pp. 4700–4708 (2017)

11. Isensee, F., Kickingereder, P., Wick, W., Bendszus, M., Maier-Hein, K.H.: No new-net. In: Crimi, A., Bakas, S., Kuijf, H., Keyvan, F., Reyes, M., van Walsum, T. (eds.) BrainLes 2018. LNCS, vol. 11384, pp. 234–244. Springer, Cham (2019). https://doi.org/10.1007/978-3-030-11726-9_21

12. Jiang, Z., Ding, C., Liu, M., Tao, D.: Two-stage cascaded U-net: 1st place solution to BraTS challenge 2019 segmentation task. In: Crimi, A., Bakas, S. (eds.) BrainLes 2019. LNCS, vol. 11992, pp. 231–241. Springer, Cham (2020). https://doi.org/10.1007/978-3-030-46640-4_22

13. Menze, B.H., et al.: The multimodal brain tumor image segmentation benchmark (BRATS). IEEE Trans. Med. Imaging 34(10), 1993 (2015)

14. Myronenko, A.: 3D MRI brain tumor segmentation using autoencoder regularization. In: Crimi, A., Bakas, S., Kuijf, H., Keyvan, F., Reyes, M., van Walsum, T. (eds.) BrainLes 2018. LNCS, vol. 11384, pp. 311–320. Springer, Cham (2019). https://doi.org/10.1007/978-3-030-11726-9_28

15. Oktay, O., et al.: Attention U-net: learning where to look for the pancreas. arXiv preprint arXiv:1804.03999 (2018)

16. Tu, Z., Bai, X.: Auto-context and its application to high-level vision tasks and 3D brain image segmentation. IEEE Trans. Pattern Anal. Mach. Intell. 32(10), 1744–1757 (2009)

17. Vaswani, A., et al.: Attention is all you need. In: Advances in Neural Information Processing Systems, pp. 5998–6008 (2017)

18. Wang, G., Li, W., Ourselin, S., Vercauteren, T.: Automatic brain tumor segmentation using cascaded anisotropic convolutional neural networks. In: Crimi, A., Bakas, S., Kuijf, H., Menze, B., Reyes, M. (eds.) BrainLes 2017. LNCS, vol. 10670, pp. 178–190. Springer, Cham (2018). https://doi.org/10.1007/978-3-319-75238-9_16

19. Yu, F., Koltun, V.: Multi-scale context aggregation by dilated convolutions. arXiv preprint arXiv:1511.07122 (2015)

20. Zhao, Y.-X., Zhang, Y.-M., Liu, C.-L.: Bag of tricks for 3D MRI brain tumor segmentation. In: Crimi, A., Bakas, S. (eds.) BrainLes 2019. LNCS, vol. 11992, pp. 210–220. Springer, Cham (2020). https://doi.org/10.1007/978-3-030-46640-4_20

21. Zhong, L., et al.: (TS)2WM: tumor segmentation and tract statistics for assessing white matter integrity with applications to glioblastoma patients. Neuroimage 223, 117368 (2020)

22. Zhou, C., Chen, S., Ding, C., Tao, D.: Learning contextual and attentive information for brain tumor segmentation. In: Crimi, A., Bakas, S., Kuijf, H., Keyvan, F., Reyes, M., van Walsum, T. (eds.) BrainLes 2018. LNCS, vol. 11384, pp. 497–507. Springer, Cham (2019). https://doi.org/10.1007/978-3-030-11726-9_44

23. Zhou, C., Ding, C., Lu, Z., Wang, X., Tao, D.: One-pass multi-task convolutional neural networks for efficient brain tumor segmentation. In: Frangi, A.F., Schnabel, J.A., Davatzikos, C., Alberola-López, C., Fichtinger, G. (eds.) MICCAI 2018. LNCS, vol. 11072, pp. 637–645. Springer, Cham (2018). https://doi.org/10.1007/978-3-030-00931-1_73

Multidimensional and Multiresolution Ensemble Networks for Brain Tumor Segmentation

Gowtham Krishnan Murugesan[1]($\boxtimes$), Sahil Nalawade[1], Chandan Ganesh[1],
Ben Wagner[1], Fang F. Yu[1], Baowei Fei[3], Ananth J. Madhuranthakam[1,2],
and Joseph A. Maldjian[1,2]

[1] Department of Radiology, University of Texas Southwestern Medical Center, Dallas, TX, USA
Gowtham.Murugesan@UTSouthwestern.edu
[2] Advanced Imaging Research Center, University of Texas Southwestern Medical Center,
Dallas, TX, USA
[3] Department of Bioengineering, University of Texas at Dallas, Dallas, TX, USA

Abstract. In this work, we developed multiple 2D and 3D segmentation models with multiresolution input to segment brain tumor components and then ensembled them to obtain robust segmentation maps. Ensembling reduced overfitting and resulted in a more generalized model. Multiparametric MR images of 335 subjects from the BRATS 2019 challenge were used for training the models. Further, we tested a classical machine learning algorithm with features extracted from the segmentation maps to classify subject survival range. Preliminary results on the BRATS 2019 validation dataset demonstrated excellent performance with DICE scores of 0.898, 0.784, 0.779 for the whole tumor (WT), tumor core (TC), and enhancing tumor (ET), respectively and an accuracy of 34.5% for predicting survival. The Ensemble of multiresolution 2D networks achieved 88.75%, 83.28% and 79.34% dice for WT, TC, and ET respectively in a test dataset of 166 subjects.

Keywords: Residual inception dense networks · Densenet-169 · Squeezenet · Survival prediction · Brain tumor segmentation

1 Introduction

Brain Tumors account for 85–90% of all primary CNS tumors. The most common primary brain tumors are gliomas, which are further classified into a high grade (HGG) and low grade gliomas (LGG) based on their histologic features. Magnetic Resonance Imaging (MRI) is a widely used modality in the diagnosis and clinical treatment of gliomas. Despite being a standard imaging modality for tumor delineation and treatment planning, brain tumor segmentation on MR images remains a challenging task due to the high variation in tumor shape, size, location, and particularly the subtle intensity changes relative to the surrounding normal brain tissue. Consequently, manual tumor contouring is performed, which is both time-consuming and subject to large inter- and

---

G. K. Murugesan and S. Nalawade—Equal Contribution.

© Springer Nature Switzerland AG 2021
A. Crimi and S. Bakas (Eds.): BrainLes 2020, LNCS 12658, pp. 448–457, 2021.
https://doi.org/10.1007/978-3-030-72084-1_40

intra-observer variability. Semi- or fully-automated brain tumor segmentation methods could circumvent this variability for better patient management (Zhuge et al. 2017). As a result, developing automated, semi-automated, and interactive segmentation methods for brain tumors has important clinical implications, but remains highly challenging. Efficient deep learning algorithms to segment brain tumors into their subcomponents may help in early clinical diagnosis, treatment planning, and follow-up of patients (Saouli et al. 2018).

The multimodal Brain Tumor Segmentation Benchmark (BRATS) dataset provided a comprehensive platform by outsourcing a unique brain tumor dataset with known ground truth segmentations performed manually by experts (Menze et al. 2014). Several advanced deep learning algorithms were developed on this unique platform provided by BRATS and benchmarked against standard datasets allowing comparisons between them. Convolutional Neural Networks (CNN)-based methods have shown advantages for learning the hierarchy of complex features and have performed the best in recent BRATS challenges. U-net (Ronneberger et al. 2015) based network architectures have been used for segmenting complex brain tumor structures. Pereira et al. Developed a 2D CNN method with two CNN architectures for HGG and LGG separately and combined the outputs in the post-processing steps (Pereira et al. 2016). Havaei et al. Developed a multi-resolution cascaded CNN architecture with two pathways, each of which takes different 2D patch sizes with four MR sequences as channels (Havaei et al. 2017). The BRATS 2018 top performer developed a 3D decoder encoder style CNN architecture with inter-level skip connections to segment the tumor (Myronenko 2018). In addition to the decoder part, a Variation Autoencoder (VAE) was included to add reconstruction loss to the model.

In this study, we propose to ensemble output from Multiresolution and Multidimensional models to obtain robust tumor segmentations. We utilized off-the-shelf model architectures (DensNET-169, SERESNEXT-101, and SENet-154) to perform segmentation using 2D inputs. We also implemented a 2D and 3D Residual Inception Densenet (RID) network to perform tumor segmentation with patch-based inputs (64 × 64 and 64 × 64 × 64). The outputs from the model trained on different resolutions and dimensions were combined to eliminate false positives and post-processed using cluster analysis to obtain the final outputs.

2   Materials and Methods

2.1   Data and Preprocessing

The BRATS 2019 dataset included a total of 335 multi-institutional subjects (Menze et al. 2014; Bakas et al. 2017a, b, c, d), consisting of 259 HGGs and 76 LGGs. The standard preprocessing steps by the BRATS organizers on all MR images included co-registration to an anatomical template (Rohlfing et al. 2010), resampling to isotropic resolution ($1 \times 1 \times 1$ mm^3), and skull-stripping (Bakas et al. 2018). Additional preprocessing steps included N4 bias field correction (Tustison et al. 2014) for removing RF inhomogeneity and normalizing the multi-parametric MR images to zero mean and unit variance.

The purpose of the survival prediction task is to predict the overall survival of the patient based on the multiparametric pre-operative MR imaging features in combination with the segmented tumor masks. Survival prediction based on only imaging-based features (with age and resection status) is a difficult task. Additional information such as histopathology, genomic information, radiotracer based imaging, and other non-MR imaging features can be used to improve the overall survival prediction. Pooya et.. al. (Mobadersany et al. 2018) reported better accuracy by combining genomic information and histopathological images to form a genomic survival convolutional neural network architecture (GSCNN model). Several studies have reported predicting overall survival for cerebral gliomas using ^{11}C-acetate and ^{18}F-FDG PET/CT scans (Tsuchida et al. 2008; Yamamoto et al. 2008; Kim et al. 2018).

2.2 Network Architecture

We trained several models to segment tumor components. All network architectures used for the segmentation task, except Residual Inception dense Network, were imported using Segmentation models, a python package (Yakubovskiy 2019). The models selected for brain tumor segmentation had different backbones (DenseNet-169 (Huang et al. 2017), SERESNEXT-101 (Chen et al. 2018) and SENet-154 (Hu et al. 2018)). The DenseNet architecture has shown promising results in medical data classification and image segmentation tasks (Islam and Zhang 2017; Chen et al. 2018; Dolz et al. 2018). The DenseNet model has advantages in feature propagation from one dense block to the next and overcomes the problem of the vanishing gradient (Huang et al. 2017). The squeeze and excitation block was designed to improve the feature propagation by enhancing the interdependencies between features for the classification task. This helps in propagating more useful features to the next block and suppressing less informative features. This network architecture was the top performer at the ILSVC 2017 classification challenge. SENet-154 and SE-ResNeXt-101 have more parameters and is computationally expensive but has shown good results on the ImageNet classification tasks (Hu et al. 2018). Three of the proposed models were ensembled to obtain the final results. All of these models from the Segmentation Models package were trained with 2D axial slices of size 240×240 (Fig. 1).

The Residual Inception Dense Network (RID) was first proposed and developed by Khened et al. for cardiac segmentation. We incorporated our implementation of the RID network in Keras with a Tensorflow backend (Fig. 2). In the DenseNet architecture, the GPU memory footprint increases with the number of feature maps of larger spatial resolution. The skip connections from the down-sampling path to the up-sampling path use element-wise addition in this model, instead of the concatenation operation in DenseNet, to mitigate feature map explosion in the up-sampling path. For the skip connections, a projection operation was performed using Batch Normalization (BN)-1 $\times$ 1-convolution-dropout to match the dimensions for element-wise addition (Fig. 3). These additions to the Densenet architecture help in reducing the parameters and the GPU memory footprint without affecting the quality of segmentation output. In addition to performing dimension reduction, the projection operation facilitates learning interactions of cross channel information (Lin et al. 2013) and faster convergence. Further, the initial layer of the RID networks includes parallel CNN branches similar to the inception module with multiple kernels of varying receptive fields. The inception module helps in

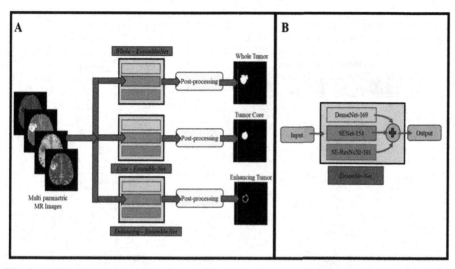

Fig. 1. A. Ensemble of Segmentation models (DenseNET-169, SERESNEXT-101 and SENet-154). B. Ensemble methodology used to combine the outputs from Segmentation Models to produce output segmentation maps

capturing view-point dependent object variability and learning relations between image structures at multiple-scales.

2.2.1 Model Training and Ensemble Methodology

All models from the Segmentation models package were trained with full resolution axial slices of size 240 × 240 as input to segment the tumor subcomponents separately. The outputs of each component from the models were combined following post-processing steps that included removing clusters of smaller size to reduce false positives. Each tumor component was then combined to form the segmentation map (Fig. 1B).

The RID model was trained on 2D input patches of size 64 × 64. For each component of the brain tumor (e.g., Whole Tumor (WT), Tumor Core (TC), and Enhancing Tumor (ET)), we trained a separate RID model with axial as well as sagittal slices as input. In addition to the six RID models, we also trained a RID with axial slices as input with patch size of 64x64 to segment TC and Edema simultaneously (TC-ED). A three-dimensional RID network model was also trained to segment ET and a multiclass TC-ED (TC-ED-3D). All models were trained with dice loss and Adam optimizers with a learning rate of 0.001 using NVIDIA Tesla P40 GPU's.

2.2.2 Ensemble Methodology

The DenseNET-169, SERESNEXT-101, and SENet-154 model outputs were first combined to form segmentation maps, as shown in Fig. 1B, which we will refer to as the Segmentation model output. Then, for each component, we combined outputs from the RID models and Segmentation models, as shown in Fig. 3.

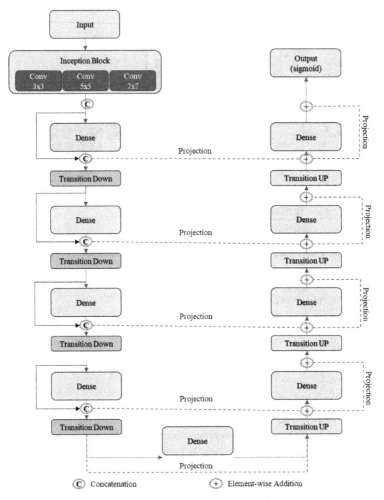

Fig. 2. Residual inception densnet architecture

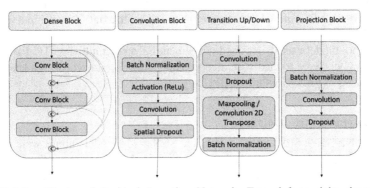

Fig. 3. Building Blocks of Residual Inception Network. From left to right, dense block, convolution block, transition block and projection block

2.2.3 Survival Prediction

The tumor segmentation maps extracted from the above methodology was used to extract texture and wavelet based features using the PyRadiomics (Van Griethuysen et al. 2017) and Pywavelets (Lee et al. 2019) packages from each tumor subcomponent for each contrast. In addition, we also added volume and surface area features of each tumor component (Feng et al. 2018), along with age. We performed feature selection based on SelectKBest features using the sklearn package (Pedregosa et al. 2011; Buitinck et al. 2013), which resulted in a reduced set of 25 features. We trained four different models, including XGBoost (XGB), K-Nearest Neighbour (KNN), Extremely randomized trees (ET), and Linear Regression (LR) models (Chen and Guestrin 2016) for the survival classification task. An ensemble of the four different models was used to form a voting

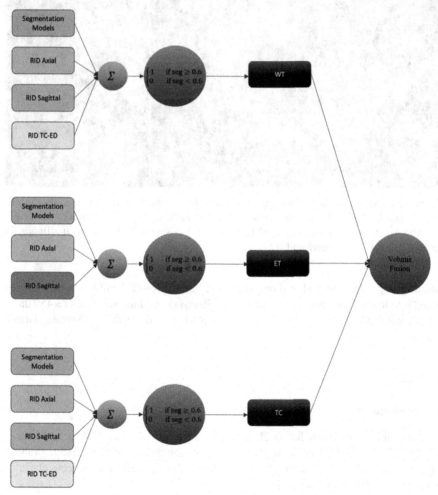

Fig. 4. An ensemble of multidimensional and multiresolution networks. Top to bottom, the ensemble for the Whole Tumor (WT), Tumor Core (TC), and Enhancing Tumor (ET), respectively.

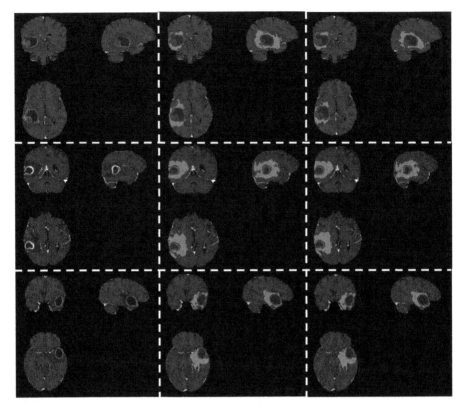

Fig. 5. Example Tumor Segmentation Performance for 3 subjects shown in each row. (a) T1-post contrast (T1C), (b) Segmentation output, (c) Overlay of segmentation output on the T1-post contrast images. Colors: Blue = Non-enhancing tumor + Necrosis, Red = Enhancing Tumor, and Green = Edema (Color figure online)

classifier to predict survival in days. These predictions for each subject were then separated into low (<300 days), medium (300–450 days) and long survivors (>450 days). Twenty-nine subjects from the validation dataset were used to validate the trained model (Fig. 4).

3 Results

3.1 Segmentation

The Ensemble of multiresolution 2D networks achieved 89.79%, 78.43% and 77.97% dice for WT, TC, and ET respectively in the validation dataset of 125 subjects (Table 1, Fig. 5) and 88.75%, 83.28% and 79.34% dice for WT, TC, and ET respectively in the test dataset of 166 subjects (Table 2).

Table 1. Validation segmentation results for multiresolution 2D ensemble model and multidimensional multiresolution ensemble model.

Models	WT	TC	ET
Multiresolution 2D Ensemble	0.892	0.776	0.783
Multidimensional and Multiresolution Ensemble	0.898	0.78	0.784

Table 2. Testing segmentation results for the multidimensional and multiresolution ensemble model.

Models	WT	TC	ET
Multidimensional and Multiresolution Ensemble	0.888	0.833	0.793

3.2 Survival Prediction

Accuracy and mean square error for overall survival prediction for the 29 subjects using a Voting Classifier were 51.7% and 117923.1, respectively (Table 3). In testing the proposed method achieved 41.1% accuracy.

Table 3. Validation survival results for the voting classifier network.

	Accuracy	Mean Squared Error
Validation	51.7%	117923.1
Testing	41.1%	446765.3

4   Discussion

We ensembled several models with multiresolution inputs to segment brain tumors. The RID network was parameter and memory efficient, and able to converge in as few as three epochs. This allowed us to train several models for ensemble in a short amount of time. The proposed methodology of combining multidimensional models improved performance and achieved excellent segmentation results, as shown in Table 1. For survival prediction, we extracted numerous features based on texture, first-order statistics, and wavelets. Efficient model based feature selection allowed us to reduce the otherwise large feature set to 25 features per subject. We trained several classical machine learning models and then combined them to improve results on the validation dataset.

5   Conclusion

We demonstrated two-dimensional multiresolution ensemble network for automated brain tumor segmentation to generate robust segmentation of tumor subcomponents. We also predicted the overall survival based on the segmented mask using an xgboost model. These may assist in diagnosis, treatment planning and therapy response monitoring of brain tumor patients with more objective and reproducible measures.

References

Bakas, S., et al.: Segmentation labels and radiomic features for the pre-operative scans of the TCGA-GBM collection. The Cancer Imaging Archive (2017a)

Bakas, S., et al.: Segmentation labels and radiomic features for the pre-operative scans of the TCGA-LGG collection. The Cancer Imaging Archive, **286** (2017b).

Bakas, S., et al.: Advancing the cancer genome atlas glioma MRI collections with expert segmentation labels and radiomic features Sci. Data, **4**, 170117 (2017c)

Bakas, S., et al.: Advancing the cancer genome atlas glioma MRI collections with expert segmentation labels and radiomic features. Sci. Data, **4**, 170117 (2017d)

Bakas, S., et al.: Identifying the best machine learning algorithms for brain tumor segmentation, progression assessment, and overall survival prediction in the BRATS challenge (2018)

Buitinck, L., et al.: API design for machine learning software: experiences from the scikit-learn project (2013)

Chen, C.F., Fan, Q., Mallinar, N., Sercu, T., Feris, R.: Big-little net: an efficient multi-scale feature representation for visual and speech recognition (2018)

Chen, L., Wu, Y., DSouza, A.M., Abidin, A.Z., Wismüller, A., Xu, C.:. MRI tumor segmentation with densely connected 3D CNN. In: Medical Imaging 2018: Image Processing. International Society for Optics and Photonics (2018)

Chen, T., Guestrin, C.: Xgboost: a scalable tree boosting system. In: Proceedings of the 22nd ACM SIGKDD International Conference on Knowledge Discovery and Data Mining. ACM (2016)

Dolz, J., Gopinath, K., Yuan, J., Lombaert, H., Desrosiers, C., Ayed, I.B.: HyperDense-Net: a hyper-densely connected CNN for multi-modal image segmentation. IEEE Trans. Med. Imaging **38**(5), 1116–1126 (2018)

Feng, X., Tustison, N., Meyer, C.: Brain tumor segmentation using an ensemble of 3d u-nets and overall survival prediction using radiomic features. In: Crimi, A., Bakas, S., Kuijf, H., Keyvan, F., Reyes, M., van Walsum, T. (eds.) BrainLes 2018. LNCS, vol. 11384, pp. 279–288. Springer, Cham (2019). https://doi.org/10.1007/978-3-030-11726-9_25

Havaei, M., et al.: Brain tumor segmentation with deep neural networks. Med. Image Anal. **35**, 18–31 (2017)

Hu, J., Shen, L., Sun, G.: Squeeze-and-excitation networks. In: Proceedings of the IEEE Conference on Computer Vision and Pattern Recognition (2018)

Huang, G., Liu, Z., Van Der Maaten, L., Weinberger, K.Q.: Densely connected convolutional networks. In: Proceedings of the IEEE Conference on Computer Vision and Pattern Recognition (2017)

Islam, J., Zhang, Y.: An ensemble of deep convolutional neural networks for Alzheimer's disease detection and classification (2017)

Kim, S., Kim, D., Kim, S.H., Park, M., Chang, J.H., Yun, M.: The roles of 11Ccetate PET/CT in predicting tumor differentiation and survival in patients with cerebral glioma. Eur. J. Nucl. Med. Mol. Imaging **45**(6), 1012–1020 (2018). https://doi.org/10.1007/s00259-018-3948-9

Lee, G.R., Gommers, R., Waselewski, F., Wohlfahrt, K., O'Leary, A.: PyWavelets: a python package for wavelet analysis. J. Open Source Softw. **4**(36), 1237 (2019)

Lin, M., Chen, Q., Yan, S.: Network in network. arXiv preprint arXiv:1312.4400

Menze, B.H., et al.: The multimodal brain tumor image segmentation benchmark (BRATS). IEEE Trans. Med. Imaging **34**(10), 1993–2024 (2014)

Mobadersany, P., et al.: Predicting cancer outcomes from histology and genomics using convolutional networks. Proc. Natl. Acad. Sci. **115**(13), E2970–E2979 (2018)

Myronenko, A.: 3D MRI brain tumor segmentation using autoencoder regularization. In: Crimi, A., Bakas, S., Kuijf, H., Keyvan, F., Reyes, M., van Walsum, T. (eds.) BrainLes 2018. LNCS, vol. 11384, pp. 311–320. Springer, Cham (2019). https://doi.org/10.1007/978-3-030-11726-9_28

Pedregosa, F., et al.: Scikit-learn: machine learning in Python. J. Mach. Learn. Res. **12**(Oct), 2825–2830 (2011)

Pereira, S., Pinto, A., Alves, V., Silva, C.A.: Brain tumor segmentation using convolutional neural networks in MRI images. IEEE Trans. Med. Imaging **35**(5), 1240–1251 (2016)

Rohlfing, T., Zahr, N.M., Sullivan, E.V., Pfefferbaum, A.: The SRI24 multichannel atlas of normal adult human brain structure. Human Brain Mapp. **31**(5), 798–819 (2010)

Ronneberger, O., Fischer, P., Brox, T.: U-net: convolutional networks for biomedical image segmentation. In: Navab, N., Hornegger, J., Wells, W.M., Frangi, A.F. (eds.) Medical Image Computing and Computer-Assisted Intervention – MICCAI 2015: 18th International Conference, Munich, Germany, October 5-9, 2015, Proceedings, Part III, pp. 234–241. Springer International Publishing, Cham (2015). https://doi.org/10.1007/978-3-319-24574-4_28

Saouli, R., Akil, M., Kachouri, R.: Fully automatic brain tumor segmentation using end-to-end incremental deep neural networks in MRI images. Comput. Methods Programs Biomed. **166**, 39–49 (2018)

Tsuchida, T., Takeuchi, H., Okazawa, H., Tsujikawa, T., Fujibayashi, Y.: Grading of brain glioma with 1–11C-acetate PET: comparison with 18F-FDG PET. Nucl. Med. Biol. **35**(2), 171–176 (2008)

Tustison, N.J., et al.: Large-scale evaluation of ANTs and FreeSurfer cortical thickness measurements. Neuroimage **99**, 166–179 (2014)

Van Griethuysen, J.J., et al.: Computational radiomics system to decode the radiographic phenotype. Cancer Res. **77**(21), e104–e107 (2017)

Yakubovskiy, P.: Segmentation models. GitHub repository (2019)

Yamamoto, Y.Y., et al.: 11 C-acetate PET in the evaluation of brain glioma: comparison with 11 C-methionine and 18 F-FDG-PET. Mol. Imaging Biol. **10**(5), 281 (2008)

Zhuge, Y., et al.: Brain tumor segmentation using holistically nested neural networks in MRI images. Med. Phys. **44**(10), 5234–5243 (2017)

Cascaded Coarse-to-Fine Neural Network for Brain Tumor Segmentation

Shuojue Yang, Dong Guo, Lu Wang, and Guotai Wang[✉]

School of Mechanical and Electrical Engineering,
University of Electronic Science and Technology of China, Chengdu, China
guotai.wang@uestc.edu.cn

Abstract. A cascaded framework of coarse-to-fine networks is proposed to segment brain tumor from multi-modality MR images into three subregions: enhancing tumor, whole tumor and tumor core. The framework is designed to decompose this multi-class segmentation into two sequential tasks according to hierarchical relationship among these regions. In the first task, a coarse-to-fine model based on Global Context Network predicts segmentation of whole tumor, which provides a bounding box of all three substructures to crop the input MR images. In the second task, cropped multi-modality MR images are fed into another two coarse-to-fine models based on NvNet trained on small patches to generate segmentation of tumor core and enhancing tumor, respectively. Experiments with BraTS 2020 validation set show that the proposed method achieves average Dice scores of 0.8003, 0.9123, 0.8630 for enhancing tumor, whole tumor and tumor core, respectively. The corresponding values for BraTS 2020 testing set were 0.81715, 0.88229, 0.83085, respectively.

Keywords: Brain tumor · Segmentation · Convolutional neural network

1 Introduction

Brain tumor is currently one of the most deadly cancers and considerable efforts have been made from academia to fight against it. Brain tumor segmentation from multi-modality Magnetic Resonance Imaging (MRI) takes an indispensable part in early diagnosis and prognosis. Manually segmenting gliomas from MR images is time-consuming and laborious. Therefore, it is necessary to develop an automatic brain tumor segmentation tool to assist clinical diagnosis. This segmentation is challenging because some characteristics of glioma MR scans, such as complicated shapes, ambiguous boundaries of adjacent regions and great variations in both location and size across patients [23], have prevented these tools from achieving a robust performance.

In recent years, Convolutional Neural Networks (CNNs) have achieved great success for medical image segmentation tasks [6], such as segmentation of fetal head [13], optic disc [19], brain tumor [23,24] and pancreas [18]. Kamnitsas

ⓒ Springer Nature Switzerland AG 2021
A. Crimi and S. Bakas (Eds.): BrainLes 2020, LNCS 12658, pp. 458–469, 2021.
https://doi.org/10.1007/978-3-030-72084-1_41

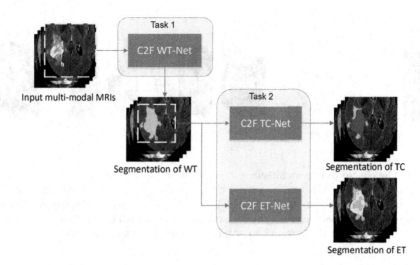

Fig. 1. The proposed cascaded framework for brain tumor segmentation. Three networks are proposed to segment whole tumor (C2F WT-Net), tumor core (C2F TC-Net) and enhancing tumor (C2F ET-Net) in sequence. C2F means that each model is composed of a coarse network and a fine network. WT-Net is used to segment the whole tumor and generates the bounding box of this region, from which TC-Net and ET-Net predict the tumor core segmentation and enhancing tumor segmentation, respectively.

et al. [11] proposed DeepMedic, a model performing well in 3D image segmentation. In spite of extracting larger contextual information, the adopted approach to focus on local patches impedes to leverage large contextual information. Since encoder-decoder framework achieved state-of-the-art performance on semantic segmentation, especially FCN [20] and U-Net [17], many works [7,9,23,26] in brain tumor segmentation have presented their networks based on encoder-decoder architecture. In Brain Tumor Segmentation (BraTS) Challenge, participants are requested to segment the tumor into three regions: whole tumor, tumor core and enhancing tumor. In BraTS 2017, Wang et al. [23] designed a cascade of three FCN-based networks corresponding to the three tumor subregions. Even though this method is complicated, it demonstrated a competitive performance in the challenge. Isensee et al. [9] fully exploited a well trained 3D U-Net with minor modifications and their model won the second place in BraTS 2018. Myronenko [15], winner in BraTS 2018, fed the network with a large patch and added a variational auto-encoder branch after the shared encoder to serve as a regularization strategy. Jiang et al. [10] designed a model consisting of two cascaded U-Nets and reduced GPU consumption by gradient checkpointing, which achieved the best performance in BraTS 2019.

In this work, we follow the cascaded framework [23] and decouple the segmentation task into two subtasks. The first subtask is to segment whole tumor while the second subtask is to segment tumor core and enhancing tumor. We

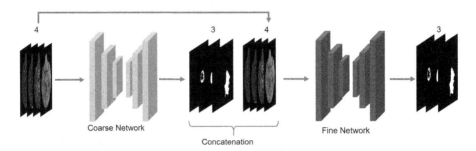

Fig. 2. The coarse-to-fine architecture consists of two sequential networks. The input of coarse network is MR images with 4 channels and the coarse prediction contains three binary segmentation results for whole tumor, tumor core and enhancing tumor, respectively. After concatenating the coarse segmentation with the multi-modal images, we fed them into the fine network to obtain more accurate segmentation.

also introduce a small-patch training strategy to improve the segmentation performance. Besides, each model in our framework is designed with a coarse-to-fine architecture and trained in an end-to-end fashion.

2   Methods

2.1   Cascaded Framework

The proposed framework is shown in Fig. 1. It consists of two sequential subtasks. The first subtask includes one network trained with a large patch size $(128 \times 128 \times 128)$ while the latter includes two networks trained with a small patch size $(96 \times 96 \times 96)$. C2F means that the model is designed under a coarse-to-fine architecture which consists of two networks (coarse network and fine network). The model (C2F WT-Net) in the first subtask is used for providing an accurate segmentation of whole tumor. In the second subtask, based on the bounding box of whole tumor predicted by WT-Net, the original multi-modal MR images are cropped into small patches to eliminate redundant information and are fed into another two networks (C2F TC-Net and C2F ET-Net) trained on small patches to predict segmentation of tumor core and enhancing tumor, respectively. The final segmentation result is a fusion of outputs of these three networks.

2.2   Coarse-to-Fine Networks

Each target-specific CNN in the cascaded framework is composed of two 3D U-Nets to construct a coarse-to-fine architecture, as shown in Fig. 2. Although these target-specific models are only required to segment one corresponding tumor subregion, we still make them to predict all the three tumor regions. Thus both the outputs of coarse network and fine network are in 3 channels. We introduce this multi-class segmentation to benefit the training process since it serves the

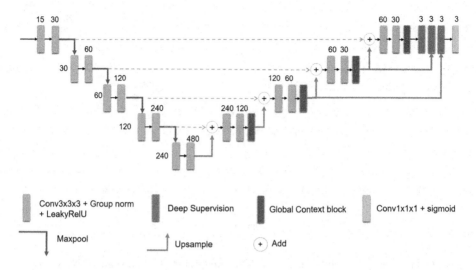

Fig. 3. The network architecture of WT-Net. Each block comprises two convolutional layers with leaky RelU and group normalization. A Global Context block is added at each decoder except the bottom one to fully exploit global context information. Deep supervision is utilized to employ multi-scale features for the final segmentation.

same function as multi-task regularization used by Myronenko [15] and Zhao et al. [26].

For each coarse-to-fine model, we fed the coarse U-Net with four modalities of images. And each coarse network outputs all three binary predictions (whole tumor, tumor core and enhancing tumor). Then we concatenate these outputs (3 channels) with multi-modality images (4 channels) to compose the inputs (7 channels) for fine network. The second network also predicts all three binary predictions, although only the task-specific one, such as the whole tumor prediction of coarse-to-fine WT-Net, would be chosen to compose final segmentation results.

Apart from the number of input channels (4 for coarse network and 7 for fine network), the two networks have the same structure. These models with a huge number of parameters are not applicable in an end-to-end training manner, thus we utilize a memory-saving technique called gradient checkpointing [21] to significantly cut down the memory consumption at the expense of training speed.

In order to expedite training process and further boost performance, our coarse-to-fine model would also calculate an extra loss between labels and coarse segmentation. Considering that there are two patch size settings for different targets, we utilize two distinctive networks. One composes coarse-to-fine WT-Net, and another one composes both coarse-to-fine TC-Net and coarse-to-fine ET-Net.

Network Architecture of WT-Net. The network of WT-Net is a Global Context Network [5], which is shown in Fig. 3. WT-Net is trained on a relatively

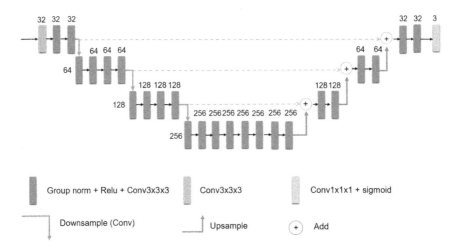

Fig. 4. The network architecture of TC-Net and ET-Net. A variant of NvNet composes TC-Net and ET-Net. The network architecture consists of a comparatively large encoder and a decoder. Both of them consist of residual blocks. We add the features from encoder to the decoder as skip-connection.

large patch size ($128 \times 128 \times 128$). This network extracts long-range dependency and inter-channel dependency by using a Global Context Block [5], which is used in the decoder at each resolution level except the bottom one. Referring to other best-performing networks [9,23,26] in medical segmentation, the Global Context Network also extensively uses residual blocks [8] to handle the potential accuracy degrading problems. Besides, a group normalization is introduced to replace the conventional batch normalization in each block. This network downsamples features in encoder part by max-pooling and restores output features' resolution in decoder part by trilinear interpolation. Finally, a deep supervision module is applied to last three decoders to encourage the model learning a better representation of both high-level semantic features and low-level locational information.

Network Architecture of TC-Net and ET-Net. TC-Net and ET-Net are trained with small patches ($96 \times 96 \times 96$) in our methodology. We introduce NvNet [15], a strong variant of U-Net to implement network for TC-Net and ET-Net. The network architecture is demonstrated in Fig. 4. Compared with original NvNet, we do not apply any auxiliary branches to regularize the encoder by multi-task learning. Our variant of NvNet encompasses many residual blocks adopting a distinctive sequence of normalization, activation function and convolution from the above Global Context Network [7].

At the beginning of the encoder, an initial convolution with 32 filters is utilized to generate more feature maps from original input images (4 or 7 channels), followed by a dropout with rate of 0.2 to defend against potential overfitting problem. Subsequently, the channels of feature maps would be doubled after

Table 1. Dice scores of our methods on BraTS 2020 online validation set. ET, WT, TC denote enhancing tumor, whole tumor and tumor core, respectively.

	Dice (%)		
	ET	WT	TC
Baseline	77.86 ± 27.42	90.49 ± 7.32	82.34 ± 18.96
Baseline + small-patch	77.63 ± 28.41	90.08 ± 7.31	84.31 ± 14.24
Baseline + coarse-to-fine	77.34 ± 28.50	90.82 ± 6.74	82.00 ± 19.19
baseline + coarse-to-fine + small-patch	78.51 ± 26.31	90.43 ± 6.81	84.98 ± 14.04
Proposed	79.61 ± 24.73	90.97 ± 5.99	86.24 ± 12.22
Ensemble	80.03 ± 24.44	91.23 ± 5.84	86.30 ± 12.44

each spatial level and three convolutions with a stride of 2 are used to replace maxpooling. Finally, a skip connection is used to add features from encoder to corresponding ones with the same resolution in decoder path.

We also adopt a method called Training with ROI Patch [23] when it comes to tumor core and enhancing tumor core. For training, we utilize a bounding box (with 5 voxels margin) of the ground truth of whole tumor to locate a volume in original input images. Then we use random crop in this volume to generate ROI patches and keep the whole ROI patch inside the bounding box so that the models focus more on the tumor region. When the bounding box of whole tumor is smaller than the patch size ($96 \times 96 \times 96$), voxels outside the bounding box would be padded with zero to the patch size. For inference, without the ground truth, we alternatively use the predicted segmentation of whole tumor to locate the bounding box (with 5 voxels margin) and the volume inside it would be fed into the network for inference.

2.3 Logarithmic Dice

In BraTS challenge, Dice Similarity Coefficient (DSC) is the dominant metric to evaluate the segmentation results submitted by participants. And most of the top methods [9,10,15] use the Dice Loss straightforwardly to optimize their models and reach a high accuracy. Inspired by previous work [25], we adopt a logarithmic Dice loss function as following:

$$\mathcal{L}(P, G) = \left(-\log \left(\frac{1}{C} \sum_c \frac{2 \sum_i p_{ci} g_{ci} + \varepsilon}{\sum_i (p_{ci} + g_{ci}) + \varepsilon} \right) \right)^{0.3} \tag{1}$$

where C means the total number of classes, p_{ci} denotes the probability of being class c for voxel i predicted by a CNN, g_{ci} is the corresponding ground truth value, and $\varepsilon = 10^{-5}$ is a small number for numerical stability.

Table 2. Hausdorff distances of our methods on BraTS 2020 online validation set. ET, WT, TC denote enhancing tumor, whole tumor and tumor core, respectively.

	Hausdorff distance (mm)		
	ET	WT	TC
Baseline	33.43 ± 100.90	5.28 ± 9.16	8.52 ± 16.32
Baseline + small patch	35.63 ± 105.38	6.20 ± 10.51	6.91 ± 11.78
Baseline + coarse-to-fine	38.46 ± 109.61	4.74 ± 6.67	12.49 ± 47.40
Baseline + coarse-to-fine + small patch	32.50 ± 100.95	5.66 ± 7.85	5.90 ± 10.21
Proposed	24.16 ± 85.52	4.40 ± 6.06	5.51 ± 9.75
Ensemble	20.53 ± 79.58	4.39 ± 6.05	5.46 ± 9.97

3   Experiments and Preliminary Results

Data and Implementation Details. We experimented with the Multi-modal Brain Tumor Segmentation Challenge (BraTS) 2020[1] [1–4,14] training and validation set. BraTS challenge focuses on evaluating methods for brain tumor segmentation with 3D MRI scans [1–4,14]. The BraTS 2020 training set comprises 369 cases (293 HGG and 76 LGG) with four 3D MRI modalities (T1, T1c, T2 and FLAIR). Participants are requested to segment gliomas into three overlapping regions: enhancing tumor (ET), tumor core (TC) which contains ET and whole tumor (WT) which includes TC. The BraTS 2020 validation set gathered images from 125 patients with brain tumor. The online evaluation platform evaluated our uploaded segmentation performance based on the Dice score, sensitivity, specificity and Hausdorff distance.

Our proposed method is implemented with Pytorch [16]. Adaptive Moment Estimation (Adam) [12] was adopted as optimizer with an $L2$ weight decay 10^{-5} to regularize our large model. We set the initial learning rate as 10^{-4}, which was reduced in each epoch as following:

$$\alpha = \alpha_0 \times \left(1 - \frac{e}{N_e}\right)^{0.9} \tag{2}$$

where α_0 denotes the initial learning rate, N_e means the number of epochs which is 300 in our experiment and e is an epoch counter.

We adopted patch size of $128 \times 128 \times 128$ for WT-Net and $96 \times 96 \times 96$ for both TC-Net and ET-Net. The batch size is 2 for all the models. Experiments were conducted on two NVIDIA GeForce RTX 2080 Ti GPUs and each network

---

[1] http://www.med.upenn.edu/cbica/brats2020.html.

Table 3. Dice and Hausdorff measurements of our method on BraTS 2020 testing set. EN, WT, TC denote enhancing tumor, whole tumor and tumor core respectively.

	Dice			Hausdorff distance (mm)		
	ET	WT	TC	ET	WT	TC
Mean	0.81751	0.88229	0.83085	15.51376	4.8308	21.46259
StdDev	0.18917	0.13259	0.25961	69.51044	6.46241	79.50443
Median	0.8591	0.92439	0.92499	1.73205	3	2.23607
25quantile	0.78803	0.87165	0.87049	1	1.73205	1.41421
75quantile	0.92285	0.95124	0.95797	2.23607	5.17187	4.21276

Table 4. Sensitivity and Specificity measurements of our method on BraTS 2020 testing set. EN, WT, TC denote enhancing tumor, whole tumor and tumor core respectively.

	Sensitivity			Specificity		
	ET	WT	TC	ET	WT	TC
Mean	0.85905	0.91058	0.85776	0.99964	0.99904	0.99957
StdDev	0.20021	0.13259	0.23182	0.0004	0.00098	0.00109
Median	0.92501	0.94343	0.93689	0.99976	0.9994	0.99983
25quantile	0.84115	0.91087	0.87252	0.99949	0.99875	0.99965
75quantile	0.96554	0.97066	0.97475	0.99993	0.9997	0.99993

was trained for 300 epochs. The training required about 40 h for WT-Net and 26 h for ET-Net and TC-Net.

We normalized each image by its intensity mean value and standard deviation as pre-processing. Besides, random crop, random elastic deformations and random mirroring were used for data augmentation. Test time augmentation was used for more robust results [22]. We sequentially flipped the input images in each dimension to produce 8 inputs and fed them into our models. We calculated the average of all 8 output probability maps and a Sigmoid function transferred it into binary maps as our final prediction.

Segmentation Results. We carried out experiments on BraTS 2020 validation set to test the usefulness of our approaches. Table 1 shows the results of different methods based on Dice metrics while Table 2 is based on Hausdorff distance.

The baseline is a single Global Context Network [5] trained with a patch size of $128 \times 128 \times 128$. It should be noticed that the baseline is trained without the cascaded framework. And this framework was only implemented in the experiments of "proposed" and "ensemble" shown in Table 1 and Table 2. We compared the baseline with or without small-patch training strategy, the baseline with or without coarse-to-fine models to investigate the effect of the training strategy and the coarse-to-fine architecture separately. In the experiments

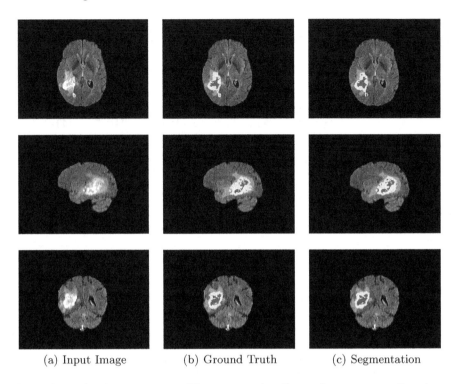

(a) Input Image   (b) Ground Truth   (c) Segmentation

Fig. 5. Example of segmentation. The green, red, yellow colors represent the edema, non-enhancing and enhancing tumor cores, respectively. (Color figure online)

of "baseline + small-patch" and "baseline + coarse-to-fine + small-patch", the cascaded framework was not introduced and we used the ground truth of whole tumor to locate the small patches for training and used the whole tumor prediction for bounding box in inference phase. The last second row in these tables shows the results achieved by our proposed method, i.e., combining the baseline with small patch, coarse-to-fine and cascaded framework. Small-patch training strategy demonstrated a significant improvement on segmenting tumor core in both Dice and Hausdorff distance but this strategy could reduce the segmentation accuracy in both whole tumor and enhancing tumor to different extents. Coarse-to-fine architecture boosted segmentation performance in whole tumor but the other regions were not improved. We further investigated the combination of small-patch training strategy and coarse-to-fine architecture. It could be observed that using this combination obtained better segmentation performance on tumor core and enhancing tumor. Then we trained these task-specific networks under the proposed cascaded framework and it reached better segmentation performance on all substructures. Finally, given 5 outputs from 5-cross validation for each subregion, we averaged the predictions to ensemble them.

Our proposed method realized a competitive performance on the leaderboard[2], which shows that we achieved Dice scores of 0.8003, 0.9123 and 0.8630 for ET, WT and TC, respectively.

We randomly selected 80% of BraTS 2020 training images as the training set and the remaining was used as local validation. Figure 5 shows an example of qualitative segmentation of images in local validation. We present the results and labels on the FLAIR image. The green, red, yellow colors represent the edema, non-enhancing and enhancing tumor cores, respectively. In Fig. 5, segmentation of whole tumor and tumor core has a high similarity with ground truth. However, the segmentation of enhancing tumor is less accurate, which is mainly because of its complex shape and scatter localization.

The proposed method was used as our entry to BraTS 2020 challenge in which a testing set containing 166 cases was provided. The results on testing set given by challenge organizers were presented in Table 3 and Table 4. They include the means, standard deviations, medians, 25 quantiles and 75 quantiles of Dice, Sensitivity, Specificity and Hausdorff measurements for each substructure. The final Dice scores of enhancing tumor, whole tumor and tumor core were 0.81715, 0.88229, 0.83085, respectively. Obviously, the Hausdorff distance of tumor core was abnormally deviated from the measurement on online validation set. This might be attributed to the poor generalization of aforementioned bounding box used to extract small patches. Therefore, it is of a great interest to find methods which are more robust to the variable accuracy of bounding box. Besides, the cascaded framework seems to have a poor efficiency. Since each target-specific model predicts all the three tumor substructures but only the target one would be used. We believe that multi-class segmentation serves as regularization and thus benefits the training process. However, the pipeline would be more efficient if TC-Net and ET-Net (the second task in Fig. 1) can further exploit the outputs of WT-Net (the first task in Fig. 1), i.e. we could combine the predictions of WT-Net with original MR images to compose the inputs fed into TC-Net and ET-Net.

4   Conclusion

We proposed a cascaded framework consisting of three coarse-to-fine models to segment gliomas from multi-modality brain MR images. We separately train these models for specific substructures and use the cascaded framework to produce the segmentation. Our small-patch training strategy can eliminate the redundant information from background and helps to improve the accuracy. And the coarse-to-fine architecture also boosts the performance. Experimental results demonstrate that our method achieved average Dice scores of 0.8003, 0.9123, 0.8630, respectively on the BraTS 2020 validation set. The counterpart values for BraTS 2020 testing set were 0.81715, 0.88229, 0.83085, respectively.

---

[2] https://www.cbica.upenn.edu/BraTS20/lboardValidation.html.

References

1. Bakas, S., Akbari, H., Sotiras, A., et al.: Advancing the cancer genome atlas glioma MRI collections with expert segmentation labels and radiomic features. Sci. Data 4(170117) (2017). https://doi.org/10.1038/sdata.2017.117

2. Bakas, S., et al.: Segmentation labels and radiomic features for the pre-operative scans of the TCGA-GBM collectio, July 2017. https://doi.org/10.7937/K9/TCIA.2017.KLXWJJ1Q

3. Bakas, S., et al.: Segmentation labels and radiomic features for the pre-operative scans of the TCGA-LGG collection, July 2017. https://doi.org/10.7937/K9/TCIA.2017.GJQ7R0EF

4. Bakas, S., Reyes, M., Jakab, A., Bauer, S., Rempfler, M., Crimi, A., et al.: Identifying the best machine learning algorithms for brain tumor segmentation, progression assessment, and overall survival prediction in the BRATS challenge. CoRR abs/1811.02629 (2018). http://arxiv.org/abs/1811.02629

5. Cao, Y., Xu, J., Lin, S., Wei, F., Hu, H.: Gcnet: non-local networks meet squeeze-excitation networks and beyond. In: 2019 IEEE/CVF International Conference on Computer Vision Workshop (ICCVW), pp. 1971–1980 (2019)

6. Shen, D., Wu, G., Suk, H.I.: Deep learning in medical image analysis. Ann. Rev. Biome. Eng. **19**, 221–248 (2017)

7. Guo, D., Wang, L., Song, T., Wang, G.: Cascaded global context convolutional neural network for brain tumor segmentation. In: Crimi, A., Bakas, S. (eds.) BrainLes 2019. LNCS, vol. 11992, pp. 315–326. Springer, Cham (2020). https://doi.org/10.1007/978-3-030-46640-4_30

8. He, K., Zhang, X., Ren, S., Sun, J.: Deep residual learning for image recognition. In: 2016 IEEE Conference on Computer Vision and Pattern Recognition (CVPR), pp. 770–778 (2016)

9. Isensee, F., Kickingereder, P., Wick, W., Bendszus, M., Maier-Hein, K.H.: No new-net. In: Crimi, A., Bakas, S., Kuijf, H., Keyvan, F., Reyes, M., van Walsum, T. (eds.) BrainLes 2018. LNCS, vol. 11384, pp. 234–244. Springer, Cham (2019). https://doi.org/10.1007/978-3-030-11726-9_21

10. Jiang, Z., Ding, C., Liu, M., Tao, D.: Two-stage cascaded U-Net: 1st place solution to BraTS challenge 2019 segmentation task. In: Crimi, A., Bakas, S. (eds.) BrainLes 2019. LNCS, vol. 11992, pp. 231–241. Springer, Cham (2020). https://doi.org/10.1007/978-3-030-46640-4_22

11. Kamnitsas, K., et al.: Deepmedic for brain tumor segmentation. In: Crimi, A., Menze, B., Maier, O., Reyes, M., Winzeck, S., Handels, H. (eds.) Brainlesion: Glioma, Multiple Sclerosis, Stroke and Traumatic Brain Injuries, pp. 138–149. Springer International Publishing, Cham (2016). https://doi.org/10.1007/978-3-319-55524-9_14

12. Kingma, D., Ba, J.: Adam: a method for stochastic optimization. In: International Conference on Learning Representations, December 2014

13. Lingyun, W., Yang, X., Li, S., Wang, T., Heng, P.A., Ni, D.: Cascaded fully convolutional networks for automatic prenatal ultrasound image segmentation, pp. 663–666, April 2017. https://doi.org/10.1109/ISBI.2017.7950607

14. Menze, B.H., et al.: The multimodal brain tumor image segmentation benchmark (brats). IEEE Trans. Med. Imaging **34**(10), 1993–2024 (2015). https://doi.org/10.1109/TMI.2014.2377694

15. Myronenko, A.: 3D MRI brain tumor segmentation using autoencoder regularization. In: Crimi, A., Bakas, S., Kuijf, H., Keyvan, F., Reyes, M., van Walsum, T. (eds.) BrainLes 2018. LNCS, vol. 11384, pp. 311–320. Springer, Cham (2019). https://doi.org/10.1007/978-3-030-11726-9_28

16. Paszke, A., et al.: Automatic differentiation in pytorch (2017)

17. Ronneberger, O., Fischer, P., Brox, T.: U-Net: convolutional networks for biomedical image segmentation. In: Navab, N., Hornegger, J., Wells, W.M., Frangi, A.F. (eds.) MICCAI 2015. LNCS, vol. 9351, pp. 234–241. Springer, Cham (2015). https://doi.org/10.1007/978-3-319-24574-4_28

18. Roth, H.R., et al.: DeepOrgan: multi-level deep convolutional networks for automated pancreas segmentation. In: Navab, N., Hornegger, J., Wells, W.M., Frangi, A.F. (eds.) MICCAI 2015. LNCS, vol. 9349, pp. 556–564. Springer, Cham (2015). https://doi.org/10.1007/978-3-319-24553-9_68

19. Sevastopolsky, A.: Optic disc and cup segmentation methods for glaucoma detection with modification of u-net convolutional neural network. Pattern Recognit. Image Anal. **27**, 618–624 (2017). https://doi.org/10.1134/S1054661817030269

20. Shelhamer, E., Long, J., Darrell, T.: Fully convolutional networks for semantic segmentation. IEEE Trans. Pattern Anal. Mach. Intell. **39**(4), 640–651 (2017)

21. Siskind, J., Pearlmutter, B.: Divide-and-conquer checkpointing for arbitrary programs with no user annotation. Optim. Methods Softw. **33** (2017). https://doi.org/10.1080/10556788.2018.1459621

22. Wang, G., Li, W., Aertsen, M., Deprest, J., Ourselin, S., Vercauteren, T.: Aleatoric uncertainty estimation with test-time augmentation for medical image segmentation with convolutional neural networks. Neurocomputing **338**, 34–45 (2019). https://doi.org/10.1016/j.neucom.2019.01.103

23. Wang, G., Li, W., Ourselin, S., Vercauteren, T.: Automatic brain tumor segmentation using cascaded anisotropic convolutional neural networks. In: Crimi, A., Bakas, S., Kuijf, H., Menze, B., Reyes, M. (eds.) BrainLes 2017. LNCS, vol. 10670, pp. 178–190. Springer, Cham (2018). https://doi.org/10.1007/978-3-319-75238-9_16

24. Wang, G., Song, T., Dong, Q., Cui, M., Huang, N., Zhang, S.: Automatic ischemic stroke lesion segmentation from computed tomography perfusion images by image synthesis and attention-based deep neural networks. Med. Image Anal. **65**, 101787 (2020). http://www.sciencedirect.com/science/article/pii/S1361841520301511

25. Wong, K.C.L., Moradi, M., Tang, H., Syeda-Mahmood, T.: 3D segmentation with exponential logarithmic loss for highly unbalanced object sizes. In: Frangi, A.F., Schnabel, J.A., Davatzikos, C., Alberola-López, C., Fichtinger, G. (eds.) MICCAI 2018. LNCS, vol. 11072, pp. 612–619. Springer, Cham (2018). https://doi.org/10.1007/978-3-030-00931-1_70

26. Zhao, Y.-X., Zhang, Y.-M., Liu, C.-L.: Bag of tricks for 3D MRI brain tumor segmentation. In: Crimi, A., Bakas, S. (eds.) BrainLes 2019. LNCS, vol. 11992, pp. 210–220. Springer, Cham (2020). https://doi.org/10.1007/978-3-030-46640-4_20

Low-Rank Convolutional Networks
for Brain Tumor Segmentation

Pooya Ashtari[1(✉)], Frederik Maes[2,3], and Sabine Van Huffel[1]

[1] Department of Electrical Engineering (ESAT),
STADIUS Center for Dynamical Systems, Signal Processing and Data Analytics,
KU Leuven, Leuven, Belgium
pooya.ashtari@esat.kuleuven.be
[2] Department of Electrical Engineering (ESAT),
Processing Speech and Images (PSI), KU Leuven, Leuven, Belgium
[3] Medical Imaging Research Center, UZ Leuven, Leuven, Belgium

Abstract. The automated segmentation of brain tumors is crucial for various clinical purposes from diagnosis to treatment planning to follow-up evaluations. The vast majority of effective models for tumor segmentation are based on convolutional neural networks with millions of parameters being trained. Such complex models can be highly prone to overfitting especially in cases where the amount of training data is insufficient. In this work, we devise a 3D U-Net-style architecture with residual blocks, in which low-rank constraints are imposed on weights of the convolutional layers in order to reduce overfitting. Within the same architecture, this helps to design networks with several times fewer parameters. We investigate the effectiveness of the proposed technique on the BraTS 2020 challenge.

Keywords: Low-rank representation · U-Net · Glioma segmentation

1 Introduction

Gliomas are brain tumors with the highest mortality rate and prevalence. They can be classified into two grades: low-grade glioma (LGG) and high-grade glioma (HGG), with the former being less aggressive than the latter. Multi-modal MRI is widely used to diagnose and assess gliomas in clinical practice. The accurate segmentation of gliomas is crucial for various clinical purposes, including diagnosis, treatment planning, image-guided surgery, and follow-up evaluations. However, manual delineation of tumors is laborious, time-consuming, and expensive especially because experts need to deal with 3D images and several modalities; therefore, accurate computer-assisted methods are needed to automatically perform this task. Despite considerable advances in medical imaging, glioma segmentation is still a challenging task since tumors can vary dramatically in shape, structure, and location across patients and over time within a specific patient. Moreover, the growing tumor mass may displace and deform the surrounding normal brain tissues, as do resection cavities that are present after surgery.

© Springer Nature Switzerland AG 2021
A. Crimi and S. Bakas (Eds.): BrainLes 2020, LNCS 12658, pp. 470–480, 2021.
https://doi.org/10.1007/978-3-030-72084-1_42

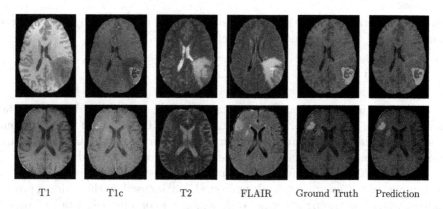

| T1 | T1c | T2 | FLAIR | Ground Truth | Prediction |

Fig. 1. The multimodal MR images along with the corresponding ground truth and prediction for a HGG (top row) and a LGG (bottom row) representative cases (green: edema, yellow: enhancing tumor, red: necrosis and non-enhancing tumor). (Color figure online)

The brain tumor segmentation challenge (BraTS) [1–4,19] aims to develop effective data-driven algorithms for brain tumor segmentation by providing a large dataset of annotated LGG and HGG 3D MRI scans, each with four MRI modalities (T1 weighted, post-contrast T1-weighted, T2-weighted, and FLAIR) rigidly co-registered, resampled to the voxel size 1 mm³, and skull-stripped. The BraTS 2020 training dataset consists of 369 cases, each of which is manually annotated by up to 4 raters who label each voxel as enhancing tumor (ET), edema (ED), necrotic and non-enhancing tumor (NCR/NET), or everything else (see Fig. 1). However, for evaluation, the 3 nested subregions, namely whole tumor (WT), tumor core (TC–i.e., the union of ED and NCR/NET), and enhancing tumor (ET), are used.

Recently, deep learning models, particularly convolutional neural networks (CNN), surpassed traditional computer vision methods for semantic segmentation. In contrast to the conventional approach based on hand-crafted features, CNNs are able to automatically learn high-level features adapted specifically to the task of brain tumor segmentation. Currently, the vast majority of effective CNNs for medical image segmentation are based on a U-Net [23] architecture with millions of trainable parameters. However, such complex models can be highly prone to overfitting especially in cases where the amount of training data is insufficient, which is usually the case for medical imaging. In this work, we introduce a new layer, called low-rank convolution, in which low-rank constraints are imposed to regularize weights and thus reduce overfitting. We make use of a 3D U-Net [5] architecture with residual modules [10] and further improve it by replacing ordinary convolution layers with low-rank ones, achieving models with several times fewer parameters than the initial ones. This leads to significantly better performance especially because the amount of training data is limited.

The rest of this paper is organized as follows: Sect. 2 briefly reviews relevant semantic segmentation techniques. Section 3 presents our approach to brain tumor segmentation using low-rank convolutional layer. Experiments are presented in Sect. 4. We conclude this paper in Sect. 5.

2   Related Work

Over the past few years, considerable research efforts have been directed to the development of fully convolutional neural networks for semantic segmentation. Encoder-decoder architectures and their variants, in particular U-Net [23], are probably the most successful ones in the segmentation of medical images.

In BraTS 2018 and 2019, all top solutions made use of such architectures in their models one way or another. Isensee et al. [13] focused on the training procedure instead of proposing a new network, winning second place in BraTS 2018 by making only minor modifications to the standard 3D U-Net [5], using additional training data, and applying a simple post-processing technique. McKinley et al. [18] proposed an architecture, in which dense blocks [12] of dilated convolutions are embedded in a shallow U-Net-style network. Following an encoder-decoder CNN architecture, Myronenko [21] won first place in BraTS 2018 by adding a branch to the encoder endpoint and taking a variational auto-encoder (VAE) approach. The winning model [15] in BraTS 2019, was based on a similar architecture but further employed a two-stage cascaded strategy.

CNNs are computationally demanding and memory-intensive. Since convolution operations comprise the vast bulk of computations in a deep CNN during both training and inference, methods have been proposed to speed up and compress convolutional layers. MobileNet [11] exploits depthwise separable filters to represent a standard convolution layer more compactly, leading to a substantial reduction in computational complexity at the cost of a small loss of accuracy. Inception [25] uses bottleneck architectures made of cheap 1×1 convolutions to limit the network size. These methods suggest new architectures by factorizing a convolution into smaller blocks.

An alternative approach is based on low-rank approximations [14] and tensor decompositions [16,17], where the weights in a convolution layer are constrained to be low-rank. One advantage of this approach is that for a fixed architecture we can easily control the number of parameters and the computational complexity of the model by adjusting the rank. Furthermore, imposing low-rank constraints can regularize the model and reduce overfitting. In addition to speed-up, Tai et al. [26] achieved significant improvements in some cases using CNNs with low-rank regularization. Note that although all the mentioned techniques are only applied to the task of image classification, they can be also deployed effectively in an encoder-decoder architecture for image segmentation. The impact of low-rank regularization on the performance is expected to be greater when less data is available for training.

3   Method

In this section, we present our approach to brain tumor segmentation. The baseline architecture used is based on a 3D U-Net (Fig. 2a) proposed in [5] and

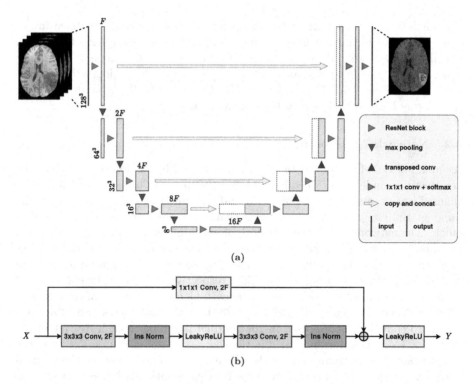

Fig. 2. The U-Net architecture (a) and the ResNet block (b) used for brain tumor segmentation.

customized for BraTS 2018 by [13], except that here convolutional blocks at each level are replaced by ResNet blocks [10]. We also introduce a new layer, called low-rank convolution, as a regularization technique to reduce overfitting. By replacing ordinary convolution layers with low-rank ones, we can achieve significantly better performance especially when the amount of training data is insufficient, where the model is more prone to overfitting. In the following, we describe the training procedure and the building blocks of our networks.

3.1 Data Preprocessing and Augmentation

The BraTS data is heterogeneous in the sense that it is multiparametric and acquired with different protocols at multiple institutions using various scanners, making intensity values nonstandardized. There is also high between-subject variability in tumors due to the presence of both low- and high-grade gliomas. To alleviate this heterogeneity and insufficiency of data, it is crucial to perform an effective preprocessing workflow before feeding the data into the network.

For each scan, we first form a 4-channel 3D image as the input, where each channel corresponds to one of the modalities (i.e. T1, post-contrast T1, T2, and FLAIR). We crop each image with a minimal box containing the whole brain region then resize it to the size $128 \times 128 \times 128$. Each channel of each image is then normalized independently using z-score to have intensities with zero mean and unit variance. Three data augmentation techniques are also utilized to reduce overfitting. Firstly, the input image is randomly flipped along the left-right axis. Secondly, we apply a random affine transform (scale $\sim U(0.9, 1.1)$, rotation $\sim U(-10, 10)$). Finally, a Gaussian noise ($\mu = 0, \sigma \sim U(0, 0.25)$) is added to intensities per-channel.

3.2   Network Architecture

As mentioned before, our network, as shown in Fig. 2a, follows a U-Net-like architecture made up of encoder and decoder parts. The network takes a 4-channel image of size $128 \times 128 \times 128$ and outputs a *probability map* with the same spatial size and with 4 channels that correspond to the 4 segmentation labels. The network has 4 levels, at each of which in the encoder (decoder) part, the input tensor is downscaled (upscaled) by a factor of two while the number of channels is doubled (halved). Downscaling and upscaling are performed via max-pooling and transposed convolution, respectively. In both the encoder and decoder, we use ResNet blocks [10], where each block is composed of convolution, Instance Normalization [27], and LeakyReLU activation layers (Fig. 2b). Two $3 \times 3 \times 3$ convolutions are used in the residual mapping of each ResNet block, and a $1 \times 1 \times 1$ convolution is used in the shortcut connection in order to match the number of input channels with the number of output channels of the residual mapping. At the decoder endpoint, a $1 \times 1 \times 1$ convolution followed by a *softmax layer* is applied to get the segmentation probability map.

3.3   Low-Rank Convolution

Convolutions form the backbone of a CNN. A typical U-Net has millions of training parameters, the majority of which are the weights that correspond to convolutional layers (this is true for any CNN with no fully-connected layers). In practice, such a complex model is very likely to overfit in particular when it comes to medical image segmentation applications, where the amount of annotated data is typically limited. Common regularization techniques like *dropout* [24] and *weight decay* can be used to mitigate this problem. However, these methods do not decrease the total number of parameters while modern CNNs are known to be heavily over-parameterized [8], i.e., the number of parameters exceeds the size of training data and what is theoretically sufficient. In this work, by imposing low-rank constraints on weights, we propose a new operation, termed Low-Rank Convolution (LRCONV), enabling the design of deep architectures with much fewer parameters but more robustness to overfitting. It is noteworthy that this idea is unrelated but complementary to other regularization techniques, such as dropout and weight decay.

Let tensor $\mathcal{X} \in \mathbb{R}^{C_{in} \times H \times W \times D}$ be the input of a 3D convolutional layer (for simplicity, we assume unit stride and dilation, and a zero-padded input) with the kernel tensor $\mathcal{V} \in \mathbb{R}^{C_{out} \times C_{in} \times H' \times W' \times D'}$ and bias $\mathbf{b} \in \mathbb{R}^{C_{out}}$. The output tensor $\tilde{\mathcal{X}} \in \mathbb{R}^{C_{out} \times H \times W \times D}$ is obtained as follows:

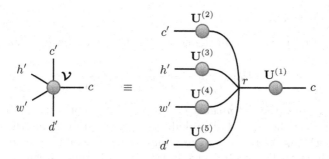

Fig. 3. Canonical Polyadic tensor network. The 5th-order weight tensor (left panel) in a convolutional layer is represented with a Canonical Polyadic tensor network (right panel). In the graphical notation of an Einstein summation, nodes and edges denote the tensors and their corresponding indices, respectively.

$$\tilde{x}_{chwd} = b_c + \sum_{c'=1}^{C_{\mathrm{in}}} \sum_{h'=1}^{H'} \sum_{w'=1}^{W'} \sum_{d'=1}^{D'} x_{c'(h+h')(w'+w)(d+d')} v_{cc'h'w'd'} \qquad (1)$$

where (H, W, D) is the resolution of the input image; C_{in} and C_{out} denote the number of channels in the input and output, respectively; and (H', W', D') is the size of the convolution kernel. To define a low-rank convolution, we can use a Canonical Polyadic [6] form and re-parameterize the weight as a sum of rank-1 tensors:

$$\mathcal{V} = \sum_{r=1}^{R} \mathbf{u}_r^{(1)} \circ \mathbf{u}_r^{(2)} \circ \mathbf{u}_r^{(3)} \circ \mathbf{u}_r^{(4)} \circ \mathbf{u}_r^{(5)}, \qquad (2)$$

where "$\circ$" denotes the vector *outer product*; $\mathbf{u}_r^{(1)} \in \mathbb{R}^{C_{\mathrm{out}}}$, $\mathbf{u}_r^{(2)} \in \mathbb{R}^{C_{\mathrm{in}}}$, $\mathbf{u}_r^{(3)} \in \mathbb{R}^{H'}$, $\mathbf{u}_r^{(4)} \in \mathbb{R}^{W'}$, and $\mathbf{u}_r^{(5)} \in \mathbb{R}^{D'}$; and R is the rank. Equivalently, the above equation can be re-written elementwise as:

$$v_{cc'h'w'd'} = \sum_{r=1}^{R} u_{cr}^{(1)} u_{c'r}^{(2)} u_{h'r}^{(3)} u_{w'r}^{(4)} u_{d'r}^{(5)}, \qquad (3)$$

where $\mathbf{U}^{(j)}$ is a *factor matrix* whose columns are $\{\mathbf{u}_1^{(j)}, \ldots, \mathbf{u}_R^{(j)}\}$. Since the Eq. (3) can be treated as an Einstein summation, we can illustrate it using a *tensor network* diagram [6] as shown in Fig. 3. In this paper, we only use the Canonical Polyadic form, but other sparsely connected tensor network like Tucker and *tensor train* can be also utilized. In a LRCONV layer, the factor matrices and the bias are the parameters to be learned. The rank R is a hyperparameter by which we can control the number of parameters although, throughout the rest of this paper, we tune the ratio $\alpha = R/C_{\mathrm{in}}$ rather than R for controlling the layer complexity. Obviously, the smaller α, the fewer parameters the layer has.

It is worth noting that our approach exploits low-rank representations to regularize a CNN before the training in contrast to methods compressing pre-trained CNNs using low-rank approximations [14] and tensor decompositions [17].

3.4   Loss Function

The loss function used to train the network is the *soft Dice loss* [20], defined as

$$\mathbf{L}_{\text{Dice}} = 1 - \frac{2\langle P, G \rangle + 1}{||P||^2 + ||G||^2 + 1} \tag{4}$$

where $\langle \cdot, \cdot \rangle$ denotes the *dot product* of tensors; $||.||$ denotes the Frobenius norm of a tensor; P is the predicted probability segmentation map (the out of the softmax layer); and G is the *one-hot binary mask* encoding the corresponding ground truth. Both P and G are $4 \times 128 \times 128 \times 128$ tensors, where the 4 channels correspond to the 4 segmentation labels. We add 1 to both the numerator and denominator (sometimes known as *additive smoothing*) to smooth the loss and avoid division by zero. Although we optimize the Dice loss obtained by the labels, e.g. enhancing tumor, edema, necrosis and non-enhancing, we also monitor the Dice for the three overlapping regions, i.e., whole tumor, tumor core, and enhancing tumor.

3.5   Optimization

All networks are trained for 50 epochs, with a batch size of one. We use Adam optimizer with initial learning rate of 10^{-4} and regularize models with ℓ_2 weight decay of 10^{-5}. The learning rate is scheduled to decrease by a factor of 5 if the validation metric sees no improvement within 5 epochs. The training set is randomly split into 80% (298 cases) used for training and the rest 20% (73 cases) used for validation.

3.6   Postprocessing

For a given test image in the inference phase, the probability map from the network output is resized to its original size. The map is then padded to have the same size as it had before cropping in the preprocessing step. We need to process the resulting probability map to obtain the final binary segmentation mask. The most trivial way is to select the label with the highest probability for each voxel. However, this does not exploit the fact that the tumor subregions, i.e., necrosis, tumor core, and whole tumor, are nested within each other. To overcome this shortcoming, we perform a hierarchical scheme, where the whole tumor is first extracted by thresholding the probability map. Having restricted to the voxels of the whole tumor, the edema channel of the probability map is then thresholded to extract the tumor core. The final threshold is applied to separate the necrosis and enhanced tumor within the tumor core. As shown in the next section, this postprocessing improves the performance significantly, with the Hausdorff distance of enhancing tumor decreasing by $\approx 15\%$.

4   Experiments and Results

All the models were implemented using PyTorch [22] and PyTorch Lighting [9] frameworks and trained on a NVIDIA P100 SXM2 GPU. We experimented with

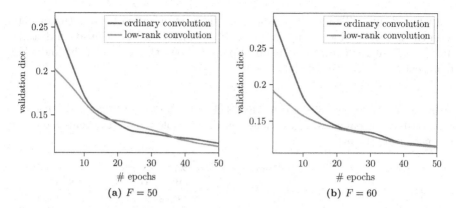

Fig. 4. Validation metric plotted against epochs for two different values of F (the initial number of feature maps). The low-rank convolutions with $\alpha = 0.5$ are used. For visualization purposes, the curves are smooth via LOESS [7].

different values for the initial number of feature maps (denoted by F in Fig. 2) and found the larger values to perform better although the GPU memory limitation did not allow us to try values greater than 60. We noticed that low-rank convolution can improve the results particularly for larger networks, roughly those with $F > 40$. We obtain the low-rank version of a U-Net by replacing $3 \times 3 \times 3$ ordinary convolutions with low-rank ones in all except the first level (the number of parameters at the first level is already relatively small). As seen in Fig. 4, the networks with low-rank convolutions converge faster, attributed to the fact that the number of training parameters of the initial networks is dramatically reduced (see Table 1), resulting in much less complex models and optimization problems. The impact of low-rank regularization on convergence and performance is more substantial for the case $F = 60$ (with 92 million parameters) compared to the case $F = 50$ (with 64 million parameters), which is somewhat expected since the former has far more parameters.

The results on the BraTS 2020 validation set (125 cases) obtained by the BraTS online evaluation framework are reported in Table 1. Our base model follows the Res-U-Net architecture (with $F = 60$ initial feature maps) described in Sect. 3.2. By postprocessing the probability maps using the hierarchical scheme described in Sect. 3.6, we achieved an improvement, particularly on the validation Dice of enhancing tumor. The results were further improved by training the low-rank version (with $\alpha = 0.5$) of the network, with the Hausdorff distance of enhancing tumor falling from 32.22 to 25.01. The low-rank network with 20.8 million parameters is far more memory efficient than the initial network with 92.2 million parameters. Finally, an ensemble of 8 models, including networks with $F \in \{30, 40, 50, 60\}$ and their low-rank versions (with $\alpha = 0.5$), was used to reduce the variance in predictions. To build the ensemble model, the probability maps were first averaged and then thresholded via the hierarchical scheme. This further improved all the scores, leading to the Dice score of over 90% for the whole tumor. The greatest improvement was observed for the tumor core, with the Dice score increasing from 77.84% to 79.1% and the Hausdorff distance dropping from 16.96 mm to 7.76 mm. Figure 1 shows the results of this model on representative HGG and LGG cases.

Table 1. The average scores on the BraTS 2020 validation set (125 cases). They are computed by the online evaluation platform (WT: whole tumor, TC: tumor core, and ET: enhancing tumor).

Method	# Params (M)	Dice (%)			Hausdorff		
		ET	WT	TC	ET	WT	TC
Res-U-Net	92.2	71.99	89.68	78.79	38.18	5.74	14.49
Res-U-Net + post	92.2	73.32	89.84	78.77	32.22	5.75	14.46
Low-rank Res-U-Net + post	20.8	74.82	89.12	77.84	25.01	7.61	16.94
Ensemble of 8 models	–	75.06	90.37	79.1	25.61	5.23	7.76

Table 2 presents the performance of the method on the BraTS 2020 test set. The average Dice scores for enhancing tumor core, whole tumor, and tumor core are 77.73%, 87.37%, and 81.24%, respectively. The corresponding values of Hausdorff distance are 16.21 mm, 6.28 mm, and 20.52 mm, respectively. Overall, the test results are consistent with those of the validation set, which reflects our model is neither biased towards the validation set nor has a high variance.

Table 2. Summary statistics of the scores on the BraTS 2020 test set (166 cases). They are computed by the online evaluation platform (WT: whole tumor, TC: tumor core, and ET: enhancing tumor).

	Dice			Hausdorff		
	ET	WT	TC	ET	WT	TC
Mean	77.73	87.37	81.24	16.21	6.28	20.52
StdDev	21.6	13.93	25.45	69.39	11.76	74.75
Median	83.97	91.58	91.05	2.0	3.25	2.83
25th quantile	75.08	86.60	82.82	1.41	2.0	1.49
75th quantile	89.59	94.54	94.92	3.0	5.72	5.34

5   Conclusion

This paper proposes a regularization technique for CNNs by re-parameterizing convolutional layers as a low-rank structure, particularly canonical polyadic form. We devised a U-Net architecture with ResNet blocks consisting of low-rank convolutions. We examined the impact of this low-rank regularization on performance, verifying its effectiveness for brain tumor segmentation in multi-modal MRI scans. The results on the BraTS 2020 data show that despite having much fewer parameters, the low-rank networks can outperform the unregularized versions especially in terms of Dice coefficients and Hausdorff distances on the enhancing tumor.

Acknowledgements. The research leading to these results has received funding from EU H2020 MSCA-ITN-2018: INtegrating Magnetic Resonance SPectroscopy and Multimodal Imaging for Research and Education in MEDicine (INSPiRE-MED), funded by the European Commission under Grant Agreement #813120. This research also received funding from the Flemish Government (AI Research Program). Sabine Van Huffel and Pooya Ashtari are affiliated to Leuven. AI - KU Leuven institute for AI, B-3000, Leuven, Belgium.

References

1. Bakas, S., et al.: Segmentation labels and radiomic features for the pre-operative scans of the TCGA-GBM collection. The cancer imaging archive. Nat. Sci. Data **4**, 170117 (2017)
2. Bakas, S., et al.: Segmentation labels and radiomic features for the pre-operative scans of the tcga-lgg collection. Cancer Imaging Archive **286**, (2017)
3. Bakas, S., et al.: Advancing the cancer genome atlas glioma MRI collections with expert segmentation labels and radiomic features. Sci. Data **4**, 170117 (2017)
4. Bakas, S., et al.: Identifying the best machine learning algorithms for brain tumor segmentation, progression assessment, and overall survival prediction in the brats challenge. arXiv preprint arXiv:1811.02629 (2018)
5. Çiçek, Ö., Abdulkadir, A., Lienkamp, S.S., Brox, T., Ronneberger, O.: 3D U-net: learning dense volumetric segmentation from sparse annotation. In: Ourselin, S., Joskowicz, L., Sabuncu, M.R., Unal, G., Wells, W. (eds.) MICCAI 2016. LNCS, vol. 9901, pp. 424–432. Springer, Cham (2016). https://doi.org/10.1007/978-3-319-46723-8_49
6. Cichocki, A., Lee, N., Oseledets, I., Phan, A.H., Zhao, Q., Mandic, D.P.: Tensor networks for dimensionality reduction and large-scale optimization: part 1 low-rank tensor decompositions. Found. Trends® Mach. Learn. **9**(4–5), 249–429 (2016)
7. Cleveland, W.S., Devlin, S.J.: Locally weighted regression: an approach to regression analysis by local fitting. J. Am. Stat. Assoc. **83**(403), 596–610 (1988)
8. Denil, M., Shakibi, B., Dinh, L., Ranzato, M., De Freitas, N.: Predicting parameters in deep learning. In: Advances in Neural Information Processing Systems, pp. 2148–2156 (2013)
9. Falcon, W.: Pytorch lightning. GitHub (2019). https://github.com/PyTorchLightning/pytorch-lightning. Cited by 3
10. He, K., Zhang, X., Ren, S., Sun, J.: Deep residual learning for image recognition. In: Proceedings of the IEEE Conference on Computer Vision and Pattern Recognition, pp. 770–778 (2016)
11. Howard, A.G., et al.: MobileNets: efficient convolutional neural networks for mobile vision applications. arXiv preprint arXiv:1704.04861 (2017)
12. Huang, G., Liu, Z., Van Der Maaten, L., Weinberger, K.Q.: Densely connected convolutional networks. In: Proceedings of the IEEE Conference on Computer Vision and Pattern Recognition, pp. 4700–4708 (2017)

13. Isensee, F., Kickingereder, P., Wick, W., Bendszus, M., Maier-Hein, K.H.: No new-net. In: Crimi, A., Bakas, S., Kuijf, H., Keyvan, F., Reyes, M., van Walsum, T. (eds.) BrainLes 2018. LNCS, vol. 11384, pp. 234–244. Springer, Cham (2019). https://doi.org/10.1007/978-3-030-11726-9_21

14. Jaderberg, M., Vedaldi, A., Zisserman, A.: Speeding up convolutional neural networks with low rank expansions. arXiv preprint arXiv:1405.3866 (2014)

15. Jiang, Z., Ding, C., Liu, M., Tao, D.: Two-stage cascaded U-net: 1st place solution to BraTS challenge 2019 segmentation task. In: Crimi, A., Bakas, S. (eds.) BrainLes 2019. LNCS, vol. 11992, pp. 231–241. Springer, Cham (2020). https://doi.org/10.1007/978-3-030-46640-4_22

16. Kim, Y.D., Park, E., Yoo, S., Choi, T., Yang, L., Shin, D.: Compression of deep convolutional neural networks for fast and low power mobile applications. arXiv preprint arXiv:1511.06530 (2015)

17. Lebedev, V., Ganin, Y., Rakhuba, M., Oseledets, I., Lempitsky, V.: Speeding-up convolutional neural networks using fine-tuned CP-decomposition. arXiv preprint arXiv:1412.6553 (2014)

18. McKinley, R., Meier, R., Wiest, R.: Ensembles of densely-connected CNNs with label-uncertainty for brain tumor segmentation. In: Crimi, A., Bakas, S., Kuijf, H., Keyvan, F., Reyes, M., van Walsum, T. (eds.) BrainLes 2018. LNCS, vol. 11384, pp. 456–465. Springer, Cham (2019). https://doi.org/10.1007/978-3-030-11726-9_40

19. Menze, B.H., et al.: The multimodal brain tumor image segmentation benchmark (BRATS). IEEE Trans. Med. Imaging 34(10), 1993–2024 (2014)

20. Milletari, F., Navab, N., Ahmadi, S.A.: V-net: fully convolutional neural networks for volumetric medical image segmentation. In: 2016 Fourth International Conference on 3D Vision (3DV), pp. 565–571. IEEE (2016)

21. Myronenko, A.: 3D MRI brain tumor segmentation using autoencoder regularization. In: Crimi, A., Bakas, S., Kuijf, H., Keyvan, F., Reyes, M., van Walsum, T. (eds.) BrainLes 2018. LNCS, vol. 11384, pp. 311–320. Springer, Cham (2019). https://doi.org/10.1007/978-3-030-11726-9_28

22. Paszke, A., et al.: PyTorch: an imperative style, high-performance deep learning library. In: Advances in Neural Information Processing Systems, pp. 8026–8037 (2019)

23. Ronneberger, O., Fischer, P., Brox, T.: U-net: convolutional networks for biomedical image segmentation. In: Navab, N., Hornegger, J., Wells, W.M., Frangi, A.F. (eds.) MICCAI 2015. LNCS, vol. 9351, pp. 234–241. Springer, Cham (2015). https://doi.org/10.1007/978-3-319-24574-4_28

24. Srivastava, N., Hinton, G., Krizhevsky, A., Sutskever, I., Salakhutdinov, R.: Dropout: a simple way to prevent neural networks from overfitting. J. Mach. Learn. Res. 15(1), 1929–1958 (2014)

25. Szegedy, C., et al.: Going deeper with convolutions. In: Proceedings of the IEEE Conference on Computer Vision and Pattern Recognition, pp. 1–9 (2015)

26. Tai, C., Xiao, T., Zhang, Y., Wang, X., et al.: Convolutional neural networks with low-rank regularization. arXiv preprint arXiv:1511.06067 (2015)

27. Ulyanov, D., Vedaldi, A., Lempitsky, V.: Instance normalization: the missing ingredient for fast stylization. arXiv preprint arXiv:1607.08022 (2016)

Automated Brain Tumour Segmentation Using Cascaded 3D Densely-Connected U-Net

Mina Ghaffari[1,2(✉)], Arcot Sowmya[2], and Ruth Oliver[1]

[1] Macquarie University, Sydney, Australia
`mina.ghaffari@hdr.mq.edu.au`
[2] University of New South Wales, Sydney, Australia

Abstract. Accurate brain tumour segmentation is a crucial step towards improving disease diagnosis and proper treatment planning. In this paper, we propose a deep-learning based method to segment a brain tumour into its subregions: whole tumour, tumour core and enhancing tumour. The proposed architecture is a 3D convolutional neural network based on a variant of the U-Net architecture of Ronneberger et al. [17] with three main modifications: (i) a heavy encoder, light decoder structure using residual blocks (ii) employment of dense blocks instead of skip connections, and (iii) utilization of self-ensembling in the decoder part of the network. The network was trained and tested using two different approaches: a multitask framework to segment all tumour subregions at the same time, and a three-stage cascaded framework to segment one subregion at a time. An ensemble of the results from both frameworks was also computed. To address the class imbalance issue, appropriate patch extraction was employed in a pre-processing step. Connected component analysis was utilized in the post-processing step to reduce the false positive predictions. Experimental results on the BraTS20 validation dataset demonstrates that the proposed model achieved average Dice Scores of 0.90, 0.83, and 0.78 for whole tumour, tumour core and enhancing tumour respectively.

Keywords: Brain tumour segmentation · Multimodal MRI · Cascaded network · Densely connected CNN

1 Introduction

Accurate and reliable brain tumour segmentation from neuroimaging scans is a critical step towards improving disease diagnosis and proper treatment planning that increases the survival chance of patients. Brain tumours are highly heterogeneous in terms of shape, size and location, which makes their segmentation challenging. In addition, brain tumours can be highly infiltrating and it may be difficult to distinguish healthy brain tissue from the tumour. Manual segmentation of brain tumours in MR images is a laborious task that is both

© Springer Nature Switzerland AG 2021
A. Crimi and S. Bakas (Eds.): BrainLes 2020, LNCS 12658, pp. 481–491, 2021.
https://doi.org/10.1007/978-3-030-72084-1_43

time-consuming and subject to rater variability. Therefore, reliable automatic segmentation of brain tumours has attracted considerable attention over the past two decades. Most recent automatic segmentation methods build on convolutional neural networks (CNNs) trained on manually annotated dataset of a large cohort of patients. The Brain Tumour Segmentation (BraTS) challenge public dataset has become the benchmark in this area [1–5,15] since 2012.

The BraTS challenge organizers provide magnetic resonance imaging (MRI) scans in four modalities: T1-weighted (T1), T1-weighted post-contrast (T1c), T2-weighted (T2), and fluid-attenuated inversion recovery (FLAIR) MRIs, with corresponding manual segmentation. The participants are required to produce segmentation masks of three glioma sub-regions that contain enhancing tumour (ET), tumour core (TC) and whole tumour (WT). The BraTS2020 dataset consists of 369, 125 and 166 cases for training, validation and test respectively.

Since the revolution of deep learning and more specifically CNN, the most successful models for BraTS are based on CNN [9]. The top ranked models submitted to recent BraTs challenges were DeepMedic, that is based on multi-scale processing [13], cascaded Fully Convolutional Networks, based on hierarchical binary segmentation [20] and U-net based models [7,11].

U-net, which was first introduced in 2015 [17], is a CNN architecture consisting of an encoder and a decoder. Due to its straightforward architecture as well as its high segmentation accuracy, U-net or its variants such as V-net [16] are used in most state-of-the-art medical image segmentation tasks [6,12,14] and it has been argued that 'a well-trained U-net is hard to beat' [11]. Hence, while using U-net as the backbone of their network, most BraTS18 and BraTS19 participants focussed on enhancing their model performance by optimizing the preprocessing step, training procedure, co training using local datasets, or applying ensemble learning. Most of the top-ranked models in recent BraTS challenges were inspired by U-net [9] including all top ranked models of BraTS19. As an example, the top-ranked model in BraTS 2019 was a two stage Cascaded U-net in which the first stage U-net predicts a coarse segmentation map and the second stage U-net provides a more accurate segmentation [7]. Zhao et al. [7] applied a bag of tricks including data sampling, random patch-size training, semi-supervised learning, self-ensembling, result fusion, and warming-up learning rate to enhance their U-net performance. McKinley et al. [7] modified their 3rd place entry to the BRaTS18: 'DeepSCAN' model, (a shallow U-net-style network with densely connected blocks of dilated convolutions and label-uncertainty loss) by adding a lightweight local attention mechanism and were ranked 3rd again in BraTS19.

In this work a modified 3D version of the U-net is utilized, and in order to enhance the model accuracy, pre- and post- processing steps as well as training procedure optimization are applied, and ensembling and data augmentation are utilized during inference time. We report preliminary results on the validation as well as test dataset of BraTS20 dataset. The results are computed online using the CBICA Image Processing Portal (https://ipp.cbica.upenn.edu). The rest of this paper is organized as follows: Methodology including the dataset preprocessing, network architecture, and post-processing are explained in Sect. 2.

Experiments and results are provided in Sect. 3, followed by a discussion and conclusion in Sect. 4.

2   Methodology

A top level diagram of the proposed method is illustrated in Fig. 1. The proposed model is composed of four different modules. In this section, each of these modules is discussed in more detail.

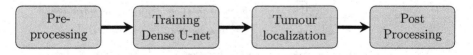

Fig. 1. Top-level schematic of the proposed model

2.1   Dataset Preprocessing

The BraTS20 training dataset consists of 369 multi-institutional pre-operative multimodal MRI scans. This dataset already has been put through various pre-processing steps by the organizers, so that all images are skull-stripped, have isotropic resolution, and are co-registered MR volumes [15]. MR scans often contain intensity non-uniformities due to magnetic field inhomogeneity, which can impact the segmentation results. To compensate for this, a bias field correction algorithm was applied using N4ITK [19]. All modalities were then normalized by subtracting the mean from each of their voxels and dividing by the standard deviation of the intensities within the brain region of that image, so that each modality has zero mean and unit variance. In order to address the small size of the dataset, data augmentation was performed using random rotation (-6 and 6 degree), scaling (0.9..1.1) and mirroring.

2.2   Network Architecture

Inspired by other work [22], we propose a modified 3D version of the well known U-net, as shown in Fig. 2. The main differences between this network and a generic U-net are threefold: (i) a heavy encoder, light decoder structure using residual blocks [10] (ii) employment of dense blocks instead of skip connections [22] and (iii) utilization of self-ensembling in the decoder part of the network.

(i) The residual block as shown in Fig. 3 consists of two $3 \times 3 \times 3$ convolutions and Group Normalization [21] with group size of 8 and Rectified Linear Unit (ReLU) activation, followed by additive identity skip connection. Residual blocks learn a non-linear residual that is added to the input and provide a deeper architecture to improve the gradient flow.

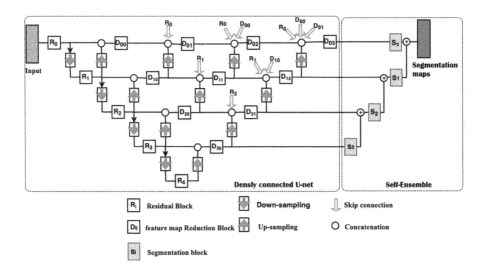

Fig. 2. Schematic visualization of the network architecture [8]

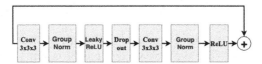

Fig. 3. Schematic illustration of the residual block used in the proposed network [8]

(ii) Dense convolutional blocks merge the feature maps at each level of the network by integrating the output of all previous convolutional blocks in the same level with the upsampled output of the corresponding lower-level dense block. This method provides the network decoder with more semantic information from the encoder feature maps. The transferred information helps the optimizer in optimizing the network more effectively [22].

(iii) Segmentation blocks were also employed for self-ensembling and deep supervision purposes [11] within the decoder of the network. These segmentation blocks make predictions at each scale of the U-net by reducing the number of feature maps at each level of the decoder to the number of feature maps at the final output layer of the network. Combining these segmentation maps helps the network converge faster by transferring more information from the earlier levels of the decoder. The network blocks structures are explained in more detail in Table 1.

The network was trained on both a multi-task framework as well as a cascaded binary segmentation task to perform sequential segmentation of the tumour components such that the output of one network is transformed and fed to the next one. Experiments were conducted to evaluate the performance of each model as well as an ensemble of both models. In the multi-task framework, the

Table 1. Structure of different blocks of the proposed network. GN stands for group normalization , Conv3 - 3 × 3 × 3 convolution, Conv1 - 1 × 1 × 1 convolution, AddId - addition of identity/skip connection

Block name	Details	Output size
R_0	Conv3, GN, Leaky ReLU, Drop out, Conv3, GN, ReLU, AddId	16 × 128 × 128 × 128
R_1	Conv3, GN, Leaky ReLU, Drop out, Conv3, GN, ReLU, AddId	32 × 64 × 64 × 64
R_2	Conv3, GN, Leaky ReLU, Drop out, Conv3, GN, ReLU, AddId	64 × 32 × 32 × 32
R_3	Conv3, GN, Leaky ReLU, Drop out, Conv3, GN, ReLU, AddId	128 × 16 × 16 × 16
R_4	Conv3, GN, Leaky ReLU, Drop out, Conv3, GN, ReLU, AddId	256 × 8 × 8 × 8
D_{00}, D_{01}, D_{02}, D_{03}	Conv3, GN, Leaky ReLU, Conv1, GN, ReLU	16 × 128 × 128 × 128
D_{10}, D_{11}, D_{12}	Conv3, GN, Leaky ReLU, Conv1, GN, ReLU	32 × 64 × 64 × 64
D_{20}, D_{21}	Conv3, GN, Leaky ReLU, Conv1, GN, ReLU	64 × 32 × 32 × 32
D_{30}	Conv3, GN, Leaky ReLU, Conv1, GN, ReLU	32 × 16 × 16 × 16
S_0	Conv3	3 × 128 × 128 × 128
S_1	Conv3, UpSampling3D	3 × 128 × 128 × 128
S_2	Conv3, UpSampling3D	3 × 64 × 64 × 64
S_3	Conv3, UpSampling3D	3 × 32 × 32 × 32

network output has three 3D channels of size 128^3, each of which corresponds to one of the three mutually inclusive tumour subregions and the network is trained to predict all tumour subregions at the same time. This multi-task learning regularizes the network by providing additional information related to the learning task. For the cascaded model, three cascaded networks were trained separately, one for each of the three tumour subregions. Each of these networks has a single 3D output of size 128^3 corresponding to one of the tumour subregions and at the time of inference, the input of each stage of the model is limited to patches containing the tumour region extracted by the previous cascaded stage. This means that the predicted tumour core is forced to lie inside the whole tumour, and the enhanced tumour core also inside the tumour core region.

2.3   Training Procedure

In the proposed method, all the four modalities were fed into the network at the same time to benefit from the information present in all. Due to memory limitations of the Graphical Processing Unit (GPU) used, the network was trained with a batch size of 1. To extract patches, each image was cropped so that the manually segmented tumour was located in the middle of each patch. Overlapping patches were extracted if the size of tumour was bigger than 128 voxels in any dimension. This strategy of extracting appropriate patches addresses the class imbalance issue to some extent, and also reduces the training time significantly. To further address the class imbalance problem, Dice Score was used as the loss function [18]; for the multi-task framework it was modified to take into account the mean value of the Dice Scores of all output segmentation maps. Dice score (DS) and multi-class Dice loss (MDS) function are expressed as:

$$DS = 2 \times \frac{Y_{true} \times Y_{pred}}{Y_{true} + Y_{pred} + \epsilon} \qquad MDL = \frac{-1}{n} \sum_{1}^{n} DS$$

where Y_{true} and Y_{pred} are the reference segmentation map (gold standard) and the predicted segmentation map for each of the output channels respectively. n is the number of channels (3 for the multi-task model), and ϵ is a small value used to avoid division by zero (we set $\epsilon = 0.00001$).

The network was trained using the Keras framework with TensorFlow backend, on an Nvidia Tesla Volta V100 GPU. The GPU memory allowed training of the network with batches of size 1 while allowing 16 filters in the highest level of the Dense U-net. The network was trained using Adam Optimizer with an initial learning rate of 5e−4, and reducing by a factor of 0.5 if the validation accuracy was not improving within the last 10 epochs. Dropout with a rate of 0.3 and also L2 norm regularization with a weight of 1e−5 were used for regularization purposes. The network was programmed to be trained for a maximum of 300 epochs or until the validation accuracy did not improve in the last 50 epochs.

2.4   Tumour Localization

To increase the prediction accuracy, a cascaded method was used in which the whole tumour region was localized using low-resolution images. To do so, the images were resized to 128^3 voxels and then the model was applied to obtain an estimate of the tumour location. The prediction map was then resized to its original size and patches of size 128^3 voxels were extracted such that the whole tumour is in the middle of that patch. If the predicted whole tumour is larger than 128^3 voxels, more overlapping patches were extracted to cover the whole tumour region. The extracted patches were then used to predict the three tumour subregions. Finally, the segmentation maps were reconstructed by zero-padding, considering the location of the extracted patches. This method of patch extraction reduces both the test time and the rate of false positives prediction.

2.5   Post-processing

To further improve the accuracy of brain tumour segmentation, a post-processing method was applied. The post-processing may be summarized as follows:

1. It was ensured that the hierarchy of the tumour subregions was respected. This means that the enhancing tumour subregion is inside the tumour core, and the tumour core is inside the whole tumour. To achieve this, any enhancing tumour voxels appearing outside the tumour core were removed, as well as any tumour core appearing outside the whole tumour.
2. Connected components of any subregion smaller than 10 voxels were removed.

3. Connected components of whole tumour that were primarily uncertain (did not include any tumour core or enhancing tumour voxels, usually in cerebellum region, and/or having higher intensities in only FLAIR or T2 modalities). Such simple post-processing worked well in discarding false positive predictions possibly due to inherent system noise. This heuristics was verified by examining validation subject having such connected components as whole tumour.
4. Considering the fact that low grade gliomas (LGG) patients may have no enhancing tumour sub-region and also inspired by other work [11], all enhancing tumour regions with less than 50 voxels were replaced by necrosis.

After these steps, the remaining tumour subregions were fused to reconstruct the final segmentation map.

3   Experiments and Results

All scans in the BraTS20 training dataset were used for training the proposed network in two different frameworks separately: cascaded binary segmentation task as well as the multi-task framework. Validation set results were obtained by using each of these models separately as well as by their ensemble, which was performed by averaging the Sigmoid probabilities of both model predictions. Thresholding and post-processing were then applied to the final prediction map. A threshold of 0.5 was chosen for whole tumour and tumour core subregion while for Enhancing tumour subregion the threshold was reduced to 0.4. This choice of threshold for enhancing tumour subregion was done based on the performance of the validation dataset and it enhanced the prediction accuracy for the subjects with under-segmented prediction. However, it did not resolve the issue for some of the LGG cases as a result of which the Hausdorff distance were not as expected. So, in our ensemble experiment the enhancing core threshold was gradually decreased to 0.3, 0.2, and 0.1 if no enhancing tumour was not detected. This also resolved the under-segmentation issue for some of those LGG cases and enhanced the Hausdorff distance of enhancing tumour subregion (Table 2). Test time augmentation was applied for enhancing the prediction accuracy in all of the experiments. This was done by flipping the input image axes, and averaging the outputs of the resulting flipped segmentation probability maps. The results are listed in Table 2. All reported values were computed by the online evaluation platform. We also trained models using 5 fold cross-validation and a qualitative example generated using the trained model is depicted in Fig. 4. Images include three sample cases from the training dataset, along with the manual annotations and the model prediction. According to the reported results, the multi-task model and the cascaded model have almost the same performance, while their ensemble resulted in a more accurate and robust model.

The final method including the ensemble model with test time augmentation was used as our entry into the BraTS 2020 challenge. For this purpose, the model was applied to the test dataset including 166 patient images. Results including the Dice Scores, sensitivities, specificities and 95 Hausdorff distances of the enhanced tumor, whole tumor and tumor core were automatically calculated by the challenge organizers after submitting to the *CBICA's* Image Processing Portal. In Table 3, the mean dice scores and 95 Hausdorff distances of ET, WT and TC for the training, validation and testing datasets are shown.

Table 2. Mean Dice Score (DSC) and Hausdorff distance (95%) (HD95), of the proposed segmentation method on BraTS 2020 validation dataset (125 cases) using the online IPP portal. ET: enhancing tumour, WT: whole tumour, TC: tumour core.

Model	DSC			DH95		
Validation	ET	WT	TC	ET	WT	TC
Multitask model	0.76 ± 0.28	0.89 ± 0.08	0.82 ± 0.17	22.3	5.89	7.16
Cascaded model	0.75 ± 0.28	0.88 ± 0.13	0.81 ± 0.19	17.17	7.16	10.4
Ensemble model	$\mathbf{0.78 \pm 0.26}$	$\mathbf{0.90 \pm 0.06}$	$\mathbf{0.82 \pm 0.17}$	**7.71**	**5.14**	**6.64**

Table 3. Mean Dice Score (DSC) and Hausdorff distance (95%) (HD95), of the proposed segmentation method (ensemble model) on BraTS 2020 test dataset (166 cases). ET: enhancing tumour, WT: whole tumour, TC: tumour core.

Model	DSC			DH95		
Validation	ET	WT	TC	ET	WT	TC
Ensemble model	**0.81**	**0.88**	**0.82**	**16.02**	**5.65**	**22.0**
StdDev	0.2	0.14	0.25	69.56	7.78	79.5
Median	0.85	0.92	0.91	1.73	3.0	2.64
25quantile	0.77	0.87	0.85	1.0	1.73	1.73
75quantile	0.90	0.95	0.95	2.45	5.83	4.66

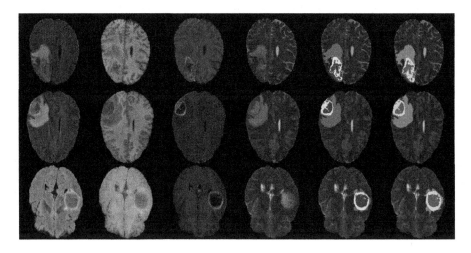

Fig. 4. Qualitative results for three patients from BraTS20 training dataset. From left to right: Flair, T1, T1c, T2, manual segmentation overlaid on T2, and model prediction.

4   Conclusion

In this paper a 3D densely connected U-net network was trained on the BraTS20 training dataset using two different approaches: a multi-task framework and a three-stage cascaded framework. The networks were trained using 3D patches of size 128^3. To address the class imbalance issue, instead of random patch selection, patches were selected by cropping images so that the manually segmented tumour was located in the middle of each patch. Brain tumour segmentation was performed in two stages: first a coarse prediction was performed on low resolution images to localize the tumour, and then fine segmentation was performed by extracting patches in the tumour region, applying the model to those patches, and finally stitching the patches together. The model predictions were then filtered through a post-processing step in which connected components analysis was used to reduce the false positive predictions. The performance of the multi-task model, the cascaded model and also ensemble of both models was evaluated using the BraTS20 validation dataset, and results were calculated on the online CBICA Image Processing Portal. The reported average Dice Scores for the ensemble model were 0.90, 0.82, and 0.78 for whole tumour, tumour core and enhancing tumour respectively. The final ensemble model was our entry into the BraTS 2020 challenge and it archived Dice Score accuracies of 0.88, 0.82, and 0.81 for whole tumour, tumour core and enhancing tumour respectively when tested on the BraTS20 testing set. We suffered from some hardware limitations while training our models, which we believe has impacted model performance. For example, the maximum duration time for the cloud GPU we were using was 48 h after which the training was terminated, and also due to GPU memory limitations we could only train with a batch size of one. We believe that with access to more powerful hardware, the model performance is likely to be further enhanced.

References

1. Bakas, S., et al.: Segmentation labels and radiomic features for the pre-operative scans of the TCGA-GBM collection. The Cancer Imaging Archive (2017). https://doi.org/10.7937/K9/TCIA.2017.KLXWJJ1Q
2. Bakas, S., et al.: Segmentation labels and radiomic features for the pre-operative scans of the TCGA-LGG collection. The Cancer Imaging Archive (2017). https://doi.org/10.7937/K9/TCIA.2017.GJQ7R0EF
3. Bakas, S., et al.: Advancing the cancer genome atlas glioma MRI collections with expert segmentation labels and radiomic features. Sci. Data 4(July), 1–13 (2017). https://doi.org/10.1038/sdata.2017.117
4. Bakas, S., et al.: Identifying the best machine learning algorithms for brain tumor segmentation, progression assessment, and overall survival prediction in the brats challenge. ArXiv abs/1811.02629 (2018)
5. Multimodal brain tumor segmentation challenge 2019 (2019). http://braintumorsegmentation.org/. Accessed 2 July 2020

6. Chen, W., Liu, B., Peng, S., Sun, J., Qiao, X.: S3D-UNet: separable 3D U-net for brain tumor segmentation. In: Crimi, A., Bakas, S., Kuijf, H., Keyvan, F., Reyes, M., van Walsum, T. (eds.) BrainLes 2018. LNCS, vol. 11384, pp. 358–368. Springer, Cham (2019). https://doi.org/10.1007/978-3-030-11726-9_32

7. Crimi, A., Eds, S.B., Goos, G.: Brainlesion: Glioma, Multiple Sclerosis, Stroke and Traumatic Brain Injuries. Lecture Notes in Computer Science (2019). https://doi.org/10.1007/978-3-030-46640-4

8. Ghaffari, M., Sowmya, A., Oliver, R., Hamey, L.: Multimodal brain tumour segmentation using densely connected 3D convolutional neural network. In: 2019 Digital Image Computing: Techniques and Applications (DICTA), pp. 1–5 (2019)

9. Ghaffari, M., Sowmya, A., Oliver, R.: Automated brain tumor segmentation using multimodal brain scans: a survey based on models submitted to the BraTS 2012–2018 challenges. IEEE Rev. Biomed. Eng. **13**, 156–168 (2020). https://doi.org/10.1109/RBME.2019.2946868

10. He, K., Sun, J.: Deep residual learning for image recognition. In: 2016 IEEE Conference on Computer Vision and Pattern Recognition (CVPR), pp. 770–778. IEEE (2016). https://doi.org/10.1109/CVPR.2016.90

11. Isensee, F., Kickingereder, P., Wick, W., Bendszus, M., Maier-Hein, K.H.: No new-net. In: Crimi, A., Bakas, S., Kuijf, H., Keyvan, F., Reyes, M., van Walsum, T. (eds.) BrainLes 2018. LNCS, vol. 11384, pp. 234–244. Springer, Cham (2019). https://doi.org/10.1007/978-3-030-11726-9_21

12. Isensee, F., et al.: nnU-Net: self-adapting framework for U-net-based medical image segmentation (2018)

13. Kamnitsas, K., et al.: DeepMedic on brain tumor segmentation. In: Crimi, A., Menze, B., Maier, O., Reyes, M., Winzeck, S., Handels, H. (eds.) BrainLes 2016. LNCS, vol. 10154, pp. 138–149. Springer, Cham (2016). https://doi.org/10.1007/978-3-319-55524-9_14

14. Li, X., Chen, H., Qi, X., Dou, Q., Fu, C.W., Heng, P.A.: H-DenseUNet: hybrid densely connected UNet for liver and tumor segmentation from CT volumes. IEEE Trans. Med. Imaging **37**(2), 2663–2674 (2018)

15. Menze, B.H., et al.: The multimodal brain tumor image segmentation benchmark (BRATS). IEEE Trans. Med. Imaging **34**(10), 1993–2024 (2015). https://doi.org/10.1109/TMI.2014.2377694

16. Milletari, F., Navab, N., Ahmadi, S.A.: V-net: fully convolutional neural networks for volumetric medical image segmentation. In: Fourth International Conference on 3D Vision (3DV) (2016)

17. Ronneberger, O., Fischer, P., Brox, T.: U-Net: convolutional networks for biomedical image segmentation. In: Medical Image Computing and Computer-Assisted Intervention (MICCAI), vol. 9351, pp. 234–241 (2015). http://arxiv.org/abs/1505.04597

18. Shen, C., et al.: On the influence of Dice loss function in multi-class organ segmentation of abdominal CT using 3D fully convolutional networks. eprint arXiv:1801.05912 (2018). http://arxiv.org/abs/1801.05912

19. Tustison, N.J., et al.: N4ITK: improved N3 bias correction. IEEE Trans. Med. Imaging **29**(6), 1310–1320 (2010)

20. Wang, G., Li, W., Ourselin, S., Vercauteren, T.: Automatic brain tumor segmentation using cascaded anisotropic convolutional neural networks. In: Crimi, A., Bakas, S., Kuijf, H., Menze, B., Reyes, M. (eds.) BrainLes 2017. LNCS, vol. 10670, pp. 178–190. Springer, Cham (2018). https://doi.org/10.1007/978-3-319-75238-9_16

21. Wu, Y., He, K.: Group normalization. Int. J. Comput. Vis. **128**(3), 742–755 (2019). https://doi.org/10.1007/s11263-019-01198-w
22. Zhou, Z., Rahman Siddiquee, M.M., Tajbakhsh, N., Liang, J.: UNet++: a nested U-net architecture for medical image segmentation. In: Stoyanov, D., et al. (eds.) DLMIA/ML-CDS -2018. LNCS, vol. 11045, pp. 3–11. Springer, Cham (2018). https://doi.org/10.1007/978-3-030-00889-5_1

Segmentation then Prediction: A Multi-task Solution to Brain Tumor Segmentation and Survival Prediction

Guojing Zhao[1,2], Bowen Jiang[1,2], Jianpeng Zhang[2(✉)], and Yong Xia[1,2(✉)]

[1] Research & Development Institute of Northwestern Polytechnical University in Shenzhen, Shenzhen 518057, China
yxia@nwpu.edu.cn

[2] National Engineering Laboratory for Integrated Aero-Space-Ground-Ocean Big Data Application Technology, School of Computer Science and Engineering, Northwestern Polytechnical University, Xi'an 710072, China
james.zhang@mail.nwpu.edu.cn

Abstract. Accurate brain tumor segmentation and survival prediction are two fundamental but challenging tasks in the computer aided diagnosis of gliomas. Traditionally, these two tasks were performed independently, without considering the correlation between them. We believe that both tasks should be performed under a unified framework so as to enable them mutually benefit each other. In this paper, we propose a multi-task deep learning model called *segmentation then prediction* (STP), to segment brain tumors and predict patient overall survival time. The STP model is composed of a segmentation module and a survival prediction module. The former uses 3D U-Net as its backbone, and the latter uses both local and global features. The local features are extracted by the last layer of the segmentation encoder, while the global features are produced by a global branch, which uses 3D ResNet-50 as its backbone. The STP model is jointly optimized for two tasks. We evaluated the proposed STP model on the BraTS 2020 validation dataset and achieved an average Dice similarity coefficient (DSC) of 0.790, 0.910, 0.851 for the segmentation of enhanced tumor core, whole tumor, and tumor core, respectively, and an accuracy of 65.5% for survival prediction.

Keywords: Brain tumor segmentation · Survival prediction · Deep learning · Joint learning · MR image generation

1 Introduction

As the most common type of primary brain tumors, gliomas can affect the brain function and be life-threatening. Accurate brain tumor segmentation and survival prediction using multi-sequence magnetic resonance (MR) imaging play a critical role in the diagnosis, treatment planning, and prognosis of this disease.

G. Zhao and B. Jiang—Contributed equally to this work.

© Springer Nature Switzerland AG 2021
A. Crimi and S. Bakas (Eds.): BrainLes 2020, LNCS 12658, pp. 492–502, 2021.
https://doi.org/10.1007/978-3-030-72084-1_44

Accurate segmentation aims to delineate the glioma and its internal structures, including the enhancing tumor and tumor core, while survival prediction aims to predict the survival time of a post-operative patient from the pre-operative MR scan. Manual segmentation of brain tumors usually requires a high degree of concentration and expertise and is time-consuming, expensive, and prone to subjective bias. Automated brain tumor segmentation can bypass these issues, and hence has been thoroughly studied in the literature. Based on the segmentation of brain tumors, automated survival prediction can assist doctors in developing a suitable treatment plan for the patient.

In recent years, deep learning-based methods have become the mainstream in the automated segmentation of gliomas. Kamnitsas *et al.* [10] proposed an ensemble of DeepMedic, FCN, and U-Net, in which the predictions made by different models are integrated to form the robust segmentation of brain tumors. Andriy [13] developed a semantic segmentation network with an encoder-decoder architecture to perform this task and adopted a variational auto-encoder (VAE) to regularize the shared encoder. Jiang *et al.* [9] proposed a novel two-stage cascaded U-Net, which is composed of a coarse segmentation network and a fine segmentation network, and won the first place in the Brain Tumor segmentation Challenge (BraTS) 2019. Zhao *et al.* [17] introduced a self-ensemble architecture, which makes predictions at each scale of U-Net and then combines them to obtain the final prediction. McKinley [11] introduced a modification of DeepSCAN, in which the batch normalization is replaced by the instance normalization and a lightweight local attention mechanism is also used.

As for survival prediction, deep learning techniques are seldom used due to the lack of sufficient training data. Instead, most survival prediction methods have three major steps, including hand-crafted feature extraction, feature dimension reduction, and classification. Li *et al.* [15] selected more than 40 features and applied a random forest to them for the prediction of survival days. Agravat *et al.* [1] extracted the statistical features and shape features, including the volume of necrosis, tumor enhancement, edema, and whole tumor, tumor range, elongation, flatness, axial length, diameter, surface area, and patient age. Based on these features, they constituted a random forest to predict the survival time, and achieved the first place in BraTS 2019 with an accuracy of 58.6% on the validation dataset. Dai *et al.* [16] extracted radiation characteristics and physical characteristics, combined them with the age feature, and adopted a regression mode to predict survival days. Feng *et al.* [7] extracted physical features, including the volume and gradient features, and predicted the survival time by applying a linear regression model to these features.

Despite their improved performance, these methods were designed either for brain tumor segmentation or for survival prediction. However, there is a strong correlation between the tumor status and survival time. Therefore, we believe both the segmentation and survival prediction tasks may mutually benefit each other, and their performance can possibly be improved together. In this paper, we propose a segmentation then prediction (STP) model for brain tumor segmentation and survival prediction. We use 3D U-Net as the segmentation backbone and use both local and global features for survival prediction. The local features are extracted by the last layer of the segmentation encoder, while the global features are produced by a global branch, which uses 3D ResNet-50 as its backbone.

In the STP model, segmentation module and local branch are trained together in an end-to-end manner. The global branch is trained based on results of segmentation module. The STP model is jointly optimized for two tasks, which improves the performance of segmentation and the accuracy of survival prediction. We evaluated the proposed STP model on the BraTS 2020 validation dataset and achieved an average Dice similarity coefficient (DSC) of 0.790, 0.910, 0.851 for the segmentation of enhanced tumor core, whole tumor, and tumor core, respectively, and an accuracy of 65.5% for survival prediction.

2   Method

Our proposed STP model containing a segmentation module and a survival prediction module is shown in Fig. 1. In STP model, we train the segmentation module in patches, which means that we usually extract the local features of tumor. We also propose a global branch to extract the global features of whole tumor to predict survival days. We show this branch in the bottom part of Fig. 1. We then delve into the details of segmentation module and survival prediction module, respectively.

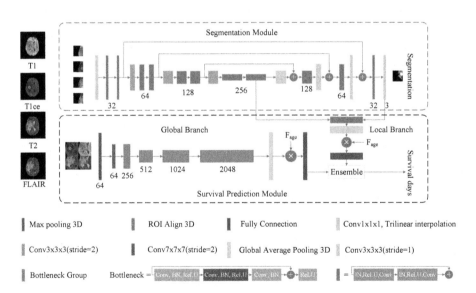

Fig. 1. Diagram of our proposed STP model, which is composed of a segmentation module (top) and a survival prediction module (bottom). The final results of prediction module are an ensemble of local branch and global branch.

2.1 Segmentation Module

As shown in the top part of Fig. 1, the segmentation module is composed of an encoder and a decoder.

The encoder takes four MR sequences as input and consists of a convolutional layer with kernel size of $3 \times 3 \times 3$ and stride of 1, eight convolutional blocks, and three downsampling layers, which are represented by yellow squares, blue squares, and orange squares, respectively, in Fig. 1. Each convolutional block contains sequentially an instance normalization layer, a ReLU layer, a convolutional layer, another instance normalization layer, another ReLU layer, another convolutional layer, and a residual connection [8]. After every two successive convolutional blocks, there is a downsampling layer, in which the kernel size is $3 \times 3 \times 3$ and the stride is 2. It should be noted that we replace the group normalization with the instance normalization due to the limited GPU memory.

The segmentation decoder has three successive pairs of upsampling layers (represented by light green squares in Fig. 1) and convolutional blocks, followed by a output layer. An upsampling layer contains convolutional kernels of size $1 \times 1 \times 1$, followed by the trilinear interpolation. After each upsampling layer, the skip connection is adopted to combine the feature maps produced by the upsampling layer with the same-size feature maps generated by the encoder. In each convolutional block, the number of channels is reduced to half of the channels in the previous block. The output layer contains $3 \times 3 \times 3$ convolutional kernels and three channels, which are expected to produce the segmentation mask of the enhancing tumor core, whole tumor, and tumor core, respectively.

2.2 Survival Prediction Module

In our STP model, the survival prediction module is composed of a local branch and a global branch. The prediction produced by these two branches are fused to generate the final result.

Local Branch. The output of segmentation module has three channels, which means probabilities of the enhancing tumor (ET), the whole tumor (WT) and the tumor core (TC), respectively. So, we get three masks for the different sub-tumors. We then get three bounding boxes, which are minimum cubes included different sub-tumors. We apply the ROI Alignment [14] to the outputs of encoder and get the features. Although three bounding boxes have different size, we finally get features with the same size as $256 \times 3 \times 3 \times 3$. We then use Global Average Pooling 3D to get the local tumor features, which size is $256 \times 1 \times 1 \times 1$. We concatenate three sub-tumor local features and reshape them into one-dimension vector and it is multiplied by the F_{age}. According to a fully connection layer, we get the result of survival prediction from local branch.

Global Branch. Our purposed global branch of survival prediction module based on 3D ResNet 50 is used to predict survival days. As shown in the bottom part of Fig. 1, the proposed global branch is composed of a convolutional layer, a max pooling layer, 4 bottleneck groups, a global average pooling layer and a fully connected layer. The convolutional layer with kernel size $7 \times 7 \times 7$ takes the combined MR image as input and its stride is 2. The kernel size of max pooling layer is $3 \times 3 \times 3$, and the stride is 2. The bottleneck groups include 3 bottlenecks, 4 bottlenecks, 6 bottlenecks, and 3 bottlenecks, respectively. There is a shortcut and three combinations of a convolutional layer, a batch normalization layer and a ReLU layer in each bottleneck. The kernel size of convolutional layer in the first and the third combination is $1 \times 1 \times 1$. The kernel size of convolution in the second combination is $3 \times 3 \times 3$. Then we use global average pooling layer to get global tumor features which are extracted from the whole tumor and F_{age} is multiplied with the global features. After a fully connection layer, we can get the survival prediction from global branch.

2.3 Hybrid Loss Function

Loss of Segmentation. For the segmentation task, the loss function is defined as follows:

$$L_{seg} = 1 - L_{Dice} + L_{BCE} \tag{1}$$

where L_{Dice} is the DSC metric, and L_{BCE} is the binary cross entropy (BCE). DSC measures the similarity between a segmentation mask S and a ground truth G as follows

$$L_{Dice} = \frac{2\,S \cap G|}{|S| + |G|} \tag{2}$$

It takes a value from the range $[0, 1]$, and a higher value represents a more accurate segmentation result. The BCE between a segmentation mask S and a ground truth G can be calculated as follows

$$L_{BCE} = \frac{1}{N} \sum_{i=1}^{n} -[G_i \log(S_i) + (1 - G_i) \log(1 - S_i))] \tag{3}$$

The value of BCE decreases if the a segmentation mask S gets closer to the ground truth G.

Loss of Survival Prediction. The mean square error (MSE) between a prediction P_i and its ground truth G_i can be calculated as follows

$$L_{MSE}(P, G) = \frac{\sum_{i=1}^{N}(P_i - G_i)^2}{N} \tag{4}$$

Obviously, a smaller MSE means a more accurate survival prediction result.

In our global branch of survival prediction module, the prediction results are classified firstly before calculating the loss function. If the categories of prediction results are inconsistent with their ground truths, the loss function is calculated for the data in inconsistent categories. When the prediction results in a batch are in correct categories, the loss function is calculated for all data in this batch. The structure of the loss function is shown as follows:

$$L_{global} = \begin{cases} L_{MSE}\left(P,G\right), & \forall class\left(P_i\right) = class\left(G_i\right) \\ L_{MSE}\left(P_{unclass}, G_{unclass}\right), & \exists class\left(P_i\right) \neq class\left(G_i\right) \end{cases} \quad (5)$$

where $P_{unclass}$ is a set of prediction results whose categories are inconsistent with their ground truths, $G_{unclass}$ is a set of their corresponding ground truths. The structure of $class(\cdot)$ is shown as follows:

$$class(P_i) = \begin{cases} 0, P_i < 10 \\ 2, P_i > 15 \\ 1, else \end{cases} \quad (6)$$

where $class(\cdot)$ is a classification of survival months, which can be divided into three categories: less than 10 months, more than or equal to 10 months and less than or equal to 15 months and greater than 15 months.

The local features from our segmentation module can just represent a part of the whole tumor. Since the predictions of local features are unconvincing, we purpose a confidence degree w_2 to represent the proportion of segmented tumor in each patch to the whole tumor. We design a loss for our local branch in survival prediction module, shown as follows:

$$L_{local} = w_2 \times L_{MSE}(P,G) \quad (7)$$

When the segmented result of whole tumor is closer to the ground truth, the value of w_2 is larger. w_2 is set as follows:

$$w_2 = \frac{\sum (S_P \cap G_p)_i}{\sum G_i} \quad (8)$$

where S_p and G_p are patches of prediction results and ground truth.

Since local features are based on the results of segmentation module,for this study, we jointly use all these metrics and propose a following hybrid loss to train our segmentation module and the local branch of our survival prediction module.

$$L_{joint} = L_{seg} + L_{local} \quad (9)$$

2.4   Testing

In the testing stage, we need two steps to get the final results. We first feed MR sequences to the segmentation module to generate a segmentation mask and local tumor features. Then we get the survival days from survival prediction module, ensemble the results of two branches to get result of survival prediction.

It should be noted that there might be no enhanced tumor core in many LGG cases. Therefore, if the enhanced tumor core we detected is smaller than 350 voxels, we believe that there might be no enhanced tumor core and hence remove those detected voxels from the segmentation result.

3 Dataset

The 2020 Brain Tumor Segmentation (BraTS 2020) challenge is focusing on the evaluation of state-of-art methods for the segmentation of brain tumors and survival prediction in multi-sequence MR scans [2–5,12]. The BraTS 2020 dataset was collected from multiple institutions with the different clinical protocols and various scanners. This dataset has been partitioned into a training dataset, a validation dataset and a testing dataset, which have 369,125 and 166 cases. Each case has four MR sequences (i.e. FLAIR, T1, T1ce, and T2) and training dataset is annotated by professional radiologists. Each annotation contains the non-enhancing tumor core, tumor core, and enhancing tumor. All MR sequences have been interpolated to the voxel size of $1 \times 1 \times 1$ mm^3 and the dimension of $155 \times 240 \times 240$, and have been made publicly available. It's noted that the annotations of validation cases and testing cases are withheld for online evaluation. There are 236 cases in the survival task. We can get some information such as age, survival days, extent of resection and grade of tumor. There are 118 GTR (i.e., Gross Total Resection) cases and 118 other extent of resection cases.

4 Experiments and Results

4.1 Data Processing

Every MR sequence is normalized by its mean and standard deviation. Simple data augmentation techniques, such as random scaling with a scaling factor from $[0.9, 1.1]$ and random axis mirror flipping, are employed to increase the number of training samples.

The survival days are converted to months of survival. Then survival months are normalization as $x = (x - x_{min})/(x_{max} - x_{min})$. Since age is related to the recovery of patients' prognosis, the age feature is purposed to multiply with feature map before the fully connection layer. The structure of age feature is shown in Eq. 10.

$$F_{age} = 1.0 - \frac{age}{100} \tag{10}$$

4.2 Experiment Settings

We aim to predict survival days and segment each brain tumor into three subregions, including the tumor core, enhancing tumor, and whole tumor (i.e. the tumor core, enhancing tumor, and non-enhancing tumor core). Our STP model is implemented in PyTorch and trained on a desktop with a NVIDIA GTX 1080Ti GPU. Our segmentation module takes four input volumes of size $80 \times 160 \times 160$, and is optimized by the Adam optimizer. We set the batch size to 1, the maximum epochs to 200 and the initial learning rate to $lr = 1e - 4$, which is decayed as $lr = lr \times (1 - epoch/epochs)^{0.9}$.

In our global branch, since we use a pretrained model by MedicalNet [6] (input channel = 1), we crop four MR images based on whole tumors which are from their segmentation result, zoom them to $56 \times 112 \times 112$ and combine them to $56 \times 224 \times 224$. The global branch is optimized by the SGD optimizer. We set the batch size to 6, the maximum epochs to 3000 and the initial learning rate to $lr = 1e - 6$.

4.3   Results of Segmentation

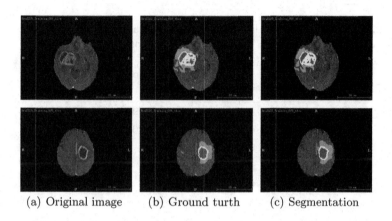

(a) Original image   (b) Ground turth   (c) Segmentation

Fig. 2. Segmentation results on BraTS 2020 training dataset.

We evaluated our method on the BraTS 2020 validation dataset. The test time augmentation (TTA) was used in this experiment. Figure 2 shows four transverse slices from a T1ce sequence, ground truth and the corresponding segmentation results produced by our model. In each result, the segmented the enhancing tumor, tumor core, and non-enhancing tumor core were highlighted in red, yellow, and green, respectively. Table 1 shows the average DSC and Hausdorff distance obtained by baseline and our STP model. Our single model achieved the averaged DSC of 0.776, 0.906, 0.830 for the the enhanced tumor core, whole tumor, and tumor core, respectively. Compared with the baseline model, our single model improves the DSC by 1.2% for the segmentation of tumor core and decrease hausdorff distance for ET and WT. Then we ensembled a set of nine our models (trained by the same hyperparameters) and postprocessing of connected component analysis to further improve the performance.

Finally, we verified the proposed STP model on the BraTS 2020 testing dataset and displayed the obtained performance in Table 2. It shows that our model achieves an average DSC of 0.780, 0.874, and 0.830 in the segmentation of enhancing tumor, whole tumor, and tumor score, respectively.

Table 1. Performance of our STP model and its baseline on BraTS 2020 validation dataset (ET: Enhancing Tumor, WT: Whole Tumor, and TC: Tumor Core).

	Dice			Hausdorff		
Model	ET	WT	TC	ET	WT	TC
Baseline	0.7741	0.9043	0.8188	32.5728	5.456	6.7842
STP	0.7756	0.9061	0.8303	26.8573	9.8114	12.7711
Ensemble of STP	0.7900	0.9096	0.8507	23.5650	5.1929	6.1816

Table 2. Performance of our BraTSeg model on BraTS 2020 testing dataset (ET: Enhancing Tumor, WT: Whole Tumor, and TC: Tumor Core).

	DSC			Hausdorff Distance		
	ET	WT	TC	ET	WT	TC
Mean	0.7798	0.8744	0.8295	22.9699	6.4236	20.2642
StdDev	0.2346	0.1611	0.2537	84.3320	12.0924	74.8721
Median	0.8545	0.9231	0.9221	1.5414	2.9142	2.2361
25quantile	0.7650	0.8753	0.8696	1	1.7321	1.4142
75quantile	0.9113	0.9505	0.9573	2.4495	5.3852	4.1231

4.4 Results of Survival Prediction

We verified our methods of survival prediction module on the BraTS 2020 validation dataset and displayed the obtained performance in Table 3. In this experiment, we use TTA by flipping MR images. Table 3 shows that our local branch of STP model achieves 55.2% accuracy, which has been a competitive results. Our global branch of STP model achieves 62.1% accuracy. Our ensemble of local and global branch achieves 65.5% accuracy in survival prediction. It means that the fusion of the local information and global information plays a positive role. The performance is generally consistent with that obtained on the validation dataset.

Table 4 shows the obtained performance of survival prediction moudle on the BraTS 2020 testing dataset. Our ensemble of local and global branch achieves 44.9% accuracy in survival prediction task.

Table 3. Performance of our models on BraTS 2020 validation dataset.

Model	Accuracy	MSE	medianSE	stdSE	SpearmanR
Local branch	0.552	114966.6	26110.94	220758	0.348
Global branch	0.621	74269.57	**25042.1**	**115255.9**	0.485
Ensemble	**0.655**	**71529.95**	25869.75	125672	**0.584**

Table 4. Performance of our model on BraTS 2020 testing dataset.

Model	Accuracy	MSE	medianSE	stdSE	SpearmanR
Ensemble	0.449	428141.393	49320.448	1307984.92	0.43

5   Conclusion

This paper proposes a STP model for brain tumor segmentation and survival prediction using multi-sequence MR scans. The performance of results on the BraTS 2020 validation dataset suggests that the dice of segmentation can be improved by local branch of survival prediction module. Meanwhile, local features produced by segmentation module can combine with global features to further improve the accuracy of survival prediction. Our future work will focus on extending the proposed model to embedded statistics feature in survival prediction branches, i.e. surface and volume characteristics.

Acknowledgement. This work was supported in part by the Science and Technology Innovation Committee of Shenzhen Municipality, China, under Grants JCYJ20180306171334997, in part by the National Natural Science Foundation of China under Grants 61771397, and in part by Seed Foundation of Innovation and Creation for Graduate Students in NPU under Grants CX2020024. We appreciate the efforts devoted by BraTS 2020 Challenge organizers to collect and share the data for comparing brain tumor segmentation algorithms for multi-sequence MR sequences.

References

1. Agravat, R.R., Raval, M.S.: Brain tumor segmentation and survival prediction. In: Crimi, A., Bakas, S. (eds.) BrainLes 2019, Part I. LNCS, vol. 11992, pp. 338–348. Springer, Cham (2020). https://doi.org/10.1007/978-3-030-46640-4_32
2. Bakas, S., et al.: Segmentation labels and radiomic features for the pre-operative scans of the TCGA-GBM collection. The cancer imaging archive (2017)
3. Bakas, S., et al.: Segmentation labels and radiomic features for the pre-operative scans of the TCGA-LGG collection. The Cancer Imaging Archive **286** (2017)
4. Bakas, S., et al.: Advancing the cancer genome atlas glioma MRI collections with expert segmentation labels and radiomic features. Sci. Data **4**, 170117 (2017)
5. Bakas, S., et al.: Identifying the best machine learning algorithms for brain tumor segmentation, progression assessment, and overall survival prediction in the brats challenge. arXiv preprint arXiv:1811.02629 (2018)
6. Chen, S., Ma, K., Zheng, Y.: Med3d: Transfer learning for 3d medical image analysis. arXiv preprint arXiv:1904.00625 (2019)
7. Feng, X., Tustison, N.J., Patel, S.H., Meyer, C.H.: Brain tumor segmentation using an ensemble of 3D U-Nets and overall survival prediction using radiomic features. Front. Comput. Neurosci. **14**, 25 (2020)
8. He, K., Zhang, X., Ren, S., Sun, J.: Deep residual learning for image recognition. In: Proceedings of the IEEE Conference on Computer Vision and Pattern Recognition, pp. 770–778 (2016)

9. Jiang, Z., Ding, C., Liu, M., Tao, D.: Two-stage cascaded U-Net: 1st place solution to BraTS challenge 2019 segmentation task. In: Crimi, A., Bakas, S. (eds.) BrainLes 2019, Part I. LNCS, vol. 11992, pp. 231–241. Springer, Cham (2020). https://doi.org/10.1007/978-3-030-46640-4_22

10. Kamnitsas, K., et al.: Ensembles of multiple models and architectures for robust brain tumour segmentation. In: Crimi, A., Bakas, S., Kuijf, H., Menze, B., Reyes, M. (eds.) BrainLes 2017. LNCS, vol. 10670, pp. 450–462. Springer, Cham (2018). https://doi.org/10.1007/978-3-319-75238-9_38

11. McKinley, R., Rebsamen, M., Meier, R., Wiest, R.: Triplanar ensemble of 3D-to-2D CNNs with label-uncertainty for brain tumor segmentation. In: Crimi, A., Bakas, S. (eds.) BrainLes 2019. LNCS, vol. 11992, pp. 379–387. Springer, Cham (2020). https://doi.org/10.1007/978-3-030-46640-4_36

12. Menze, B.H., et al.: The multimodal brain tumor image segmentation benchmark (BRATS). IEEE Trans. Med. Imaging **34**(10), 1993–2024 (2014)

13. Myronenko, A.: 3D MRI brain tumor segmentation using autoencoder regularization. In: Crimi, A., Bakas, S., Kuijf, H., Keyvan, F., Reyes, M., van Walsum, T. (eds.) BrainLes 2018, Part II. LNCS, vol. 11384, pp. 311–320. Springer, Cham (2019). https://doi.org/10.1007/978-3-030-11726-9_28

14. Ramien, G.N., Jaeger, P.F., Kohl, S.A.A., Maier-Hein, K.H.: Reg R-CNN: lesion detection and grading under noisy labels. In: Greenspan, H., et al. (eds.) CLIP/UNSURE -2019. LNCS, vol. 11840, pp. 33–41. Springer, Cham (2019). https://doi.org/10.1007/978-3-030-32689-0_4

15. Sun, L., Zhang, S., Chen, H., Luo, L.: Brain tumor segmentation and survival prediction using multimodal MRI scans with deep learning. Front. Neuurosci. **13**, 810 (2019)

16. Wang, S., Dai, C., Mo, Y., Angelini, E., Guo, Y., Bai, W.: Automatic brain tumour segmentation and biophysics-guided survival prediction. In: Crimi, A., Bakas, S. (eds.) BrainLes 2019, Part II. LNCS, vol. 11993, pp. 61–72. Springer, Cham (2020). https://doi.org/10.1007/978-3-030-46643-5_6

17. Zhao, Y.-X., Zhang, Y.-M., Liu, C.-L.: Bag of tricks for 3D MRI brain tumor segmentation. In: Crimi, A., Bakas, S. (eds.) BrainLes 2019, Part I. LNCS, vol. 11992, pp. 210–220. Springer, Cham (2020). https://doi.org/10.1007/978-3-030-46640-4_20

Enhancing MRI Brain Tumor Segmentation with an Additional Classification Network

Hieu T. Nguyen[1,2(✉)], Tung T. Le[1,3], Thang V. Nguyen[1],
and Nhan T. Nguyen[1]

[1] Medical Imaging Department, Vingroup Big Data Institute (VinBigdata),
Hanoi, Vietnam
[2] Hanoi University of Science and Technology (HUST), Hanoi, Vietnam
hieu.nt170073@sis.hust.edu.vn
[3] University of Engineering and Technology (UET), VNU, Hanoi, Vietnam

Abstract. Brain tumor segmentation plays an essential role in medical image analysis. In recent studies, deep convolution neural networks (DCNNs) are extremely powerful to tackle tumor segmentation tasks. We propose in this paper a novel training method that enhances the segmentation results by adding an additional classification branch to the network. The whole network was trained end-to-end on the Multimodal Brain Tumor Segmentation Challenge (BraTS) 2020 training dataset. On the BraTS's test set, it achieved an average Dice score of 80.57%, 85.67% and 82.00%, as well as Hausdorff distances (95%) of 14.22, 7.36 and 23.27, respectively for the enhancing tumor, the whole tumor and the tumor core.

Keywords: Deep learning · Brain tumor segmentation · FPN · U-Net

1 Introduction

Gliomas are the most common primary brain malignancies, with different degrees of aggressiveness, variable prognosis and various heterogeneous histological sub-regions [1–3]. One objective of The Brain Tumor segmentation (BraTS) challenge is to identify state-of-the-art machine learning methods for segmentation of brain tumors in magnetic resonance imaging (MRI) scans [4,19]. One MRI data sample of BraTS consists of a native T1-weighted scan (T1), a post-contrast T1-weighted scan (T1Gd), a native T2-weighted scan (T2), and a T2 Fluid Attenuated Inversion Recovery (T2-FLAIR) scan. However, each tumor-region-of-interest (TRoI) is visible in one pulse. Specifically, the whole tumor is visible in T2-FLAIR, the tumor core is visible in T2, and the enhancing tumor is visible in T1Gd.

This work was done when Hieu Nguyen and Tung Le were AI Interns at Medical Imaging Department, Vingroup Big Data Institute (VinBigdata).
H. T. Nguyen and T. T. Le—These authors share first authorship on this work.

© Springer Nature Switzerland AG 2021
A. Crimi and S. Bakas (Eds.): BrainLes 2020, LNCS 12658, pp. 503–513, 2021.
https://doi.org/10.1007/978-3-030-72084-1_45

An accurate deep learning segmentation model not only can save time for neuroradiologists but also can provide a reliable result for further tumor analysis. Recently, deep learning approaches have consistently surpassed traditional computer vision methods [6,11,22,24,27]. Specifically, convolutional neural networks (CNN) are able to learn deep representative features to generate accurate segmentation mask both in 2D and 3D medical images.

The BraTS 2020 training dataset, which comprises 369 cases for training and 125 cases for validation, is manually annotated by both clinicians and board-certified radiologists. Each tumor is segmented into enhancing tumor, peritumoral edema, and the necrotic and non-enhancing tumor core. To evaluate the segmentation performance, various metrics are used: Dice score, Hausdorff distance (95%), sensitivity and specificity.

Since the introduction of U-Net [23] in 2015, various types of U-shape DCNN have been proposed and gained significant results in medical image segmentation tasks. In BraTS 2017, Kamnitsas et al. [10], who was the winner of the segmentation challenge, explored Ensembles of Multiple Models and Architecture (EMMA) for robust performance by combining several DCNNs including DeepMedic [11], 3D FCN [17] and 3D U-Net [5]. In BraTS 2018, Myronenko [21], who won segmentation track, utilized asymmetrically large encoder to extract deep image features, and the decoder part reconstructs dense segmentation masks. The authors also added the variational autoencoder (VAE) branch in order to regularize the network. In BraTS 2019, Jiang et al. [9], who recently achieved the highest score on the private test set, deployed two-stage cascaded U-Net which basically stacked 2 U-Net networks together. In the first stage, they used a variant of U-Net to train a coarse prediction. In the next stage, they increased the network capacity by using 2 decoders simultaneously. The model was trained in an end-to-end manner and achieved the best result.

Contribution. Through exploratory model analysis after training, we notice that deep learning segmentation models sometimes make false positive predictions. To bridge the gap between segmentation model efficiency and avoid these problems, we proposed a novel end-to-end training method by combining both segmentation and classification. The classification branch helps to predict whether a mask slice contains region of interest as well as to regularize the segmentation branch. We explored this approach with 2 architectures which are variant of nested U-Net [28] and Bi-directional Feature Pyramid Network (BiFPN) [25]. Our method achieved Dice score of 80.57%, 85.67% and 82.00% respectively for the enhancing tumor, the whole tumor and the tumor core on the test dataset of the 2020 BraTS challenge[1].

2   Methods

In this section, we describe the proposed approach in which two different models, BiFPN and Nested U-Net, are leveraged as the base segmentation architecture,

---

[1] https://www.med.upenn.edu/cbica/brats2020/data.html.

enhanced by a classifier head. While the segmentation head largely relies on local features to segment tumor area, the classification branch leverages global features of the whole slice as well as neighbors slices to aid segmentation task. The main advantage of classification head is that it significantly reduces false positive regions since minute, high intensity regions of enhancing tumor are often confused with other non-tumor, high intensity regions. In addition, to tackle small batch size problem when using batch-norm, we deploy Group Normalization [26] with number groups of 8 instead.

2.1 Bi-directional Feature Pyramid Network

In this approach, an encoder-decoder based network is leveraged [12,28] with an additional classification branch to further enhance segmentation results. The classification head is placed at the end of the encoder to classify whether an image slice has tumor region. In the following subsections, we describe the details of the encoder and decoder parts (see Fig. 1).

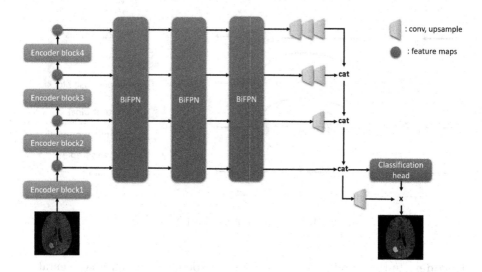

Fig. 1. Overview of the BiFPN architecture with classifier.

Encoder. For the encoder part, we exploited residual block [7] for features extraction with the number of channels being doubled after each convolutional layer of stride 2, which results in multi-scale features maps for the latter part. There are four scales of feature maps, where the smallest one were 16 times smaller than the input image (see Table 1 for the details of the feature extractor). In order to combine features of multiple sizes, we adapted the BiFPN layer from EfficientDet architecture [25] (see Fig. 2), which was an improved version of the Feature Pyramid networks [14]. We used three consecutive BiFPN layers with feature dimensions of 256, as deeper networks did not improve performance.

Table 1. Details of the feature extractor, where *conv3* is $3 \times 3 \times 3$ convolution and GN denotes group norm. Note that output shape of encoder blocks correspond to input image of shape $4 \times 128 \times 128 \times 96$

Block	Details	Repeat	Output size
Encoder block1	(conv3 stride2, GN, dropout, ReLU, conv3 stride1, GN, dropout, ReLU) + conv3 stride2	1	$16 \times 64 \times 64 \times 48$
Encoder block2	(conv3 stride2, GN, dropout, ReLU, conv3 stride1, GN, dropout, ReLU) + conv3 stride2	1	$32 \times 32 \times 32 \times 24$
Encoder block3	(conv3 stride2, GN, dropout, ReLU, conv3 stride1, GN, dropout, ReLU) + conv3 stride2	1	$64 \times 16 \times 16 \times 12$
Encoder block4	(conv3 stride2, GN, dropout, ReLU, conv3 stride1, GN, dropout, ReLU) + conv3 stride2	1	$128 \times 8 \times 8 \times 6$

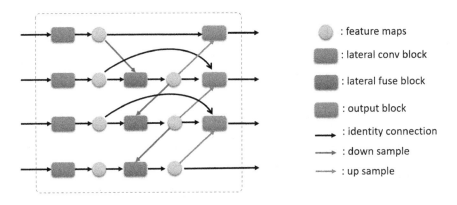

Fig. 2. Illustration of the BiFPN layer.

Decoder. In the decoder part, we followed the design of semantic segmentation branch of the Panoptic Feature Pyramid networks [12]. Each feature map from the BiFPN layers was put into a number of up-sample blocks, depending on the spatial size. Each up-sample block consists of a $3 \times 3 \times 3$ convolution, group-norm and ReLU, followed by $2\times$ trilinear interpolation, and the feature dimension is fixed to 256. Due to GPU memory constraint, all feature maps were up-sampled to a common size, which is half of the input image size, then were concatenated before putting into the final upsample block, which has a $1 \times 1 \times 1$ convolution with 3 filters corresponding to three classes of tumor regions, and subsequently a $2\times$ trilinear interpolation layer.

2.2   Nested U-Net

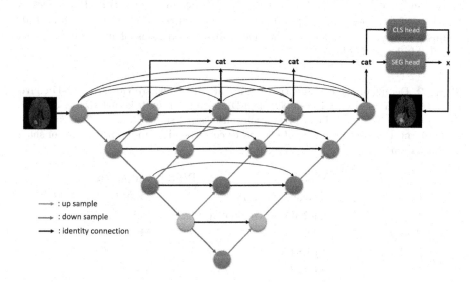

Fig. 3. Overview of the nested UNET architecture with an additional classifier.

Skip Pathways. According to Zhou et al. [28], nested U-Net (UNet++) proposed dense convolution block whose number of convolution layers depend on the pyramid level. Therefore, they re-designed skip pathways to bring the semantic level of the encoder feature maps closer to that of the feature maps awaiting in the decoder. Figure 3 clarifies how the feature maps travel through the top skip pathway of UNet++.

Deep Supervision. In order to take advantage of lower feature maps, we used deep supervision [13] wherein the final segmentation map was selected from all segmentation branches averaged. Instead of using another layer before upsampling to the map size, our final prediction mask upsamples directly from last layer. Each layer contains ReLU activation, followed by 2 convolution layers.

2.3   Classification Branch

In all segmentation architectures, the concatenated feature maps before the final blocks were used as the input for the classification head. While the feature maps for classifier in UNET++ have the same spatial size as the input image, its counterpart in BiFPN are only half of the input, leading to an additional upsample layer in the classification branch of the BiFPN. The classification branch includes: (1) a $3 \times 3 \times 3$ convolution block to reduce feature channels; (2) a global average pooling layer that averages feature maps over frontal and sagittal axes to produce slice-wise feature maps along axial axis; (3) a transpose convolution block which is used to upsample axial axis of the feature maps to match input image size (for BiFPN only); (4) a sequence of several BiLSTM layers [8] which leverages inter-slice dependence from both directions; and (5) a final fully connected layer to classify whether each slice has regions of interested classes. (see Table 2 and Table 3)

Table 2. Details of the classification branch in BiFPN, where *conv1d1* is 1-D convolution with kernel size of 1, *conv3d3* is 3-D convolution with kernel size of $3 \times 3 \times 3$, *tconv3d3* is 3-D transpose convolution with kernel size of $3 \times 3 \times 3$, GN denotes group-norm. Here output shape of each layer corresponding to input features of shape $1024 \times 64 \times 64 \times 48$.

Names	Details	Repeat	Output size
Conv block	conv3d3, GN, ReLU	1	$512 \times 64 \times 64 \times 48$
Pool	Global average pooling	1	512×48
TransConv block	tconv3d3, GN, ReLU	1	512×96
BiLSTM	Bi-LSTM with dropout	2	1024×96
FC	conv1d1	1	3×96

Table 3. Details of the classification branch in U-NET++. Here output shape of each layer corresponding to input features of shape $128 \times 128 \times 128 \times 128$.

Names	Details	Repeat	Output size
Conv block	conv3d3, BN, ReLU	1	$256 \times 128 \times 128 \times 128$
Pool	Global average pooling	1	256×128
BiLSTM	Bi-LSTM with dropout	3	512×128
FC	conv1d1	1	3×128

2.4   Losses

Segmentation Loss

Dice loss. Dice loss originates from Sørensen–Dice coefficient, which is a statistic developed in 1940s to gauge the similarity between two samples. It was brought in V-Net paper [20]. The Dice Similarity Coefficient (DSC) measures the the degree of overlap between the prediction map and ground truth based on dice coefficient, which is a quantity ranging between 0 and 1 which we aim to maximize. The Dice loss is calculated as in Eq. 1.

$$\mathcal{L}_{dice} = 1 - \frac{2\sum_{i}^{N} p_i g_i}{\sum_{i}^{N} p_i^2 + \sum_{i}^{N} g_i^2 + \epsilon}, \tag{1}$$

where p_i is predicted voxels, and g_i is ground truth. The sums run over N voxels and we add a small $\epsilon = 1e-5$ to avoid zero denominator.

Focal Loss. To deal with large class imbalance in the segmentation problem, we also used focal loss [15] to penalize more on wrongly segmented regions

$$\mathcal{L}_{focal}(p_t) = -\alpha_t(1 - p_t)^\gamma \log(p_t). \tag{2}$$

We directly optimized the label regions (whole tumor, tumor core, and enhancing tumor) with the losses.

Classification Loss. For the classification branch, we used **focal loss** and standard **binary cross entropy loss**.

Finally, we summed over all the losses to obtain the final loss.

$$\mathcal{L}_{total} = \mathcal{L}_{focal\_seg} + \mathcal{L}_{dice} + \mathcal{L}_{focal\_cls} + \mathcal{L}_{BCE}. \tag{3}$$

3 Experiments

3.1 Data Pre-processing and Augmentation

For preprocessing, we cropped out zero-intensity regions in order to reduce the image size as well as discard the out-of-interest regions in training (see Fig. 4). To prevent the network from overfitting, we executed several types of data augmentation. Firstly, we applied random flip with probability of 0.5 in every spatial axis. Then, we applied a random scale intensity shift of input in the range of $[0.9, 1.1]$. Finally, we applied random intensity shift of input between offset $[-0.1, 0.1]$. Both these shift augmentations were applied with the probability of 0.8. We also applied random crop data with the size of $128 \times 128 \times 128$ and $128 \times 128 \times 96$ voxels due to memory limitation.

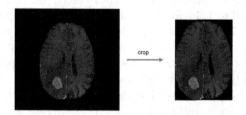

Fig. 4. Crop out zero-intensity region.

3.2   Training Details

Thanks to Pytorch 1.6, we took advantage of automatic mixed precision to save GPU memory. Adam optimizer was exploited to update model's parameters. Moreover, instead of using fixed learning rate during training, we deployed 2 learning rate schedulers (see below) which were cosine learning rate scheduler [18] and polynomial learning rate scheduler [16]. Both of the two settings have achieved the same level of performance after 200 epochs with base learning rate of $1e-3$ and 10 epochs of warming up.

Cosine Learning Rate Scheduler. Ignoring the warmup stage and assuming the total number of batches is T, initial learning rate η, the learning rate η_t at batch t is computed as in Eq. 4.

$$\eta_t = \frac{1}{2}(1 + \cos(\frac{t\pi}{T}))\eta. \tag{4}$$

Polynomial Learning Rate Scheduler. Ignoring the warm-up stage, $\eta_0 = 1e-3$ is initial learning rate, e is epoch counter, N_e is total number of epochs, the learning rate η_t at batch t is computed as in Eq. 5.

$$\eta_t = \eta_0 \times (1 - \frac{e}{N_e})^{0.9}. \tag{5}$$

The cosine learning rate scheduler was used to train in BiFPN model and the polynomial learning rate scheduler was used in nested Unet model. Each network was trained from scratch without neither pretrained weight nor external data on two NVIDIA Tesla V100 32GB RAM. While BiFPN takes a batch of 4 samples, each of shape $4 \times 128 \times 128 \times 96$, UNET++ takes a batch of 2 samples with size of $4 \times 128 \times 128 \times 128$.

3.3   Inference

To achieve a robust prediction, we applied test time augmentations (TTA) for every model before averaging them. The augmentations were different flipping in each axis. Finally, we averaged the output before put them through sigmoid function. Decision threshold for both classification and segmentation of three classes was set at 0.5. The negative predictions at slice level of the classification head were used to exclude predicted segmentation regions. While removal of small tumor regions is a simpler approach to address false positive predictions, this method is highly sensitive to volume threshold and can mistakenly exclude the actual tumor which is minute. The classification head provides a more robust solution which can significantly reduce false positive rate, at the same time less likely to miss tiny tumor.

4   Results

In this section we show the performance of the two architectures with and without the classification head (see Table 4) in terms of Dice score and Hausdorff distance (95%). The results indicate that classification head improves both architectures by a significant margin. Ensemble of 5 models of the same architecture trained on 5-folds cross-validation gives marginal improvement, while ensemble of the two architectures gives more considerable enhancement. The results of ensemble of 10 models on the testing set are given in Table 5.

Table 4. Mean Dice score and Hausdorff distance of the proposed method on BraTS 2020 validation set.

Method	Dice score			Hausdorff distance (95%)		
Validation	ET	WT	TC	ET	WT	TC
Best single Unet++ w/o cls	0.7029	0.8967	0.8239	42.2474	7.4907	9.0179
Best single Unet++ w. cls	0.7742	0.8940	0.8241	35.4246	8.4361	10.4074
Ensemble of 5-fold Unet++ w/o cls	0.7017	0.8953	0.8239	47.1436	5.8179	11.0075
Ensemble of 5-fold Unet++ w. cls	0.7841	0.8960	0.8233	35.4841	5.0862	10.0780
Best single BiFPN w/o cls	0.7480	0.8896	0.8400	31.1209	5.8924	6.9682
Best single BiFPN w. cls	0.7729	0.8881	0.8373	21.5720	6.9531	6.5573
Ensemble of 5-fold BiFPN w/o cls	0.7471	0.8915	0.8371	28.9473	5.9362	6.8706
Ensemble of 5-fold BiFPN w. cls	0.7774	0.8914	0.8380	24.6944	5.9834	6.8527
Ensemble of 10 models	**0.7843**	**0.8999**	**0.8422**	**24.0235**	**5.6808**	**9.5663**

Table 5. Mean Dice score and Hausdorff distance of the proposed method on BraTS 2020 testing set.

Method	Dice score			Hausdorff distance (95%)		
Testing	ET	WT	TC	ET	WT	TC
Ensemble of 10 models	0.80569	0.85671	0.81997	14.21938	7.35549	23.27358

5   Conclusion

In this work, we described a novel training method for brain tumor segmentation from multimodal 3D MRIs. Our results on BraTS 2020 indicated that our model is able to achieve an extremely competitive segmentation result. On the BraTS 2020 test set, the proposed method obtained an average Dice score of 80.57%, 85.67% and 82.00%, as well as Hausdorff distances (95%) of 14.22%, 7.36% and 23.27%, for the enhancing tumor, the whole tumor and the tumor core. In the future, we plan to focus on investigation of new methods for improving small region segmentation as well as classification performance of the network.

Acknowledgments. This work was highly supported by Medical Imaging Department at Vingroup Big Data Institute (VinBigdata).

References

1. Bakas, S., et al.: Segmentation labels and radiomic features for the pre-operative scans of the TCGA-LGG collection. The cancer imaging archive **286** (2017)
2. Bakas, S., et al.: Segmentation labels and radiomic features for the pre-operative scans of the TCGA-GBM collection. The cancer imaging archive. Nat Sci Data **4**, 170117 (2017)
3. Bakas, S., et al.: Advancing the cancer genome atlas glioma MRI collections with expert segmentation labels and radiomic features. Sci. Data **4**, 170117 (2017)
4. Bakas, S., et al.: Identifying the best machine learning algorithms for brain tumor segmentation, progression assessment, and overall survival prediction in the brats challenge. arXiv preprint arXiv:1811.02629 (2018)
5. Çiçek, Ö., Abdulkadir, A., Lienkamp, S.S., Brox, T., Ronneberger, O.: 3D U-Net: learning dense volumetric segmentation from sparse annotation. In: Ourselin, S., Joskowicz, L., Sabuncu, M.R., Unal, G., Wells, W. (eds.) MICCAI 2016, Part II. LNCS, vol. 9901, pp. 424–432. Springer, Cham (2016). https://doi.org/10.1007/978-3-319-46723-8_49
6. Havaei, M., et al.: Brain tumor segmentation with deep neural networks. Med. Image Anal. **35**, 18–31 (2017)
7. He, K., Zhang, X., Ren, S., Sun, J.: Deep residual learning for image recognition. In: IEEE CVPR, pp. 770–778 (2016). https://doi.org/10.1109/CVPR.2016.90
8. Hochreiter, S., Schmidhuber, J.: Long short-term memory. Neural Comput. **9**(8), 1735–1780 (1997)
9. Jiang, Z., Ding, C., Liu, M., Tao, D.: Two-stage cascaded U-Net: 1st place solution to BraTS challenge 2019 segmentation task. In: Crimi, A., Bakas, S. (eds.) BrainLes 2019, Part I. LNCS, vol. 11992, pp. 231–241. Springer, Cham (2020). https://doi.org/10.1007/978-3-030-46640-4_22
10. Kamnitsas, K., et al.: Ensembles of multiple models and architectures for robust brain tumour segmentation. In: Crimi, A., Bakas, S., Kuijf, H., Menze, B., Reyes, M. (eds.) BrainLes 2017. LNCS, vol. 10670, pp. 450–462. Springer, Cham (2018). https://doi.org/10.1007/978-3-319-75238-9_38
11. Kamnitsas, K., et al.: Efficient multi-scale 3D CNN with fully connected CRF for accurate brain lesion segmentation. Med. Image Anal. **36**, 61–78 (2017)
12. Kirillov, A., Girshick, R., He, K., Dollár, P.: Panoptic feature pyramid networks. In: Proceedings of the IEEE Conference on Computer Vision and Pattern Recognition, pp. 6399–6408 (2019)
13. Lee, C.Y., Xie, S., Gallagher, P., Zhang, Z., Tu, Z.: Deeply-supervised nets. In: Artificial Intelligence and Statistics, pp. 562–570 (2015)
14. Lin, T.Y., Dollár, P., Girshick, R., He, K., Hariharan, B., Belongie, S.: Feature pyramid networks for object detection. In: Proceedings of the IEEE Conference on Computer Vision and Pattern Recognition, pp. 2117–2125 (2017)
15. Lin, T.Y., Goyal, P., Girshick, R., He, K., Dollár, P.: Focal loss for dense object detection. In: Proceedings of the IEEE International Conference on Computer Vision, pp. 2980–2988 (2017)
16. Liu, W., Rabinovich, A., Berg, A.C.: Parsenet: Looking wider to see better. arXiv preprint arXiv:1506.04579 (2015)
17. Long, J., Shelhamer, E., Darrell, T.: Fully convolutional networks for semantic segmentation. In: Proceedings of the IEEE Conference on Computer Vision and Pattern Recognition, pp. 3431–3440 (2015)

18. Loshchilov, I., Hutter, F.: Sgdr: Stochastic gradient descent with warm restarts. arXiv preprint arXiv:1608.03983 (2016)
19. Menze, B.H., et al.: The multimodal brain tumor image segmentation benchmark (BRATS). IEEE Trans. Med. Imaging **34**(10), 1993–2024 (2014)
20. Milletari, F., Navab, N., Ahmadi, S.A.: V-net: Fully convolutional neural networks for volumetric medical image segmentation. In: 2016 Fourth International Conference on 3D Vision (3DV), pp. 565–571. IEEE (2016)
21. Myronenko, A.: 3D MRI brain tumor segmentation using autoencoder regularization. In: Crimi, A., Bakas, S., Kuijf, H., Keyvan, F., Reyes, M., van Walsum, T. (eds.) BrainLes 2018, Part II. LNCS, vol. 11384, pp. 311–320. Springer, Cham (2019). https://doi.org/10.1007/978-3-030-11726-9_28
22. Pereira, S., Pinto, A., Alves, V., Silva, C.A.: Brain tumor segmentation using convolutional neural networks in MRI images. IEEE Trans. Med. Imaging **35**(5), 1240–1251 (2016)
23. Ronneberger, O., Fischer, P., Brox, T.: U-net: Convolutional networks for biomedical image segmentation. arXiv preprint arXiv:1505.04597 (2015)
24. Shen, H., Wang, R., Zhang, J., McKenna, S.J.: Boundary-aware fully convolutional network for brain tumor segmentation. In: Descoteaux, M., Maier-Hein, L., Franz, A., Jannin, P., Collins, D.L., Duchesne, S. (eds.) MICCAI 2017, Part II. LNCS, vol. 10434, pp. 433–441. Springer, Cham (2017). https://doi.org/10.1007/978-3-319-66185-8_49
25. Tan, M., Pang, R., Le, Q.V.: Efficientdet: Scalable and efficient object detection. arXiv preprint arXiv:1911.09070 (2019)
26. Wu, Y., He, K.: Group normalization. In: Ferrari, V., Hebert, M., Sminchisescu, C., Weiss, Y. (eds.) ECCV 2018, Part XIII. LNCS, vol. 11217, pp. 3–19. Springer, Cham (2018). https://doi.org/10.1007/978-3-030-01261-8_1
27. Zhao, X., Wu, Y., Song, G., Li, Z., Zhang, Y., Fan, Y.: A deep learning model integrating FCNNs and CRFs for brain tumor segmentation. Med. Image Anal. **43**, 98–111 (2018)
28. Zhou, Z., Rahman Siddiquee, M.M., Tajbakhsh, N., Liang, J.: UNet++: a nested U-Net architecture for medical image segmentation. In: Stoyanov, D., et al. (eds.) DLMIA/ML-CDS -2018. LNCS, vol. 11045, pp. 3–11. Springer, Cham (2018). https://doi.org/10.1007/978-3-030-00889-5_1

Self-training for Brain Tumour Segmentation with Uncertainty Estimation and Biophysics-Guided Survival Prediction

Chengliang Dai[1(✉)], Shuo Wang[1(✉)], Hadrien Raynaud[1], Yuanhan Mo[1], Elsa Angelini[2], Yike Guo[1], and Wenjia Bai[1,3]

[1] Data Science Institute, Imperial College London, London, UK
{c.dai,shuo.wang}@imperial.ac.uk
[2] ITMAT Data Science Group, Imperial College London, London, UK
[3] Department of Brain Sciences, Imperial College London, London, UK

Abstract. Gliomas are among the most common types of malignant brain tumours in adults. Given the intrinsic heterogeneity of gliomas, the multi-parametric magnetic resonance imaging (mpMRI) is the most effective technique for characterising gliomas and their sub-regions. Accurate segmentation of the tumour sub-regions on mpMRI is of clinical significance, which provides valuable information for treatment planning and survival prediction. Thanks to the recent developments on deep learning, the accuracy of automated medical image segmentation has improved significantly. In this paper, we leverage the widely used attention and self-training techniques to conduct reliable brain tumour segmentation and uncertainty estimation. Based on the segmentation result, we present a biophysics-guided prognostic model for the prediction of overall survival. Our method of uncertainty estimation has won the second place of the MICCAI 2020 BraTS Challenge.

Keywords: Brain imaging · Deep learning · Tumour segmentation · Radiomics

1 Introduction

Gliomas are the most common type of malignant brain tumours, accounting for more than 60% of all brain tumors in adults [5]. Sub-regions such as necrosis (NCR), non-enhancing tumour (NET), enhancing tumour (ET) and peritumoural edema (ED) with varying biological properties coexist within the tumour, which can be captured by multi-parametric magnetic resonance imaging (mpMRI) scans. Accurate segmentation of these sub-regions on mpMRI is of clinical importance, especially for preoperative diagnosis, treatment planning and survival prediction. However, due to the highly heterogeneous shape and

---

C. Dai and S. Wang—Contributed equally.

© Springer Nature Switzerland AG 2021
A. Crimi and S. Bakas (Eds.): BrainLes 2020, LNCS 12658, pp. 514–523, 2021.
https://doi.org/10.1007/978-3-030-72084-1_46

appearance, accurate segmentation of the tumour sub-regions remains challenging and requires expertise from experienced radiologists.

Thanks to the recent development of deep learning techniques and the availability of more open medical image datasets, automatic segmentation and analysis of brain tumour has drawn a lot of attention in the recent years. Comparing to segmenting the medical images manually, a well-trained convolutional neural network (CNN) model can finish the segmentation task much faster with acceptable accuracy. However, there are still a few challenges including limited manually-annotated training data, variations in image acquisition protocols and MRI scanners etc.

Multimodal Brain Tumour Segmentation Challenge (BraTS) has been organised for many years [1–3,14] to encourage the researchers to propose their most advanced methods for automatic segmentation and survival prediction. The BraTS datasets consist of mpMRI scans for glioblastoma (GBM/HGG) and low grade glioma (LGG). The modalities include T1-weighted scan (T1), post-contrast T1-weighted scan (T1Gd), T2-weighted scan (T2) and T2 Fluid Attenuated Inversion Recovery (T2-FLAIR) scan. These scans were acquired preoperatively with different clinical protocols from multiple institutions and annotated by experienced radiologists.

Apart from challenges for image segmentation, another challenge lies in building a robust prognostic model taking accout of the image features. Data-driven radiomics approach has demonstrated promising results in prognostic modelling while the reproducibility and explainability are still questionable [17]. To integrate physiological knowledge into the prognostic modelling, biophysics-guided prognostic models were proposed to provide robust survival prediction [12,13,18].

In this paper, we tested our methods for segmentation and uncertainty estimation on the BraTS 2020 dataset and investigate the influence of attention units, loss function and post-processing on segmentation and uncertainty estimation performance. Based on the segmentation results, we also present a biophysics-guided model for survival prediction.

2   Methods

In this section, we first present our segmentation and uncertainty estimation models. Based on the tumour sub-region segmentation, we extract the tumour invasiveness feature from biophysics and develop a concise prognostic model.

2.1   Tumour Sub-region Segmentation

Background: Winning Methods in BraTS 2018 and 2019. UNet and UNet-like models are adopted by most of the top participants in BraTS 2018 and 2019. The method described in [15] won the first place in BraTS 2018 with a UNet architecture plus an additional decoder branch derived from a generic variational autoencoder (VAE). Given this modified network structure, the loss function used by [15] consists of a soft Dice loss for the segmenter branch, the KL divergence loss and reconstruction loss for the reconstruction branch. The second place of BraTS 2018 used a vanilla UNet with minor modifications [6] tailored for the BraTS dataset. The method [7] that won the first place of BraTS 2019 adopted a 2-stage training scheme. A UNet for generating coarse segmentation maps is used in the first stage. The coarse segmentation maps are then combined with the training data to refine the segmentation result from a network that is similar to the one used in [15].

Proposed Network Architectures. Although the vanilla UNet has shown great potential in various medical imaging segmentation challenges including winning the second place in BraTS 2018 [6], we found that it requires a series of optimization to the model and to the dataset in order to maximum the performance of UNet.

The attention-gated mechanism [16] has shown very promising performance in BraTS 2019 as demonstrated in [18] and it has also been proved to be effective in improving the network performance across different tasks [16] thanks to the gating mechanism of using the large spatial context (high-level features) to disambiguate task-irrelevant feature content in low-level features. Therefore, we adopted the attention UNet model (UNet-AT) shown in Fig. 1 for the challenge, which leverages the ability of attention-gated blocks to concentrate more on semantic features to achieve a better segmentation performance. Moreover, there is no need for putting too much effort to tuning the hyperparameter and pre-processing the dataset.

Instance normalisation was empirically chosen for the UNet-AT. Leaky-ReLU with a leakiness of 0.01 as the activation function.

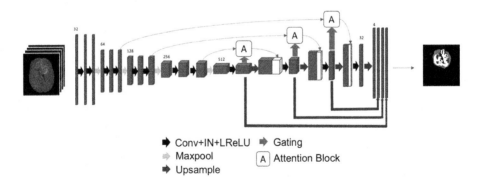

Fig. 1. 3D UNet with attention gates for tumour segmentation.

Image Pre-processing and Augmentation. We only applied z-score normalisation onto the raw images as it was suggested in [8] and some other works that different image normalisation methods have not shown significant impact

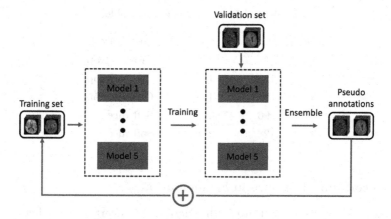

Fig. 2. Self-training scheme.

on segmentation performance. Random image rotation and horizontal flipping were performed on the fly for data augmentation during training.

Ensemble of Network Models. We trained a number of different models (Table 1) for building the ensemble, using different loss functions (cross entropy or soft Dice). Due to the constraint of GPU memory, it was not possible for us to use the whole 3D volume as the input to our networks. To tackle this issue, we applied patch extraction to the training data to create the training samples. Different patch sizes were used for the models for ensemble. The patches were randomly extracted with a 50% chance from background and the other 50% chance from any of the tumour sub-regions. Adam optimizer was used for training the models with the learning rate set to 10^{-4} and weight decay set to 10^{-5}.

Self-training. To increase the size of the training set, we adopted a simple self-training scheme shown in Fig. 2. We first trained 5 models for 20,000 iterations on the training set. The trained models were used to generate the pseudo-annotations of the validation set. Then we combined the validation set and the training set to further train the models for another 20,000 iterations just to improve the quality of the pseudo-annotations. Finally the models were trained on the training set and the improved pseudo-annotations for 20,000 more iterations. As a result, all the models were trained for 60,000 iterations in total and it took approximately 40 h to train a model on Nvidia GPU P100.

Post-processing. We empirically adopted some automatic post-processing methods to further improve accuracy of the prediction from ensemble of the models, including removing small isolated whole/enhancing tumour (smaller than 10 voxels) from the prediction, adjusting the size of tumour core in line with the size of enhancing tumour. The post-processing significantly improved the Dice score and Hausdorff distance of enhancing tumour and tumour core.

Table 1. List of models for ensemble.

Model	Patch size	Batch size	Loss function
1	96	4	Cross entropy
2	128	4	Cross entropy
3	128	4	Cross entropy
4	96	4	Soft dice
5	128	2	Soft dice

2.2 Uncertainty Measure in Segmentation

With the segmentation result, we further performed uncertainty prediction using the same models. The uncertainty maps were normalised from 0 to 100 for the following nested sub-regions: Whole Tumor (WT), Tumor Core (TC) and Enhancing Tumor (ET). To generate each uncertainty map, we take the maximum probability from our 5 models for each voxel in each glioma sub-region. The uncertainty maps are independent and for each one of them, we consider the voxels to be either part of the given nested sub-region or not, by looking at our segmentation map which was generated with the method described earlier. If a voxel belongs to the nested sub-region, we invert its probability. We multiply the probabilities by 100 to get into the 0–100 range and round them to get our uncertainty maps.

Once we have the initial uncertainty maps, we apply double thresholds to filter the result. We define low and high probability thresholds for each nested sub-region: WT(0.1, 0.3), TC(0.2, 0.3) ET(0.3, 0.5). For each voxel that belongs to a nested sub-region we set the uncertainty to 0 when the maximum probability from our 5 models is higher than the corresponding high threshold. For each voxel which does not belong to the nested sub-region, we set its uncertainty to 0 when the maximum probability is lower than the low threshold. By doing so, we can adjust the uncertainty of the 2 regions of each uncertainty maps (nested sub-region and its inverse) independently. The thresholds were determined empirically based on the probability distribution of our models for each glioma sub-region.

2.3 Invasiveness Feature Extraction

Based on the segmentation, we constructed a tumour structure map with four discrete values (1 for NCR and NET, 2 for ET, 3 for ED and 4 for normal tissue) for each patient. The spatial distribution of tumour sub-regions provides the information of tumour heterogeneity and tumour invasiveness [9]. Instead of data-driven radiomics features, we considered the biophysics modelling of tumour growth [4]. The relative invasiveness coefficient (RIC) is of particular interest, which is defined as the ratio between the hypoxic tumour core and infiltration front [12].

We adopted the biophysics-guided invasiveness feature proposed in [18], which assumes the tumour core as the hypoxic region and the whole tumour as the infiltration front, respectively. This biophysics-guided feature yielded better performance than data-driven radiomics approach in prognostic modelling. RIC derived from the tumour structure map was used to describe tumour invasiveness. The characteristic extent of each ROI is calculated from the minimum volume ellipsoid. In this study, the ratio of the first semi-axis length between TC and WT was calculated as the RIC (Fig. 3).

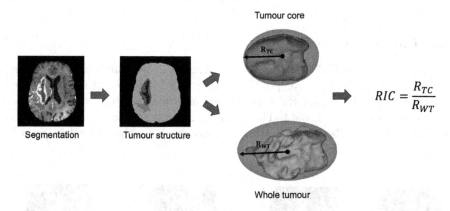

Fig. 3. Illustration of the calculation of relative invasiveness coefficient (RIC).

2.4 Prognostic Model

We used a linear regression model with age and RIC as the two predictors. Classification and regression metrics were used to evaluate the prognostic performance. For classification, overall survival time was quantitised into three survival categories: short survival (<10 months), intermediate survival (10–15 months) and long survival (>15 months). The 3-class accuracy metric wass evaluated. Evaluation metrics for regression include the mean squared error (MSE), median squared error (mSE) and Spearman correlation coefficient ρ.

3 Results

3.1 Segmentation Performance

The segmentation performance on BraTS 2020 validation dataset is reported in Table 2, in terms of Dice score and Hausdorff distance. It is quite clear that the post-processing helps to improve the model performance on BraTS dataset quite significantly and this is partially due to the evaluation metrics used by the BraTS organiser. The result on BraTS 2020 test dataset is given in Table 3.

Table 2. Segmentation result on validation set.

Model	Dice_ET	Dice_WT	Dice_TC	Hausdorff_ET	Hausdorff_WT	Hausdorff_TC
Ensemble	0.75	0.90	0.83	35.39	4.87	5.97
Post- processing	**0.79**	**0.90**	**0.84**	**26.55**	**4.19**	**5.93**

Table 3. Segmentation result on test set.

Model	Dice_ET	Dice_WT	Dice_TC	Hausdorff_ET	Hausdorff_WT	Hausdorff_TC
Post- processing	0.80	0.88	0.80	26.55	14.03	25.49

3.2 Uncertainty Prediction

We observed a trade-off between the performance of Dice area under the ROC curve (AUC) and filtered true positive/negative (FTP/FTN) AUC, so we tuned our hyperparameter to get the best overall performance. The result on the validation set is given in Table 4. The qualitative result of uncertainty estimation is given in Fig. 4. The uncertainty estimation result on the test set is given in Table 5.

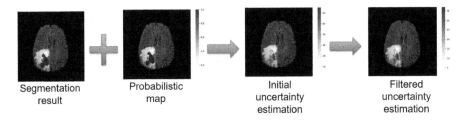

| Segmentation result | Probabilistic map | Initial uncertainty estimation | Filtered uncertainty estimation |

Fig. 4. The uncertainty estimation of the tumour core extracted from one of the scans in validation set. It is clear that the uncertainty of the edge of the tumour is refined after the filtering

3.3 Survival Prediction

The training set includes 107 patients with Gross Tumour Resection (GTR). We first trained the models on the full training set and evaluated on the training set. This overestimates the real performance so we also use internal cross-validation to assess the generalisation performance. The training set was split into five folds where models were trained on four folds and evaluated on the hold-out fold.

The performance on the external validation set with 29 patients were evaluated on the organiser's online platform. Finally, we submitted trained model for ranking on the external test set including 107 patients. The overall performance of our model is summarised in Table 6.

Table 4. Uncertainty prediction on the validation set.

DICE_AUC_WT	DICE_AUC_TC	DICE_AUC_ET
0.9213	0.8518	0.7955
FTP_RATIO_AUC_WT	FTP_RATIO_AUC_TC	FTP_RATIO_AUC_ET
5.49e−07	0.0137	0.0011
FTN_RATIO_AUC_WT	FTN_RATIO_AUC_TC	FTN_RATIO_AUC_ET
0.0047	0.0014	0.0009

Table 5. Uncertainty prediction on the test set.

DICE_AUC_WT	DICE_AUC_TC	DICE_AUC_ET
0.8948	0.8112	0.7962
FTP_RATIO_AUC_WT	FTP_RATIO_AUC_TC	FTP_RATIO_AUC_ET
4.50e−07	0.0057	0.0011
FTN_RATIO_AUC_WT	FTN_RATIO_AUC_TC	FTN_RATIO_AUC_ET
0.0073	0.0022	0.0014

4   Discussion

For tumour segmentation task, [6] demonstrated that the generic UNet with a set of carefully selected hyperparameters and a pre-processing method achieved the state-of-the-art performance on brain tumour segmentation. Our experiments results further confirm this. However, [6] adopted co-training scheme to further improve the UNet performance on segmenting enhancing tumour and our models also encountered some difficulties when segmenting enhancing tumour for some subjects. Including networks with different or more complex structure in our ensembles has been demonstrated to be effective to alleviate this issue to some extent.

The performance of our method on test dataset did not get into top three in the end. Judging from the Dice score and Hausdorff distance we achieved, we suspect including 6 models in the ensemble may not necessary and may introduce a higher false negative rate in the prediction of whole tumour.

For survival prediction, our invasiveness model achieved similar performance among the training, validation and test set. This suggests the robustness and generalisation ability of classical linear regression models. It is noted that we only use age and RIC as the predictors. This linear regression also provides an explainable prognostic model which is essential for clinical practice. Although more complex features can be integrated into a radiomics model, it did not outperform the simple model on unseen sets [18] and lack of interpretability.

Table 6. Performance of survival prediction on the training, validation and test set.

Dataset	Accuracy	MSE	mSE	Spearman ρ
Training	0.45	94737	25875	0.43
Internal cross-validation	0.44	99016	26497	0.40
Independent validation	0.45	87008	32400	0.29
Test	0.49	416229	65025	0.34

Although the accuracy of our models are not as high as the top 3 teams, we find our model provides a similar MSE, mSE and Spearman ρ metrics on the validation learderboard. This is as expected because we formulate the survival prediction as a regression problem and the accuracy is not the optimisation goal. Moreover, the model assumption of linear regression might oversimplify the survival prediction problem, especially for those subjects at tail regions of the survival distribution.

Another limitation of current invasiveness feature is the definition of tumor boundaries. Although the boundaries of whole tumor and tumor core revealed on the four structural modalities of MRI (i.e. T1-Gd, T1, T2 and T2-FLAIR) can be used to calculate the invasiveness, more advance MRI modalities including perfusion weighted imaging (PWI) and diffusion tensor imaging (DTI) could be used for better assessment of tumour heterogeneity and invasiveness [9–11].

5   Conclusion

We have developed a deep learning framework for automatic brain tumour segmentation, uncertainty estimation and a biophysics-guided prognostic model for survival prediction. Our framework performs well across all three tasks in BraTS 2020 and shows potential to be further improved in the future.

Acknowledgement. This work was supported by the SmartHeart EPSRC Programme Grant (EP/P001009/1) and the NIHR Imperial Biomedical Research Centre (BRC). We gratefully acknowledge the support of NVIDIA Corporation with the donation of the GPU used for this challenge.

References

1. Bakas, S., et al.: Segmentation labels and radiomic features for the pre-operative scans of the TCGA-GBM collection. The cancer imaging archive (2017)
2. Bakas, S., et al.: Advancing the cancer genome atlas glioma MRI collections with expert segmentation labels and radiomic features. Sci. Data **4**, 170117 (2017)
3. Bakas, S., Reyes, M., Jakab, A., Bauer, S., Rempfler, M., et al.: Identifying the Best Machine Learning Algorithms for Brain Tumor Segmentation, Progression Assessment, and Overall Survival Prediction in the BraTS Challenge. arXiv:1811.02629 (2018)

4. Baldock, A.L., Ahn, S., Rockne, R., Johnston, S., Neal, M., et al.: Patient-specific metrics of invasiveness reveal significant prognostic benefit of resection in a predictable subset of gliomas. PLoS One **9**(10), e99057 (2014)

5. Hanif, F., Muzaffar, K., Perveen, K., Malhi, S.M., Simjee, S.U.: Glioblastoma multiforme: a review of its epidemiology and pathogenesis through clinical presentation and treatment. Asian Pac. J. Cancer Prev. APJCP **18**(1), 3 (2017)

6. Isensee, F., Kickingereder, P., Wick, W., Bendszus, M., Maier-Hein, K.H.: No newnet. In: Crimi, A., Bakas, S., Kuijf, H., Keyvan, F., Reyes, M., van Walsum, T. (eds.) BrainLes 2018, Part II. LNCS, vol. 11384, pp. 234–244. Springer, Cham (2019). https://doi.org/10.1007/978-3-030-11726-9_21

7. Jiang, Z., Ding, C., Liu, M., Tao, D.: Two-stage cascaded U-Net: 1st place solution to BraTS challenge 2019 segmentation task. In: Crimi, A., Bakas, S. (eds.) BrainLes 2019, Part I. LNCS, vol. 11992, pp. 231–241. Springer, Cham (2020). https://doi.org/10.1007/978-3-030-46640-4_22

8. Kamnitsas, K., et al.: Ensembles of multiple models and architectures for robust brain tumour segmentation. In: Crimi, A., Bakas, S., Kuijf, H., Menze, B., Reyes, M. (eds.) BrainLes 2017. LNCS, vol. 10670, pp. 450–462. Springer, Cham (2018). https://doi.org/10.1007/978-3-319-75238-9_38

9. Li, C., Wang, S., Liu, P., Torheim, T., Boonzaier, N.R., et al.: Decoding the interdependence of multiparametric magnetic resonance imaging to reveal patient subgroups correlated with survivals. Neoplasia **21**(5), 442–449 (2019)

10. Li, C., et al.: Multi-parametric and multi-regional histogram analysis of MRI: modality integration reveals imaging phenotypes of glioblastoma. Eur. Radiol. **29**(9), 4718–4729 (2019). https://doi.org/10.1007/s00330-018-5984-z

11. Li, C., et al.: Intratumoral heterogeneity of glioblastoma infiltration revealed by joint histogram analysis of diffusion tensor imaging. Neurosurgery **85**, 524–534 (2018)

12. Li, C., Wang, S., Yan, J.L., Torheim, T., Boonzaier, N.R., et al.: Characterizing tumor invasiveness of glioblastoma using multiparametric magnetic resonance imaging. J. Neurosurg. **1**, 1–8 (2019)

13. Mang, A., Bakas, S., Subramanian, S., Davatzikos, C., Biros, G.: Integrated biophysical modeling and image analysis: application to neuro-oncology. Annu. Rev. Biomed. Eng. **22**, 309–341 (2020)

14. Menze, B.H., et al.: The multimodal brain tumor image segmentation benchmark (BRATS). IEEE Trans. Med. Imaging **34**(10), 1993–2024 (2014)

15. Myronenko, A.: 3D MRI brain tumor segmentation using autoencoder regularization. In: Crimi, A., Bakas, S., Kuijf, H., Keyvan, F., Reyes, M., van Walsum, T. (eds.) BrainLes 2018, Part II. LNCS, vol. 11384, pp. 311–320. Springer, Cham (2019). https://doi.org/10.1007/978-3-030-11726-9_28

16. Schlemper, J., et al.: Attention gated networks: learning to leverage salient regions in medical images. Med. Image Anal. **53**, 197 (2019)

17. Scialpi, M., Bianconi, F., Cantisani, V., Palumbo, B.: Radiomic machine learning: is it really a useful method for the characterization of prostate cancer? Radiology **291**(1), 269 (2019)

18. Wang, S., Dai, C., Mo, Y., Angelini, E., Guo, Y., Bai, W.: Automatic brain tumour segmentation and biophysics-guided survival prediction. In: Crimi, A., Bakas, S. (eds.) BrainLes 2019, Part II. LNCS, vol. 11993, pp. 61–72. Springer, Cham (2020). https://doi.org/10.1007/978-3-030-46643-5_6

Author Index

Printed in the United States
by Baker & Taylor Publisher Services